Integral and Integro-Differential Equations

This book provides a comprehensive study of numerical techniques for solving integral and integro-differential equations using wavelet-based approximation methods. It combines both theoretical insights and practical applications, focusing on integer- and fractional-order equations, including those with weakly singular kernels. Starting with key definitions and theorems from integral equations and fractional calculus, this book establishes a clear mathematical framework. It then introduces wavelet-based schemes for approximating solutions, with a particular focus on convergence, stability, and error analysis. Each chapter is enriched with numerical examples, graphs, and tables that demonstrate the accuracy and computational efficiency of the proposed methods.

Features

- Employs wavelet approximation methods to solve a wide range of integral and integro-differential equations
- Expands the scope of applicable equations by covering both classical and fractional-order models
- Illustrates the effectiveness of wavelet methods in handling weakly singular kernels through worked-out examples
- Provides rigorous theoretical validation for the accuracy and reliability of each method.

This book is intended for postgraduate students, researchers, and professionals in mathematics, engineering, and applied sciences who seek modern computational techniques for solving integral and integro-differential equations.

Integral and Integro-Differential Equations

Wavelet-Based Numerical Methods

Santanu Saha Ray and Srikanta Behera

CRC Press
Taylor & Francis Group
Boca Raton London New York

CRC Press is an imprint of the
Taylor & Francis Group, an **informa** business
A CHAPMAN & HALL BOOK

First edition published 2026
by CRC Press
2385 NW Executive Center Drive, Suite 320, Boca Raton FL 33431

and by CRC Press
4 Park Square, Milton Park, Abingdon, Oxon, OX14 4RN

CRC Press is an imprint of Taylor & Francis Group, LLC

ISBN: 9781032586120 (hbk)
ISBN: 9781032591940 (pbk)
ISBN: 9781003453451 (ebk)

DOI: 10.1201/9781003453451

Typeset in Times
by codeMantra

Contents

Author Biography

Santanu Saha Ray is currently a Professor in the Department of Mathematics, Indian Institute of Engineering Science and Technology, Shibpur, India. He was also a Professor and former Head of the Department of Mathematics, National Institute of Technology, Rourkela, India. He was elected Fellow of the Institute of Mathematics and Its Applications, United Kingdom, in 2018. Dr. Saha Ray is enlisted in the list of "WORLD RANKING OF TOP 2% SCIENTISTS" from India as per a subject-wise analysis conducted by a team of scientists at Stanford University, led by Dr. John PA Ioannidis. Recently, he was elected for Full Membership in Sigma Xi, The Scientific Research Honour Society, USA. Dr. Saha Ray has more than 23 years of teaching experience at the undergraduate and postgraduate levels in National Institute of Technology, Rourkela and two private engineering institutes in Kolkata, West Bengal. He has more than 22 years of research experience in various fields of applied mathematics. He has published many peer-reviewed research papers in numerous fields and various international SCI journals of repute. He was awarded an IOP Publishing Top Cited Author Award 2018 from India in the field of physics, published across the whole IOP Publishing portfolio, for 3 years (2015–2017), using citations recorded in Web of Science. He has contributed papers on several topics such as fractional calculus, mathematical modelling, mathematical physics, stochastic modelling, integral equations, and wavelet methods.

Srikanta Behera is an Assistant Professor in the Department of Mathematics, Kalinga Institute of Industrial Technology (KIIT), Bhubaneswar, Odisha, India. Dr. Behera obtained his PhD in 2022 from National Institute of Technology, Rourkela, Odisha, India and his MSc (Master of Science) degree in 2015 from Maharaja Sriram Chandra Bhanja Deo University (North Orissa University), Baripada, Odisha, India. He completed his BSc (Bachelor of Science) degree in mathematics (honors) at Gangadhar Meher Autonomous College (Sambalpur University), Sambalpur, Odisha, India, in 2013. Dr. Behera has 6 years of teaching experience at the undergraduate and postgraduate levels. He has also about 7 years of research experience in various fields of applied mathematics. He has published 12 research papers in numerous fields and various international SCI journals of repute, including *Applied Mathematics and Computation*, *Journal of Computational and Applied Mathematics*, *Computational and Applied Mathematics*, *Mathematics and Computers in Simulation*, *Mediterranean Journal of Mathematics*, *Engineering Computations*, *Iranian Journal of Science and Technology*, *Transactions A Science*, and *Mathematical Methods in the Applied Sciences*. He has contributed papers on several topics such as fractional calculus, numerical solution of integral and integro-differential equations, and wavelet methods.

Preface

Integral equation has been regarded as one of the most effective tools in applied mathematics. We adopted different approximation techniques to solve both linear and nonlinear integral and integro-differential equations of both integer and fractional orders. The main objective of this study is to classify the approached techniques and estimate their accuracy and efficiency. Moreover, the concept of fractional calculus has been discussed in our work. Fractional calculus has many applications in the field of science and engineering such as electromagnetics, viscoelasticity, fluid mechanics, electrochemistry, and signal processing. It plays an important role in the integration and differentiation of functions with non-integer orders.

This book deals with the development of operational matrix method and collocation method based on different types of one-dimensional and two-dimensional wavelets for solving linear and nonlinear integral and integro-differential equations. The purpose is to analyse the accuracy and efficiency of the wavelet methods of solving the proposed equations.

The preliminary concept of integral equations and integro-differential equations along with different kinds of kernels and the introductory concept of fractional calculus are described in Chapter 1. In Chapter 2, basic definitions and properties of different types of wavelets such as Legendre wavelet, Bernoulli wavelet, Jacobi wavelet, Gegenbauer wavelet, Cosine and Sine (CAS) wavelet, Müntz–Legendre wavelet, Euler wavelet, Taylor wavelet, and Laguerre wavelet are discussed.

In Chapter 3, a novel technique based on Bernoulli wavelets is proposed to solve two-dimensional Fredholm integral equation of second kind. Bernoulli wavelets are created by dilation and translation of Bernoulli polynomials. Furthermore, a collocation scheme based on Laguerre wavelets is introduced for solving the linear and nonlinear Fredholm integral equations as well as the linear system of Fredholm integral equations with weakly singular logarithmic kernel. The approached techniques are used to transform the respective proposed equations into a system of algebraic equations. Moreover, the numerical results of the several experiments are also presented in both tabular and graphical forms to describe the accuracy and efficiency of the approached method.

In Chapter 4, two numerical methods, namely, Gegenbauer wavelet method and CAS wavelet method, are introduced for the numerical solution of two-dimensional Volterra integral equation. In addition, a collocation-based method based on the Jacobi wavelets method is studied for the solution of system of two-dimensional Volterra integral equations. Finally, a brief discussion including numerical problems is presented in order to show the accuracy and effectiveness of the proposed schemes.

In Chapter 5, an effective approach is proposed to obtain the approximate solutions of linear and nonlinear two-dimensional Volterra integro-differential equations. The operational matrices of integration, differentiation, and product based on two-dimensional Bernoulli wavelets are constructed. By utilizing the properties and matrices of wavelets along with the collocation point, the two-dimensional linear and nonlinear Volterra integro-differential equations are reduced into the system of linear and nonlinear algebraic equations, respectively. The convergence analysis and error analysis are extensively studied by the help of two-dimensional wavelets approximation. A comparison of error values and some figures obtained by the proposed wavelets are presented in order to justify the error analysis of the proposed method.

In Chapter 6, an operational matrix scheme based on two-dimensional wavelets is introduced for the solution of weakly singular linear partial integro-differential equations and nonlinear partial integro-differential equations. By implementing two-dimensional wavelets approximations and its operational matrices of integration and differentiation along with collocation points, the linear and nonlinear weakly singular PIDEs are reduced into the system of linear and nonlinear algebraic equations, respectively. Some numerical examples are included to establish the accuracy of the proposed scheme via Bernoulli wavelet approximation and Legendre wavelet approximation, respectively.

In Chapter 7, an effective numerical framework is developed to obtain the solution of the Pantograph Volterra delay-integro-differential equation. By utilizing the Müntz–Legendre wavelet–based operational matrices along with the collocation points, the one-dimensional Pantograph Volterra delay-integro-differential equation is reduced into an explicit system of algebraic equations. Moreover, a numerical operational matrix approach based on Euler wavelet is proposed to solve the nonlinear Pantograph Volterra delay-integro-differential equation of fractional order. Additionally, some numerical problems are solved to justify the applicability and validity of the presented techniques.

In Chapter 8, the main objective is to establish a fractional-order operational matrix method based on Euler wavelet for solving linear Volterra–Fredholm integro-differential equations with weakly singular kernels. Again, an approximation technique based on Taylor wavelet is employed for the solution of linear and nonlinear fractional-order Volterra integro-differential equation with weakly singular kernels. The results of numerical experiments are also reported in both graphical and tabular forms to demonstrate the efficiency and validity of the proposed techniques.

In Chapter 9, a numerical operational matrix approach based on the Taylor wavelet method is proposed for solving the linear and nonlinear Volterra-Fredholm integro-differential equations. Several theorems are presented to establish the convergence and error analysis of the proposed method. Additionally, several numerical problems are included to justify the efficiency and validity of the presented technique.

1 Preliminaries of Integral Equations and Fractional Calculus

1.1 INTRODUCTION

Integral equations (IEs) are one of the most significant mathematical tools in both pure and applied mathematics. It has a wide range of physical applications. Integral transforms are widely used mathematical techniques for solving advanced science and engineering problems that can be expressed mathematically as differential equations, partial differential equations (PDEs), ordinary differential equations (ODEs), IEs, partial integro-differential equations (PIDEs), integro-differential equations (IDEs), and so on. The theory of Fourier integral introduced IEs for the first time. In 1825, Abel, an Italian mathematician, developed the first integral equation with respect to the well-known tautochrone problem. The problem is closely associated with defining a curve along which a heavy particle slides without friction to its lowest location, or more specifically, a curve along which the time of descent is a function of its initial state. The actual development of the theory of IEs began with the works of the Italian mathematician V. Volterra (1896) and the Swedish mathematician I. Fredholm (1900). Initial value problems for a finite interval $[\alpha, \beta]$ are used to construct IEs in the fields of physics, chemistry, biology, and engineering. IEs are also extremely useful tools in many fields of pure analysis, such as functional analysis and stochastic processes. A large class of initial and boundary value problems can be converted to Volterra or Fredholm IEs. Mathematical physics models, such as diffraction problems, scattering in quantum mechanics, conformal mapping, and water waves, also contributed to the creation of IEs. Many other applications in science and engineering are described by IEs or IDEs. Volterra's population growth model, biological species living together, propagation of stocked fish in a new lake, the heat transfer, and the heat radiation are among the many areas that are described by integral equations. Many scientific problems give rise to IEs with logarithmic kernels. IEs often arise in electrostatic, low-frequency electromagnetic problems, electromagnetic scattering problems, and propagation of acoustical and elastical waves.

1.2 INTEGRAL EQUATION AND ITS CLASSIFICATION

An IE is an equation in which the unknown function $y(x)$ to be determined appears under the integral sign. A typical form of an IE for $\alpha \le x \le \beta$, $\alpha \le t \le \beta$ is of the following form:

DOI: 10.1201/9781003453451-1

$$y(x) = f(x) + \int_{\alpha}^{\beta} \tilde{K}(x,t)y(t)\,dt, \quad x \in [\alpha, \beta], \tag{1.1}$$

where $y(x)$ is the unknown function to be determined, $f(x)$ is the known function, $\tilde{K}(x,t)$ is the kernel of the IE, and α, β are constants. In terms of x and t, the functions specified above can be real or complex valued. Only real-valued functions are discussed in our study.

An IE can be distinguished as a linear or nonlinear integral equation. An IE is called linear if only linear operations are performed in it upon the unknown function. An IE that is not linear is known as nonlinear IE. IEs are usually categorized into two parts: Volterra integral equations (VIEs) and Fredholm integral equations (FIEs). These IEs are also classified as homogeneous or non-homogeneous IEs.

1.2.1 Volterra Integral Equation

The standard form of linear VIEs is given by the following form:

$$h(x)y(x) = f(x) + \lambda \int_{\alpha}^{x} \tilde{K}(x,t)y(t)\,dt, \tag{1.2}$$

where $y(x)$ is the unknown function; $h(x), f(x)$, and $\tilde{K}(x,t)$ are known functions; α is a constant; and λ is a non-zero real or complex parameter. Equation (1.2) is known as a nonlinear VIE if the power of unknown function that appears under the integral sign, i.e., $y(t)$, is more than 1.

The VIEs are again classified into different kinds based on the value of $h(x)$:

- If $h(x) = 0$, Eq. (1.2) becomes

$$f(x) + \lambda \int_{\alpha}^{x} \tilde{K}(x,t)y(t)\,dt = 0, \tag{1.3}$$

  and the IE is known as VIE of the first kind.
- If $h(x) = 1$, Eq. (1.2) becomes

$$y(x) = f(x) + \lambda \int_{\alpha}^{x} \tilde{K}(x,t)y(t)\,dt, \tag{1.4}$$

  and the IE is known as VIE of the second kind.
- If $f(x) = 0$ and $h(x) = 1$, Eq. (1.2) becomes

$$y(x) = \lambda \int_{\alpha}^{x} \tilde{K}(x,t)y(t)\,dt, \tag{1.5}$$

  and the IE is known as homogeneous VIE of the second kind.

1.2.2 Fredholm Integral Equation

The general form of linear FIEs is given by the following form:

$$h(x)y(x) = f(x) + \lambda \int_{\alpha}^{\beta} \tilde{K}(x,t)y(t)\,dt, \tag{1.6}$$

where $y(x)$ is the unknown function; $h(x), f(x)$, and $\tilde{K}(x,t)$ are known functions; α, β are constants; and λ is a non-zero real or complex parameter. Equation (1.6) is known as a nonlinear FIE if the power of unknown function that appears under the integral sign, i.e., $y(t)$, is more than 1.

The FIEs are again classified into different kinds based on the value of $h(x)$:

- If $h(x) = 0$, Eq. (1.6) becomes

$$f(x) + \lambda \int_{\alpha}^{\beta} \tilde{K}(x,t)y(t)\,dt = 0, \tag{1.7}$$

and the IE is known as FIE of the first kind.

- If $h(x) = 1$, Eq. (1.6) becomes

$$y(x) = f(x) + \lambda \int_{\alpha}^{\beta} \tilde{K}(x,t)y(t)\,dt, \tag{1.8}$$

and the IE is known as FIE of the second kind.

- If $f(x) = 0$ and $h(x) = 1$, Eq. (1.6) becomes

$$y(x) = \lambda \int_{\alpha}^{\beta} \tilde{K}(x,t)y(t)\,dt, \tag{1.9}$$

and the IE is known as homogeneous FIE of the second kind.

1.2.3 Volterra–Fredholm Integral Equation

The Volterra–Fredholm integral equation (VFIE) is the combination of disjoint Volterra and Fredholm integrals, which appears in one IE. The most general form of VFIE is defined as follows:

$$y(x) = f(x) + \lambda_1 \int_{\alpha}^{x} \tilde{K}_1(x,t)y(t)\,dt + \lambda_2 \int_{\alpha}^{\beta} \tilde{K}_2(x,t)y(t)\,dt, \tag{1.10}$$

where $y(x)$ is the unknown function to be determined; $f(x)$ is the known function; and λ_1, λ_2 are non-zero real or complex parameters, respectively. $\tilde{K}_1(x,t)$ and $\tilde{K}_2(x,t)$ are the kernels of VFIE.

1.2.4 Singular Integral Equation

An integral with infinite limits or when the kernel of the integral becomes unbounded at one or more points within the integration interval is known as a singular integral equation (SIE).

For example,

$$y(x) = 1 + x^2 + \frac{1}{2}\int_{-\infty}^{\infty} (x+t)y(t)\,dt, \tag{1.11}$$

$$y(x) = 1 - 2\sqrt{x} - \int_{0}^{x} \frac{1}{\sqrt{x-t}}\,y(t)\,dt, \tag{1.12}$$

are known as SIEs.

1.3 SYSTEMS OF INTEGRAL EQUATION

A system of IEs is a set of two or more IEs in two or more unknown functions. Many scientific applications utilize systems of Fredholm and Volterra integral equations of the first and second kinds, which can be solved using various numerical methods. The system of IEs is categorized into the following types:

- Systems of Volterra integral equation
- Systems of Fredholm integral equation.

1.3.1 Systems of Volterra Integral Equation

Systems of VIEs appear in many scientific applications, including species propagation and Volterra population growth models. The standard system of VIEs of the second kind is defined as follows:

$$y_1(x) = g_1(x) + \int_0^x (k_1(x,t)y_1(t) + \bar{k}_1(x,t)y_2(t))dt,$$

$$y_2(x) = g_2(x) + \int_0^x (k_2(x,t)y_1(t) + \bar{k}_2(x,t)y_2(t))dt,$$

$$\vdots$$

where the kernel functions $k_i(x,t), i = 1,2,\ldots, \bar{k}_j(x,t), j = 1,2,\ldots$ and the functions $g_i(x), i = 1,2,\ldots$ are known real-valued functions. The functions $y_i(x), i = 1,2,\ldots$ are unknown functions to be determined.

For example,

$$y_1(x) = x^2 - \frac{x^5}{12} + \int_0^x ((x-t)^2 y_1(t) + (x-t) y_2(t))dt,$$

$$y_2(x) = x^3 + \frac{x^5}{60} + \int_0^x ((x-t)^2 y_1(t) - (x-t) y_2(t))dt.$$

1.3.2 Systems of Fredholm Integral Equation

The standard system of FIEs of the second kind is defined as follows:

$$y_1(x) = g_1(x) + \int_a^b (k_1(x,t) y_1(t) + \bar{k}_1(x,t) y_2(t))dt,$$

$$y_2(x) = g_2(x) + \int_a^b (k_2(x,t) y_1(t) + \bar{k}_2(x,t) y_2(t))dt,$$

$$\vdots$$

where the kernel functions $k_i(x,t), i = 1,2,\ldots, \bar{k}_j(x,t), j = 1,2,\ldots$ and the functions $g_i(x), i = 1,2,\ldots$ are known real-valued functions. The functions $y_i(x), i = 1,2,\ldots$ are unknown functions to be determined.

For example,

$$y_1(x) = x + \sin x - \frac{\pi^2}{4} - \frac{\pi}{2} x + \int_0^{\frac{\pi}{2}} ((1+xt) y_1(t) + (1-xt) y_2(t))dt,$$

$$y_2(x) = x - \cos x - \frac{\pi^2}{4} + \frac{\pi}{2} x + \int_0^{\frac{\pi}{2}} ((1-xt) y_1(t) + (1+xt) y_2(t))dt.$$

1.4 INTEGRO-DIFFERENTIAL EQUATION

Vito Volterra investigated population growth in the early 1900s, which led to the development of new types of equations known as integro-differential equations (IDEs). In this type of equations, the unknown function $y(x)$ occurs in one side as an ordinary derivative, and it appears on the other side under the integral sign. The IDE can be easily observed as an intermediate stage when a differential equation is converted into an IE. IDEs are categorized into the following types:

- Volterra integro-differential equations (VIDEs)
- Fredholm integro-differential equations (FIDEs)

- Volterra–Fredholm integro-differential equations (VFIDEs)
- Partial integro-differential equations (PIDEs)
- Pantograph Volterra delay integro-differential equations (PVDIDEs).

1.4.1 Volterra Integro-Differential Equation

The standard form of VIDE is given by the following equation:

$$y^{(n)}(x) = f(x) + \int_{\alpha}^{x} \tilde{K}(x,t)y(t)\,dt, \tag{1.13}$$

where $\tilde{K}(x,t)$ is the kernel function, $f(x)$ is the known function, and n is the order of the ordinary derivative of the unknown function $y(x)$. Notice that because this kind of equations contain ordinary derivatives, then initial conditions should be prescribed depending on the order of the derivative involved. For example,

$$y''(x) = x + \int_{0}^{x} (x+2t)y(t)\,dt, \quad y(0) = 0, \quad y'(0) = 1,$$

is the second-order VIDE.

1.4.2 Fredholm Integro-Differential Equation

The standard form of FIDE is given by the following equation:

$$y^{(n)}(x) = f(x) + \int_{\alpha}^{\beta} \tilde{K}(x,t)y(t)\,dt, \tag{1.14}$$

where $\tilde{K}(x,t)$ is the kernel function, $f(x)$ is the known function, and n is the order of the ordinary derivative of the unknown function $y(x)$. Notice that because this kind of equations contain ordinary derivatives, then initial conditions should be prescribed depending on the order of the derivative involved. For example,

$$y'(x) = 1 + \int_{0}^{1} (xt)y(t)\,dt, \quad y(0) = 1,$$

is the first-order FIDE.

1.4.3 Volterra–Fredholm Integro-Differential Equation

The standard form of VFIDE is given by the following equation:

$$y^{(n)}(x) = f(x) + \int_{\alpha}^{x} \tilde{K}_1(x,t)y(t)\,dt + \int_{\alpha}^{\beta} \tilde{K}_2(x,t)y(t)\,dt, \tag{1.15}$$

where $\tilde{K}_1(x,t)$ and $\tilde{K}_2(x,t)$ are the kernel functions, $f(x)$ is the known function, and n is the order of the ordinary derivative of the unknown function $y(x)$. Notice that because this kind of equations contain ordinary derivatives, then initial conditions should be prescribed depending on the order of the derivative involved. For example,

$$y'(x) = 1 + \int_0^x (x-t)y(t)\,dt + \int_0^1 (xt)y(t)\,dt, \quad y(0) = 1,$$

is the first-order VFIDE.

1.4.4 Partial Integro-Differential Equation

PIDE is an equation in which the unknown function $y(x)$ occurs in one side as a partial derivative, and it appears on the other side under the integral sign. In our study, we consider two types of PIDES:

- **Linear Volterra Weakly Singular PIDEs:** The form of linear Volterra weakly singular PIDE is given by

$$y_{xt}(x,t) + y_t(x,t) = y(x,t) + h(x,t) + \int_0^x \int_0^t \Phi(x,z)\tilde{H}(y(z,s))dsdz, \quad (x,t) \in \Delta, \qquad (1.16)$$

  having the following boundary conditions:

$$y(x,0) = y_0(x),$$

$$y(0,t) = y_0(t).$$

  where $\tilde{H}(y(x,t)), h(x,t), y_0(x), y_0(t)$ are known functions and $y(x,t)$ is the unknown function. In Eq. (1.16), $\Delta = [0,1]\times[0,1]$, $\Phi(x,z) = \frac{1}{(x-z)^\theta}$, $0 < \theta < 1$ is the kernel function which has a weak singularity at $z = x$. Moreover, $\tilde{H}$ is a differential operator that is considered as $\tilde{H}(y(x,t)) = y_{tt}(x,t)$, and $\tilde{H}(y(x,t)) = y_{xt}(x,t)$ for linear case as the first kind and the second kind, respectively.
- **Nonlinear Volterra Weakly Singular PIDEs:** The form of nonlinear Volterra weakly singular PIDE is given by the following equation:

$$y_{xt}(x,t) + y_t(x,t) = y(x,t) + h(x,t) + \int_0^x \int_0^t \Phi(x,z)\tilde{H}(y(z,s))dsdz, \quad (x,t) \in \Delta, \qquad (1.17)$$

  having the following boundary conditions:

$$y(x,0) = y_0(x),$$

$$y(0,t) = y_0(t).$$

  In Eq. (1.17), $\tilde{H}(y(x,t))$, $h(x,t)$, $y_0(x)$, $y_0(t)$ are known functions and $y(x,t)$ is an unknown function that should be determined. The domain $\Delta = [0,1]\times[0,1]$, $\Phi(x,z) = \frac{1}{(x-z)^\theta}$, $0 < \theta < 1$, and the differential operator

$\tilde{H}$ is considered as $\tilde{H}(y(x,t)) = [y(x,t)]^{\rho}$, and $\tilde{H}(y(x,t)) = [\eta_{i_1,i_2} y(x,t)]^{\rho}$ for the nonlinear case as the first kind and the second kind, respectively, where ρ is a positive number and η_{i_1,i_2} is a differential operator.

1.4.5 Pantograph Volterra Delay Integro-Differential Equation

In 1971, Ockendon and Tayler introduced the term pantograph in their study of delay differential equations. The pantograph equations are usually referred to as IDEs with proportional delays. In our study, we consider two types of PVDIDES:

- **Integer-Order Pantograph Volterra Delay-Integro-Differential Equation:** The form of integer-order PVDIDE is given by the following equation:

$$y'(x) = u(x)y(x) + v(x)y(qx) + f(x) + \int_0^x k_1(x,\eta)y(\eta)\,d\eta + \int_0^{qx} k_2(x,\eta)y(\eta)\,d\eta, \quad (1.18)$$

where $y(0) = 0$, $x \in \tilde{I} := [0,T]$, and qx $(0 < q < 1)$ is the proportional delays.

In Eq. (1.18), the known functions $u(x)$, $v(x)$, and $f(x)$ are defined on $\tilde{I}$, and $y(x)$ is the unknown function to be determined. Here, $k_1(x,\eta)$ and $k_2(x,\eta)$ are the kernel functions defined on $D_1 := \{(x,\eta) : 0 < x < T, 0 < \eta < x\}$ and $D_2 := \{(x,\eta) : 0 < x < T, 0 < \eta < qx\}$, respectively.

- **Fractional-Order Pantograph Volterra Delay-Integro-Differential Equation:** The form of fractional-order PVDIDE is given by the following equation:

$$\begin{cases} {}_0^C D_x^{\alpha} y(x) = a(x)y(x) + b(x)y(qx) + g(x) + \int_0^x k_1(x,s)y(s)\,ds + \int_0^{qx} k_2(x,s)y(s)\,ds \\ y(0) = 0 \end{cases}, \quad (1.19)$$

where $x \in \tilde{I} := [0,T]$ and qx is the proportional delays $(0 < q < 1)$.

Here $a(x)$, $b(x)$, and $g(x)$ are known functions defined on $\tilde{I}$, and $y(x)$ is the unknown function to be determined. The kernel functions $k_1(x,s)$ and $k_2(x,s)$ are defined on $\Omega_1 := \{(x,s): 0 < x < T, 0 < s < x\}$ and $\Omega_2 := \{(x,s): 0 < x < T, 0 < s < qx\}$, respectively, and ${}_0^C D_x^{\alpha}$ denotes the Caputo fractional derivative of fractional order α, $0 < \alpha \le 1$.

1.5 TYPES OF KERNELS

The special cases of the kernel functions of an IE are classified as follows:

- Symmetric kernel
- Separable or degenerate kernel
- Iterated kernel
- Weakly singular kernel.

1.5.1 Symmetric Kernel

A complex kernel $\tilde{K}(x,t)$ is symmetric if

$$\tilde{K}(x,t) = \tilde{K}^*(t,x),$$

where the asterisk denotes the complex conjugate.

Similarly, a real kernel $\tilde{K}(x,t)$ is symmetric if

$$\tilde{K}(x,t) = \tilde{K}(t,x).$$

1.5.2 Separable or Degenerate Kernel

A kernel function $\tilde{K}(x,t)$ is called separable or degenerate if it can be expressed as the sum of a finite number of terms, each of which is the product of a function of x and t only, i.e.,

$$\tilde{K}(x,t) = \sum_{l=0}^{n} f_l(x) g_l(t).$$

Furthermore, a kernel function $\tilde{K}(x,t)$ is called non-degenerate if it cannot be separated as the function of x and the function of t. For example, e^{xt}, $\sqrt{x+t}$ are the non-degenerate kernels.

1.5.3 Iterated Kernel

Consider FIE of the second kind:

$$y(x) = f(x) + \lambda \int_{\alpha}^{\beta} \tilde{K}(x,t) y(t)\, dt.$$

Then, the iterated kernels $\tilde{K}_n(x,t)$, $n = 1,2,3,\ldots$ are defined as follows:

$$\tilde{K}_1(x,t) = \tilde{K}(x,t),$$

$$\tilde{K}_n(x,t) = \int_{\alpha}^{\beta} \tilde{K}(x,z) \tilde{K}_{n-1}(x,t)\, dt,\ n = 2,3,\ldots.$$

1.5.4 Weakly Singular Kernel

A kernel $\tilde{K}(x,t)$ is said to be weakly singular kernel if it is of the following forms:

1. $\tilde{K}(x,t) = \dfrac{1}{|x-t|^{\gamma}}$, $0 < \gamma < 1$, and it is known as weakly singular algebraic kernel.
2. $\tilde{K}(x,t) = \log|x-t|$, and it is known as weakly singular logarithmic kernel.

1.6 APPLICATION OF INTEGRAL EQUATION

The study of IEs is essential in both the pure and practical fields of mathematics. The progress of research has resulted in the formulation of numerous physical laws, and when restated mathematically, frequently appear as differential equations. Differential equations may be used to explain engineering problems mathematically; hence, they play a significant role in the solution of practical problems. For example, Newton's law, which states that the rate of change of a particle's momentum is equal to the force acting on it, can be expressed mathematically as a differential equation. Similarly, difficulties appearing in electric circuits, chemical kinetics, and heat transport in a medium can all be theoretically described as differential equations. These differential equations can be converted into VIEs and FIEs. Applications of some special IEs are discussed in the following:

- **Abel's Integral Equation:** $\left(f(x) = \int_0^x k(x,t)y(t)\,dt \right)$

  The above equation derives from the problem of finding the path of a particle that is confined to move in a vertical plane under gravity.
- **Vandrey's Equation:** $\left(y(x) = f(x) - \frac{\lambda}{\pi}\int_0^L k(x,t)y(t)\,dt; \quad 0 \le x \le L \right)$

  The above equation appears in fluid dynamics when computing the pressure distribution on the surface of a moving body of revolution in a fluid.
- **Transverse Oscillations of a Homogenous Elastic Bar:**

  Consider a homogeneous elastic bar with linear mass density d and its axis coincides with the segment $(0, l)$ of the t-axis when the bar is in its state of rest. It is clamped at the end $t = 0$, is free at the end $t = l$, and is forced to perform simple harmonic oscillations with period $2\pi/\omega$, and ω is the angular frequency (Figure 1.1).

  The problem is to find the deflection $y(t)$ that is parallel to the y-axis and satisfies the system of equations

$$\frac{d^4y}{dt^4} - Q^4 y = 0, \quad Q = \frac{\omega^2 d}{EI},$$

$$y(0) = y'(0) = 0,$$

$$y''(l) = y'''(l) = 0,$$

where EI is the bending rigidity of the bar.

The above differential equation with initial conditions can be reduced to $y''(0) = c_2$, $y'''(0) = c_3$, and then we obtain the required IE as follows:

$$\varphi(t) = Q^4\left(\frac{t^2}{2!}c_2 + \frac{t^3}{3!}c_3\right) + Q^4\int_0^t \frac{(t-s)^3}{3!}\varphi(s)\,ds,$$

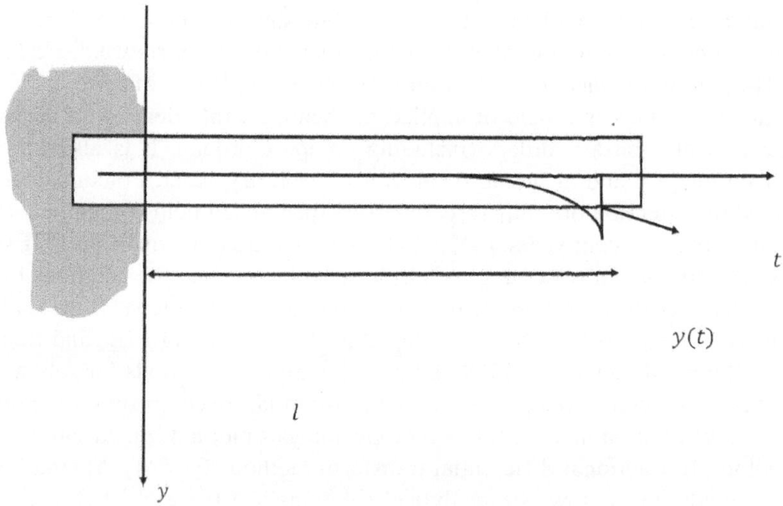

FIGURE 1.1 Transverse oscillations of a homogenous elastic bar.

where $\varphi(t) = \frac{d^4 y}{dt^4}$.

Then the solution $y(t)$ of the above differential equation is given by

$$y(t) = \int_0^t \frac{(t-s)^3}{3!} y(s)\,ds + \frac{t^2}{2!}c_2 + \frac{t^3}{3!}c_3.$$

From the given conditions, we can determine the constants c_2 and c_3. As a result, we can simply solve a given differential equation with initial conditions by transforming it into an IE.

1.7 FRACTIONAL CALCULUS

Fractional calculus is used in a wide range of engineering and science fields, including electromagnetics, viscoelasticity, fluid mechanics, electrochemistry, biological population models, optics, and signal processing. It has been used to simulate physical and engineering processes that are better explained by fractional differential equations (FDEs). The fractional derivative models are used to accurately represent systems that require correct damping modelling. In these disciplines, different analytical and numerical approaches, as well as their applicability to new problems, have been suggested in recent years. Mathematical simulations of real-life problems usually result in FDEs and various other problems such as physical phenomena of fluid dynamics, quantum mechanics, electricity, and ecological systems. In 1695, Leibniz, one of the founders of standard calculus, proposed the fractional calculus for the first time. Because of Liouville, the primary effort to provide a logical definition has been established. Many mathematicians later made significant contributions towards fractional calculus, including Abel, Caputo, Euler, Fourier, Grünwald, Hadamard, Hardy, Holmgren, Laplace, Leibniz, Letnikov, Riemann, Riesz, and Weyl.

Fractional calculus is a field of applied mathematics that deals with derivatives and integrals of arbitrary orders (including complex orders). It is also known as generalized integral and differential calculus of arbitrary order. Fractional calculus has gained the attention of many researchers in specialised fields of applied science and engineering in recent years. FDEs and fractional integral equations (FIEs) have been used to simulate a wide range of physical problems, and the solutions to these equations have been the source of numerous physical phenomena. Both analytical and numerical methods have been developed for the solution of FIEs and fractional integro-differential equations (FIDEs). Several analytical methods for solving FIEs and FIDEs have been developed, such as the Adomian decomposition method [1], homotopy perturbation method [2], homotopy analysis method [3], variational iteration method [4], fractional differential transform method (FDTM) [5], and Laplace transform method [6]. However, analytical solutions to FIEs and FIDEs are sometimes difficult to obtain. Because of this, establishing numerical solutions to these problems has become a challenging task for many researchers.

Various numerical schemes have been introduced for solving the FDEs, FIEs, and FIDEs. Different wavelet techniques such as Euler wavelet [7,8], Taylor wavelet [9], Haar wavelet [10], Riesz wavelet [11], Legendre wavelet [12], Bernoulli wavelet [13], and second-kind Chebyshev wavelet [14] have been applied to solve FIDEs. In a previous study [15], Lepik has used the Haar wavelet method for solving FIE. Alqahtani et al. [16] have applied hybrid contraction technique for solving the Volterra FIE. In another study [17], Abdeljawad et al. have introduced fixed point scheme for numerical solution of nonlinear IEs and FDEs. Recently, Mohammad et al. [18] have employed framelet expansion method to fractional nonlinear Volterra–Fredholm integral equations. The existence and uniqueness of solutions for fractional Volterra–Fredholm integro-differential equations have been discussed in previous works [19,20]. The FDEs have been solved by many numerical methods, viz., B-spline wavelet collocation method [21], variational iteration method [22], Legendre multiwavelet Galerkin method [23], and also Chebyshev wavelets [24].

1.7.1 Fractional Integration and Derivative

Fractional calculus plays an important role in the integration and differentiation of functions with non-integer orders [25,26]. After Leibniz, Euler identified the difficulty with a derivative of non-integer order in 1738 [27,28]. In 1822, Fourier proposed an integral representation to define the derivative, and his interpretation can be considered the first definition for the derivative of arbitrary (positive) order. In 1826, Abel solved an integral equation related to the tautochrone problem, which is regarded as the first application of fractional calculus. In 1832, Liouville proposed a formulation for the differentiation of exponential function. This is referred to as the first Liouville definition. The second definition proposed by Liouville is stated in terms of an integral and is now known as the Liouville version for the integration of non-integer order. In 1967, Caputo developed a definition for dealing with difficulties comprising a fractional differential equation with initial conditions. The concept introduced by Caputo inverts the order of integral and derivative operators with the non-integer order derivative of the Riemann–Liouville equation.

In fractional calculus, different types of fractional integrals and derivatives have been used such as

- Riemann–Liouville fractional integral
- Riemann–Liouville fractional derivative
- Caputo fractional derivative
- Grünwald–Letnikov fractional derivative.

1.7.1.1 Riemann–Liouville Fractional Integral

The Riemann–Liouville (RL) fractional integral operator $\tilde{I}^{\alpha}$ of order $\alpha \geq 0$ is defined as follows [29]:

$$\tilde{I}^{\alpha} y(x) = \frac{1}{\Gamma(\alpha)} \int_{0}^{x} (x-z)^{\alpha-1} y(z)\,dz, \quad x > 0, \alpha \in \mathbb{R}^{+}, \tag{1.20}$$

$$\tilde{I}^{0} y(x) = y(x), \tag{1.21}$$

where $\Gamma(.)$ is the well-known gamma function and is defined as follows:

$$\Gamma(\alpha) = \int_{0}^{\infty} e^{-x} x^{\alpha-1}\,dx \text{ and for real } \alpha, \Gamma(\alpha) = (\alpha-1)!.$$

Let $\Omega = [a,b]$, $(-\infty < a < b < \infty)$ be a finite interval on the real axis $\mathbb{R}$. The RL fractional integrals $\tilde{I}_{a+}^{\alpha} y$ and $\tilde{I}_{b-}^{\alpha} y$ of order $\alpha \geq 0$ are defined by [30]

$$\tilde{I}_{a+}^{\alpha} y(x) = \frac{1}{\Gamma(\alpha)} \int_{a}^{x} \frac{y(z)}{(x-z)^{1-\alpha}}\,dz, \quad x > a, \alpha \in \mathbb{R}^{+}, \tag{1.22}$$

and

$$\tilde{I}_{b-}^{\alpha} y(x) = \frac{1}{\Gamma(\alpha)} \int_{x}^{b} \frac{y(z)}{(z-x)^{1-\alpha}}\,dz, \quad x < b, \alpha \in \mathbb{R}^{+}. \tag{1.23}$$

Equations (1.22) and (1.23) are known as left-sided and right-sided RL fractional integrals, respectively.

The integral operator $\tilde{I}^{\alpha}$ has the following properties:

- $\tilde{I}^{\alpha}(\lambda_1 y(x) + \lambda_2 y(x)) = \lambda_1 \tilde{I}^{\alpha}(y(x)) + \lambda_2 \tilde{I}^{\alpha}(y(x))$,
- $\tilde{I}^{\alpha_1} \tilde{I}^{\alpha_2} y(x) = \tilde{I}^{\alpha_1+\alpha_2} y(x)$,

- $\tilde{I}^{\alpha_2}\tilde{I}^{\alpha_1}y(x) = \tilde{I}^{\alpha_1}\tilde{I}^{\alpha_2}y(x)$,
- $\tilde{I}^{\alpha}x^{\gamma} = \dfrac{\Gamma(\gamma+1)}{\Gamma(\gamma+\alpha+1)}x^{\gamma+\alpha}, \quad \gamma > -1,$

where λ_1, λ_2 are constants.

1.7.1.2 Riemann–Liouville Fractional Derivative

The RL fractional derivative of order $\alpha \geq 0$ is defined as follows:

$$D^{\alpha}y(x) = D^{n}\tilde{I}^{n-\alpha}y(x)$$

$$= \frac{d^n}{dx^n}\left[\frac{1}{\Gamma(n-\alpha)}\int_0^x \frac{y(z)}{(x-z)^{n-\alpha-1}}dz\right], \quad n-1 < \alpha \leq n, n \in \mathbb{N}. \tag{1.24}$$

The left RL fractional derivative can be defined by

$$D_{a+}^{\alpha}y(x) = \frac{1}{\Gamma(n-\alpha)}\frac{d^n}{dx^n}\left[\int_a^x (x-z)^{n-\alpha-1}y(z)\,dz\right]. \tag{1.25}$$

The right RL fractional derivative can be defined by

$$D_{b-}^{\alpha}y(x) = \frac{1}{\Gamma(n-\alpha)}\left(-\frac{d^n}{dx^n}\right)\left[\int_x^b (z-x)^{n-\alpha-1}y(z)\,dz\right]. \tag{1.26}$$

For $\alpha, \beta \geq 0$, the derivative operator D^{α} has the following properties:

- $(D_{a+}^{\alpha}(z-a)^{\beta-1})(x) = \dfrac{\Gamma(\beta)}{\Gamma(\beta-\alpha)}(x-a)^{\beta-\alpha-1}$,
- $(D_{b-}^{\alpha}(b-z)^{\beta-1})(x) = \dfrac{\Gamma(\beta)}{\Gamma(\beta-\alpha)}(b-x)^{\beta-\alpha-1}$,
- If $\beta = 1$, then the RL fractional derivatives of a constant are, in general, not equal to zero:

$$(D_{a+}^{\alpha}1)(x) = \frac{(x-a)^{-\alpha}}{\Gamma(1-\alpha)} \text{ and } (D_{b-}^{\alpha}1)(x) = \frac{(b-x)^{-\alpha}}{\Gamma(1-\alpha)}, (0 < \alpha < 1).$$

1.7.1.3 Caputo Fractional Derivative

The Caputo fractional derivative operator ${}_0^C D_x^{\alpha}$ of order $\alpha \geq 0$ is defined as follows [31]:

$${}_0^C D_x^{\alpha}y(x) = \frac{1}{\Gamma(n-\alpha)}\int_0^x (x-z)^{n-\alpha-1}y^{(n)}(z)\,dz, \quad n-1 < \alpha < n, n \in \mathbb{N}, \tag{1.27}$$

$$ {}_0^C D_x^{\alpha} y(x) = y^{(n)}(x), \quad \alpha = n, \tag{1.28} $$

where n is an integer.

The Caputo fractional derivative has the following two basic properties:

- $ {}_0^C D_x^{\alpha} x^{\gamma} = \dfrac{\Gamma(1+\gamma)}{\Gamma(1+\gamma-\alpha)} x^{\gamma-\alpha}, \quad 0 < \alpha < \gamma + 1, \quad \gamma > -1,$
- $\tilde{I}^{\alpha} D^{\alpha} y(x) = y(x) - \displaystyle\sum_{j=0}^{n-1} y^{j}(0^{+}) \frac{x^{j}}{j!}, \quad n-1 < \alpha \le n, n \in \mathbb{N}.$

Let $[a,b]$ be a finite interval of the real line $\mathbb{R}$, and let $D_{a+}^{\alpha}[y(z)](x) = (D_{a+}^{\alpha} y)(x)$ and $D_{b-}^{\alpha}[y(z)](x) = (D_{b-}^{\alpha} y)(x)$ be the RL fractional derivative of order α. The fractional derivative $(D_{a+}^{\alpha} y)(x)$ and $(D_{b-}^{\alpha} y)(x)$ of order α on $[a,b]$ are defined via the above RL fractional derivatives by [30]

$$ (D_{a+}^{\alpha} y)(x) = \left(D_{a+}^{\alpha} \left[y(z) - \sum_{j=0}^{n-1} \frac{y^{j}(a)}{j!} (z-a)^{j} \right] \right)(x), \quad n-1 < \alpha \le n, n \in \mathbb{N}, \tag{1.29} $$

and

$$ (D_{b-}^{\alpha} y)(x) = \left(D_{b-}^{\alpha} \left[y(z) - \sum_{j=0}^{n-1} \frac{y^{j}(a)}{j!} (b-z)^{j} \right] \right)(x), \quad n-1 < \alpha \le n, n \in \mathbb{N}. \tag{1.30} $$

Equations (1.29) and (1.30) are known as left-sided and right-sided Caputo fractional derivatives of order α, respectively.

1.7.1.4 Grünwald–Letnikov Fractional Derivative

Anton Karl Grünwald (1838–1920) of Prague initially developed the Grünwald–Letnikov (GL) fractional derivative in 1867, and Aleksey Vasilievich Letnikov (1837–1888) of Moscow in 1868. The usual differentiation of a function $y(x)$ of order $n \in \mathbb{N}$ is given by

$$ y^{(n)}(x) = \lim_{h \to 0} \frac{(\Delta_h^n y)(x)}{h^n}, \tag{1.31} $$

where $(\Delta_h^n y)(x)$ is a finite difference of order n of a function $y(x)$ with a step $h \in \mathbb{R}$.

Equation (1.31) is used to define a fractional derivative by directly replacing $n \in \mathbb{N}$ by $\alpha > 0$. Thus, h^n is replaced by h^{α}, while the finite difference $(\Delta_h^n y)(x)$ is replaced by the difference $(\Delta_h^{\alpha} y)(x)$ of a fractional order α defined by the following infinite series, i.e.,

$$ (\Delta_h^n y)(x) = \sum_{r=0}^{\infty} (-1)^r \binom{\alpha}{r} y(x - rh), \quad x, h \in \mathbb{R}; \alpha > 0, \tag{1.32} $$

where $\begin{pmatrix}\alpha\\ r\end{pmatrix}$ are the binomial coefficients. For $h > 0$, the difference is called left-sided difference, and for $h < 0$, the difference is called a right-sided difference.

For the function $y(x)$ defined on finite interval $[a,b]$, we have the following finite series:

$$({}_{a+}\Delta_h^\alpha y)(x) = \sum_{r=0}^{\left[\frac{x-a}{h}\right]} (-1)^r \begin{pmatrix}\alpha\\ r\end{pmatrix} y(x-rh), \quad x \in \mathbb{R}; h > 0, \alpha > 0 \tag{1.33}$$

and

$$({}_{b-}\Delta_h^\alpha y)(x) = \sum_{r=0}^{\left[\frac{b-x}{h}\right]} (-1)^r \begin{pmatrix}\alpha\\ r\end{pmatrix} y(x+rh), \quad x \in \mathbb{R}; h > 0, \alpha > 0. \tag{1.34}$$

The left-sided and right-sided Grünwald–Letnikov fractional derivative of order $\alpha > 0$ on a finite interval $[a,b]$ are defined as follows [30]:

$$D_{a+}^\alpha y(x) = \lim_{h\to 0} \frac{({}_{a+}\Delta_h^\alpha y)(x)}{h^\alpha}$$

$$= \lim_{h\to 0} \frac{1}{h^\alpha} \sum_{r=0}^{\left[\frac{x-a}{h}\right]} (-1)^r \begin{pmatrix}\alpha\\ r\end{pmatrix} y(x-rh)$$

and

$$D_{b-}^\alpha y(x) = \lim_{h\to 0} \frac{({}_{b-}\Delta_h^\alpha y)(x)}{h^\alpha}$$

$$= \lim_{h\to 0} \frac{1}{h^\alpha} \sum_{r=0}^{\left[\frac{x-a}{h}\right]} (-1)^r \begin{pmatrix}\alpha\\ r\end{pmatrix} y(x+rh).$$

REFERENCES

1. Momani S., and Noor M.A., 2006, "Numerical methods for fourth-order fractional integro-differential equations", *Applied Mathematics and Computation*, **182**(1), pp. 754–760.
2. Das P., Rana S., and Ramos H., 2019, "Homotopy perturbation method for solving Caputo-type fractional-order Volterra-Fredholm integro-differential equations", *Computational and Mathematical Methods*, **1**(5), p. e1047.
3. Noeiaghdam S., Fariborzi Araghi M.A., and Abbasbandy S., 2019, "Finding optimal convergence control parameter in the homotopy analysis method to solve integral equations based on the stochastic arithmetic", *Numerical Algorithms*, **81**(1), pp. 237–267.
4. Ahmood W.A., and Al-Anni M.K., 2018, "Solutions of partial fractional integro-differential equations by using multi-dimensional modification of He's variational iteration method", *International Journal of Pure and Applied Mathematics*, **118**(20), pp. 3517–3535.

5. Arikoglu A., and Ozkol I., 2009, "Solution of fractional integro-differential equations by using fractional differential transform method", *Chaos, Solitons & Fractals*, **40**(2), pp. 521–529.
6. Usta F., Budak H., and Sarikaya M., 2018, "Yang-Laplace transform method Volterra and Abel's integro-differential equations of fractional order", *International Journal of Nonlinear Analysis and Applications*, **9**(2), pp. 203–214.
7. Behera S., and Saha Ray S., 2021, "Euler wavelets method for solving fractional-order linear Volterra–Fredholm integro-differential equations with weakly singular kernels", *Computational and Applied Mathematics*, **40**(6), pp. 1–30.
8. Behera S., and Saha Ray S., 2022, "An efficient numerical method based on Euler wavelets for solving fractional order pantograph Volterra delay-integro-differential equations", *Journal of Computational and Applied Mathematics*, **406**, p. 113825.
9. Behera S., and Saha Ray S., 2022, "A wavelet-based novel technique for linear and nonlinear fractional Volterra–Fredholm integro-differential equations", *Computational and Applied Mathematics*, **41**(2), pp. 1–28.
10. Amin R., Shah K., Asif M., Khan I., and Ullah F., 2021, "An efficient algorithm for numerical solution of fractional integro-differential equations via Haar wavelet", *Journal of Computational and Applied Mathematics*, **381**, p. 113028.
11. Mohammad M., and Trounev A., 2020, "Implicit Riesz wavelets based-method for solving singular fractional integro-differential equations with applications to hematopoietic stem cell modelling", *Chaos, Solitons & Fractals*, **138**, p. 109991.
12. Singh A.K., and Mehra M., 2021, "Wavelet collocation method based on Legendre polynomials and its application in solving the stochastic fractional integro-differential equations", *Journal of Computational Science*, **51**, p. 101342.
13. Wang J., Xu T.Z., Wei Y.Q., and Xie J.Q., 2018, "Numerical simulation for coupled systems of nonlinear fractional order integro-differential equations via wavelets method", *Applied Mathematics and Computation*, **324**, pp. 36–50.
14. Zhu L., and Fan Q., 2012, "Solving fractional nonlinear Fredholm integro-differential equations by the second kind Chebyshev wavelet", *Communications in Nonlinear Science and Numerical Simulation*, **17**(6), pp. 2333–2341.
15. Lepik Ü., 2009, "Solving fractional integral equations by the Haar wavelet method", *Applied Mathematics and Computation*, **214**(2), pp. 468–478.
16. Alqahtani B., Aydi H., Karapınar E., and Rakočević V., 2019, "A solution for Volterra fractional integral equations by hybrid contractions", *Mathematics*, **7**(8), p. 694.
17. Abdeljawad T., Agarwal R.P., Karapınar E., and Kumari P.S., 2019, "Solutions of the nonlinear integral equation and fractional differential equation using the technique of a fixed point with a numerical experiment in extended b-metric space", *Symmetry*, **11**(5), p. 686.
18. Mohammad M., and Trounev A., 2020, "Fractional nonlinear Volterra–Fredholm integral equations involving Atangana–Baleanu fractional derivative: framelet applications", *Advances in Difference Equations*, **2020**(1), pp. 1–15.
19. Hamoud A., and Ghadle K., 2018, "Existence and uniqueness of solutions for fractional mixed Volterra-Fredholm integro-differential equations", *Indian Journal of Mathematics*, **60**(3), pp. 375–395.
20. Hamoud A., 2020, "Existence and uniqueness of solutions for fractional neutral volterra-fredholm integro differential equations", *Advances in the Theory of Nonlinear Analysis and its Application*, **4**(4), pp. 321–331.
21. Li X., 2012, "Numerical solution of fractional differential equations using cubic B-spline wavelet collocation method", *Communications in Nonlinear Science and Numerical Simulation*, **17**(10), pp. 3934–3946.
22. He J.H., 1998, "Approximate analytical solution for seepage flow with fractional derivatives in porous media", *Computer Methods in Applied Mechanics and Engineering*, **167**(1–2), pp. 57–68.

23. Mohamed M.A., and Torky M.S., 2014, "Approximate solution of fractional nonlinear partial differential equations by the Legendre multiwavelet Galerkin method", *Journal of Applied Mathematics*, **2014**, pp. 1–12.
24. Yuanlu L.I., 2010, "Solving a nonlinear fractional differential equation using Chebyshev wavelets", *Communications in Nonlinear Science and Numerical Simulation*, **15**(9), pp. 2284–2292.
25. Milici C., Draganescu G., and Tenreiro Machado J.A., 2019, *Introduction to Fractional Differential Equations*, Springer International Publishing, Cham, Switzerland.
26. Saha Ray S., 2020, *Nonlinear Differential Equations in Physics*, Springer Nature, Singapore.
27. Miller K.S., and Ross B., 1993, *An Introduction to the Fractional Calculus and Fractional Differential Equations*, Wiley, Hoboken, NJ.
28. Machado J.T., Kiryakova V., and Mainardi F., 2011, "Recent history of fractional calculus", *Communications in Nonlinear Science and Numerical Simulation*, **16**(3), pp. 1140–1153.
29. Podlubny I., 1998, *Fractional Differential Equations: An Introduction to Fractional Derivatives, Fractional Differential Equations, to Methods of Their Solution and Some of Their Applications*, Elsevier, Amsterdam, Netherlands.
30. Kilbas A.A., Srivastava H.M., and Trujillo J.J., 2006. *Theory and Applications of Fractional Differential Equations*, p. 204, Elsevier, Amsterdam, Netherlands.
31. Caputo M., 1967, "Linear models of dissipation whose Q is almost frequency independent-II", *Geophysical Journal International*, **13**(5), pp. 529–539.

2 Wavelet Methods

2.1 INTRODUCTION

Wavelet analysis has attracted much attention towards the signal processing. It has been successfully applied in many applications such as image analysis, signal analysis, and communications systems [1,2]. In the 1980s, wavelet theory emerged as an essential mathematical tool that extends the ideas of the traditional Fourier theory (1800s). It has been used for analysing, visualizing, and manipulating complex non-stationary data, for which the traditional Fourier methods cannot be applied directly. Wavelet theory can be viewed as a modern improvement and extension of the Fourier theory. Many researchers have made significant contribution in the field of wavelet analysis [3–6]. In general, wavelet is a complex valued function defined on some space. However, we describe a wavelet as a real-valued function defined on the real line. It may be viewed as a function of time t or a function of spatial variable x.

Like the Fourier analysis, wavelet analysis deals with expansion of functions in terms of a set of basis functions. Wavelet analysis expands functions not in terms of trigonometric polynomials but in terms of wavelets. These wavelets are generated in the form of translations and dilations of a fixed function called the mother wavelet. The main objective of wavelet analysis is to define these wavelet basis functions and find efficient methods for their computation.

In our study, we consider a smooth mother wavelets whose set of discrete translations and dilations form an orthonormal basis for $L^2(\mathbb{R})$, where $\mathbb{R}$ is the real numbers and L^2 is the set of all functions f, defined on $\mathbb{R}$, that satisfy

$$\int_{-\infty}^{\infty} |f(x)|^2 \, dx \le \infty.$$

In fact, we look for such "waves" that generate $L^2(\mathbb{R})$, and these waves should decay to zero at $\pm\infty$, and for all practical purpose, the decay should be very fast. That is, we look for small waves, or "wavelets", to generate $L^2(\mathbb{R})$. For this purpose, we prefer a single function ψ that generates all of $L^2(\mathbb{R})$. The wavelet ψ decays very quickly, and therefore to cover the whole real line, we shift it along $\mathbb{R}$. For computational efficiency, we have used integral powers of 2 for frequency partitioning. That is, consider the small waves

$$\psi(2^j x - k), \quad j,k \in \mathbb{Z}.$$

$\psi(2^j x - k)$ is obtained from a single wavelet function (called mother wavelet) $\psi(x)$ by a binary dilation (dilation by 2^j) and a dyadic translation (of $k/2^j$). Hence, if a function $\psi \in L^2(\mathbb{R})$, then all of the functions, $\psi_{j,k}$, are defined by

$$\psi_{j,k}(x) = 2^{j/2}\psi(2^j x - k), \quad j,k \in \mathbb{Z}.$$

DOI: 10.1201/9781003453451-2

2.2 WAVELET TRANSFORMS

In the early 1980s, the concept of "ondelettes" or "wavelets" started to appear more frequently. This innovative concept can be interpreted as a formulation of many theories originating from various fields such as mathematics, physics, and engineering. Wavelets were initially introduced by Morlet and Grossmann [7,8] in the early 1980s. In wavelet analysis, various wavelet transforms have been introduced such as Morlet and Grossmann [8] for formulation of continuous wavelet transform (CWT), Daubechies [9] for proposal of orthogonal compactly supported wavelets, and Meyer [10] and Mallat [11] for multi-resolution analysis using wavelet transform.

In 1982, Jean Morlet, a French geophysical engineer, introduced the wavelet transform concept by developing a new mathematical tool for seismic wave analysis. In Morlet's study, signals have distinct time and frequency characteristics, but their high-frequency components have a shorter time duration than their low-frequency components. For improving the time resolution for the high-frequency transients and frequency resolution for the low-frequency components, Morlet et al. first proposed the concept of wavelets as a family of functions constructed from translations and dilations of a single function called the "mother wavelet" $\psi(x)$. In the discrete domain, if a as a scaling parameter and b as translation parameter fluctuate continuously, we have the following family of continuous wavelets:

$$\psi_{a,b}(x) = |a|^{\frac{-1}{2}} \psi\left(\frac{x-b}{a}\right), \quad a, b \in \mathbb{R}, a \neq 0. \tag{2.1}$$

If $|a| < 1$, the wavelet (2.1) is a compressed form of the mother wavelet that corresponds to higher frequencies. Similarly, when $|a| > 1$, $\psi_{a,b}(x)$ has a larger time-width than $\psi(x)$ and corresponds to lower frequencies. As a result, wavelets have time-widths that are proportional to their frequencies. This is the primary reason for the Morlet wavelets' effectiveness in signal processing and time-frequency signal analysis. The resolution of wavelets at various scales varies in the time and frequency domains, as defined by the Heisenberg uncertainty principle. The resolution in the time domain becomes finer as the scale a decreases, whereas the resolution in the frequency domain becomes coarser.

If we restrict the parameters a and b to discrete values as $a = a_0^{-k}$, $b = nb_0a_0^{-k}$, $a_0 > 1$, $b_0 > 0$, and n and k are positive integers, from Eq. (2.1), we have the following family of discrete wavelets:

$$\psi_{k,n}(x) = |a_0|^{\frac{k}{2}} \psi(a_0^k x - nb_0), \quad n, k \in \mathbb{Z}^+. \tag{2.2}$$

2.2.1 Orthonormal Wavelets

The orthonormal wavelets with good time-frequency localization have been established to play a vital role in wavelet theory and have a wide range of applications. In general, the theory of wavelets instigates with a single function $\psi \in L^2(\mathbb{R})$, and a family of functions $\psi_{k,n}$ is constructed from this single function ψ by the operation of binary dilation (i.e., dilation by 2^k) and dyadic translation of $n2^{-k}$ so that

$$\psi_{k,n}(x) = 2^{\frac{k}{2}}\psi\left(2^k\left(x - \frac{n}{2^k}\right)\right), \quad k, n \in \mathbb{Z},$$

$$= 2^{\frac{k}{2}}\psi(2^k x - n),$$

where the coefficient $2^{\frac{k}{2}}$ is defined for orthonormality.

Definition 2.2.1.1 (Orthonormal Wavelet)

A wavelet $\psi \in L^2(\mathbb{R})$ is called an orthonormal wavelet, if the family of functions $\psi_{k,n}$, generated from ψ, is an orthogonal basis of $L^2(\mathbb{R})$; i.e.,

$$\langle \psi_{i,j}, \psi_{k,n} \rangle = \delta_{i,k}\delta_{j,n}, \quad i, j, k, n \in \mathbb{Z}.$$

Definition 2.2.1.2 (Semi-Orthogonal Wavelet)

A wavelet $\psi \in L^2(\mathbb{R})$ is called a semi-orthogonal wavelet, if the family $\{\psi_{k,n}\}$ satisfies the following condition:

$$\langle \psi_{i,j}, \psi_{k,n} \rangle = 0, \quad i \neq k, \quad i, j, k, n \in \mathbb{Z}.$$

2.3 MULTIRESOLUTION ANALYSIS

In 1989, Stephane Mallat and Yves Meyer proposed the concept of multiresolution analysis (MRA). The basic idea of MRA is to represent a function as the limit of a series of consecutive approximations, each of which is a "smoother" version of the original function. The successive approximations correlate to different resolutions, giving rise to the term MRA, which refers to a formal technique for constructing orthogonal wavelet bases using a specific set of rules and procedures. It also allows for the creation of scaling functions and scaling filters, which are subsequently utilized to construct wavelets and efficient numerical methods. In applications, it is an effective mathematical framework for hierarchical decomposition of a signal or an image into components of different scales represented by a sequence of function spaces on $\mathbb{R}$.

Any wavelet, orthogonal or semi-orthogonal, generates a direct sum decomposition of $L^2(\mathbb{R})$. For each $j \in \mathbb{Z}$, let us consider the closed subspaces

$$U_j = \ldots \oplus V_{j-2} \oplus V_{j-1}, \quad j \in \mathbb{Z},$$

of $L^2(\mathbb{R})$. A set of subspaces $\{U_j\}_{j\in\mathbb{Z}}$ is said to be MRA of $L^2(\mathbb{R})$ if it possesses the following properties:

1. $U_j \subset U_{j+1}$, $j \in \mathbb{Z}$,
2. $\bigcup_{j\in\mathbb{Z}} U_j$ is dense in $L^2(\mathbb{R})$,

3. $\bigcap_{j\in\mathbb{Z}} U_j = \varphi$,
4. $U_{j+1} = U_j \oplus V_j$,
5. $f(x) \in U_j \Leftrightarrow f(x) \in U_{j+1}, \quad \forall j \in \mathbb{Z}$.

Properties (2)–(5) state that $\{U_j\}_{j\in\mathbb{Z}}$ is a nested sequence of subspaces that effectively covers $L^2(\mathbb{R})$. That is, every square integrable function can be approximated as closely as desired by a function that belongs to at least one of the subspaces U_j. A function $\phi \in L^2(\mathbb{R})$ is called a scaling function if it generates the nested sequence of subspaces U_j and satisfies the dilation equation, namely,

$$\phi(x) = \sum_r p_r \phi(ax - r). \tag{2.3}$$

The coefficients $\{p_r\}$ are square-summable real or complex numbers and a being any rational number.

For each scale j, since $U_j \subset U_{j+1}$, there exists a unique orthogonal complementary subspace V_j of U_j in U_{j+1}. This subspace $V_j = \mathrm{Span}\{\psi_{j,k} \mid \psi_{j,k} = \psi(2^j x - k)\}$ is called wavelet subspace and is generated by $\psi_{j,k} = \psi(2^j x - k)$, where $\psi \in L^2$ is called the wavelet. From the above discussion, these results follow easily

- $U_{j_1} \bigcap U_{j_2}$, $j_1 > j_2$,
- $V_{j_1} \bigcap V_{j_2} = 0$, $j_1 \neq j_2$,
- $U_{j_1} \bigcap V_{j_2} = 0$, $j_1 \leq j_2$.

In order to solve integral and integro-differential equations by numerical methods, the approximated solution can be determined by wavelets of different resolutions. The operational matrix constructed from an integral operator can be sparsified using wavelet-based techniques to attain an arbitrary degree of solution accuracy. The main advantage of wavelet-based methods is its ability to transform the given integral and integro-differential equations to a system of linear or nonlinear algebraic equations that can be solved by numerical methods.

2.4 TYPES OF WAVELETS

2.4.1 Legendre Wavelets

Legendre wavelets (LWs) $\psi^L_{\hat{n},m}(x) = \psi^L(k, \hat{n}, m, x)$ have four arguments $\hat{n} = 2n - 1$, $n = 1, 2, \ldots, 2^{k-1}$, $k \in \mathbb{Z}^+$, $m = 0, 1, \ldots, M - 1$, where x is the normalized time and m is the order of Legendre polynomials. LWs are defined on $[0, 1)$ as follows [12]:

$$\psi^L_{\hat{n},m}(x) = \begin{cases} \sqrt{m + \frac{1}{2}}\, 2^{\frac{k}{2}} L_m(2^k x - 2n + 1), & \frac{\hat{n} - 1}{2^k} \leq x < \frac{\hat{n} + 1}{2^k} \\ 0, & \text{otherwise}, \end{cases} \tag{2.4}$$

where $L_0(x) = 1$,

$$L_1(x) = x,$$

$$L_{m+1}(x) = \frac{(2m+1)}{(m+1)} x L_m(x) - \frac{m}{(m+1)} L_{m-1}(x), \quad m = 1,2,3,\ldots.$$

The two-dimensional LWs can be defined over the domain $[0,1) \times [0,1)$ as follows:

$$\psi^L_{n,m,n',m'}(x,t) = \begin{cases} \sqrt{m+\frac{1}{2}} \cdot \sqrt{m'+\frac{1}{2}} 2^{\frac{k_1+k_2}{2}} L_m(2^{k_1}x - 2n+1) \cdot & \frac{n-1}{2^{k_1-1}} \le x < \frac{n}{2^{k_1-1}}, \\ L_{m'}(2^{k_2}t - 2n'+1), & \frac{n'-1}{2^{k_2-1}} \le t < \frac{n'}{2^{k_2-1}}. \\ 0, & \text{otherwise}, \end{cases} \tag{2.5}$$

where $n = 1,2,\ldots,2^{k_1-1}$, $n' = 1,2,\ldots,2^{k_2-1}$, $m = 0,1,2,\ldots,M-1$, and $m' = 0,1,2,\ldots,M'-1$. Here k_1 and k_2 are any positive integers. L_m and $L_{m'}$ are Legendre polynomials of order m and m', respectively.

2.4.2 Bernoulli Wavelets

Bernoulli wavelets (BWs) $\psi^B_{n,m}(x) = \psi^B(k,n,m,x)$ have four arguments; defined on $[0,1)$ by [13]

$$\psi^B_{n,m}(x) = \begin{cases} 2^{\frac{k-1}{2}} \mathcal{B}_m(2^{k-1}x - n+1), & \frac{n-1}{2^{k-1}} \le x < \frac{n}{2^{k-1}}, \\ 0, & \text{otherwise}, \end{cases} \tag{2.6}$$

with

$$\mathcal{B}_m(x) = \begin{cases} 1, & m = 0, \\ \frac{1}{\sqrt{\frac{(-1)^{m-1}(m!)^2}{(2m!)^2}\widehat{\alpha}_{2m}}} \widehat{\beta}_m(x), & m > 0. \end{cases}$$

where $n = 1,2,\ldots,2^{k-1}$, k is any positive integer, $m = 0,1,\ldots,M-1$ ($M > 0$ is a fixed integer) is the order for the Bernoulli polynomial and x is the normalized time.

The coefficient $\frac{1}{\sqrt{\frac{(-1)^{m-1}(m!)^2}{(2m!)^2}\widehat{\alpha}_{2m}}}$ is for orthonormality, the dilation parameter is $\alpha = 2^{-(k-1)}$, and the translation parameter is $\gamma = (n-1)2^{-(k-1)}$. Here, $\widehat{\beta}_m(x)$ are the well-known mth-order Bernoulli polynomials which can be defined by [14]

$$\widehat{\beta}_m(x)=\sum_{i=0}^{m}\binom{m}{i}\widehat{\alpha}_{m-i}x^i, \tag{2.7}$$

where $\widehat{\alpha}_i$, $i=0,1,\ldots,m$, are Bernoulli numbers.

The first few Bernoulli numbers are as follows:

$$\widehat{\alpha}_0=1,\quad \widehat{\alpha}_1=-\frac{1}{2},\quad \widehat{\alpha}_2=\frac{1}{6},\quad \widehat{\alpha}_3=0,\quad \widehat{\alpha}_4=-\frac{1}{30},\quad \widehat{\alpha}_5=0,\quad \widehat{\alpha}_6=\frac{1}{42}$$

with $\widehat{\alpha}_{2i+1}=0,\quad i=1,2,\ldots.$

The first few Bernoulli polynomials are as follows:

$$\widehat{\beta}_0(x)=1,\quad \widehat{\beta}_1(x)=x-\frac{1}{2},\quad \widehat{\beta}_2(x)=x^2-x+\frac{1}{6},\quad \widehat{\beta}_3(x)=x^3-\frac{3}{2}x^2+\frac{1}{2}x.$$

These Bernoulli polynomials satisfy the following formula [15]:

$$\int_0^1 \widehat{\beta}_m(x)\widehat{\beta}_n(x)\,dx=(-1)^{m-1}\frac{n!m!}{(n+m)!}\widehat{\beta}_{n+m},\quad m,n\ge 1. \tag{2.8}$$

The two-dimensional BWs can be defined over the domain $[0,1)\times[0,1)$ as follows:

$$\psi^{B}_{n,m,n',m'}(x,t)=\begin{cases} 2^{\frac{k_1+k_2}{2}}\tilde{\beta}_m(2^{k_1-1}x-n+1). & \text{for } \dfrac{n-1}{2^{k_1-1}}\le x<\dfrac{n}{2^{k_1-1}},\\ \tilde{\beta}_{m'}(2^{k_2-1}t-n'+1), & \dfrac{n'-1}{2^{k_2-1}}\le t<\dfrac{n'}{2^{k_2-1}},\\ 0, & \text{otherwise,}\end{cases} \tag{2.9}$$

where

$$\tilde{\beta}_m(x)=\begin{cases} 1, & m=0\\ \dfrac{1}{\sqrt{\dfrac{(-1)^{m-1}(m!)^2}{(2m)!}\beta_{2m}}}\beta_m(x), & m>0\end{cases}$$

and

$$\tilde{\beta}_{m'}(t)=\begin{cases} 1, & m'=0\\ \dfrac{1}{\sqrt{\dfrac{(-1)^{m'-1}(m'!)^2}{(2m')!}\beta_{2m'}}}\beta_{m'}(t), & m'>0,\end{cases}$$

where $n = 1,2,\ldots,2^{k_1-1}$, $n' = 1,2,\ldots,2^{k_2-1}$, $m = 0,1,2,\ldots,M-1$, and $m' = 0,1,2,\ldots,M'-1$. Here k_1 and k_2 are any positive integers. $\widehat{\beta}_m$ and $\widehat{\beta}_{m'}$ are Bernoulli polynomials of order m and m', respectively.

2.4.2.1 Properties of Bernoulli Polynomials

The followings are the properties of Bernoulli polynomials [14]:

1. $\beta_m(1-x) = (-1)^m \beta_m(x), \quad m \in \mathbb{Z}^+$.
2. $\beta_m'(x) = m\beta_{m-1}(x), \quad m \in \mathbb{Z}^+$.
3. $\int_0^1 |\beta_m(x)| dx < 16\frac{m!}{(2\pi)^{m+1}}, \quad m \geq 0$.
4. $\int_a^x \beta_m(t)\, dt = \frac{\beta_{m+1}(x) - \beta_{m+1}(a)}{m+1}$.
5. $\sup_{x\in[0,1]} |\beta_{2m}(x)| = |\alpha_{2m}|$.
6. $\sup_{x\in[0,1]} |\beta_{2m+1}(x)| \leq \frac{2m+1}{4}|\alpha_{2m}|$.

2.4.2.2 Properties of Bernoulli Numbers

The following properties satisfy the sequence of Bernoulli numbers $(\alpha_m)_{m\in N}$ [14]:

1. $\alpha_{2m+1} = 0, \quad \alpha_{2m} = \beta_{2m}(1)$.
2. $\beta_m(\frac{1}{2}) = (2^{1-m} - 1)\alpha_m$.
3. $\alpha_n = -\frac{1}{n+1}\sum_{k=0}^{n-1}\binom{n+1}{k}\alpha_k$.

2.4.3 Gegenbauer Wavelets

Gegenbauer polynomials, which is also known as ultra-spherical harmonic polynomials of order $m \in \mathbb{Z}^+$, are defined as $\mathcal{G}_m^\lambda(x)$ for $\lambda > -\frac{1}{2}$ (λ is the known ultra-spherical parameter) on the interval [−1, 1] and given by [16]

$$\mathcal{G}_0^\lambda(x) = 1,$$

$$\mathcal{G}_1^\lambda(x) = 2\lambda x,$$

$$\mathcal{G}_{m+1}^\lambda(x) = \frac{1}{m+1}(2(m+\lambda)x\, \mathcal{G}_m^\lambda(x) - (m+2\lambda-1)\, \mathcal{G}_{m-1}^\lambda(x)), \quad m = 1,2,3,\ldots.$$

Gegenbauer polynomials satisfy the orthogonality on $[-1,1]$ with the weight function $\tilde{w}(x)=(1-x^2)^{\lambda-\frac{1}{2}}$ as follows:

$$\int_{-1}^{1}(1-x^2)^{\lambda-\frac{1}{2}}\mathcal{G}_m^{\lambda}(x)\mathcal{G}_n^{\lambda}(x)\,dx=\begin{cases}\tilde{L}_m^{\lambda}, & m=n,\\ 0, & m\neq n,\end{cases}$$

where

$$\tilde{L}_m^{\lambda}=\begin{cases}\dfrac{\pi 2^{1-2\lambda}\Gamma(m+2\lambda)}{m!(m+\lambda)(\Gamma(\lambda))^2}, & \lambda\neq 0,\\ \dfrac{2\pi}{m^2}, & \lambda=0,\\ \pi, & \lambda=0,\, m=0\end{cases}$$

is the normalizing factor.

This following inequality holds for Gegenbauer polynomials [17]:

$$\left|\mathcal{G}_m^{\lambda}(\cos\theta)\right|(\sin\theta)^{\lambda}<\frac{2^{1-\lambda}\Gamma\left(m+\dfrac{3\lambda}{2}\right)}{\Gamma(\lambda)\Gamma\left(1+m+\dfrac{\lambda}{2}\right)},\quad 0\leq\theta\leq\pi. \tag{2.10}$$

The Gegenbauer polynomial is the extended forms of the Chebyshev and Legendre polynomials.

For $\lambda=0$, the first kind Chebyshev polynomials are obtained as follows:

$$T_m(x)=\frac{m}{2}\lim_{\lambda\to 0}\frac{\mathcal{G}_m^{\lambda}(x)}{\lambda},\quad m\geq 1.$$

For $\lambda=1$, the second kind Chebyshev polynomials are obtained as follows:

$$U_m(x)=\mathcal{G}_m^{1}(x).$$

For $\lambda=\frac{1}{2}$, the Legendre polynomials are obtained as follows:

$$L_m(x)=\mathcal{G}_m^{\frac{1}{2}}(x).$$

Gegenbauer wavelets (GWs) $\psi_{n,m}^{G}(x)=\psi^{G}(k,n,m,x)$ are defined on the interval $[0,1)$ as follows:

$$\psi_{n,m}^{G}(x)=\begin{cases}\dfrac{2^{\frac{k}{2}}}{\sqrt{L_m^{\lambda}}}\mathcal{G}_m^{\lambda}(2^k x-2n+1), & \dfrac{n-1}{2^{k-1}}\leq x<\dfrac{n}{2^{k-1}},\\ 0, & \text{otherwise,}\end{cases} \tag{2.11}$$

where $n = 1, 2, ..., 2^{k-1}$, $k \in \mathbb{Z}^{+}$, $m = 0, 1, ..., M-1$ ($M > 0$ is a fixed integer) is the order for the Gegenbauer polynomial and x is the normalized time. $\mathcal{G}_m^{\lambda}(2^k x - 2n + 1)$ are Gegenbauer polynomials of degree m, which are orthogonal with respect to the weight function $\tilde{w}_{n,k}(x) = \tilde{w}(2^k x - 2n + 1) = (1 - (2^k x - 2n + 1)^2)^{\lambda - \frac{1}{2}}$ on $[-1, 1]$.

The two-dimensional GWs can be defined over the domain $[0,1) \times [0,1)$ as follows:

$$\psi^{G}_{n,m,n',m'}(x,t) = \begin{cases} \dfrac{2^{\frac{k_1+k_2}{2}}}{\sqrt{L_m^{\lambda}}\sqrt{L_{m'}^{\lambda}}} \mathcal{G}_m^{\lambda}(2^{k_1} x - 2n + 1). & \text{for } \dfrac{n-1}{2^{k_1-1}} \le x < \dfrac{n}{2^{k_1-1}}, \\ \mathcal{G}_{m'}^{\lambda}(2^{k_2} t - 2n' + 1), & \dfrac{n'-1}{2^{k_2-1}} \le t < \dfrac{n'}{2^{k_2-1}}, \\ 0, & \text{otherwise,} \end{cases} \tag{2.12}$$

where $k_1, k_2 \in \mathbb{Z}^{+}$, $n = 1, 2, ..., 2^{k_1-1}$, $n' = 1, 2, ..., 2^{k_2-1}$, $m = 0, 1, ..., M-1$, and $m' = 0, 1, ..., M'-1$. $\mathcal{G}_m^{\lambda}$ and $\mathcal{G}_{m'}^{\lambda}$ are Gegenbauer polynomials of order m and m', respectively, and $\psi^{G}_{n,m,n',m'}(x,t)$ forms a basis for $L^2(\mathbb{R}^2)$.

2.4.4 Cosine and Sine Wavelets

Cosine and Sine wavelets (CASWs) are defined on the interval $[0,1)$ by [18]

$$\psi^{\text{CAS}}_{n,m}(x) = \begin{cases} 2^{\frac{k}{2}} \text{CAS}_m(2^k x - n), & \dfrac{n}{2^k} \le x < \dfrac{n+1}{2^k}, \\ 0, & \text{otherwise,} \end{cases} \tag{2.13}$$

where $n = 1, 2, ..., 2^k - 1$, $k \in \mathbb{Z}^{+}$, $m = -M, -M+1, ..., M$ and $\text{CAS}_m(x) = \cos(2m\pi x) + \sin(2m\pi x)$.

The two-dimensional CASWs are defined as follows:

$$\psi^{\text{CAS}}_{n,m,n',m'}(x,t) = \begin{cases} 2^{\frac{k_1+k_2}{2}} \text{CAS}_{m_1}(2^{k_1} x - n).\text{CAS}_{m'}(2^{k_2} t - n'), & \dfrac{n}{2^{k_1}} \le x < \dfrac{n+1}{2^{k_1}}, \\ & \dfrac{n'}{2^{k_2}} \le t < \dfrac{n'+1}{2^{k_2}} \\ 0, & \text{otherwise,} \end{cases} \tag{2.14}$$

where $k_1, k_2 \in \mathbb{Z}^{+}$, $n = 0, 1, ..., 2^{k_1} - 1$, $n' = 0, 1, ..., 2^{k_2} - 1$, $m = -M, ..., M$, and $m' = -M', ..., M'$. $\psi^{\text{CAS}}_{n,m,n',m'}(x,t)$ forms a basis for $L^2(\mathbb{R}^2)$.

2.4.5 Jacobi Wavelets

The Jacobi polynomials, associated with the real parameters $\delta, \eta \ge -1$, $\wp_m^{(\delta,\eta)}(x)$, $x \in [-1, 1]$, are defined by [19]

$$\wp_0^{(\delta,\eta)}(x) = 1,$$

$$\wp_1^{(\delta,\eta)}(x) = \frac{1}{2}(\delta+\eta+2)x + \frac{1}{2}(\delta-\eta),$$

$$\wp_{m+1}^{(\delta,\eta)}(x) = (a_m^{(\delta,\eta)}x - b_m^{(\delta,\eta)})\wp_m^{(\delta,\eta)}(x) - c_m^{(\delta,\eta)}\wp_{m-1}^{(\delta,\eta)}(x),\ m \geq 1,$$

where m is the degree of the Jacobi polynomial and

$$a_m^{(\delta,\eta)} = \frac{(2m+\delta+\eta+1)(2m+\delta+\eta+2)}{2(m+1)(m+\delta+\eta+1)},$$

$$b_m^{(\delta,\eta)} = \frac{(2m+\delta+\eta+1)(\eta^2-\delta^2)}{2(m+1)(m+\delta+\eta+1)(2m+\delta+\eta)},$$

$$c_m^{(\delta,\eta)} = \frac{(2m+\delta+\eta+2)(m+\delta)(m+\eta)}{(m+1)(m+\delta+\eta+1)(2m+\delta+\eta)}.$$

The Jacobi polynomials are orthogonal with respect to the weight function $w^{(\delta,\eta)}(x) = (1-x)^\delta(1+x)^\eta$, $\delta,\eta \geq -1$ over the interval $[-1,1]$, and

$$\int_{-1}^{1} (1-x)^\delta(1+x)^\eta \wp_m^{(\delta,\eta)}(x)\wp_n^{(\delta,\eta)}(x)dx = \begin{cases} \dfrac{1}{g_m^{(\delta,\eta)}}, & m = n, \\ 0, & m \neq n, \end{cases} \tag{2.15}$$

where

$$g_m^{(\delta,\eta)} = \frac{(2m+\delta+\eta+1)\Gamma(m+1)\Gamma(m+\delta+\eta+1)}{2^{(\delta+\eta)}\Gamma(m+\delta+1)\Gamma(m+\eta+1)}.$$

The change of variables $x \mapsto 2x-1$ gives Jacobi polynomials on $[0,1]$ for the weight function $w^{(\delta,\eta)}(x) = x^\eta(1-x)^\delta$. They are defined by the orthogonality conditions

$$\int_0^1 x^\eta(1-x)^\delta \wp_m^{(\delta,\eta)}(x)x^m\,dx = 0, \tag{2.16}$$

where $m = 0,1,\ldots,M-1$.

The Jacobi wavelets (JWs) $\psi_{n,m}^J(x) = \psi^J(k,n,m,x)$ have four arguments and are defined on $[0,1)$ by [20]

$$\psi_{n,m}^J(x) = \begin{cases} \sqrt{g_m^{(\delta,\eta)}}\,2^{\frac{k-1}{2}}\wp_m^{(\delta,\eta)}(2^k x - 2n + 1), & \dfrac{(n-1)}{2^{k-1}} \leq x < \dfrac{n}{2^{k-1}}, \\ 0, & \text{otherwise} \end{cases} \tag{2.17}$$

where $n = 1,2,\ldots,2^{k-1}$, $m = 0,1,\ldots,M-1$, k is any positive integer and x is the normalized time. Here, the coefficient $\sqrt{g_m^{(\delta,\eta)}}$ is for orthonormality.

The JWs form an orthonormal basis for $L^2([0,1))$, i.e.,

$$<\psi^J_{n,m}(x),\psi^J_{n',m'}(x)>=\int_0^1 \varpi_n(x)\psi^J_{n,m}(x)\psi^J_{n',m'}(x)\,dx=\begin{cases}1, & (n,m)=(n',m'),\\ 0, & (n,m)\neq(n',m'),\end{cases} \tag{2.18}$$

where

$$\varpi_n(x)=\begin{cases}(2n-2^k x)^{\delta}(2^k x-2n+2)^{\eta}, & \frac{(n-1)}{2^{k-1}}\le x<\frac{n}{2^{k-1}}\\ 0, & \text{otherwise.}\end{cases}$$

In the domain $[0,1)\times[0,1)$, the two-dimensional JWs are defined as follows:

$$\psi^J_{n,m,n',m'}(x,t)=\begin{cases}\sqrt{g_m^{(\delta,\eta)}}\sqrt{g_{m'}^{(\delta,\eta)}}\,2^{\frac{k_1-1}{2}}2^{\frac{k_2-1}{2}} & \frac{n-1}{2^{k_1-1}}\le x<\frac{n}{2^{k_1-1}},\\ \wp_m^{(\delta,\eta)}(2^{k_1}x-2n+1)\wp_{m'}^{(\delta,\eta)}(2^{k_2}t-2n'+1), & \frac{n'-1}{2^{k_2-1}}\le t<\frac{n'}{2^{k_2-1}},\\ 0 & \text{otherwise,}\end{cases} \tag{2.19}$$

where $n=1,2,\dots,2^{k_1-1}$, $n'=1,2,\dots,2^{k_2-1}$, $m=0,1,\dots,M-1$, and $m'=0,1,\dots,M'-1$. Here, k_1 and k_2 are any positive integers. $\wp_m^{(\delta,\eta)}$ and $\wp_{m'}^{(\delta,\eta)}$ are Jacobi polynomials of order m and m', respectively.

2.4.6 Müntz–Legendre Wavelets

The Müntz–Legendre wavelets (MLWs) $\psi^{\mathrm{ML}}_{n,m}(x)=\psi^{\mathrm{ML}}(k,\hat{n},m,x)$ have four arguments, namely, $\hat{n}=n-1$, $n=1,2,\dots,2^{k-1}$, $k\in\mathbb{Z}^+$, and $m=0,1,\dots,M-1$, and $M\in\mathbb{Z}^+$ is the order for Mu¨ntz–Legendre polynomial (MLP) and x is the normalized time. The MLWs is defined on $[0,1)$ as follows:

$$\psi^{\mathrm{ML}}_{n,m}(x)=\begin{cases}\sqrt{(2\tau_m+1)}\,2^{\frac{k-1}{2}}\tilde{L}_m(2^{k-1}x-n+1), & \frac{n-1}{2^{k-1}}\le x<\frac{n}{2^{k-1}},\\ 0, & \text{otherwise.}\end{cases} \tag{2.20}$$

The co-efficient $\sqrt{(2\tau_m+1)}$, $m=0,1,\dots,M-1$, is for orthonormality.

Here, $\tilde{L}_m(x)$ are the MLPs which are defined on [0,1] by [21]

$$\tilde{L}_m(x)=\sum_{k=0}^{m}\kappa_{k,m}x^{\tau_k},$$

where

$$\kappa_{l,m} = \frac{\prod_{j=0}^{m-1} (\tau_l + \tau_j + 1)}{\prod_{j=0, j \neq l}^{m} (\tau_l - \tau_j)}, \quad m \in N_0.$$

The MLPs satisfy the following orthogonality condition:

$$\int_0^1 \tilde{L}_n(x) \tilde{L}_m(x) dx = \frac{\delta_{n,m}}{(2\tau_m + 1)}, \quad (m \geq n),$$

where $\delta_{n,m}$ is the Kronecker symbol.

These polynomials also satisfy the following recursive formula [22]:

$$\tilde{L}_m(x) = \tilde{L}_{m-1}(x) - (\tau_m + \tau_{m-1} + 1) x^{\tau_m} \int_x^1 t^{-\tau_m - 1} \tilde{L}_{m-1}(t) dt, \quad x \in (0,1].$$

In this chapter, it has been chosen that $\tau_l = l\gamma$, where γ is a real constant.

2.4.7 Euler Wavelets

The Euler wavelets (EWs) $\psi^E_{n,m}(x) = \psi(k, n, m, x)$ have four arguments: $n = 1, 2, \ldots, 2^{k-1}$, k is any positive integer, m is the degree of the Euler polynomials (EPs), and x is the normalized time. The EW is defined on $[0,1)$ for $m = 0, 1, \ldots, M-1$ ($M > 0$ is a fixed integer) as follows [23]:

$$\psi^E_{n,m}(x) = \begin{cases} 2^{\frac{k-1}{2}} \bar{E}_m(2^{k-1}x - n + 1), & \frac{n-1}{2^{k-1}} \leq x < \frac{n}{2^{k-1}}, \\ 0, & \text{otherwise} \end{cases} \tag{2.21}$$

with

$$\bar{E}_m(x) = \begin{cases} 1, & m = 0, \\ \dfrac{1}{\sqrt{\dfrac{2(-1)^{m-1}(m!)^2}{(2m)!} E_{2m+1}(0)}} E_m(x), & m > 0. \end{cases} \tag{2.22}$$

The coefficient $\dfrac{1}{\sqrt{\dfrac{2(-1)^{m-1}(m!)^2}{(2m)!} E_{2m+1}(0)}}$ is for normality. The functions $E_m(x)$ are the well-known EPs of order m, which can be constructed from the following generating functions [24]:

$$\frac{2e^{xs}}{e^s+1}=\sum_{m=0}^{\infty}E_m(x)\frac{s^m}{m!}. \tag{2.23}$$

The Euler basis polynomials of the first kind of degree m can be defined from following relation [25]:

$$\sum_{l=0}^{m}\binom{m}{l}E_l(x)+E_m(x)=2x^m, \tag{2.24}$$

where $\binom{m}{l}$ is a binomial coefficient. The first few EPs are as follows:

$$E_0(x)=1,\quad E_1(x)=x-\frac{1}{2},\quad E_2(x)=x^2-x,\quad E_3(x)=x^3-\frac{3}{2}x^2+\frac{1}{4},\ldots.$$

These polynomials satisfy the following formula:

$$\int_0^1 E_m(x)E_n(x)\,dx=(-1)^{n-1}\frac{m!(n+1)!}{(m+n+1)!}E_{m+n+1}(0),\quad m,n\geq 1, \tag{2.25}$$

and the EPs form a complete basis over the interval [0,1].

Taking $x=\frac{1}{2}$ and normalizing by 2^n gives the Euler number, $E_n=2^nE_n(1/2)$. The first few Euler numbers are as follows:

$$E_0=1,\quad E_1=-\frac{1}{2},\quad E_3=\frac{1}{4},\quad E_5=-\frac{1}{2},\ldots.$$

2.4.8 Taylor's Wavelets

Taylor's wavelets (TWs) $\psi^T_{n,m}(x)$ are defined by the following equation [26]:

$$\psi^T_{n,m}(x)=\begin{cases}2^{\frac{k-1}{2}}\tilde{T}_m(2^{k-1}x-n+1), & \frac{n-1}{2^{k-1}}\leq x<\frac{n}{2^{k-1}},\\ 0, & \text{otherwise}\end{cases} \tag{2.26}$$

with

$$\tilde{T}_m(x)=\sqrt{2m+1}.T_m(x)$$

where $n=1,2,\ldots,2^{k-1}$, $k\in\mathbb{Z}^+$, $m=0,1,\ldots,M-1$ ($M>0$ is a fixed integer) is the degree of the Taylor polynomials (TPs), and x is the normalized time. Here, $T_m(x)=x^m$ are the well-known TPs of degree m.

2.4.9 Laguerre Wavelets

The Laguerre wavelets (LaWs) $\psi_{n,m}^{\text{LaW}}(x)$ are defined by the following equation [27]:

$$\psi_{n,m}^{\text{LaW}}(x)=\begin{cases}\dfrac{2^{\frac{k}{2}}}{(m!)}\widehat{L}_m(2^k x-2n+1), & \dfrac{n-1}{2^{n-1}}\le x<\dfrac{n}{2^{k-1}},\\ 0, & \text{otherwise}\end{cases} \tag{2.27}$$

where $n=1,2,\ldots,2^{k-1}$, $k\in\mathbb{Z}^+$, $m=0,1,\ldots,M-1$ ($M>0$ is a fixed integer) is the degree of the Laguerre polynomials (LPs), and x is the normalized time. Here $\widehat{L}_m(x)$ are the well-known LPs of order m with respect to weight function $\tilde{w}(x)=e^{-x}$ which are defined on the interval $[0,\infty]$, and can be determined with the aid of the following recurrence formulae:

$$\widehat{L}_0(x)=1,$$

$$\widehat{L}_1(x)=1-x,$$

$$\widehat{L}_m(x)=\frac{(2m-1-x)\widehat{L}_{m-1}(x)-(m-1)\widehat{L}_{m-2}(x)}{m}.$$

REFERENCES

1. Chui C.K., Lemm J.M., and Sedigh S., 1992, *An Introduction to Wavelets*, vol. **1**, Academic Press, Massachusetts.
2. Chui C.K., 1997, *Wavelets: A Mathematical Tool for Signal Analysis*, Society for Industrial and Applied Mathematics, Philadelphia, PA.
3. Morlet J., Arens G., Fourgeau I., and Giard D., 1982, "Wave propagation and sampling theory", *Geophiysics*, **47**, pp. 203–236.
4. Mallat S.G., 1986, *Multiresolution Approximation and Wavelets, Preprint GRASP Lab*, Department of Computer and Information Science, University of Pennsylvania, Philadelphia, PA.
5. Daubechies I., 1993, "Orthonormal bases of compactly supported wavelets II. Variations on a theme", *SIAM Journal on Mathematical Analysis*, **24**(2), pp. 499–519.
6. Meyer Y., 1993, *Wavelets: Algorithms and Applications*, SIAM, Philadelphia, PA.
7. Morlet J., Arens G., Fourgeau E., and Giard D., 1982, "Wave propagation and sampling theory—Part II: Sampling theory and complex waves", *Geophysics*, **47**(2), pp. 222–236.
8. Grossmann A., and Morlet J., 1984, "Decomposition of Hardy functions into square integrable wavelets of constant shape", *SIAM Journal on Mathematical Analysis*, **15**(4), pp. 723–736.
9. Daubechies I., 1992, *Ten Lectures on Wavelets*, Society for Industrial and Applied Mathematics, Philadelphia, PA.
10. Meyer Y., 1986, "Principe d'incertitude, bases hilbertiennes et algebres d'operateurs", *Séminaire Bourbaki*, **28**, pp. 209–223.
11. Mallat S.G., 1989, "Multiresolution approximations and wavelet orthonormal bases of $L^2(R)$", *Transactions of the American Mathematical Society*, **315**(1), pp. 69–87.

12. Sahu P.K., and Saha Ray S., 2015, "Legendre wavelets operational method for the numerical solutions of nonlinear Volterra integro-differential equations system", *Applied Mathematics and Computation*, **256**, pp. 715–723.
13. Sahu P.K., and Saha Ray S., 2017, "A new Bernoulli wavelet method for accurate solutions of nonlinear fuzzy Hammerstein–Volterra delay integral equations", *Fuzzy Sets and Systems*, **309**, pp. 131–144.
14. Kouba O., 2013, "Lecture Notes: Bernoulli Polynomials and Applications", arXiv:1309.7560v1 mah.CA.
15. Saha Ray S., 2016, *Numerical Analysis with Algorithms and Programming*, CRC Press, Taylor and Francis Group, Boca Raton, New York.
16. Çelik İ., 2018, "Generalization of Gegenbauer wavelet collocation method to the generalized Kuramoto–Sivashinsky equation", *International Journal of Applied and Computational Mathematics*, **4**(5), p. 111, https://doi.org/10.1007/s40819-018-0546-2.
17. Kumar S., Pandey P., and Das S., 2019, "Gegenbauer wavelet operational matrix method for solving variable-order non-linear reaction–diffusion and Galilei invariant advection–diffusion equations", *Computational and Applied Mathematics*, **38**(4), pp. 1–22.
18. Yousefi S., and Banifatemi A., 2006, "Numerical solution of Fredholm integral equations by using CAS wavelets", *Applied Mathematics and Computation*, **183**(1), pp. 458–463.
19. Bhrawy A.H., and Zaky M.A., 2017, "Numerical simulation of multi-dimensional distributed-order generalized Schrödinger equations", *Nonlinear Dynamics*, **89**(2), pp. 1415–1432.
20. Zaky M.A., Ameen I.G., and Abdelkawy M.A., 2017, "A new operational matrix based on Jacobi wavelets for a class of variable-order fractional differential equations", *Proceedings of the Romanian Academy. Series A*, **18**(4), pp. 315–322.
21. Rahimkhani P., and Ordokhani Y., 2018, "Numerical solution a class of 2D fractional optimal control problems by using 2D Müntz-Legendre wavelets", *Optimal Control Applications and Methods*, **39**(6), pp. 1916–1934.
22. Mokhtary P., Ghoreishi F., and Srivastava H.M., 2016, "The Müntz-Legendre Tau method for fractional differential equations", *Applied Mathematical Modelling*, **40**(2), pp. 671–684.
23. Wang Y., and Zhu L., 2017, "Solving nonlinear Volterra integro-differential equations of fractional order by using Euler wavelet method", *Advances in Difference Equations*, 2017(1), pp. 1–16.
24. He Y., 2015, "Some new results on products of Apostol–Bernoulli and Apostol–Euler polynomials", *Journal of Mathematical Analysis and Applications*, **431**(1), pp. 34–46.
25. Wang Y., Zhu L., and Wang Z., 2018, "Fractional-order Euler functions for solving fractional integro-differential equations with weakly singular kernel", *Advances in Difference Equations*, **254**(1), pp. 1–13.
26. Keshavarz E., and Ordokhani Y., 2019, "A fast numerical algorithm based on the Taylor wavelets for solving the fractional integro-differential equations with weakly singular kernels", *Mathematical Methods in the Applied Sciences*, **42**(13), pp. 4427–4443.
27. Sahu P.K., and Saha Ray S., 2018, "Comparison on wavelets techniques for solving fractional optimal control problems", *Journal of Vibration and Control*, **24**(6), pp. 1185–1201.s

3 Numerical Solutions of Two-Dimensional Fredholm Integral Equations of the Second Kind

3.1 INTRODUCTION

In recent years, integral equations (IEs) appear in a variety of mathematical problems and also play an important role in different areas of science and engineering [1–3]. A computational technique to solve an IE is a stupendous work in scientific research. An IE inevitably appears in various fields of applications such as continuum mechanics, potential theory, geophysics, quantum mechanics, optimization, and fluid mechanics [4]. IEs can also be applied to various fields such as ionospheric problems [5], scattering theory on planar curves [6], heat transfer problems [7], transport theory [8], and electromagnetics [9]. Usually, IEs fall under two main classes, namely, Volterra and Fredholm integral equations, out of which the Fredholm integral equation (FIE) is one of the most essential integral equations [10]. The method of successive approximations [11], quadrature rule [12], Lagrange interpolation [13], Chebyshev polynomials [14], Legendre wavelet method [15], Haar wavelet method [16], Galerkin method [17], and collocation method [18] have been applied to solve IE numerically.

Weakly singular IEs have significant applications in several fields of engineering and science, such as electrostatic and low-frequency electromagnetic [19], conformal mapping [20], and surface water waves [21]. Many numerical techniques have been introduced to solve the weakly singular FIEs such as hybrid collocation methods [22], Petrov–Galerkin methods [23], high-order collocation methods [24], iterated fast multiscale Galerkin methods [25], Sinc-collocation methods [26], and piecewise polynomial collocation and Galerkin methods [27].

Numerical solutions of FIEs have also been discussed by several authors such as Saha Ray and Behera [28], Bahmanpour et al. [29], Bellour et al. [30], and Babolian et al. [31]. In recent years, two-dimensional (2D) FIEs have been emerged from modelling of many physical phenomena in applied sciences and engineering processes. Recently, Asl et al. [32] introduced the numerical solution of 2D FIEs using triangular function. Gupta et al. [33] presented a 2D Chebyshev wavelet method to compute the numerical solution of Riesz time-fractional Camassa–Holm equation and

DOI: 10.1201/9781003453451-3

also proposed a new method based on the Legendre wavelet to solve the nonlinear time-fractional sKdV equation [34]. Liu et al. [35] have solved the 2D FIEs of the second kind by using the Barycentric Lagrange function. In an earlier study [36], Maleknejad et al. have applied Sinc collocation method for the numerical solution of the first kind FIE. Sahu et al. [37] have employed the B-spline wavelet method for solving the Fredholm Hammerstein IE. In another study [38], Shoukralla established a numerical method for FIEs of the first kind with weakly singular logarithmic kernels. Khuri and Wazwaz [39] have developed Adomianm decomposition methods for solving logarithmic FIEs. In a study [40], Gao and Jiang have introduced the trigonometric Hermite wavelet scheme for approximating the IEs with weakly singular kernels. Babolian et al. [41] have studied the solution of linear system of Fredholm integral equations (SFIEs) by using the Adomian decomposition method. Huang et al. [42] have utilized the Taylor expansion method for constructing the numerical solution of linear SFIEs. In their studies [43,44], Maleknejad et al. implemented the Taylor-Series expansion method and block-pulse functions method to determine the approximate solutions of a linear SFIE of the second kind. Rashidinia et al. [45] have considered the Sinc approximation technique for solving the second-kind SFIEs.

3.2 OUTLINE OF THE PRESENT STUDY

In this chapter, a novel technique based on Bernoulli wavelets (BWs) has been proposed to solve 2D-FIE of the second kind. BWs have been created by the dilation and translation of Bernoulli polynomials. For this purpose, we introduce properties of BWs and Bernoulli polynomials. Then, this method has been used to transform a 2D FIE into an algebraic equation. Furthermore, the convergence and error analyses of proposed scheme are examined. Finally, some illustrations are included to demonstrate the efficiency and accuracy of the proposed method.

In addition, a wavelet-based collocation scheme has been introduced for solving the linear and nonlinear FIEs as well as the linear SFIEs with a weakly singular logarithmic kernel. For the numerical solution of FIEs, all the functions have been approximated with respect to the Laguerre wavelets (LaWs). Then, the function approximations are utilized to reduce the solutions of linear and nonlinear FIEs to systems of linear and nonlinear algebraic equations, respectively. Moreover, the numerical results of the several experiments have also been presented in both tabular and graphical forms to describe the accuracy and efficiency of the approached method, and also, the approximate results and absolute error norms are compared with the other existing methods to demonstrate the validity of the presented scheme.

3.3 TWO-DIMENSIONAL FREDHOLM INTEGRAL EQUATION

In this chapter, a linear 2D-FIE of the following form has been considered:

$$u(x,t) = f(x,t) + \int_0^a \int_0^b l(x,t,s,y)u(s,y)\,ds\,dy, \quad (x,t) \in \Omega, \tag{3.1}$$

where $u(x,t)$ is an unknown function defined on Ω, where $\Omega = [0,a]\times[0,b]$ and the function $l(x,t,s,y)$, called the kernel of integral equation, and $f(x,t)$ are known functions defined on $\Omega\times\Omega$ and Ω, respectively.

3.3.1 Function Approximation by BWs

A truncated expansion of the function $g(x,t)$ in $L^2([0,1)\times[0,1))$ with respect to 2D BWs is as follows:

$$g(x,t) \approx \bar{g}(x,t) = \sum_{n=1}^{2^{k_1-1}} \sum_{n'=1}^{2^{k_2-1}} \sum_{m=0}^{M-1} \sum_{m'=0}^{M'-1} w_{n,m,n',m'} \cdot \tilde{\psi}_{n,m,n',m'}(x,t) = W^T \cdot \tilde{\Psi}(x,t), \quad (3.2)$$

where W and $\tilde{\Psi}(x,t)$ are $(2^{k_1-1}2^{k_2-1}MM' \times 1)$ vectors given by

$$W = [w_{1,0,1,0},\ldots,w_{1,0,1,M'-1},\ldots,w_{1,0,2^{k_2-1},M'-1},\ldots,w_{2^{k_1-1},M-1,2^{k_2-1},M'-1}]^T, \quad (3.3)$$

$$\tilde{\Psi}(x,t) = [\tilde{\psi}_{1,0,1,0}(x,t),\ldots,\tilde{\psi}_{1,0,1,M'-1}(x,t),\ldots,\tilde{\psi}_{1,0,2^{k_2-1},M'-1}(x,t),\ldots,\tilde{\psi}_{2^{k_1-1},M-1,2^{k_2-1},M'-1}(x,t)]^T. \quad (3.4)$$

where $\tilde{\psi}_{n,m,n',m'}(x,t)$, $n = 1,2,\ldots,2^{k_1-1}$, $n' = 1,2,\ldots,2^{k_2-1}$, $m = 0,1,\ldots,M-1$, and $m' = 0,1,\ldots,M'-1$ is the 2D BWs defined as follows:

$$\tilde{\psi}_{n,m,n',m'}(x,t) = \begin{cases} 2^{\frac{k_1+k_2}{2}} \tilde{\beta}_m(2^{k_1-1}x - n + 1).\tilde{\beta}_{m'}(2^{k_2-1}t - n' + 1), & \text{for } \frac{n-1}{2^{k_1-1}} \le x < \frac{n}{2^{k_1-1}}, \ \frac{n'-1}{2^{k_2-1}} \le t < \frac{n'}{2^{k_2-1}} \\ 0, & \text{otherwise,} \end{cases} \quad (3.5)$$

where

$$\tilde{\beta}_m(x) = \begin{cases} 1, & m = 0 \\ \dfrac{1}{\sqrt{\dfrac{(-1)^{m-1}(m!)^2}{(2m)!}\beta_{2m}}} \beta_m(x), & m > 0 \end{cases}$$

and

$$\tilde{\beta}_{m'}(t) = \begin{cases} 1, & m' = 0 \\ \dfrac{1}{\sqrt{\dfrac{(-1)^{m'-1}(m'!)^2}{(2m')!}\beta_{2m'}}} \beta_{m'}(t), & m' > 0. \end{cases}$$

Here k_1 and k_2 are any positive integers. $\tilde{\beta}_m$ and $\tilde{\beta}_{m'}$ are Bernoulli polynomials of order m and m', respectively.

The $2^{k_1-1}2^{k_2-1}MM'$-vector W contains the coefficients $w_{n,m,n',m'}$ that is defined as follows:

$$w_{n,m,n',m'} = \frac{<\tilde{\psi}_{n,m}(x), <g(x,t), \tilde{\psi}'_{n',m'}(t)>>}{<\tilde{\psi}_{n,m}(x), \tilde{\psi}_{n,m}(x)>.<\tilde{\psi}'_{n',m'}(t), \tilde{\psi}'_{n',m'}(t)>},$$

where $<.,.>$ denotes the inner product and

$$<\tilde{\psi}_{n,m}(x), \tilde{\psi}_{n,m}(x)> = \int_{\frac{n-1}{2^{k_1-1}}}^{\frac{n}{2^{k_1-1}}} \tilde{\psi}_{n,m}(x)\cdot\tilde{\psi}_{n,m}(x)\,dx = 1,$$

$$<\tilde{\psi}'_{n',m'}(t), \tilde{\psi}'_{n',m'}(t)> = \int_{\frac{n'-1}{2^{k_2-1}}}^{\frac{n'}{2^{k_2-1}}} \tilde{\psi}'_{n',m'}(t), \tilde{\psi}'_{n',m'}(t)\,dt = 1.$$

We have

$$w_{n,m,n',m'} = \int_{\frac{n-1}{2^{k_1-1}}}^{\frac{n}{2^{k_1-1}}} \tilde{\psi}_{n,m}(x) \int_{\frac{n'-1}{2^{k_2-1}}}^{\frac{n'}{2^{k_2-1}}} g(x,t)\cdot\tilde{\psi}'_{n',m'}(t)\,dt\,dx. \tag{3.6}$$

Similarly, the kernel function $l(x,t,s,y)$ can be expressed with respect to 2D BWs as follows:

$$l(x,t,s,y) = \Psi^T(x,t)\,L\,\Psi(s,y),$$

where L is the $(2^{k_1-1}2^{k_2-1}MM' \times 2^{k_1-1}2^{k_2-1}MM')$ coefficient matrix with components as follows:

$$L_{i,j} = <\tilde{\psi}_{n,m}(x), <\tilde{\psi}'_{n',m'}(t), <\tilde{\psi}_{\bar{n},\bar{m}}(x), <l(x,t,s,y), \tilde{\psi}'_{\bar{n}',\bar{m}'}(y)>>>>,$$

in which

$$i = [[(n-1)2^{k_1-1} + (n'-1)]M + m]M' + (m'+1),$$

$$j = [[(\bar{n}-1)2^{k_1-1} + (\bar{n}'-1)]M + \bar{m}]M' + (\bar{m}'+1),$$

for $n,\bar{n} = 1,2,\ldots,2^{k_1-1}$, $n',\bar{n}' = 1,2,\ldots,2^{k_2-1}$, $m,\bar{m} = 0,1,\ldots,M-1$, and $m',\bar{m}' = 0,1,\ldots,M'-1$.

3.3.2 Implementation of Proposed Method

In this section, we solve the 2D-FIE numerically using the proposed 2D BWs.

Consider the following 2D-FIE:

$$u(x,t) = f(x,t) + \int_0^1\int_0^1 l(x,t,s,y)u(s,y)\,ds\,dy, \tag{3.7}$$

where l and f are known functions and u is unknown function. Approximating functions u, f, and l with respect to 2D BWs yields

$$u(x,t) \approx W^T\tilde{\Psi}(x,t) = \tilde{\Psi}^T(x,t)W, \tag{3.8}$$

$$f(x,t) \approx F^T\tilde{\Psi}(x,t) = \tilde{\Psi}^T(x,t)F, \tag{3.9}$$

$$l(x,t,s,y) \approx \tilde{\Psi}^T(x,t)L\tilde{\Psi}(s,y) = \tilde{\Psi}^T(s,y)L^T\tilde{\Psi}(x,t), \tag{3.10}$$

where $\tilde{\Psi}(x,t)$ is defined in Eq. (3.4), and $2^{k_1-1}2^{k_2-1}MM'$-vectors W and F are 2D BWs coefficients of $u(x,t)$ and $f(x,t)$, respectively. Furthermore, the $(2^{k_1-1}2^{k_2-1}MM' \times 2^{k_1-1}2^{k_2-1}MM')$-matrix L is 2D BWs coefficients of $l(x,t,s,y)$.

Using Eqs. (3.8)–(3.10), the integral part of Eq. (3.7) can be expressed as follows:

$$\begin{aligned}\int_0^1\int_0^1 l(x,t,s,y)u(s,y)\,ds\,dy &\approx \int_0^1\int_0^1 \tilde{\Psi}^T(x,t)L\tilde{\Psi}(s,y)\tilde{\Psi}^T(s,y)W\,ds\,dy\\ &= \tilde{\Psi}^T(x,t)L\int_0^1\int_0^1 \tilde{\Psi}(s,y)\tilde{\Psi}^T(s,y)\,ds\,dy\,W \\ &= \tilde{\Psi}^T(x,t)LIW.\end{aligned} \tag{3.11}$$

Substituting Eqs. (3.8)–(3.11) in Eq. (3.7), we get

$$(I - L)W = F. \tag{3.12}$$

Equation (3.12) is a linear system of $2^{k_1-1}2^{k_2-1}MM'$-algebraic equations. The $2^{k_1-1}2^{k_2-1}MM'$ components of F are unknown. By solving the above system (3.12), W can be obtained immediately. Hence, an approximate solution of Eq. (3.7) can be obtained as follows:

$$u(x,t) \approx W^T\tilde{\Psi}(x,t).$$

3.3.3 Convergence and Error Analyses

In the present analysis, we establish that the expansion of any continuous function $g(x,t) \in [0,1)\times[0,1)$ with respect to 2D BWs, as given in Eq. (3.2), converges uniformly to g.

Theorem 3.3.1: Convergence Analysis

Let $f(x,t)$ be a continuous function on $[0,1)\times[0,1)$ and $|f(x,t)|\leq \mathcal{K}$, and then the BW expansion of $f(x,t)$ defined in Eq. (3.2) converges uniformly and also

$$|w_{n,m,n',m'}| < \mathcal{K}\frac{A_1A_2}{2^{\frac{k_1-1}{2}}2^{\frac{k_2-1}{2}}}\frac{16m!}{(2\pi)^{m+1}}\frac{16m'!}{(2\pi)^{m'+1}},$$

where $A_1 = \dfrac{1}{\sqrt{\dfrac{(-1)^{m-1}(m!)^2}{(2m)!}\alpha_{2m}}}$ and $A_2 = \dfrac{1}{\sqrt{\dfrac{(-1)^{m'-1}(m'!)^2}{(2m')!}\alpha_{2m'}}}$.

Proof:

Any function $f(x,t)\in L^2([0,1)\times[0,1))$ can be expressed as the 2D BWs as follows:

$$f(x,t) = \sum_{n=1}^{2^{k_1-1}}\sum_{n'=1}^{2^{k_2-1}}\sum_{m=0}^{M-1}\sum_{m'=0}^{M'-1} w_{n,m,n',m'}\cdot\tilde{\psi}_{n,m,n',m'}(x,t)$$

$$= \sum_{n=1}^{2^{k_1-1}}\sum_{n'=1}^{2^{k_2-1}}\sum_{m=0}^{M-1}\sum_{m'=0}^{M'-1} w_{n,m,n',m'}\cdot\tilde{\psi}_{n,m}(x)\cdot\tilde{\psi}'_{n',m'}(t),$$

where the coefficients $w_{n,m,n',m'}$ can be determined as follows:

$$w_{n,m,n',m'} = << f(x,t),\tilde{\psi}_{n,m}(x)>,\tilde{\psi}'_{n',m'}(t)>.$$

Now for $m,m'>0$,

$$w_{n,m,n',m'} = << f(x,t),\tilde{\psi}_{n,m}(x)>,\tilde{\psi}'_{n',m'}(t)>$$

$$= \int_0^1\left[\int_0^1 f(x,t)\tilde{\psi}_{n,m}(x)\,dx\right]\tilde{\psi}'_{n',m'}(t)\,dt$$

$$= \int_{\frac{n'-1}{2^{k_2-1}}}^{\frac{n'}{2^{k_2-1}}}\left[\int_{\frac{n-1}{2^{k_1-1}}}^{\frac{n}{2^{k_1-1}}} f(x,t)\tilde{\psi}_{n,m}(x)\,dx\right]\tilde{\psi}'_{n',m'}(t)\,dt$$

$$= 2^{\frac{k_1-1}{2}}2^{\frac{k_2-1}{2}}A_1A_2\int_{\frac{n'-1}{2^{k_2-1}}}^{\frac{n'}{2^{k_2-1}}}\left[\int_{\frac{n-1}{2^{k_1-1}}}^{\frac{n}{2^{k_1-1}}} f(x,t)\tilde{\beta}_m(2^{k_1-1}x-n+1)\,dx\right]\tilde{\beta}_{m'}(2^{k_2-1}t-n'+1)\,dt.$$

Now, changing the variable $2^{k_1-1}x-n+1=z_1$, we have

$$w_{n,m,n',m'} = \frac{1}{2^{\frac{k_1-1}{2}}} 2^{\frac{k_2-1}{2}} A_1 A_2 \int_{\frac{n'-1}{2^{k_2-1}}}^{\frac{n'}{2^{k_2-1}}} \left[\int_0^1 f\left(\frac{z_1+n-1}{2^{k_1-1}}, t\right) \tilde{\beta}_m(z_1)\, dz_1 \right] \tilde{\beta}_{m'}(2^{k_2-1}t - n' + 1)\, dt.$$

Similarly, changing the variable for t as $2^{k_2-1}t - n' + 1 = z_2$, we have

$$w_{n,m,n',m'} = \frac{A_1}{2^{\frac{k_1-1}{2}}} \frac{A_2}{2^{\frac{k_2-1}{2}}} \int_0^1 \left[\int_0^1 f\left(\frac{z_1+n-1}{2^{k_1-1}}, \frac{z_2+n'-1}{2^{k_2-1}}\right) \tilde{\beta}_m(z_1)\, dz_1 \right] \tilde{\beta}_{m'}(z_2)\, dz_2$$

$$\leq \frac{A_1}{2^{\frac{k_1-1}{2}}} \frac{A_2}{2^{\frac{k_2-1}{2}}} \int_0^1 \left[\int_0^1 f\left(\frac{z_1+n-1}{2^{k_1-1}}, \frac{z_2+n'-1}{2^{k_2-1}}\right) | \tilde{\beta}_m(z_1) |\, dz_1 \right] \beta_{m'}(z_2) |\, dz_2$$

$$\leq \frac{A_1}{2^{\frac{k_1-1}{2}}} \frac{A_2}{2^{\frac{k_2-1}{2}}} \mathcal{K} \left(\int_0^1 | \tilde{\beta}_m(z_1) |\, dz_1 \right) \left(\int_0^1 | \tilde{\beta}_{m'}(z_2) |\, dz_2 \right)$$

$$\leq \frac{A_1}{2^{\frac{k_1-1}{2}}} \frac{A_2}{2^{\frac{k_2-1}{2}}} \mathcal{K} \frac{16m!}{(2\pi)^{m+1}} \frac{16m'!}{(2\pi)^{m'+1}},$$ using the property of Bernoulli polynomials.

This implies that the series $\sum_{n=1}^{2^{k_1-1}} \sum_{m=0}^{M-1} \sum_{n'=1}^{2^{k_2-1}} \sum_{m'=0}^{M'-1} w_{n,m,n',m'}$ is absolutely convergent, and hence, the series $\sum_{n=1}^{2^{k_1-1}} \sum_{m=0}^{M-1} \sum_{n'=1}^{2^{k_2-1}} \sum_{m'=0}^{M'-1} w_{n,m,n',m'} \tilde{\psi}_{n,m,n',m'}(x,t)$ is uniformly convergent [46].

Theorem 3.3.2: Error Analysis

Let $f''(x,t) = \sum_{n=1}^{2^{k_1-1}} \sum_{m=0}^{M-1} \sum_{n'=1}^{2^{k_2-1}} \sum_{m'=0}^{M'-1} w_{n,m,n',m'} \tilde{\psi}_{n,m,n',m'}(x,t)$ be the truncated series, and then, the truncated error $\rho_{n,m,n',m'}(x,t)$ can be defined as follows:

$$\left\| \rho_{n,m,n',m'}(x,t) \right\|_2^2 \leq \sum_{n=2^{k_1-1}+1}^{\infty} \sum_{m=M}^{\infty} \sum_{n'=2^{k_2-1}+1}^{\infty} \sum_{m'=M'}^{\infty} \left(\frac{A_1}{2^{\frac{k_1-1}{2}}} \frac{A_2}{2^{\frac{k_2-1}{2}}} \mathcal{K} \frac{16m!}{(2\pi)^{m+1}} \frac{16m'!}{(2\pi)^{m'+1}} \right)^2.$$

Proof:

Any function $f(x,t) \in L^2([0,1) \times [0,1))$ can be expressed by the BWs as follows:

$$f(x,t)=\sum_{n=1}^{\infty}\sum_{m=0}^{\infty}\sum_{n'=1}^{\infty}\sum_{m'=0}^{\infty}w_{n,m,n',m'}\tilde{\psi}_{n,m,n',m'}(x,t).$$

If $f''(x,t)$ be the truncated series, and then truncated term can be calculated as follows:

$$\rho_{n,m,n',m'}(x,t)=f(x,t)-f''(x,t)=\sum_{n=2^{k_1-1}+1}^{\infty}\sum_{m=M}^{\infty}\sum_{n'=2^{k_2-1}+1}^{\infty}\sum_{m'=M'}^{\infty}w_{n,m,n',m'}\tilde{\psi}_{n,m,n',m'}(x,t).$$

Now,

$$\begin{aligned}\left\|\rho_{n,m,n',m'}(x,t)\right\|_2^2&=\left\|\sum_{n=2^{k_1-1}+1}^{\infty}\sum_{m=M}^{\infty}\sum_{n'=2^{k_2-1}+1}^{\infty}\sum_{m'=M'}^{\infty}w_{n,m,n',m'}\tilde{\psi}_{n,m,n',m'}(x,t)\right\|_2^2\\&=\int_0^1\int_0^1\left|\sum_{n=2^{k_1-1}+1}^{\infty}\sum_{m=M}^{\infty}\sum_{n'=2^{k_2-1}+1}^{\infty}\sum_{m'=M'}^{\infty}w_{n,m,n',m'}\tilde{\psi}_{n,m,n',m'}(x,t)\right|^2dx\,dt\\&\le\sum_{n=2^{k_1-1}+1}^{\infty}\sum_{m=M}^{\infty}\sum_{n'=2^{k_2-1}+1}^{\infty}\sum_{m'=M'}^{\infty}\left|w_{n,m,n',m'}\right|^2\int_0^1\int_0^1\left|\tilde{\psi}_{n,m,n',m'}(x,t)\right|^2dx\,dt\\&\le\sum_{n=2^{k_1-1}+1}^{\infty}\sum_{m=M}^{\infty}\sum_{n'=2^{k_2-1}+1}^{\infty}\sum_{m'=M'}^{\infty}\left(\frac{A_1}{2^{\frac{k_1-1}{2}}}\frac{A_2}{2^{\frac{k_2-1}{2}}}\mathcal{K}\frac{16m!}{(2\pi)^{m+1}}\frac{16m'!}{(2\pi)^{m'+1}}\right)^2.\end{aligned}$$

3.3.4 Numerical Experiments

In the present section, the proposed scheme as described in Section 3.3 has been successfully applied to some problems. Let $\bar{f}(x,t)$ and $f(x,t)$ are the approximate and exact solutions of the integral equation, respectively. For comparison of the obtained results with exact solution, the error function $e(x,t)$ is calculated using the following formula:

$$e(x,t)=\left|f(x,t)-\bar{f}(x,t)\right|,\quad\frac{n-1}{2^{k_1-1}}\le x<\frac{n}{2^{k_1-1}},\ \frac{n'-1}{2^{k_2-1}}\le t<\frac{n'}{2^{k_2-1}}.\tag{3.13}$$

Example 3.3.1

Let us consider the following linear FIE [47]:

$$u(x,t)=f(x,t)+\int_0^1\int_0^1 4(x+2xt-4sy+3y)u(s,y)\,dy\,ds,\quad 0\le x,t\le 1,\tag{3.14}$$

where $f(x,t)=\begin{cases}-\frac{11}{2}xt+4t-\frac{1}{4}x-\frac{34}{9}, & 0\le x<\frac{1}{2}, 0\le t\le 1,\\ \frac{3}{2}xt+2t-\frac{9}{4}x-\frac{43}{9}, & \frac{1}{2}\le x\le 1, 0\le t\le 1,\\ 0, & \text{otherwise,}\end{cases}$

with the exact solution

$$u(x,t)=\begin{cases} -3xt+4t+x-1, & 0\le x<\dfrac{1}{2}, 0\le t\le 1,\\ 4xt+2t-x-2, & \dfrac{1}{2}\le x\le 1, 0\le t\le 1,\\ 0, & \text{otherwise.}\end{cases}$$

Choosing $k_1 = 2$, $k_2 = 1$, and $M = M' = 2$, we have

$$L=\begin{bmatrix} 3 & 1.1547 & -0.5773 & -0.3333 & 1 & 0 & -0.5773 & -0.3333\\ 0.2887 & 0 & 0 & 0 & 0.2887 & 0 & 0 & 0\\ 0.5773 & 0 & 0 & 0 & 0.5773 & 0 & 0 & 0\\ 0.1667 & 0 & 0 & 0 & 0.1667 & 0 & 0 & 0\\ 5 & 1.1547 & -0.5773 & -0.3333 & 3 & 0 & -0.5773 & -0.3333\\ 0.8660 & 0 & 0 & 0 & 0.8660 & 0 & 0 & 0\\ 0.5773 & 0 & 0 & 0 & 0.5773 & 0 & 0 & 0\\ 0.1667 & 0 & 0 & 0 & 0.1667 & 0 & 0 & 0 \end{bmatrix}$$

and

$$F=[-1.7874\ \ 0.5358-0.3062-0.1620-3.4668\ \ 0.6379-0.1530\ \ 0.0042]^T.$$

Solving linear system Eq. (3.12), the unknown vector W is obtained as follows:

$$W=[0.6187\ \ 0.6634-0.0510-0.0884-0.1768\ \ 1.0206\ \ 0.1021\ \ 0.1178]^T,$$

which renders the numerical solution of Eq. (3.14) by the proposed method.

The values of error for $k_1 = 2, k_2 = 2, M = 3, M' = 3$ and $k_1 = 3, k_2 = 3, M = 3, M' = 3$ are shown in Table 3.1. The absolute error obtained by BWs and LWs is presented in Table 3.2.

TABLE 3.1
Numerical Results of Error Function $e(x,t)$ for Example 3.3.1

	Error function $e(x,t)$	
(x,t)	$k_1 = 2, k_2 = 2, M = 3, M' = 3$	$k_1 = 3, k_2 = 3, M = 3, M' = 3$
(0.0,0.0)	1.5099E-14	1.9984E-15
(0.1,0.1)	1.26565E-14	1.77636E-15
(0.2,0.2)	9.17322E-15	1.44329E-15
(0.3,0.3)	4.996E-15	2.498E-16
(0.4,0.4)	1.11022E-16	7.77156E-16
(0.5,0.5)	5.77316E-15	4.44089E-16
(0.6,0.6)	1.13243E-14	1.23512E-15
(0.7,0.7)	1.77636E-14	1.9984E-15
(0.8,0.8)	2.57572E-14	1.5543E-15
(0.9,0.9)	3.37508E-14	3.9968E-15

TABLE 3.2
Comparison of Error Function $e(x,t)$ for Example 3.3.1

Error Function $e(x,t)$

(x,t)	BWs ($k_1 = 3, k_2 = 2$ $M = 3, M' = 3$)	LWs ($k_1 = 2, k_2 = 2,$ $M = 3, M' = 3$)	(x,t)	BWs ($k_1 = 3, k_2 = 2$ $M = 3, M' = 3$)	LWs ($k_1 = 2, k_2 = 2,$ $M = 3, M' = 3$)
(0.1,0.1)	1.77636E-15	1.26565E-14	(0.6,0.1)	4.44089E-16	8.88178E-16
(0.1,0.2)	1.66533E-15	1.21014E-14	(0.6,0.2)	2.22045E-16	1.33227E-15
(0.1,0.3)	1.44329E-15	1.15741E-14	(0.6,0.3)	1.55431E-15	3.9968E-15
(0.1,0.4)	1.33227E-15	1.11022E-14	(0.6,0.4)	1.88738E-15	6.43929E-15
(0.1,0.6)	8.88178E-16	1.11022E-14	(0.6,0.6)	1.23512E-15	1.13243E-14
(0.1,0.7)	6.66134E-16	1.04361E-14	(0.6,0.7)	7.77156E-16	1.42109E-14
(0.1,0.8)	2.22045E-15	1.02141E-14	(0.6,0.8)	3.88578E-15	1.62093E-14
(0.1,0.9)	2.22045E-15	9.32587E-15	(0.6,0.9)	3.77476E-15	1.86517E-14
(0.2,0.1)	1.55431E-15	1.00475E-14	(0.7,0.1)	4.44089E-16	1.33227E-15
(0.2,0.2)	1.44329E-15	9.17322E-15	(0.7,0.2)	0	4.44089E-15
(0.2,0.3)	7.21645E-16	8.43769E-15	(0.7,0.3)	2.44249E-15	7.10543E-15
(0.2,0.4)	7.77156E-16	7.43849E-15	(0.7,0.4)	2.66454E-15	9.88098E-15
(0.2,0.6)	1.33227E-15	6.43929E-15	(0.7,0.6)	1.60982E-15	1.53211E-14
(0.2,0.7)	1.55431E-15	5.55112E-15	(0.7,0.7)	1.9984E-15	1.77636E-14
(0.2,0.8)	1.9984E-15	4.66294E-15	(0.7,0.8)	4.44089E-15	2.10942E-14
(0.2,0.9)	2.22045E-15	3.55271E-15	(0.7,0.9)	4.44089E-15	2.35367E-14
(0.3,0.1)	1.11022E-15	7.60503E-15	(0.8,0.1)	2.66454E-15	4.44089E-15
(0.3,0.2)	1.19349E-15	6.32827E-15	(0.8,0.2)	1.77636E-15	7.77156E-15
(0.3,0.3)	2.498E-16	4.996E-15	(0.8,0.3)	3.55271E-15	1.06581E-14
(0.3,0.4)	2.22045E-16	3.77476E-15	(0.8,0.4)	4.21885E-15	1.39888E-14
(0.3,0.6)	6.66134E-16	1.9984E-15	(0.8,0.6)	3.33067E-16	1.95399E-14
(0.3,0.7)	8.88178E-16	4.44089E-16	(0.8,0.7)	4.44089E-16	2.28706E-14
(0.3,0.8)	6.66134E-16	4.44089E-16	(0.8,0.8)	1.55431E-15	2.57572E-14
(0.3,0.9)	0	2.22045E-15	(0.8,0.9)	2.22045E-15	2.93099E-14
(0.4,0.1)	9.99201E-16	5.10703E-15	(0.9,0.1)	1.77636E-15	7.10543E-15
(0.4,0.2)	1.11022E-15	3.45557E-15	(0.9,0.2)	1.11022E-15	1.06581E-14
(0.4,0.3)	9.99201E-16	1.83187E-15	(0.9,0.3)	3.9968E-15	1.39888E-14
(0.4,0.4)	7.77156E-16	1.11022E-16	(0.9,0.4)	4.66294E-15	1.77636E-14
(0.4,0.6)	4.44089E-16	1.77636E-15	(0.9,0.6)	2.22045E-16	2.28706E-14
(0.4,0.7)	2.22045E-16	3.77476E-15	(0.9,0.7)	0	2.66454E-14
(0.4,0.8)	1.33227E-15	5.32907E-15	(0.9,0.8)	2.66454E-15	3.04201E-14
(0.4,0.9)	1.11022E-15	6.88338E-15	(0.9,0.9)	3.9968E-15	3.37508E-14

Example 3.3.2

Consider the following 2D FIE [47]:

$$u(x,t) = f(x,t) + 4\int_0^1\int_0^1 e^{(x+t+s+y)}u(s,y)\,dy\,ds, \quad 0 \le x,t \le 1, \tag{3.15}$$

where $f(x,t) = 2e^{(x+t+2)} - e^{(x+t+4)}$, with the exact solution $u(x,t) = e^{(x+t)}$.

The values of error for $k_1 = 2, k_2 = 2, M = 3, M' = 3$ and $k_1 = 3, k_2 = 3, M = 3, M' = 3$ are shown in Table 3.3. The absolute error obtained by BWs and LWs is presented in Table 3.4.

TABLE 3.3
Numerical Results of Error Function $e(x,t)$ for Example 3.3.2

	Error function $e(x,t)$	
(x,t)	$k_1 = 2, k_2 = 2, M = 3, M' = 3$	$k_1 = 3, k_2 = 3, M = 3, M' = 3$
(0.0,0.0)	2.59313E-3	2.90172E-4
(0.1,0.1)	1.02467E-3	9.2713E-5
(0.2,0.2)	9.4357E-4	1.32517E-4
(0.3,0.3)	9.8417E-4	1.80565E-4
(0.4,0.4)	1.50149E-3	1.56037E-4
(0.6,0.6)	2.78534E-3	2.52022E-4
(0.7,0.7)	2.56488E-3	3.60219E-4
(0.8,0.8)	2.67526E-3	4.90827E-4
(0.9,0.9)	4.08148E-3	4.24153E-4

TABLE 3.4
Comparison of Error Function $e(x,t)$ for Example 3.3.2

Error Function $e(x,t)$

(x,t)	BWs ($k_1 = 3, k_2 = 3$, $M = 3, M' = 3$)	LWs ($k_1 = 2, k_2 = 2$, $M = 2, M' = 2$)	(x,t)	BWs ($k_1 = 3, k_2 = 3$, $M = 3, M' = 3$)	LWs ($k_1 = 2, k_2 = 2$, $M = 2, M' = 2$)
(0.1,0.1)	0.0000927137	0.00102467	(0.6,0.1)	0.000152859	0.00168939
(0.1,0.2)	8.71633E-6	0.000993112	(0.6,0.2)	0.0000143708	0.00163736
(0.1,0.3)	0.000130538	0.000223236	(0.6,0.3)	0.00021522	0.000368054
(0.1,0.4)	4.78198E-6	0.000135885	(0.6,0.4)	7.88412E-6	0.000224036
(0.1,0.6)	0.000152859	0.00168939	(0.6,0.6)	0.000252022	0.00278534
(0.1,0.7)	0.0000143708	0.00163736	(0.6,0.7)	0.0000236934	0.00269956
(0.1,0.8)	0.00021522	0.000368054	(0.6,0.8)	0.000354838	0.000606819
(0.1,0.9)	7.88412E-6	0.000224036	(0.6,0.9)	0.0000129988	0.000369373
(0.2,0.1)	8.71647E-6	0.000993112	(0.7,0.1)	0.000014371	0.00163736
(0.2,0.2)	0.000132517	0.000943567	(0.7,0.2)	0.000218484	0.00155568
(0.2,0.3)	8.47128E-6	0.000076427	(0.7,0.3)	0.0000139668	0.000126007
(0.2,0.4)	0.000144805	0.0000380331	(0.7,0.4)	0.000238742	0.0000627059

(Continued)

TABLE 3.4 (*Continued*)
Comparison of Error Function $e(x,t)$ for Example 3.3.2

Error Function $e(x,t)$

(x,t)	BWs ($k_1=3, k_2=3, M=3, M'=3$)	LWs ($k_1=2, k_2=2, M=2, M'=2$)	(x,t)	BWs ($k_1=3, k_2=3, M=3, M'=3$)	LWs ($k_1=2, k_2=2, M=2, M'=2$)
(0.2,0.6)	0.000014371	0.00163736	(0.7,0.6)	0.0000236938	0.00269956
(0.2,0.7)	0.000218484	0.00155568	(0.7,0.7)	0.000360219	0.00256488
(0.2,0.8)	0.0000139668	0.000126007	(0.7,0.8)	0.0000230273	0.00020775
(0.2,0.9)	0.000238742	0.0000627059	(0.7,0.9)	0.00039362	0.000103385
(0.3,0.1)	0.000130537	0.000223236	(0.8,0.1)	0.00021522	0.000368054
(0.3,0.2)	8.47129E-6	0.000076427	(0.8,0.2)	0.0000139668	0.000126007
(0.3,0.3)	0.000180565	0.000984171	(0.8,0.3)	0.000297702	0.00162262
(0.3,0.4)	0.0000291906	0.00122314	(0.8,0.4)	0.0000481271	0.00201661
(0.3,0.6)	0.00021522	0.000368054	(0.8,0.6)	0.000354838	0.000606819
(0.3,0.7)	0.0000139668	0.000126007	(0.8,0.7)	0.0000230274	0.00020775
(0.3,0.8)	0.000297702	0.00162262	(0.8,0.8)	0.000490827	0.00267526
(0.3,0.9)	0.0000481271	0.00201661	(0.8,0.9)	0.0000793482	0.00332483
(0.4,0.1)	4.78198E-6	0.000135885	(0.9,0.1)	7.88412E-6	0.000224036
(0.4,0.2)	0.000144804	0.0000380331	(0.9,0.2)	0.000238742	0.0000627059
(0.4,0.3)	0.0000291908	0.00122314	(0.9,0.3)	0.0000481275	0.00201661
(0.4,0.4)	0.000156037	0.00150149	(0.9,0.4)	0.000257262	0.00247555
(0.4,0.6)	7.88412E-6	0.000224036	(0.9,0.6)	0.0000129988	0.000369373
(0.4,0.7)	0.000238742	0.0000627059	(0.9,0.7)	0.000393619	0.000103385
(0.4,0.8)	0.0000481275	0.00201661	(0.9,0.8)	0.0000793488	0.00332483
(0.4,0.9)	0.000257262	0.00247555	(0.9,0.9)	0.000424153	0.00408148

Example 3.3.3

Let us consider the following 2D FIE [47]:

$$u(x,t) = f(x,t) + \int_0^1 \int_0^1 \frac{8}{3}\cos(x+t+s+y)u(s,y)\,dy\,ds, \quad 0 \le x,t \le 1, \qquad (3.16)$$

where $f(x,t) = \frac{1}{3}\cos(x+t+4) - \frac{2}{3}\cos(x+t+2)$, with the exact solution $u(x,t) = \cos(x+t)$.

The values of error for $k_1 = 2, k_2 = 1, M = 2, M' = 2$ and $k_1 = 3, k_2 = 3, M = 3, M' = 3$ are shown in Table 3.5. The absolute error obtained by BWs and LWs is presented in Table 3.6.

TABLE 3.5
Numerical Results of Error Function $e(x,t)$ for Example 3.3.3

	Error function $e(x,t)$	
(x,t)	$k_1 = 2, k_2 = 1, M = 2, M' = 2$	$k_1 = 3, k_2 = 3, M = 3, M' = 3$
(0.0,0.0)	9.4328E-2	2.78E-5
(0.1,0.1)	3.00614E-2	1.73E-5
(0.2,0.2)	9.2977E-3	3.17E-5
(0.3,0.3)	2.61016E-2	5.70E-5
(0.4,0.4)	2.41668E-2	5.02E-5
(0.5,0.5)	1.52434E-2	2.326E-4
(0.6,0.6)	1.41977E-2	6.89E-5
(0.7,0.7)	4.8267E-3	9.13E-5
(0.8,0.8)	5.1996E-3	9.33E-5
(0.9,0.9)	7.9410E-3	7.15E-5

TABLE 3.6
Comparison of Error Function $e(x,t)$ for Example 3.3.3

Error Function $e(x,t)$					
(x,t)	BWs ($k_1 = 2, k_2 = 2, M = 3, M' = 3$)	LWs ($k_1 = 2, k_2 = 2, M = 2, M' = 2$)	(x,t)	BWs ($k_1 = 2, k_2 = 2, M = 3, M' = 3$)	LWs ($k_1 = 2, k_2 = 2, M = 2, M' = 2$)
(0.1,0.1)	0.000017338	0.00103335	(0.6,0.1)	0.0000491503	0.000400895
(0.1,0.2)	7.97448E-7	0.00835868	(0.6,0.2)	4.61989E-6	0.00617806
(0.1,0.3)	0.000036114	0.0082053	(0.6,0.3)	0.0000676654	0.00579575
(0.1,0.4)	2.56544E-6	0.00115102	(0.6,0.4)	1.57525E-6	0.000797485
(0.1,0.6)	0.0000491503	0.000400895	(0.6,0.6)	0.0000689288	0.000329708
(0.1,0.7)	4.61989E-6	0.00617806	(0.6,0.7)	8.90611E-6	0.00248483
(0.1,0.8)	0.0000676654	0.00579575	(0.6,0.8)	0.0000826499	0.00196719
(0.1,0.9)	1.57525E-6	0.000797485	(0.6,0.9)	1.99393E-7	0.000248702
(0.2,0.1)	7.97448E-7	0.00835868	(0.7,0.1)	4.61989E-6	0.00617806
(0.2,0.2)	0.0000316768	0.0168699	(0.7,0.2)	0.0000700697	0.0111311
(0.2,0.3)	4.51957E-6	0.0161781	(0.7,0.3)	2.77836E-6	0.00987331
(0.2,0.4)	0.0000436336	0.00671785	(0.7,0.4)	0.0000723034	0.00321696
(0.2,0.6)	4.61989E-6	0.00617806	(0.7,0.6)	8.90611E-6	0.00248483
(0.2,0.7)	0.0000700697	0.0111311	(0.7,0.7)	0.0000913071	0.00266713
(0.2,0.8)	2.77836E-6	0.00987331	(0.7,0.8)	3.56905E-7	0.00115118
(0.2,0.9)	0.0000723034	0.00321696	(0.7,0.9)	0.0000832707	0.00107155
(0.3,0.1)	0.000036114	0.0082053	(0.8,0.1)	0.0000676654	0.00579575
(0.3,0.2)	4.51957E-6	0.0161781	(0.8,0.2)	2.77836E-6	0.00987331
(0.3,0.3)	0.0000570684	0.0153824	(0.8,0.3)	0.0000856942	0.00855236
(0.3,0.4)	9.7579E-6	0.00634024	(0.8,0.4)	0.0000111103	0.00269922
(0.3,0.6)	0.0000676654	0.00579575	(0.8,0.6)	0.0000826499	0.00196719
(0.3,0.7)	2.77836E-6	0.00987331	(0.8,0.7)	3.56905E-7	0.00115118

(Continued)

TABLE 3.6 (*Continued*)
Comparison of Error Function $e(x,t)$ for Example 3.3.3

Error Function $e(x,t)$

(x,t)	BWs ($k_1=2,k_2=2, M=3,M'=3$)	LWs ($k_1=2,k_2=2, M=2,M'=2$)	(x,t)	BWs ($k_1=2,k_2=2, M=3,M'=3$)	LWs ($k_1=2,k_2=2, M=2,M'=2$)
(0.3,0.8)	0.0000856942	0.00855236	(0.8,0.8)	0.0000933391	0.000371612
(0.3,0.9)	0.0000111103	0.00269922	(0.8,0.9)	9.74243E-6	0.00160265
(0.4,0.1)	2.56544E-6	0.00115102	(0.9,0.1)	1.57525E-6	0.000797485
(0.4,0.2)	0.0000436336	0.00671785	(0.9,0.2)	0.0000723034	0.00321696
(0.4,0.3)	9.7579E-6	0.00634024	(0.9,0.3)	0.0000111103	0.00269922
(0.4,0.4)	0.0000502044	0.00167943	(0.9,0.4)	0.0000693932	0.00143907
(0.4,0.6)	1.57525E-6	0.000797485	(0.9,0.6)	1.99393E-7	0.000248702
(0.4,0.7)	0.0000723034	0.00321696	(0.9,0.7)	0.0000832707	0.00107155
(0.4,0.8)	0.0000111103	0.00269922	(0.9,0.8)	9.74243E-6	0.00160265
(0.4,0.9)	0.0000693932	0.00143907	(0.9,0.9)	0.0000715922	0.000846382

Example 3.3.4

Consider the following linear 2D IE [48]:

$$u(x,t)=f(x,t)+\int_0^1\int_0^1 \frac{x}{1+t}.(1+s+y)u(s,y)\,ds\,dy,\quad 0\le x,t<1 \tag{3.17}$$

where $f(x,t)=\dfrac{1}{(1+x+t)}-\dfrac{x}{(1+t)}$, with the exact solution $u(x,t)=\dfrac{1}{(1+x+t)}$.

The values of error for $k_1=2,k_2=2,M=3,M'=3$ and $k_1=3,k_2=3,M=3,M'=3$ are shown in Table 3.7. The absolute error obtained by BWs and LWs is presented in Table 3.8.

TABLE 3.7
Numerical Results of Error Function $e(x,t)$ for Example 3.3.4

	Error function $e(x,t)$	
(x,t)	$k_1=2,k_2=2,M=3,M'=3$	$k_1=3,k_2=3,M=3,M'=3$
(0.0,0.0)	0.00597348	0.00105277
(0.1,0.1)	0.00159457	0.00018650
(0.2,0.2)	0.00074599	0.00017298
(0.3,0.3)	0.00066119	0.00007547
(0.4,0.4)	0.00066119	0.00004570
(0.5,0.5)	0.00052639	0.00007955
(0.6,0.6)	0.00016018	0.00001742
(0.7,0.7)	0.00009325	0.00001864
(0.8,0.8)	0.00008715	0.00001135
(0.9,0.9)	0.00008649	0.00000754

TABLE 3.8
Comparison of Error Function $e(x,t)$ for Example 3.3.4

Error Function $e(x,t)$

(x,t)	BWs ($k_1=3,k_2=3,$ $M=3,M'=3$)	LWs ($k_1=2,k_2=2,$ $M=2,M'=2$)	(x,t)	BWs ($k_1=3,k_2=3,$ $M=3,M'=3$)	LWs ($k_1=2,k_2=2,$ $M=2,M'=2$)
(0.1,0.1)	0.00159457	0.000192228	(0.6,0.1)	0.000428617	0.00010709
(0.1,0.2)	0.00108562	0.00822131	(0.6,0.2)	0.000314449	0.00302934
(0.1,0.3)	0.000151863	0.00709289	(0.6,0.3)	0.000022102	0.0027258
(0.1,0.4)	0.000286385	0.00136154	(0.6,0.4)	0.0000650398	0.000501718
(0.1,0.6)	0.000428617	0.00010709	(0.6,0.6)	0.000160184	0.0000993573
(0.1,0.7)	0.000314449	0.00302934	(0.6,0.7)	0.000122534	0.00143456
(0.1,0.8)	0.000022102	0.0027258	(0.6,0.8)	3.7016E-6	0.00132158
(0.1,0.9)	0.0000650398	0.000501718	(0.6,0.9)	0.0000208944	0.000240677
(0.2,0.1)	0.00108562	0.00822131	(0.7,0.1)	0.000314449	0.00302934
(0.2,0.2)	0.000745993	0.0134445	(0.7,0.2)	0.000230544	0.00532349
(0.2,0.3)	0.000157193	0.0113416	(0.7,0.3)	0.0000361801	0.00469366
(0.2,0.4)	0.000250309	0.00328634	(0.7,0.4)	0.0000680133	0.00155756
(0.2,0.6)	0.000314449	0.00302934	(0.7,0.6)	0.000122534	0.00143456
(0.2,0.7)	0.000230544	0.00532349	(0.7,0.7)	0.0000932514	0.00263311
(0.2,0.8)	0.0000361801	0.00469366	(0.7,0.8)	0.0000116918	0.00238239
(0.2,0.9)	0.0000680133	0.00155756	(0.7,0.9)	0.0000250714	0.000849611
(0.3,0.1)	0.000151863	0.00709289	(0.8,0.1)	0.000022102	0.0027258
(0.3,0.2)	0.000157193	0.0113416	(0.8,0.2)	0.0000361801	0.00469366
(0.3,0.3)	0.000661194	0.00963791	(0.8,0.3)	0.000211483	0.00415525
(0.3,0.4)	0.000613445	0.0030227	(0.8,0.4)	0.000206247	0.00145234
(0.3,0.6)	0.000022102	0.0027258	(0.8,0.6)	3.7016E-6	0.00132158
(0.3,0.7)	0.0000361801	0.00469366	(0.8,0.7)	0.0000116918	0.00238239
(0.3,0.8)	0.000211483	0.00415525	(0.8,0.8)	0.0000871538	0.00216114
(0.3,0.9)	0.000206247	0.00145234	(0.8,0.9)	0.0000876375	0.000800298
(0.4,0.1)	0.000286385	0.00136154	(0.9,0.1)	0.0000650398	0.000501718
(0.4,0.2)	0.000250309	0.00328634	(0.9,0.2)	0.000680133	0.00155756
(0.4,0.3)	0.000613445	0.00303227	(0.9,0.3)	0.000206247	0.00145234
(0.4,0.4)	0.0005588	0.00130677	(0.9,0.4)	0.000197414	0.000535063
(0.4,0.6)	0.0000650398	0.000501718	(0.9,0.6)	0.0000208944	0.000240677
(0.4,0.7)	0.0000680133	0.00155756	(0.9,0.7)	0.0000250714	0.000849611
(0.4,0.8)	0.000206247	0.00145234	(0.9,0.8)	0.0000876375	0.000800298
(0.4,0.9)	0.000197414	0.000535063	(0.9,0.9)	0.0000864926	0.000266516

3.4 APPLICATION OF TWO-DIMENSIONAL HOMOGENEOUS FREDHOLM INTEGRAL EQUATION ASSOCIATED WITH KARHUNEN–LOÈVE EXPANSION

In this section, numerical approximation of a 2D homogeneous FIE of the second kind associated with the Karhunen–Loève (K–L) expansion of random fields has been studied.

3.4.1 Preliminaries Concept of K–L Expansion

Let $k : \tilde{\Omega} \times \tilde{\Omega} \to \mathbb{R}$ be a symmetric and nonnegative-definite covariance kernel, where $\tilde{\Omega}$ be a compact subset of $\mathbb{R}^d, d = 1,2,3$ [49], k admits the spectral decomposition

$$l(X,Y) = \sum_{i=1}^{\infty} \lambda_i u_i(X) u_i(Y), \tag{3.18}$$

where the nonnegative, monotonically decreasing eigenvalues λ_i and the orthonormal eigenfunctions $u_i (i \geq 1)$ are the solutions of the homogeneous FIE

$$\int_{\tilde{\Omega}} k(X,Y) u(Y) dY = \lambda u(X), \quad X \in \tilde{\Omega}. \tag{3.19}$$

Furthermore, let $\tilde{Y} : \tilde{\Omega} \times D \to \mathbb{R}$ be a second-order random field with expectation

$$E[\tilde{Y}(X;\varpi)] = \int_D \tilde{Y}(X,\varpi) d\vartheta,$$

and two-point covariance

$$k(X,Y) = E[(\tilde{Y}(X;\varpi) - E[\tilde{Y}(X;\varpi)])(\tilde{Y}(Y;\varpi) - E[\tilde{Y}(Y;\varpi)])],$$

where D represents the set of outcomes and ϑ is a probability measure. If the covariance kernel $k(X,Y)$ satisfies the above assumptions, then $\tilde{Y}$ can be written as follows:

$$\tilde{Y}(X;\varpi) = E[\tilde{Y}(X;\varpi)] + \sum_{i=1}^{\infty} \sqrt{\lambda_i} u_i(X) \zeta_i(\varpi),$$

in which, $\zeta_i(\varpi)$ are independent and identically distributed random variables. This representation is known as the K–L expansion.

3.4.2 Implementation of the Fredholm Integral Eigenvalue Problem

Let us consider the following 2D homogeneous FIE of the second kind [50]:

$$\int_0^1 \int_0^1 k(x_1, x_2, y_1, y_2) u(y_1, y_2) dy_1 \, dy_2 = \lambda u(x_1, x_2), \tag{3.20}$$

where $u(y_1, y_2)$ is the unknown function and $k(x_1, x_2, y_1, y_2)$ is the kernel function defined on D and D^2, respectively, where $D = [0,1] \times [0,1]$.

Approximating functions u and k with respect to 2D BWs gives

$$u(x_1, x_2) \approx W^T \tilde{\Psi}(x_1, x_2) = \tilde{\Psi}^T(x_1, x_2) W, \tag{3.21}$$

$$k(x_1, x_2, y_1, y_2) \approx \tilde{\Psi}^T(x_1, x_2) K \tilde{\Psi}(y_1, y_2) = \tilde{\Psi}^T(y_1, y_2) K^T \tilde{\Psi}(x_1, x_2), \tag{3.22}$$

where $\tilde{\Psi}(x_1,x_2)$ is defined in Eq. (3.4) and $2^{k_1-1}2^{k_2-1}MM'$-vector W is the 2D BWs coefficients of $u(x_1,x_2)$. Moreover, the $(2^{k_1-1}2^{k_2-1}MM' \times 2^{k_1-1}2^{k_2-1}MM')$-matrix K is the 2D BWs coefficients of $k(x_1,x_2,y_1,y_2)$.

Using Eq. (3.22), the integral part of Eq. (3.20) can be expressed as follows:

$$\begin{aligned}\int_0^1\int_0^1 k(x_1,x_2,y_1,y_2)u(y_1,y_2)\,dy_1\,dy_2 &\approx \int_0^1\int_0^1 \tilde{\Psi}^T(x_1,x_2)K\Psi(y_1,y_2)\tilde{\Psi}^T(y_1,y_2)W\,dy_1\,dy_2 \\ &\approx \int_0^1\int_0^1 \tilde{\Psi}^T(x_1,x_2)K\tilde{\Psi}(y_1,y_2)\tilde{\Psi}^T(y_1,y_2)W\,dy_1\,dy_2 \\ &= \tilde{\Psi}^T(x_1,x_2)K\int_0^1\int_0^1 \tilde{\Psi}(y_1,y_2)\tilde{\Psi}^T(y_1,y_2)\,dy_1\,dy_2 W \\ &= \tilde{\Psi}^T(x_1,x_2)KIW.\end{aligned} \tag{3.23}$$

Now substituting Eqs. (3.21)–(3.23) in Eq. (3.20), we get

$$(K-\lambda I)W = 0. \tag{3.24}$$

Equation (3.24) is a linear system of $2^{k_1-1}2^{k_2-1}MM'$-algebraic equations. By solving the above system (3.24), W can be obtained immediately. Hence, an approximate solution of Eq. (3.20) can be obtained as follows:

$$u(x_1,x_2) \approx W^T\Psi(x_1,x_2).$$

3.4.3 Numerical Experiments

Example 3.4.1

Let us consider the example of separable kernels: a non-smooth, exponential kernel [49,50],

$$k(x_1,x_2;y_1,y_2) = \sigma^2\exp\left(-\frac{|x_1-y_1|}{\eta} - \frac{|x_2-y_2|}{\eta}\right), \tag{3.25}$$

with $\sigma^2 = 1$ and $\eta = 0.1$. The parameters η and σ^2 are the correlation scale and variance, respectively.

The eigenvalues of this covariance function are as follows:

$$\lambda_i = \frac{2\eta\sigma}{\eta^2\gamma_i^2+1}, \quad i = 1,2,\ldots.$$

Moreover, the parameters $\gamma_1,\gamma_2,\ldots$ satisfy $(\eta^2\gamma^2-1)\sin(\gamma) = 2\eta\gamma\cos(\gamma)$.

TABLE 3.9
Comparison of Error Function $\tilde{e}$ for Example 3.4.1

	Error Function $e(x,t)$	
Sl. No.	$k_1 = 1, k_2 = 1, M = 2, M' = 2$	$k_1 = 1, k_2 = 1, M = 3, M' = 3$
1	1.63645E-3	9.57854E-4
2	1.55556E-3	5.49155E-4
3	3.4936E-3	4.5E-3
4	1.86118E-3	1.86118E-3
5		3.95694E-3
6		1.03012E-3
7		1.62595E-3
8		1.62595E-3
9		5.73482E-4

Now we have applied the proposed technique to solve the homogeneous Fredholm integral eigenvalue problem. Furthermore, we compare the eigenvalues of the given problem with the eigenvalues of the covariance function. The error function $\tilde{e}$ is calculated using the following formula:

$$\tilde{e} = \left|\tilde{\lambda}_i - \lambda_i\right|. \tag{3.26}$$

The values of $\tilde{e}$ are computed for different values of k_1, k_2, M, and M' as demonstrated in Table 3.9. The results obtained in Table 3.9 clearly manifests that our proposed BW method is quite efficient for solving the 2D homogeneous Fredholm eigenvalue problem associated with K–L expansion.

3.5 FREDHOLM INTEGRAL EQUATION WITH A WEAKLY SINGULAR LOGARITHMIC KERNEL

The main objective of this analysis is to establish a wavelet-based collocation method for solving the linear and nonlinear FIEs with a weakly singular logarithmic kernel. The FIEs are considered as follows:

$$u(x) - \int_0^1 \tilde{K}(x,t)[u(t)]^{\alpha}\, dt = f(x), \tag{3.27}$$

where $f(x), u(x) \in L^2[0,1)$ are known and unknown functions, respectively. The kernel function $\tilde{K}(x,t)$ which takes the form $\ln|x-t|$ is defined in $L^2([0,1)\times[0,1))$. In this chapter, Eq. (3.27) is considered as linear FIE for $\alpha = 1$ and nonlinear FIE for $\alpha > 1$.

Furthermore, a linear SFIE of the second kind of the following form has been considered:

$$\sum_{i=1}^{p}\left(u_i(x) - f_i(x) - \int_0^1 \tilde{K}_i(x,t)u_i(t)\, dt \right) = 0, \quad p \in \mathbb{Z}^+, \tag{3.28}$$

where $u_i(x)$ and $f_i(x)$ are unknown and known functions defined as $L^2[0,1)$ for $1 \le i \le p$. The kernel functions $\tilde{K}_i(x,t)$, $1 \le i \le p$ are the weakly singular logarithmic kernels defined in $L^2([0,1) \times [0,1))$.

3.5.1 Function Approximation by LaWs

The function $g(x) \in L^2[0,1)$ can be approximated with respect to the LaWs as follows:

$$g(x) \approx \sum_{n=1}^{2^{k-1}} \sum_{m=0}^{M-1} \tilde{g}_{n,m} \tilde{\psi}_{n,m}(x) = \tilde{G}^T \tilde{\Psi}(x), \tag{3.29}$$

where $\tilde{G}$ and $\tilde{\Psi}(x)$ are $(2^{k-1}M \times 1)$-dimensional vectors given by

$$\tilde{G} = [\tilde{g}_{1,0}, \tilde{g}_{1,1}, \ldots, \tilde{g}_{1,M-1}, \ldots, \tilde{g}_{2^{k-1},0}, \tilde{g}_{2^{k-1},1}, \ldots, \tilde{g}_{2^{k-1},M-1}]^T \tag{3.30}$$

and

$$\tilde{\Psi}(x) = [\tilde{\psi}_{1,0}(x), \tilde{\psi}_{1,1}(x), \ldots, \tilde{\psi}_{1,M-1}(x), \ldots, \tilde{\psi}_{2^{k-1},0}(x), \tilde{\psi}_{2^{k-1},1}(x), \ldots \tilde{\psi}_{2^{k-1},M-1}(x)]^T. \tag{3.31}$$

Here, $\tilde{\psi}_{n,m}(x)$ are the LaWs defined on $[0,1)$ as follows [51]:

$$\tilde{\psi}_{n,m}(x) = \begin{cases} \dfrac{2^{\frac{k}{2}}}{(m!)} \widehat{L}_m(2^k x - 2n + 1), & \dfrac{n-1}{2^{k-1}} \le x < \dfrac{n}{2^{k-1}}, \\ 0, & \text{otherwise}, \end{cases} \tag{3.32}$$

where $n = 1, 2, \ldots, 2^{k-1}$, $k \in \mathbb{Z}^+$, $m = 0, 1, \ldots, M-1$ ($M > 0$ is a fixed integer) is the degree of the Laguerre polynomials (LPs), and t is the normalized time. Here, $\widehat{L}_m(x)$ are the well-known LPs of order m with respect to weight function $\tilde{w}(x) = e^{-x}$ which are defined on the interval $[0, \infty]$, and can be determined with the aid of the following recurrence formulae:

$$\widehat{L}_0(x) = 1,$$

$$\widehat{L}_1(x) = 1 - x,$$

$$\widehat{L}_m(x) = \frac{(2m-1-x)\widehat{L}_{m-1}(x) - (m-1)\widehat{L}_{m-2}(x)}{m}.$$

Using Eq. (3.29), the value of $\tilde{G}$ can be calculated as follows:

$$\tilde{G}^T = \left(\int_0^1 g(x) \tilde{\Psi}^T(x) dx \right) . \tilde{D}^{-1}, \tag{3.33}$$

where $\tilde{D}$ is a $(2^{k-1}M \times 2^{k-1}M)$-dimensional matrix and is given by

$$\tilde{D} = \int_0^1 \tilde{\Psi}(x)\tilde{\Psi}^T(x)\,dx. \tag{3.34}$$

For $k = 2$, $M = 3$, we have

$$\tilde{D} = \begin{bmatrix} 2 & 2 & \frac{7}{6} & 0 & 0 & 0 \\ 2 & \frac{8}{3} & \frac{11}{6} & 0 & 0 & 0 \\ \frac{7}{6} & \frac{11}{6} & \frac{163}{120} & 0 & 0 & 0 \\ 0 & 0 & 0 & 2 & 2 & \frac{7}{6} \\ 0 & 0 & 0 & 2 & \frac{8}{3} & \frac{11}{6} \\ 0 & 0 & 0 & \frac{7}{6} & \frac{11}{6} & \frac{163}{120} \end{bmatrix}.$$

Similarly, $\tilde{K}(x,t)$ can be approximated as follows:

$$\tilde{K}(x,t) = \tilde{\Psi}^T(x).\tilde{\Pi}.\tilde{\Psi}(t), \tag{3.35}$$

where $\tilde{\Pi}$ is a $(2^{k-1}M \times 2^{k-1}M)$-dimensional matrix and is defined as follows:

$$\tilde{\Pi} = \tilde{D}^{-1}.\left(\int_0^1 \tilde{\Psi}(x).\left(\int_0^1 \tilde{K}(x,t)\tilde{\Psi}^T(t)dt\right)dx\right).\tilde{D}^{-1}.$$

3.5.2 Implementation of the Proposed Method

In this work, the linear and nonlinear FIEs and also the linear SFIEs have been solved by implementing the wavelets technique.

3.5.2.1 Linear FIE

In the linear case, let $\alpha = 1$ in Eq. (3.27). Therefore, we have

$$u(x) - \int_0^1 \tilde{K}(x,t)u(t)\,dt = f(x). \tag{3.36}$$

To demonstrate the proposed scheme for solving Eq. (3.36), consider the approximation of the following functions:

$$u(x) \approx \tilde{\Psi}^T(x)\tilde{U}_1 = \tilde{U}_1^T\tilde{\Psi}(x), \tag{3.37}$$

$$f(x) \approx \tilde{\Psi}^T(x)\tilde{F} = \tilde{F}^T\tilde{\Psi}(x), \tag{3.38}$$

$$\tilde{K}(x,t) \approx \tilde{\Psi}^T(x)\tilde{\Pi}\tilde{\Psi}(t) = \tilde{\Psi}^T(t)\tilde{\Pi}^T\tilde{\Psi}(x). \tag{3.39}$$

Using Eq. (3.39), the integral part of Eq. (3.36) can be expressed as follows:

$$\begin{aligned}\int_0^1 \tilde{K}(x,t)u(t)\,dt &= \int_0^1 \tilde{\Psi}^T(x)\tilde{\Pi}\tilde{\Psi}(t)\tilde{\Psi}^T(t)\tilde{U}_1\,dt \\ &= \tilde{\Psi}^T(x)\tilde{\Pi}\int_0^1 \tilde{\Psi}(t)\tilde{\Psi}^T(t)\,dt.\tilde{U}_1 \\ &= \tilde{\Psi}^T(x)\tilde{\Pi}.\tilde{D}.\tilde{U}_1 \\ &= \tilde{\Psi}^T(x)\tilde{U}_2,\end{aligned} \tag{3.40}$$

where $\tilde{U}_2 = \tilde{\Pi}.\tilde{D}.\tilde{U}_1$.

Now by plugging Eqs. (3.37)–(3.40) into Eq. (3.36), it follows

$$\tilde{\Psi}^T(x)\tilde{U}_1 - \tilde{\Psi}^T(x)\tilde{U}_2 = \tilde{\Psi}^T(x)\tilde{F}. \tag{3.41}$$

Collocating Eq. (3.41) at $2^{k-1}M$-points x_j yields

$$\tilde{\Psi}^T(x_j)\tilde{U}_1 - \tilde{\Psi}^T(x_j)\tilde{U}_2 = \tilde{\Psi}^T(x_j)\tilde{F}, \tag{3.42}$$

where x_j are the collocation points given as follows:

$$x_j = \frac{2j-1}{2.2^{k-1}.M}, \quad j = 1,2,\ldots,2^{k-1}.M.$$

By solving above algebraic system of Eq. (3.42), the value of $\tilde{U}_1$ can be obtained, and therefore, the linear FIE Eq. (3.36) can be numerically solved by substituting $\tilde{U}_1$ in Eq. (3.37).

3.5.2.2 Nonlinear FIE

For the nonlinear case, let $\alpha > 1$ in Eq. (3.27). Therefore, we have

$$u(x) - \int_0^1 \tilde{K}(x,t)[u(t)]^\alpha\,dt = f(x). \tag{3.43}$$

Now, in the case of $\alpha = 2$, it follows

$$
\begin{aligned}
u^2(t) &= (\tilde{U}_1^T\tilde{\Psi}(t)).(\tilde{U}_1^T\tilde{\Psi}(t)) \\
&= \tilde{U}_1^T\tilde{\Psi}(t)\tilde{\Psi}^T(t)\tilde{U}_1 \\
&= \tilde{U}_1^T\overline{U}_1\tilde{\Psi}(t) \\
&= \tilde{U}_3^T\tilde{\Psi}(t),
\end{aligned}
\tag{3.44}
$$

where $\tilde{U}_3 = \overline{U}_1^T.\tilde{U}_1$.

Now by plugging Eqs. (3.37)–(3.40) and Eq. (3.44) into Eq. (3.43), it follows

$$
\tilde{\Psi}^T(x)\tilde{U}_1 - \tilde{\Psi}^T(x)\tilde{U}_3 = \tilde{\Psi}^T(x)\tilde{F}. \tag{3.45}
$$

Collocating Eq. (3.45) at $2^{k-1}M$-points x_j yields

$$
\tilde{\Psi}^T(x_j)\tilde{U}_1 - \tilde{\Psi}^T(x_j)\tilde{U}_3 = \tilde{\Psi}^T(x_j)\tilde{F}, \tag{3.46}
$$

where x_j are the collocation points given as follows:

$$
x_j = \frac{2j-1}{2.2^{k-1}.M}, \quad j = 1,2,\ldots,2^{k-1}.M.
$$

By solving above algebraic system of Eq. (3.46), the value of $\tilde{U}_1$ can be obtained, and therefore, the linear FIE Eq. (3.43) can be numerically solved by substituting $\tilde{U}_1$ in Eq. (3.37).

3.5.2.3 Systems of Linear FIE

To discuss the proposed method for solving Eq. (3.28), consider the approximation of the following functions:

$$
u_i(x) \approx \tilde{\Psi}^T(x)\tilde{U}_i = \tilde{U}_i^T\tilde{\Psi}(x), \tag{3.47}
$$

$$
f_i(x) \approx \tilde{\Psi}^T(x)\tilde{F}_i = \tilde{F}_i^T\tilde{\Psi}(x), \tag{3.48}
$$

$$
\tilde{K}_i(x,t) \approx \tilde{\Psi}^T(x)\tilde{\Pi}_i\Psi(t) = \tilde{\Psi}^T(t)\tilde{\Pi}_i{}^T\tilde{\Psi}(x). \tag{3.49}
$$

Using Eq. (3.49), the integral part of Eq. (3.28) for $1 \le i \le p$ can be expressed as follows:

$$
\begin{aligned}
\int_0^1 \tilde{K}_i(x,t)u_i(t)\,dt &= \int_0^1 \tilde{\Psi}^T(x)\tilde{\Pi}_i\tilde{\Psi}(t)\tilde{\Psi}^T(t)\tilde{U}_i\,dt \\
&= \tilde{\Psi}^T(x)\tilde{\Pi}_i\int_0^1 \Psi(t)\Psi^T(t)\,dt.\tilde{U}_i \\
&= \tilde{\Psi}^T(x)\tilde{\Pi}_i.\tilde{D}.\tilde{U}_i \\
&= \tilde{\Psi}^T(x)\hat{U}_i,
\end{aligned}
\tag{3.50}
$$

where $\hat{U}_i = \tilde{\Pi}_i.\tilde{D}.\tilde{U}_i,\ 1 \le i \le p$.

Now by plugging Eqs. (3.47)–(3.50) into Eq. (3.28), it follows

$$\sum_{i=1}^{p}\left(\tilde{\Psi}^T(x)\tilde{U}_i - \tilde{\Psi}^T(x)\hat{U}_i - \tilde{\Psi}^T(x)\tilde{F}_i\right) = 0. \tag{3.51}$$

Collocating Eq. (3.51) at $2^{k-1}M$-points x_j yields

$$\sum_{i=1}^{p}\left(\tilde{\Psi}^T(x_j)\tilde{U}_i - \tilde{\Psi}^T(x_j)\hat{U}_i - \tilde{\Psi}^T(x_j)\tilde{F}_i\right) = 0, \tag{3.52}$$

where x_j are the collocation points given as follows:

$$x_j = \frac{2j-1}{2.2^{k-1}.M}, \quad j = 1,2,\ldots,2^{k-1}.M.$$

By solving the above algebraic system of Eq. (3.52), the value of $\tilde{U}_i$ can be obtained, and therefore, the linear SFIE Eq. (3.28) can be numerically solved by substituting $\tilde{U}_i$ in Eq. (3.47).

3.5.3 Convergence and Error Estimation

Theorem 3.5.1: Convergence Analysis

Let $u(x)$ be a continuous function in $L^2([0,1))$ and $|u(x)| \le \Im$, where $\Im$ is a real and positive constant. Then, the LaWs expansion of $u(x)$ converges uniformly to $u(x)$, and also [52]

$$\left|\tilde{g}_{n,m}\right| \le \frac{2\Im\sqrt{e}}{2^{\frac{k}{2}}(m!)(m-1)!}.$$

Proof:

Let us consider the function $u(x) \in L^2[0,1)$ which is defined as follows:

$$u(x) = \sum_{n=1}^{2^{k-1}}\sum_{m=0}^{M-1} \tilde{g}_{n,m}\tilde{\psi}_{n,m}(x).$$

Then, the LaWs coefficients can be calculated as follows:

$$\tilde{g}_{n,m} =< u(x), \tilde{\psi}_{n,m}(x) >$$

$$= \int_0^1 u(x)\tilde{\psi}_{n,m}(x)\,dx$$

$$= \frac{2^{\frac{k}{2}}}{(m!)} \int_{\frac{n-1}{2^{k-1}}}^{\frac{n}{2^{n-1}}} u(x)\widehat{L}_m(2^k x - 2n + 1)\,dx.$$

Now, changing the variable $2^k x - 2n + 1 = z$, it follows

$$= \frac{2^{\frac{k}{2}}}{(m!)} \frac{1}{2^k} \int_0^1 u\left(\frac{z + 2n - 1}{2^k}\right) \widehat{L}_m(z)\,dz.$$

This implies

$$\left|\tilde{g}_{n,m}\right| = \left| \frac{1}{(m!)} \frac{1}{2^{\frac{k}{2}}} \int_0^1 u\left(\frac{z + 2n - 1}{2^k}\right) \widehat{L}_m(z)\,dz \right|$$

$$\leq \frac{1}{(m!)} \frac{1}{2^{\frac{k}{2}}} \int_0^1 \left| u\left(\frac{z + 2\tilde{p} - 1}{2^k}\right) \right| \left|\widehat{L}_m(z)\right| dz \tag{3.53}$$

$$\leq \frac{1}{(m!)} \frac{1}{2^{\frac{k}{2}}} \Im \int_0^1 \left|\widehat{L}_m(z)\right| dz.$$

From classical global uniform estimates, it follows [53]

$$\left|\widehat{L}_m^{\alpha}(x)\right| \leq \frac{(\alpha + 1)m}{m!} e^{\frac{x}{2}}, \quad \alpha \geq 0,\ x \geq 0,\ m = 0, 1, 2, \ldots.$$

Here $\alpha = 0$, and therefore,

$$\left|\widehat{L}_m(x)\right| \leq \frac{1}{(m-1)!} e^{\frac{x}{2}}, \quad x \geq 0,\ m = 0, 1, 2, \ldots.$$

Now, the integral part of Eq. (3.53) becomes

$$\int_0^1 \left|\hat{L}_m(z)\right| dz \le \int_0^1 \frac{1}{(m-1)!} e^{\frac{z}{2}}\, dz$$

$$= \frac{1}{(m-1)!} \int_0^1 e^{\frac{z}{2}}\, dz$$

$$= \frac{2\sqrt{e}}{(m-1)!}.$$

This implies

$$\left|\tilde{g}_{n,m}\right| \le \frac{2\mathfrak{I}\sqrt{e}}{2^{\frac{k}{2}}(m!)(m-1)!}.$$

Hence, the series $\sum_{n=1}^{2^{k-1}} \sum_{m=0}^{M-1} \tilde{g}_{n,m}$ is absolutely convergent, and it follows that the series $\sum_{n=1}^{2^{k-1}} \sum_{m=0}^{M-1} \tilde{g}_{n,m} \tilde{\psi}_{n,m}(x)$ is uniformly convergent [54].

Theorem 3.5.2: Error Estimation

Let $\tilde{u}_{2^{k-1}M}(x)$ be the 1D LaWs expansion of the real sufficiently smooth function $u(x) \in L^2([0,1))$ and u is M times differentiable. Then [55],

$$\left\| u(x) - \tilde{u}_{2^{k-1}M}(x) \right\|_2 \le \frac{2\tilde{\aleph}}{M!\,2^{M(k+1)}},$$

where $\tilde{\aleph} = \max_{x\in[0,1]} \left|u^{(M)}(x)\right|$.

Proof:

Assume that $u(x)$ be a sufficiently smooth function on $L^2([0,1))$ and $\tilde{u}_{2^{k-1}M}(x)$ is the best approximation of u from $\Lambda_{2^{k-1}M}$, where

$$\Lambda_{2^{k-1}M} = \operatorname{Span}\left\{ \tilde{\psi}_{n,m}(x) : n = 1,2,\ldots,2^{k-1}, m = 0,1,\ldots,M-1 \right\}.$$

Now dividing the closed interval [0,1] into 2^{k-1}-subintervals $\tilde{I}_n = \left[\frac{n-1}{2^{k-1}}, \frac{n}{2^{k-1}}\right]$ with $n = 1,2,\ldots,2^{k-1}$.

Then, for any $P_{2^{k-1}M}[u] \in \Lambda_{2^{k-1}M}$, the following inequality is held:

$$\left\| u(x) - \tilde{u}_{2^{k-1}M}(x) \right\|_2 \le \left\| u(x) - P_{2^{k-1}M}[u] \right\|_2, \tag{3.54}$$

where $P_{2^{k-1}M}[u]$ is the interpolating polynomial of u at the Chebyshev nodes in the interval $\tilde{I}_n$. According to [56], the interpolation error is

$$\begin{aligned} \left| u(x) - P_{2^{k-1}M}[u] \right| &\le \frac{1}{2^{k-1}M!} \left(\frac{|\tilde{I}_n|}{2} \right)^M \max_{t \in \tilde{I}_n} \left| u^{(M)}(x) \right| \\ &\le \frac{2\tilde{\aleph}}{M! 2^{M(k+1)}}. \end{aligned} \tag{3.55}$$

In particular, from Eq. (3.54),

$$\begin{aligned} \left\| u(x) - \tilde{u}_{2^{k-1}M}(x) \right\|_2^2 &\le \left\| u(x) - P_{2^{k-1}M}[u] \right\|_2^2 \\ &= \int_0^1 \left| u(x) - P_{2^{k-1}M}[u] \right|^2 dx \\ &\le \int_0^1 \left(\frac{2\tilde{\aleph}}{M! 2^{M(k+1)}} \right)^2 dx \\ &= \left(\frac{2\tilde{\aleph}}{M! 2^{M(k+1)}} \right)^2. \end{aligned}$$

Therefore,

$$\left\| u(x) - \tilde{u}_{2^{k-1}M}(x) \right\|_2 \le \frac{2\tilde{\aleph}}{M! 2^{M(k+1)}}. \tag{3.56}$$

Hence, the theorem is proved. ■

3.5.4 Numerical Experiments

To justify the applicability of the presented scheme, some numerical examples are solved. The absolute error definition and L^2-error norm are defined as follows to verify the efficiency and validity of the proposed technique:

$$\rho(x) = \left| u(x) - \tilde{u}(x) \right|$$

and

$$L^2_{\text{error}} = \left[\int_0^1 |u(x) - \tilde{u}(x)|^2 \, dx \right]^{\frac{1}{2}},$$

where $\tilde{u}(x)$ is the approximate solution and $u(x)$ is the exact solution.

Moreover, the accuracy of the presented scheme has been demonstrated by the maximum absolute error (MAE) formula, which is given by

$$\text{MAE} = \underset{1 \le l \le 2^{k-1}M}{\text{Max}} |u(x_l) - \tilde{u}(x_l)|, \quad x_l = \frac{2.l - 1}{2.2^{k-1}M}, \ l = 1, 2, \ldots, 2^{k-1}M.$$

Example 3.5.1

Consider the following linear FIE with weakly singular logarithmic kernel [57]:

$$u(x) - \int_0^1 \ln|x - t| u(t) \, dt = f(x), \quad 0 \le x \le 1, \tag{3.57}$$

where

$$f(x) = 4x^2 - \frac{5}{3} + x\ln(x) + \frac{x}{2} - x^3 \ln(x) - x\ln(1 - x) + x^3 \ln(1 - x).$$

The exact solution is $u(x) = 3x^2 - 1$.

For different values of k and M, the absolute error values $\rho(x)$ obtained by Laguerre wavelets method (LaWM) are tabulated in Tables 3.10 and 3.11. Table 3.12 displays the comparison of approximate solutions between LaWM and Legendre wavelets method (LWM) [57]. Table 3.13 shows the MAE norms obtained by LaWM for $k = 2,3,4$ and $M = 3,4$. The numerical solutions and error graphs by LaWM are plotted in Figures 3.1–3.8.

TABLE 3.10
Absolute Errors $\rho(x)$ Obtained by LaWM for Example 3.5.1 with $k = 2,3,4$ and $M = 3$

	$\rho(x)$		
x	$k = 2, M = 3$	$k = 3, M = 3$	$k = 4, M = 3$
0.1	5.31797E-14	2.60902E-14	5.62883E-14
0.2	3.94129E-14	2.88658E-15	5.78426E-14
0.3	2.36578E-14	2.44249E-15	4.9849E-14
0.4	5.77316E-15	4.66294E-15	2.60902E-14
0.5	3.55271E-15	1.21014E-14	3.21743E-13
0.6	7.54952E-15	2.09832E-14	3.03785E-14
0.7	1.02141E-14	1.87073E-14	6.09512E-14
0.8	1.11022E-14	2.64233E-14	2.64233E-14
0.9	1.02141E-14	2.4647E-14	1.11466E-13

TABLE 3.11
Absolute Errors $\rho(x)$ Obtained by LaWM for Example 3.5.1 with $k = 2,3,4$ and $M = 4$

	$\rho(x)$		
x	$k = 2, M = 4$	$k = 3, M = 4$	$k = 4, M = 4$
0.1	5.54479E-12	1.49647E-12	6.00742E-12
0.2	2.5584E-12	4.7593E-12	1.94633E-12
0.3	1.26088E-12	4.86833E-13	2.10587E-12
0.4	1.46216E-12	1.96465E-12	1.41442E-13
0.5	1.10973E-11	4.34985E-12	4.76508E-12
0.6	6.55476E-12	2.31881E-12	1.70813E-12
0.7	3.45723E-12	1.42064E-12	1.74805E-13
0.8	1.24767E-12	9.0703E-12	3.15081E-12
0.9	6.31273E-13	2.09388E-12	1.02831E-11

TABLE 3.12
Comparison of Approximate Solutions for Example 3.5.1

		$k = 2, M = 3$		$k = 2, M = 4$	
x	Exact Solution	LaWM ($\varepsilon_0 = 10^{-14}$)	LWM [57] ($\varepsilon_0 = 10^{-5}$)	LaWM ($\varepsilon_0 = 10^{-12}$)	LWM [57] ($\varepsilon_0 = 10^{-4}$)
0.1	−0.97	−0.9699999999	−0.9699999998	−0.9700000000	−0.9700000000
0.2	−0.88	−0.8799999999	−0.8799999990	−0.8800000000	−0.8800000000
0.3	−0.73	−0.7299999999	−0.7299999990	−0.7300000000	−0.7300000000
0.4	−0.52	−0.5199999999	−0.5199999998	0.5200000000	0.5200000000
0.5	−0.25	−0.2499999999	−0.2500000013	0.2500000001	0.2500000001
0.6	0.08	0.0800000000	0.0799999994	0.0800000000	0.0800000004
0.7	0.47	0.4700000000	0.4700000001	0.4700000000	0.4699999999
0.8	0.92	0.9200000000	0.9200000003	0.9200000000	0.9199999998
0.9	1.43	1.4300000000	1.4300000001	1.4300000000	1.4300000000

TABLE 3.13
MAE for Example 3.5.1

k, M	MAE	k, M	MAE
$k = 2, M = 3$	5.51781E-14	$k = 2, M = 4$	8.05933E-12
$k = 3, M = 3$	3.11973E-14	$k = 3, M = 4$	1.22569E-11
$k = 4, M = 3$	4.71179E-13	$k = 4, M = 4$	1.22665E-11

Figures 3.1–3.4 manifest the numerical results by LaWM for $k = 2,3$ and $M = 3,4$.

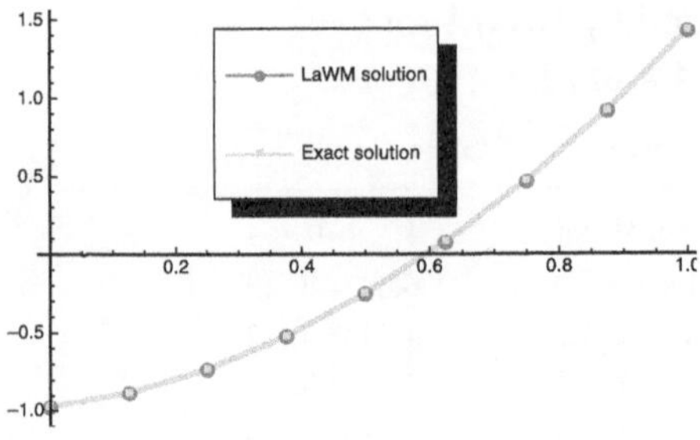

FIGURE 3.1 Numerical results by LaWM for $k = 2, M = 3$.

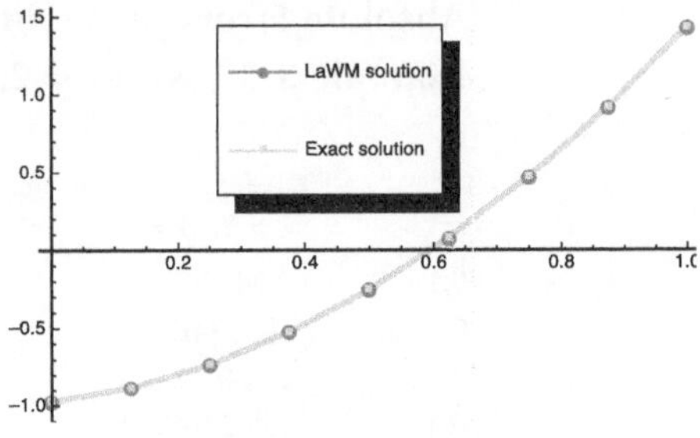

FIGURE 3.2 Numerical results by LaWM for $k = 3, M = 3$.

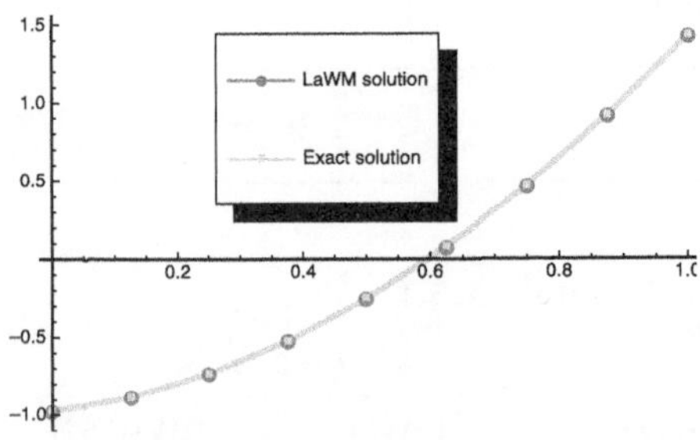

FIGURE 3.3 Numerical results by LaWM for $k = 2, M = 4$.

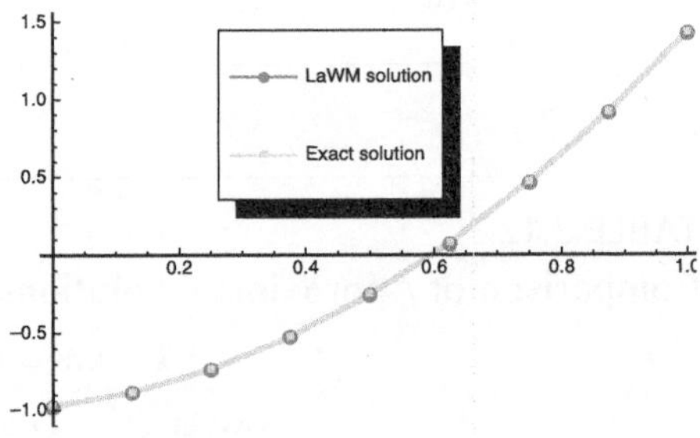

FIGURE 3.4 Numerical results by LaWM for $k = 3, M = 4$.

Figures 3.5–3.8 represent the error distribution by LaWM for $k = 2,3$ and $M = 3,4$.

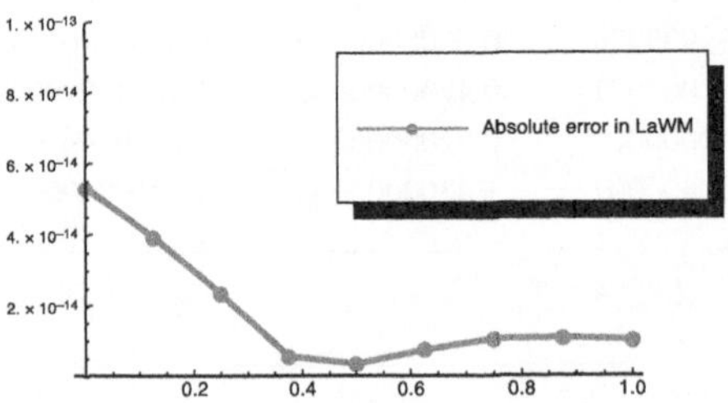

FIGURE 3.5 Absolute error norm by LaWM for $k = 2, M = 3$.

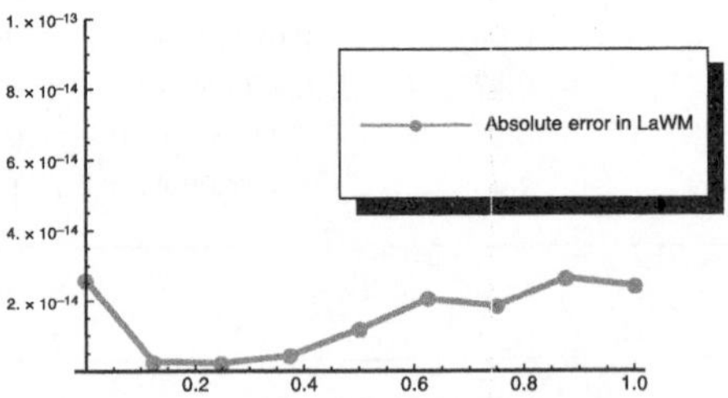

FIGURE 3.6 Absolute error norm by LaWM for $k = 3, M = 3$.

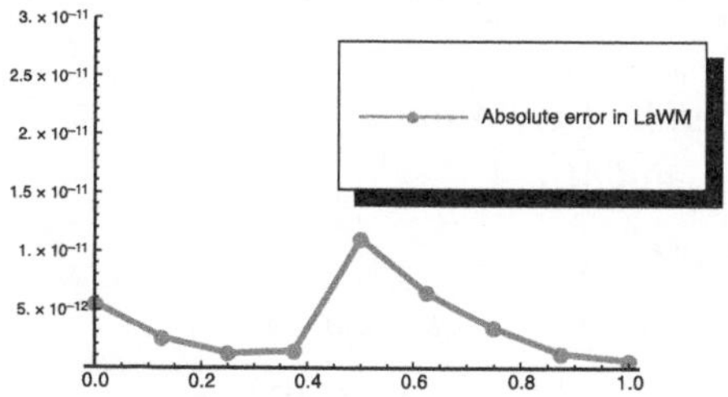

FIGURE 3.7 Absolute error norm by LaWM for $k = 2, M = 4$.

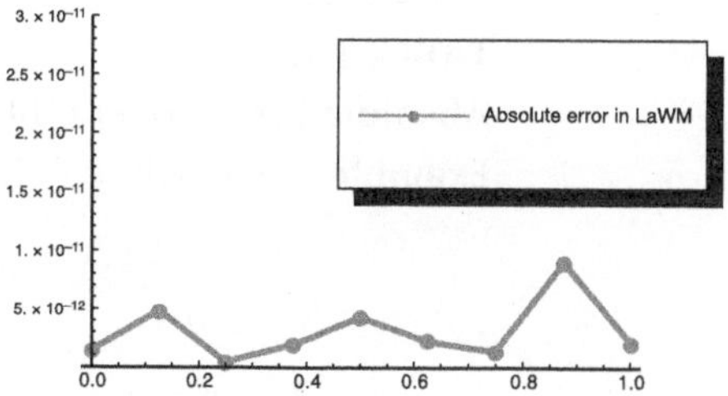

FIGURE 3.8 Absolute error norm by LaWM for $k = 3, M = 4$.

Example 3.5.2

Consider the following nonlinear FIE with a weakly singular logarithmic kernel [57]:

$$u(x) - \int_0^1 \ln|x-t|[u(t)]^3\,dt = f(x), \quad 0 \le x \le 1, \tag{3.58}$$

where

$$f(x) = \left(x-\frac{1}{2}\right)^{\frac{2}{3}} + \frac{1}{3}\left(x^2 - x + \frac{1}{3}\right) - x\left(\frac{x^2}{3} - \frac{x}{2} + \frac{1}{4}\right)\ln\left(\frac{x}{1-x}\right) - \frac{1}{12}\ln(1-x).$$

The exact solution is $u(x) = \left(x-\frac{1}{2}\right)^{\frac{2}{3}}$.

For different values of k and M, the absolute error values $\rho(x)$ obtained by LaWM are tabulated in Tables 3.14 and 3.15. Table 3.16 shows the comparison of absolute error norms between LaWM and LWM [57]. The numerical solutions and error graphs by LaWM are plotted in Figures 3.9–3.16.

TABLE 3.14
Absolute Errors $\rho(x)$ Obtained by LaWM for Example 3.5.2 with $k = 2,3,4$ and $M = 3$

	$\rho(x)$		
x	$k = 2, M = 3$	$k = 3, M = 3$	$k = 4, M = 3$
0.1	3.44395E-03	1.20961E-04	1.30483E-05
0.2	5.55769E-03	1.67374E-04	1.96117E-05
0.3	1.74484E-03	2.187E-03	7.6299E-05
0.4	9.204E-03	1.11536E-03	1.37564E-03
0.5	5.16467E-02	3.24746E-02	2.04544E-02
0.6	9.53835E-03	1.21017E-03	1.37219E-03
0.7	2.22849E-03	2.16831E-03	7.47683E-05
0.8	5.0912E-03	1.59741E-04	2.0267E-05
0.9	3.41613E-03	1.15728E-04	1.32268E-05

TABLE 3.15
Absolute Errors $\rho(x)$ Obtained by LaWM for Example 3.5.2 with $k = 2,3,4$ and $M = 4$

	$\rho(x)$		
x	$k = 2, M = 4$	$k = 2, M = 4$	$k = 4, M = 4$
0.1	2.75448E-03	8.48869E-06	7.6468E-07
0.2	6.07429E-04	1.90853E-05	1.27059E-06
0.3	3.04036E-03	1.72148E-03	5.57974E-06
0.4	3.01606E-03	1.89001E-03	1.08069E-03
0.5	3.44143E-02	2.1653E-02	1.36371E-02
0.6	3.23003E-03	1.87754E-03	1.07827E-03
0.7	2.92471E-03	1.7008E-03	5.24824E-06
0.8	6.93995E-04	2.24905E-05	1.09802E-06
0.9	2.62563E-03	7.79234E-06	7.73698E-07

TABLE 3.16
Comparison of Absolute Errors for Example 3.5.2

	$k = 2, M = 3$		$k = 4, M = 3$	
x	LaWM	LWM [57]	LaWM	LWM [57]
0.1	3.44395E-03	3.464146741E-03	1.30483E-05	1.554874885E-05
0.2	5.55769E-03	5.431141791E-03	1.96117E-05	1.916614782E-05
0.3	1.74484E-03	1.816516290E-03	7.6299E-05	7.605330014E-05
0.4	9.204E-03	9.274629359E-03	1.37564E-03	1.374712585E-03
0.5	5.16467E-02	5.072718022E-02	2.04544E-02	2.045453913E-02
0.8	5.0912E-03	5.431138603E-03	2.0267E-05	1.947048811E-05
0.9	3.41613E-03	3.464070958E-03	1.32268E-05	1.269229914E-05

Figures 3.9–3.12 display the numerical results by LaWM for $k = 2,3$ and $M = 3,4$.

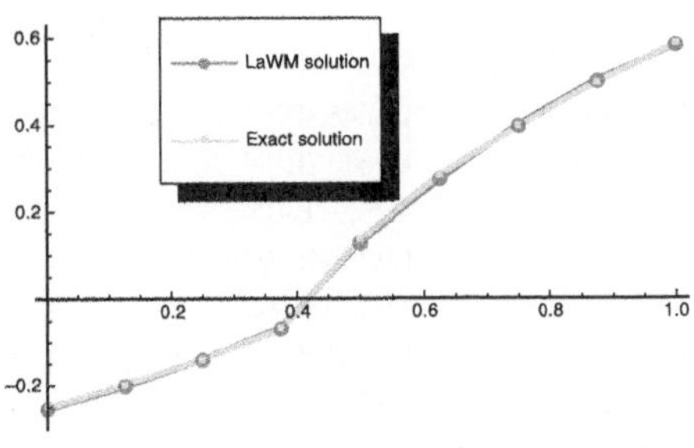

FIGURE 3.9 Numerical results by LaWM for $k = 2, M = 3$.

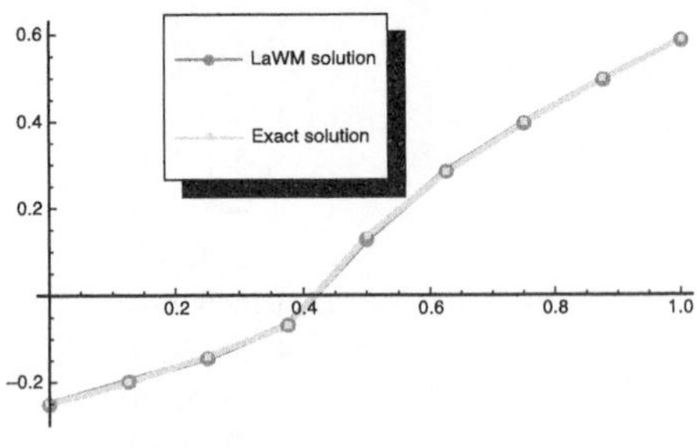

FIGURE 3.10 Numerical results by LaWM for $k = 3, M = 3$.

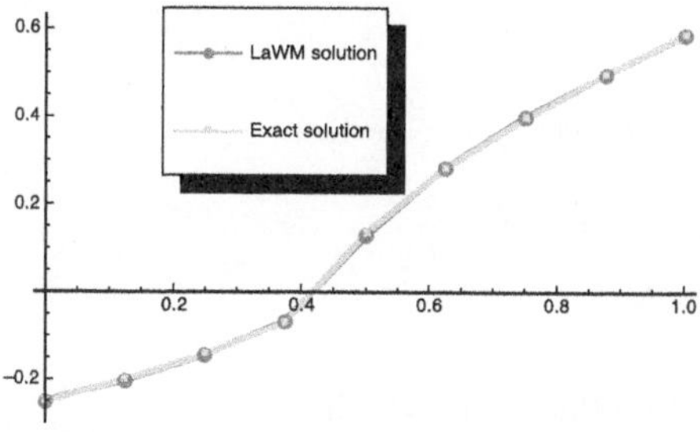

FIGURE 3.11 Numerical results by LaWM for $k = 2, M = 4$.

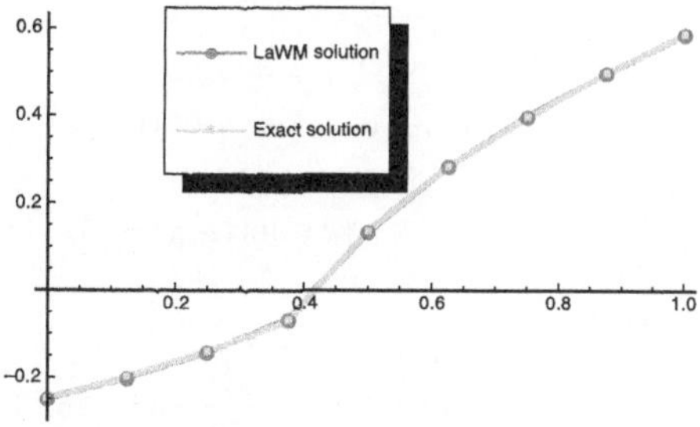

FIGURE 3.12 Numerical results by LaWM for $k = 3, M = 4$.

Figures 3.13–3.16 represent the error distribution by LaWM for $k = 2,3$ and $M = 3,4$.

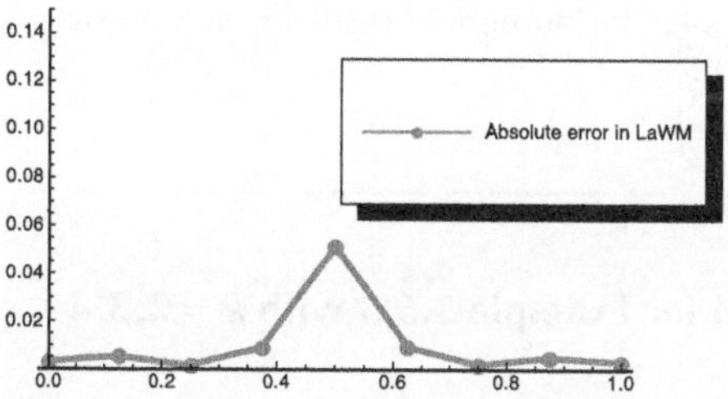

FIGURE 3.13 Absolute error norm by LaWM for $k = 2, M = 3$.

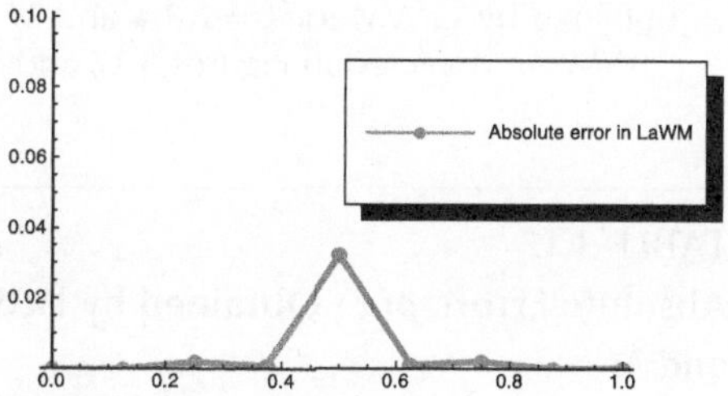

FIGURE 3.14 Absolute error norm by LaWM for $k = 3, M = 3$.

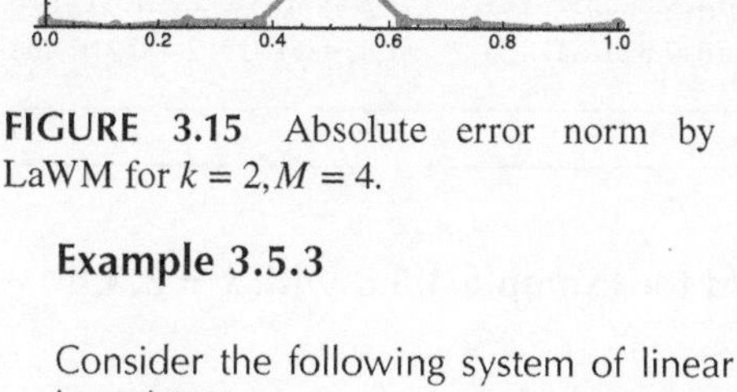

FIGURE 3.15 Absolute error norm by LaWM for $k = 2, M = 4$.

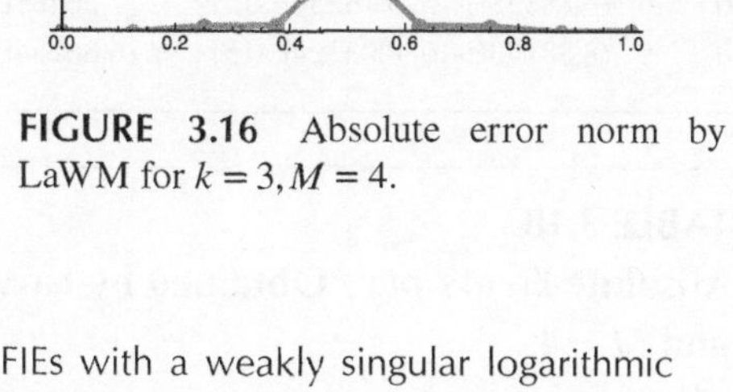

FIGURE 3.16 Absolute error norm by LaWM for $k = 3, M = 4$.

Example 3.5.3

Consider the following system of linear FIEs with a weakly singular logarithmic kernel [58]:

$$\begin{aligned} u_1(x) - \int_0^1 2\ln|x-t| u_1(t)\,dt - \int_0^1 \ln|x-t| u_2(t)\,dt &= f_1(x), \\ u_2(x) - \int_0^1 \ln|x-t| u_1(t)\,dt - \int_0^1 2\ln|x-t| u_2(t)\,dt &= f_2(x), \end{aligned} \tag{3.59}$$

where

$$f_1(x) = x + x^2 - \frac{1}{36}[-31 - 60x - 12x^2 + 66\ln(1-x) - 54x^2\ln(1-x)$$
$$-12x^3\ln(1-x) + 54x^2\ln(x) + 12x^3\ln(x)]$$

and

$$f_2(x) = x - x^2 - \frac{1}{36}[-23 - 48x + 12x^2 + 42\ln(1-x) - 54x^2\ln(1-x)$$
$$+12x^3\ln(1-x) + 54x^2\ln(x) - 12x^3\ln(x)].$$

The exact solutions are $u_1(x) = x + x^2$ and $u_2(x) = x - x^2$.

For different values of k and M, the absolute error values $\rho(x)$ obtained by LaWM are tabulated in Tables 3.17 and 3.18. Table 3.19 shows the MAE norms obtained by LaWM for $k = 2,3,4$ and $M = 3,4$. The numerical results obtained by LaWM are depicted in Figures 3.17 and 3.18.

TABLE 3.17
Absolute Errors $\rho(x)$ Obtained by LaWM for Example 3.5.3 with $k = 2,3,4$ and $M = 3$

	$\rho(x)$		
x	$k = 2, M = 3$	$k = 3, M = 3$	$k = 4, M = 3$
0.1	(9.71445E-16, 3.66374E-15)	(7.46625E-15, 4.2466E-15)	(9.10383E-14, 4.88082E-14)
0.2	(2.2482E-15, 2.19269E-15)	(2.41474E-15, 1.33227E-15)	(6.40231E-14, 4.28546E-14)
0.3	(5.60663E-15, 9.4369E-16)	(4.17444E-14, 1.51268E-14)	(2.06501E-14, 1.06304E-14)
0.4	(9.10383E-15, 5.55112E-17)	(3.10862E-15, 1.30174E-14)	(5.21805E-15, 1.82354E-14)
0.5	(2.04281E-14, 1.99285E-14)	(2.90878E-14, 1.67644E-14)	(1.52434E-13, 5.4956E-15)
0.6	(1.64313E-14, 1.11022E-14)	(1.85407E-14, 7.35523E-15)	(1.83187E-14, 2.08999E-14)
0.7	(1.19904E-14, 6.21725E-15)	(1.64313E-14, 3.96072E-14)	(9.39249E-14, 2.59792E-14)
0.8	(6.88338E-15, 5.16254E-15)	(2.26485E-14, 9.67559E-14)	(7.9492E-14, 5.20139E-14)
0.9	(8.88178E-16, 8.03524E-15)	(6.66134E-16, 7.82707E-15)	(1.47438E-13, 2.45221E-14)

TABLE 3.18
Absolute Errors $\rho(x)$ Obtained by LaWM for Example 3.5.3 with $k = 2,3,4$ and $M = 4$

	$\rho(t)$		
x	$k = 2, M = 4$	$k = 3, M = 4$	$k = 4, M = 4$
0.1	(2.45856E-12, 1.61263E-12)	(7.0513E-13, 2.67367E-12)	(6.69319E-12, 2.80045E-12)
0.2	(1.20978E-12, 9.36695E-13)	(8.36303E-13, 8.25812E-13)	(2.57999E-12, 1.43019E-13)

(Continued)

TABLE 3.18 (*Continued*)
Absolute Errors $\rho(x)$ Obtained by LaWM for Example 3.5.3 with $k = 2,3,4$ and $M = 4$

	$\rho(t)$		
x	$k = 2, M = 4$	$k = 3, M = 4$	$k = 4, M = 4$
0.3	(1.84242E-13, 7.55174E-13)	(1.22069E-13, 1.75843E-12)	(3.48199E-12, 2.16571E-12)
0.4	(1.73506E-12, 4.57467E-13)	(3.27649E-12, 2.54632E-12)	(1.04206E-12, 3.84959E-12)
0.5	(2.00608E-11, 1.1896E-12)	(1.16657E-11, 1.45517E-12)	(7.37632E-13, 1.07858E-12)
0.6	(1.02913E-11, 2.24193E-12)	(1.89115E-12, 3.13344E-12)	(7.02771E-12, 2.27726E-12)
0.7	(4.71179E-12, 2.01167E-12)	(8.92642E-12, 2.80434E-12)	(4.57345E-12, 4.65711E-13)
0.8	(2.00107E-12, 1.07253E-12)	(2.61093E-11, 3.97324E-12)	(3.29137E-12, 2.06524E-12)
0.9	(8.38218E-13, 1.72085E-15)	(8.80851E-13, 2.7959E-12)	(4.84102E-12, 7.13055E-12)

TABLE 3.19
MAE for Example 3.5.3

k, M	MAE	k, M	MAE
$k = 2, M = 3$	(1.70974E-14, 1.23235E-14)	$k = 2, M = 4$	(1.33928E-11, 2.11822E-12)
$k = 3, M = 3$	(4.61853E-14, 1.07692E-13)	$k = 3, M = 4$	(3.42024E-11, 6.75335E-12)
$k = 4, M = 3$	(4.13655E-13, 1.95836E-13)	$k = 4, M = 4$	(2.30838E-11, 7.61391E-12)

Figures 3.17–3.20 display the numerical results by LaWM for $k = 3,4$ and $M = 3,4$.

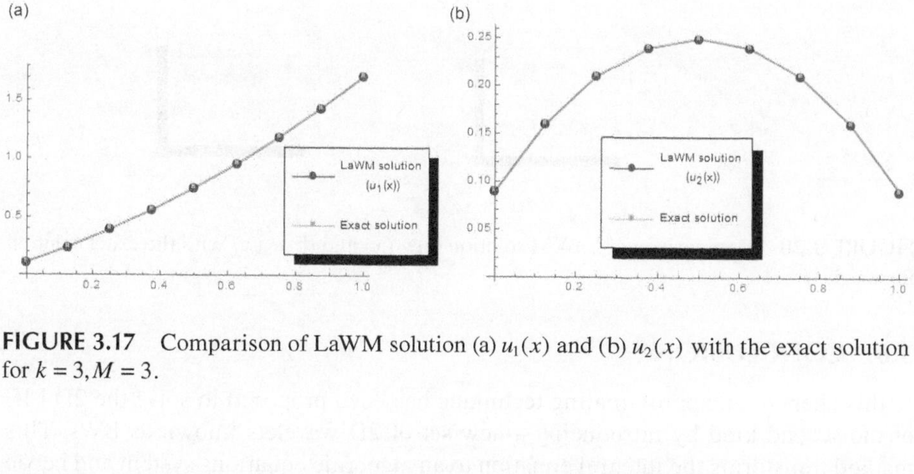

FIGURE 3.17 Comparison of LaWM solution (a) $u_1(x)$ and (b) $u_2(x)$ with the exact solution for $k = 3, M = 3$.

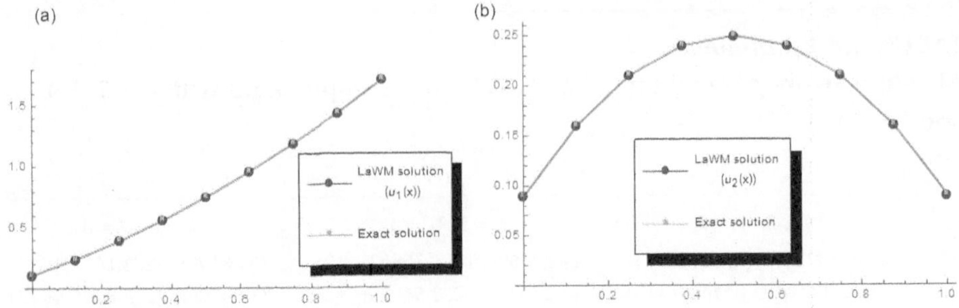

FIGURE 3.18 Comparison of LaWM solution (a) $u_1(x)$ and (b) $u_2(x)$ with the exact solution for $k=4, M=3$.

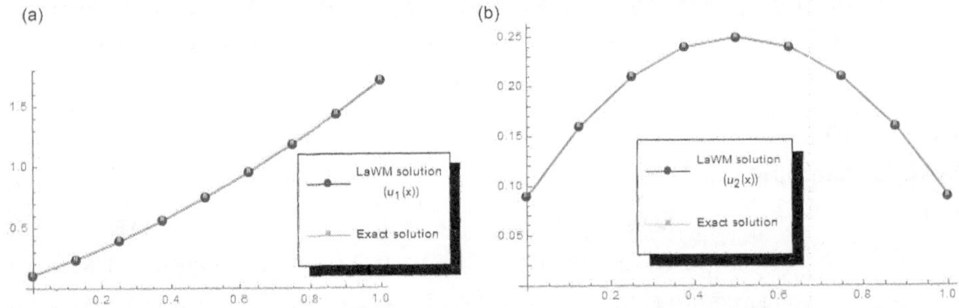

FIGURE 3.19 Comparison of LaWM solution (a) $u_1(x)$ and (b) $u_2(x)$ with the exact solution for $k=3, M=4$.

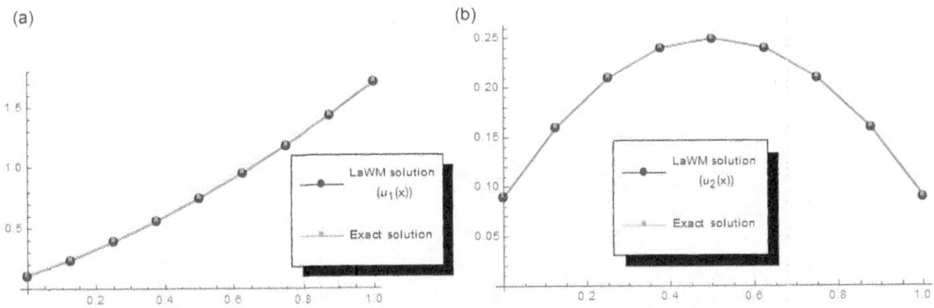

FIGURE 3.20 Comparison of LaWM solution (a) $u_1(x)$ and (b) $u_2(x)$ with the exact solution for $k=4, M=4$.

3.6 CONCLUSION

In this chapter, an approximating technique has been proposed to solve the 2D FIE of the second kind by introducing a new set of 2D wavelets known as BWs. This method transforms the integral equation to an algebraic equations system and hence results in a desirable accurate approximate solution. Furthermore, the uniform convergence and error estimation of the function expansion with respect to 2D BWs are presented. Moreover, the absolute error can be minimized by increasing the order of Bernoulli polynomials. The error analyses for BWs and LWs are also presented in this chapter. The numerical illustrations show efficiency and accuracy of the proposed method.

In addition, a wavelet-based collocation technique has been developed for the approximate solution of the linear and nonlinear FIEs and second-kind SFIEs with weakly singular logarithmic kernels. The proposed method transforms the linear and nonlinear FIEs into linear and nonlinear algebraic systems of equations, respectively, and also converts the second-kind linear SFIEs into a system of algebraic equations. By solving the resulting algebraic system of equations, a desirable accurate approximate solution has been obtained. Moreover, the error estimation and convergence analysis for the proposed method have been discussed. The accuracy and applicability of the approached technique have been demonstrated through the numerical experiments. Furthermore, the numerical results obtained by LaWM have been compared with other existing methods to display the efficiency of the presented scheme.

REFERENCES

1. Wazwaz A.M., 2011, *Linear and Nonlinear Integral Equations Methods and Applications*, Higher Education Press, Beijing and Springer-Verlag, Berlin.
2. Atkinson K.E., 1997, "The numerical solution of boundary integral equations", *Institute of Mathematics and Its Applications Conference Series*, **63**, pp. 223–260.
3. Rahman M., 2007, *Integral Equations and Their Applications*, WIT Press, Southampton.
4. Saha Ray S., and Sahu P. K., 2013, "Numerical methods for solving Fredholm integral equations of second kind", *Abstract and Applied Analysis*, **2013**(426916), p. 17.
5. Parand K., and Delkhosh M., 2017, "Solving the nonlinear Schlomilch's integral equation arising in ionospheric problems", *Afrika Matematika*, **28**(3–4), pp. 459–480.
6. Bremer J., 2012, "A fast direct solver for the integral equations of scattering theory on planar curves with corners", *Journal of Computational Physics*, **231**(4), pp. 1879–1899.
7. Kulish V.V., and Novozhilov V.B., 2003, "Integral equation for the heat transfer with the moving boundary", *Journal of Thermophysics and Heat Transfer*, **17**(4), pp. 538–540.
8. Rammer J., 2018, *Quantum Transport Theory*, CRC Press, Boca Raton, FL.
9. Roth T.E., and Chew W.C., 2020, "Stability analysis and discretization of A-Φ time domain integral equations for multiscale electromagnetics", *Journal of Computational Physics*, **408**, pp. 1–29.
10. Wazwaz A.M., 1997, *A First Course in Integral Equations*, World Scientific, Singapore and River Edge, NJ.
11. Norkin B.V., 2004, "Method of Successive approximation for solving integral equations of the theory of risk processes", *Cybernetics and Systems Analysis*, **40**(4), pp. 517–526.
12. Ioakimidis N.I., 1982, "On the Quadrature methods for the numerical solution of singular integral equations", *Journal of Computational and Applied Mathematics*, **8**(2), pp. 81–86.
13. Rashed M.T., 2004, "Lagrange interpolation to compute the numerical solutions to differential, integral and integro-differential equations", *Applied Mathematics and Computations*, **151**(3), pp. 869–878.
14. Avazzadeh Z., and Heydari M., 2012, "Chebyshev polynomials for solving two dimensional linear and nonlinear integral equations of the second kind", *Computational and Applied Mathematics*, **31**(1), pp. 127–142.

15. Wang X.T., and Li Y.M., 2012, "Numerical solution of Fredholm integral equations of the second kind by General Legendre wavelets", *International Journal of Innovative computing, Information and Control*, **8**, pp. 799–805.
16. Babolian E., and Shahsavaran A., 2009, "Numerical solution of nonlinear Fredholm integral equations of the second kind using Haar wavelets", *Journal of Computational and Applied Mathematics*, **225**(1), pp. 87–95.
17. Thomas K.S., 1981, "Galerkin methods for singular integral equations", *Mathematics of Computation*, **36**(153), pp. 193–205.
18. Saeed R.K., and Hassan J.S., 2014, "Solving singular integral equations by using Collocation method", *Mathematical Sciences Letters*, **3**(3), pp. 185–187.
19. Mei K., and Van Bladel J., 1963, "Low-frequency scattering by rectangular cylinders", *IEEE Transactions on Antennas and Propagation*, **11**(1), pp. 52–56.
20. Symm G.T., 1966, "An integral equation method in conformal mapping", *Numerische Mathematik*, **9**(3), pp. 250–258.
21. Chakrabarti A., Manam S.R., and Banerjea S., 2003, "Scattering of surface water waves involving a vertical barrier with a gap", *Journal of Engineering Mathematics*, **45**(2), pp. 183–194.
22. Cao Y., Huang M., Liu L., and Xu Y., 2007, "Hybrid collocation methods for Fredholm integral equations with weakly singular kernels", *Applied Numerical Mathematics*, **57**(5–7), pp. 549–561.
23. Chen W.S., and Lin W., 2001, "Galerkin trigonometric wavelet methods for the natural boundary integral equations", *Applied Mathematics and Computation*, **121**(1), pp. 75–92.
24. Dominguez V., 2003, "High-order collocation and quadrature methods for some logarithmic kernel integral equations on open arcs", *Journal of Computational and Applied Mathematics*, **161**(1), pp. 145–159.
25. Long G., Nelakanti G., and Zhang X., 2012, "Iterated fast multiscale Galerkin methods for Fredholm integral equations of second kind with weakly singular kernels", *Applied Numerical Mathematics*, **62**(3), pp. 201–211.
26. Okayama T., Matsuo T., and Sugihara M., 2010, "Sinc-collocation methods for weakly singular Fredholm integral equations of the second kind", *Journal of Computational and Applied Mathematics*, **234**(4), pp. 1211–1227.
27. Pedas A., and Vainikko G., 2006, "Smoothing transformation and piecewise polynomial projection methods for weakly singular Fredholm integral equations", *Communications on Pure & Applied Analysis*, **5**(2), p. 395.
28. Saha Ray S., and Behera S., 2019, "A novel technique based on Bernoulli wavelets for numerical solutions of two-dimensional Fredholm integral equation of second kind", *Engineering Computations*, **36**(6), pp. 1798–1819.
29. Bahmanpour M., Kajani M.T., and Maleki M., 2019, "Solving Fredholm integral equations of the first kind using Müntz wavelets", *Applied Numerical Mathematics*, **143**, pp. 159–171.
30. Bellour A., Sbibih D., and Zidna A., 2016, "Two cubic spline methods for solving Fredholm integral equations", *Applied Mathematics and Computation*, **276**, pp. 1–11.
31. Babolian E., and Shahsavaran A., 2009, "Numerical solution of nonlinear Fredholm integral equations of the second kind using Haar wavelets", *Journal of Computational and Applied Mathematics*, **225**(1), pp. 87–95.
32. Asl E.H., and Saberi-Nadjafi J., 2016, "Solving linear two dimensional Fredholm integral equation system by Triangular functions", *International Journal of Applied Mathematical Research*, **5**(4), pp. 187–191.
33. Gupta A.K., and Saha Ray S., 2017, "Two dimensional Chebyshev wavelet method for Camassa-Holm equation with Riesz fractional derivative describing propagation of shallow water waves", *Fundamenta Informatica*, **151**(1–4), pp. 77–89.

34. Gupta A.K., and Saha Ray S., 2017, "Two-dimensional Legendre wavelet method for travelling wave solution of time-fractional generalized seventh order KdV equation", *Computers and Mathematics with Applications*, **73**, pp. 1118–1133.
35. Liu H., Huang J., and Pan Y., 2017, "Numerical solution of two dimensional Fredholm integral equations of the second kind by the Barycentric Lagrange functions", *Journal of Applied Mathematics and Physics*, **5**, pp. 259–266.
36. Maleknejad K., Rostami Y., and Kalalagh H.S., 2016, "Numerical solution for first kind Fredholm integral equations by using sinc collocation method", *International Journal of Applied Physics and Mathematics*, **6**(3), p. 120.
37. Sahu P.K., Ranjan A.K., and Saha Ray S., 2018, "B-spline wavelet method for solving Fredholm Hammerstein integral equation arising from chemical reactor theory", *Nonlinear Engineering*, **7**(3), pp. 163–169.
38. Shoukralla E.S., 2020, "A numerical method for solving Fredholm integral equations of the first kind with logarithmic kernels and singular unknown functions", *International Journal of Applied and Computational Mathematics*, **6**(6), pp. 1–14.
39. Khuri S.A., and Wazwaz A.M., 1996, "The decomposition method for solving a second kind Fredholm integral equation with a logarithmic kernel", *International Journal of Computer Mathematics*, **61**(1–2), pp. 103–110.
40. Gao J., and Jiang Y.L., 2008, "Trigonometric Hermite wavelet approximation for the integral equations of second kind with weakly singular kernel", *Journal of Computational and Applied Mathematics*, **215**(1), pp. 242–259.
41. Babolian E., Biazar J., and Vahidi A.R., 2004, "The decomposition method applied to systems of Fredholm integral equations of the second kind", *Applied Mathematics and Computation*, **148**(2), pp. 443–452.
42. Huang Y., Fang M., and Li X.F., 2009, "Approximate solution of a system of linear integral equations by the Taylor expansion method", *International Journal of Computer Mathematics*, **86**(5), pp. 924–937.
43. Maleknejad K., Aghazadeh N., and Rabbani M., 2006, "Numerical solution of second kind Fredholm integral equations system by using a Taylor-series expansion method", *Applied Mathematics and Computation*, **175**(2), pp. 1229–1234.
44. Maleknejad K., Shahrezaee M., and Khatami H., 2005, "Numerical solution of integral equations system of the second kind by block–pulse functions", *Applied Mathematics and Computation*, **166**(1), pp. 15–24.
45. Rashidinia J., and Sajjadian M., 2018, "The impact of two transformations on the solutions of second kind Fredholm integral equations system", *International Journal of Applied and Computational Mathematics*, **4**(3), pp. 1–12.
46. Liu N., and Lin E.B., 2010, "Legendre wavelet method for numerical solution of partial differential equations", *Numerical Methods for Partial Differential Equations: An International Journal*, **26**(1), pp. 81–94.
47. Roodaki M., and Jafari Behbahani Z., 2015, "Two dimensional Legendre wavelets and their application to integral equations", *Indian Journal of Scince and Technology*, **8**(2), pp. 157–164.
48. Derili H., Sohrabi S., and Arzhang A., 2012, "Two dimensional wavelets for numerical solutions of integral equation", *Mathematical Sciences*, **6**(1), pp. 1–4.
49. Azevedo J.S., Wisniewski F., and Oliveira S.P., 2018, "A Galerkin method with two-dimensional Haar basis functions for the computation of the Karhunen–Loève expansion", *Computational and Applied Mathematics*, **37**(2), pp. 1825–1846.
50. Oliveira S.P., and Azevedo J.S., 2015, "Numerical approximation of 2D Fredholm integral eigenvalue problems by orthogonal wavelets", *Applied Mathematics and Computation*, **267**, pp. 517–528.
51. Sahu P.K., and Saha Ray S., 2018, "Comparison on wavelets techniques for solving fractional optimal control problems", *Journal of Vibration and Control*, **24**(6), pp. 1185–1201.

52. Maleknejad K., and Dehkordi M.S., 2021, "Numerical solutions of two-dimensional nonlinear integral equations via Laguerre Wavelet method with convergence analysis", *Applied Mathematics-A Journal of Chinese Universities*, **36**(1), pp. 83–98.
53. Michalska M., and Szynal J., 2001, "A new bound for the Laguerre polynomials", *Journal of Computational and Applied Mathematics*, **133**(1–2), pp. 489–493.
54. Liu N., and Lin E.B., 2010, "Legendre wavelet method for numerical solutions of partial differential equations", *Numerical Methods for Partial Differential Equations: An International Journal*, **26**(1), pp. 81–94.
55. Behera S., and Saha Ray S., 2022, "A wavelet-based novel technique for linear and nonlinear fractional Volterra–Fredholm integro-differential equations", *Computational and Applied Mathematics*, **41**(2), pp. 1–28.
56. Stewart G.W., 1993, *Afternotes on Numerical Analysis*, University of Maryland, College Park.
57. Adibi H., and Assari P., 2011, "On the numerical solution of weakly singular Fredholm integral equations of the second kind using Legendre wavelets", *Journal of Vibration and Control*, **17**(5), pp. 689–698.
58. Chakraborty S., Kant K., and Nelakanti G., 2021, "Approximation methods for system of linear Fredholm integral equations of second kind", *Applied Mathematics and Computation*, **403**, p. 126173.

4 Numerical Solutions of Two-Dimensional Volterra Integral Equations

4.1 INTRODUCTION

The integral equation is a valuable part of the analysis theory, which plays an important role in the numerical analysis. In 1825, Abel, an Italian mathematician, first proposed an integral equation in connection with the famous tautochrone problem. The two-dimensional (2D) integral equations have been used to study the dynamic properties of various models arising in engineering and mathematics [1,2]. Integral equations can also be used in various applications such as biomechanics, control, electrical engineering, heat and mass transfer, and oscillation theory [3]. Volterra integral equations (VIEs) have been applied in several areas, including mathematical physics [4], heat conduction problem [5], diffusion problems [6], electroelastic [7], plasma physics [8], contact problems [9], and the unsteady poiseuille flow in a pipe [10].

Wavelet analysis is an emerging area in the field of engineering and applied mathematics. It has been successfully applied in many applications such as image analysis, signal analysis, and communications systems [11,12].

Several studies have been done on developing and accessing numerical methods for solving one-dimensional integral equations of the second kind [13–16]. For the 2D case, some work has been done to solve the integral equations [17–19]. Mckee et al. [20] revealed that a class of nonlinear telegraph equations can be represented as 2D VIEs. Mirzaee et al. [21] have used a hybrid approach for approximate solution of nonlinear partial mixed integro-differential equations of fractional order. For nonlinear VIEs, the method of Haar wavelets was introduced in an earlier study [22]. Assari et al. presented a meshless local discrete Galerkin (MLDG) scheme and a meshless local discrete collocation (MLDC) scheme for the solution of nonlinear 2D VIEs and 2D singular integral equations with logarithmic kernels, respectively [23,24]. Brunner et al. [18] have solved 2D VIEs by collocation and iterated collocation method. Sahu et al. [25] introduced the Legendre wavelets operational method for the solution of nonlinear Volterra integro-differential equation. Assari [26] has proposed a meshless local discrete Galerkin scheme for solving the 2D integral equations. Khajehnasiri et al. [27] presented an approach for 2D nonlinear partial VIE by using the Bernoulli wavelet. Maleknejad et al. [28] applied radial basis functions (RBFs) for solving a class of nonlinear 2D Volterra–Fredholm integral equations.

DOI: 10.1201/9781003453451-4

The RBFs have been applied for the numerical solution of the 2D logarithmic integral equations [29] and nonlinear VIEs of the second kind [30].

In recent years, the system of IEs has been numerically solved by using the Haar functions method [31], Adomian method [32], homotopy perturbation method [33], differential transform method [34], Taylor expansion method [35], and Galerkin method [36]. Maleknejad et al. [37] have solved the system of IEs of the first kind by applying an operational matrix with block pulse functions. Jafarian et al. [38] introduced the Bernstein collocation method for the numerical solution of linear integral equations system. Hasan et al. [39] developed a new fixed-point method for solving a system of Volterra–Fredholm integral equations of the second kind. Babolian et al. [40] applied the Adomian decomposition method for solving the system of linear VIEs. The system of linear Fredholm integral equations has been solved by Javidi [41] by employing the modified homotopy perturbation method. Golbabai et al. [42] introduced an RBF network for the solution of the system of nonlinear integral equations. An analytical method has been presented by Shidfar et al. [43] to solve a system of integral equations.

4.2 OUTLINE OF THE PRESENT STUDY

In this chapter, two numerical methods, known as Gegenbauer wavelet method (GWM) and Cosine and Sine wavelet method (CASWM), have been introduced. The different collocation points are used for these two different wavelet methods along with the Gaussian quadrature method to reduce the given integral equations to the system of algebraic equations. The convergence and error analyses for both GWM and CASWM have also been presented in this chapter. Furthermore, some examples are provided to describe the efficiency and validity of the proposed techniques through the comparison of errors for two proposed methods.

In addition, a collocation-based method has been studied for the solution of system of 2D VIEs. The proposed approach is based on the Jacobi wavelet method (JWM) collocation procedure. In addition, the GWM has been implemented to show the comparison of error norms obtained by the proposed method. This proposed approach has been applied to reduce the system of VIEs into a system of algebraic equations. The convergence and error analyses of the proposed method are also discussed. A brief discussion, including numerical problems, is presented in order to show the accuracy and effectiveness of the proposed scheme.

4.3 TWO-DIMENSIONAL NONLINEAR VOLTERRA INTEGRAL EQUATIONS

In this chapter, the nonlinear 2D VIE of the second kind has been considered as follows:

$$u(x,t) = h(x,t) + \int_0^t \int_0^x \tilde{k}(x,t,y,z,u(y,z))\,dy\,dz, \quad (x,t) \in \Omega = [0,1) \times [0,1), \tag{4.1}$$

where $u(x,t)$ is an unknown function and the functions $h(x,t)$ and $\tilde{k}(x,t,y,z,u)$ are the given continuous functions defined on Ω and $\Omega \times \Omega$, respectively.

4.3.1 Function Approximation by GWs

A continuous function $h(x,t)$ defined over $L^2(\Omega)$ may be expanded as follows:

$$h(x,t) \approx \sum_{n=1}^{2^{k_1-1}} \sum_{n'=1}^{2^{k_2-1}} \sum_{m=0}^{M-1} \sum_{m'=0}^{M'-1} d_{n,m,n',m'} \cdot \psi_{n,m,n',m'}(x,t) = \Psi^T(x)\Phi_1\Psi(t), \tag{4.2}$$

where $\Psi(x)$ and $\Psi(t)$ are vectors of dimension $2^{k_1-1}M \times 1$ and $2^{k_2-1}M' \times 1$, respectively, given by

$$\Psi(x) = [\psi_{1,0}(x),...,\psi_{1,M-1}(x),\psi_{2,0}(x),...,\psi_{2,M-1}(x),...,\psi_{2^{k_1-1},0}(x),...,\psi_{2^{k_1-1},M-1}(x)]^T, \tag{4.3}$$

$$\Psi(t) = [\psi_{1,0}(t),...,\psi_{1,M'-1}(t),\psi_{2,0}(t),...,\psi_{2,M'-1}(t),...,\psi_{2^{k_2-1},0}(t),...,\psi_{2^{k_2-1},M'-1}(t)]^T, \tag{4.4}$$

where $\psi_{n,m}(x)$, $n = 1,2,...,2^{k_1-1}$, $m = 0,1,...,M-1$ is the GWs defined as follows:

$$\psi_{n,m}(x) = \begin{cases} \dfrac{2^{\frac{k}{2}}}{\sqrt{L_m^\lambda}} \mathcal{G}_m^\lambda(2^k x - 2n + 1), & \dfrac{n-1}{2^{k-1}} \le x < \dfrac{n}{2^{k-1}}, \\ 0, & \text{otherwise.} \end{cases} \tag{4.5}$$

Here, $\Phi_1 = [\tilde{\varphi}_{p,q}]$ is the coefficient matrix of dimension $(2^{k_1-1}M \times 2^{k_2-1}M')$ whose elements can be calculated from

$$\tilde{\varphi}_{p,q} = < \psi_{n,m}(x), < h(x,t), \psi_{n',m'}(t) >_{w_{n,k}} >_{w_{n,k}},$$

where

$$p = (n-1)M + m + 1,$$

$$q = (n'-1)M' + m' + 1,$$

for $n = 1,2,...,2^{k_1-1}$, $n' = 1,2,...,2^{k_2-1}$, $m = 0,1,...,M-1$, and $m' = 0,1,...,M'-1$.

4.3.1.1 GWM for Nonlinear Two-Dimensional VIEs

Consider the nonlinear VIEs given in Eq. (4.1). By using Eq. (4.2), $u(x,t)$ and $h(x,t)$ can be approximated as follows [44]:

$$u(x,t) = \Psi^T(x)U\Psi(t) \tag{4.6}$$

and

$$h(x,t) = \Psi^T(x)\Phi_1\Psi(t). \tag{4.7}$$

From Eqs. (4.1), (4.6), and (4.7), it can be written as follows:

$$\Psi^T(x)U\Psi(t) = \Psi^T(x)\Phi_1\Psi(t) + \int_0^t\int_0^x \tilde{k}(x,t,y,z,\Psi^T(y)U\Psi(z))\,dy\,dz, \quad 0 \le x,t \le 1. \tag{4.8}$$

We now collocate Eq. (4.8) at $2^{k_1-1}M \times 2^{k_2-1}M'$ points (x_p, t_q) as follows:

$$\Psi^T(x_p)U\Psi(t_q) = \Psi^T(x_p)\Phi_1\Psi(t_q) + \int_0^{t_q}\int_0^{x_p} \tilde{k}(x_p,t_q,y,z,\Psi^T(y)U\Psi(z))\,dy\,dz, \tag{4.9}$$

where the collocation points x_p and t_q are zeros of shifted Chebyshev first-kind polynomials:

$$x_p = \frac{\cos[\pi + \left(\frac{2p-1}{2(2^{k_1-1}M)}\right)\pi] + 1}{2}, \quad p = 1,2,\ldots,2^{k_1-1}M,$$

$$t_q = \frac{\cos[\pi + \left(\frac{2q-1}{2(2^{k_2-1}M')}\right)\pi] + 1}{2}, \quad q = 1,2,\ldots,2^{k_2-1}M'.$$

The intervals $[0,t_q]$ and $[0,x_p]$ are each converted into interval $[-1,1]$ by the following transformations:

$$\tau = \frac{2}{t_q}z - 1, \quad \alpha = \frac{2}{x_p}y - 1,$$

to use the Gaussian quadrature formula for Eq. (4.9).

Let

$$\tilde{H}(x_p,t_q,\alpha,\tau) = \tilde{k}\left(x_p,t_q,\frac{x_p}{2}(\alpha+1),\frac{t_q}{2}(\tau+1),\Psi^T\left(\frac{x_p}{2}(\alpha+1)\right)U\Psi\left(\frac{t_q}{2}(\tau+1)\right)\right).$$

Equation (4.9) may then be approximately written as follows:

$$\Psi^T(x_p)U\Psi(t_q) \approx \Psi^T(x_p)\Phi_1\Psi(t_q) + \frac{x_p t_q}{4}\int_{-1}^{1}\int_{-1}^{1} \tilde{H}(x_p,t_q,\alpha,\tau)\,d\alpha\,d\tau.$$

From the Gaussian quadrature formula, we get

$$\Psi^T(x_p)U\Psi(t_q) \approx \Psi^T(x_p)\Phi_1\Psi(t_q) + \frac{x_p t_q}{4}\sum_{\tilde{p}=1}^{n_1}\sum_{\tilde{q}=1}^{n_2} \tilde{w}_{1p}\tilde{w}_{2q}\tilde{H}(x_p,t_q,\alpha_{\tilde{q}},\tau_{\tilde{p}}), \tag{4.10}$$

where $\tau_{\tilde{p}}$ and $\alpha_{\tilde{q}}$ are zeros of the Legendre polynomial of degrees n_1 and n_2, respectively, and $\tilde{w}_{1p}, \tilde{w}_{2q}$ are the corresponding weights. With the help of $2^{k_1-1}M \times 2^{k_2-1}M'$ nonlinear equations from Eq. (4.10), elements of U in Eq. (4.6) can be obtained by using Newton's iterative method.

4.3.1.2 Convergence Analysis

In the present analysis, we established that the expansion of any continuous function $h(x,t) \in L^2(\Omega)$ with respect to 2D GWs, as given in Eq. (4.2), converges uniformly to h.

Theorem 4.3.1

Suppose that $h(x,t) \in L^2(\Omega)$ is a real-valued function and $|h(x,t)| \le \aleph_1$, where $\aleph_1$ is a real and positive constant. Then, the GWs expansion of $h(x,t)$ converges uniformly to $h(x,t)$.

Proof:

The GWs coefficients can be determined as follows:

$$d_{n,m,n',m'} = << h(x,t), \psi_{n,m}(x) >, \psi_{n',m'}(t) >$$

$$= \int_0^1 \int_0^1 h(x,t)\psi_{n,m}(x)\psi_{n',m'}(t)\tilde{w}(x)\tilde{w}(t)\,dx\,dt$$

$$= \frac{2^{\frac{k_1+k_2}{2}}}{\sqrt{L_m^\lambda}\sqrt{L_{m'}^\lambda}} \int_{\frac{n'-1}{2^{k_2-1}}}^{\frac{n'}{2^{k_2-1}}} \left[\int_{\frac{n-1}{2^{k_1-1}}}^{\frac{n}{2^{k_1-1}}} h(x,t)\mathcal{G}_m^\lambda(2^{k_1}x-2n+1)\tilde{w}(2^{k_1}x-2n+1)\,dx \right]$$

$$.\mathcal{G}_{m'}^\lambda(2^{k_2}t-2n'+1)\tilde{w}(2^{k_2}t-2n'+1)dt.$$

Now, changing the variable $2^{k_1}x-2n+1=\cos\theta_1$, it follows

$$= \frac{2^{\frac{k_1+k_2}{2}}}{\sqrt{L_m^\lambda}\sqrt{L_{m'}^\lambda}} \frac{1}{2^{k_1}} \int_{\frac{n'-1}{2^{k_2-1}}}^{\frac{n'}{2^{k_2-1}}} \left[\int_0^\pi h\left(\frac{\cos\theta_1+2n-1}{2^{k_1}}, t\right)\mathcal{G}_m^\lambda(\cos\theta_1)\tilde{w}(\cos\theta_1)(\sin\theta_1)\,d\theta_1 \right]$$

$$.\mathcal{G}_{m'}^\lambda(2^{k_2}t-2n'+1)\tilde{w}(2^{k_2}t-2n'+1)dt.$$

Similarly, changing the variable $2^{k_2}t-2n'+1=\cos\theta_2$, it follows

$$= \frac{2^{\frac{k_1+k_2}{2}}}{\sqrt{L_m^\lambda}\sqrt{L_{m'}^\lambda}} \frac{1}{2^{k_1+k_2}} \int_0^\pi \left[\int_0^\pi h\left(\frac{\cos\theta_1+2n-1}{2^{k_1}}, \frac{\cos\theta_2+2n'-1}{2^{k_2}}\right)\mathcal{G}_m^\lambda(\cos\theta_1)\tilde{w}(\cos\theta_1)(\sin\theta_1)\,d\theta_1 \right]$$

$$.\mathcal{G}_{m'}^\lambda(\cos\theta_2)\tilde{w}(\cos\theta_2)(\sin\theta_2)d\theta_2$$

$$= \frac{2^{\frac{k_1+k_2}{2}}}{\sqrt{L_m^\lambda}\sqrt{L_{m'}^\lambda}} \frac{1}{2^{k_1+k_2}} \int_0^\pi \int_0^\pi h\left(\frac{\cos\theta_1+2n-1}{2^{k_1}}, \frac{\cos\theta_2+2n'-1}{2^{k_2}}\right)\mathcal{G}_m^\lambda(\cos\theta_1)(1-\cos^2\theta_1)^{\lambda-\frac{1}{2}}(\sin\theta_1)\,d\theta_1$$

$$.\mathcal{G}_{m'}^\lambda(\cos\theta_2)(1-\cos^2\theta_2)^{\lambda-\frac{1}{2}}(\sin\theta_2)d\theta_2$$

$$= \frac{2^{\frac{k_1+k_2}{2}}}{\sqrt{L_m^\lambda}\sqrt{L_{m'}^\lambda}} \frac{1}{2^{k_1+k_2}} \int_0^\pi \int_0^\pi h\left(\frac{\cos\theta_1 + 2n-1}{2^{k_1}}, \frac{\cos\theta_2 + 2n'-1}{2^{k_2}}\right) \mathcal{G}_m^\lambda(\cos\theta_1)(\sin\theta_1)^{2\lambda}\, d\theta_1$$

$$.\mathcal{G}_{m'}^\lambda(\cos\theta_2)(\sin\theta_2)^{2\lambda}\, d\theta_2.$$

This implies

$$\left|d_{n,m,n',m'}\right| = \left| \frac{2^{\frac{k_1+k_2}{2}}}{\sqrt{L_m^\lambda}\sqrt{L_{m'}^\lambda}} \frac{1}{2^{k_1+k_2}} \int_0^\pi \int_0^\pi h\left(\frac{\cos\theta_1 + 2n-1}{2^{k_1}}, \frac{\cos\theta_2 + 2n'-1}{2^{k_2}}\right) \mathcal{G}_m^\lambda(\cos\theta_1)(\sin\theta_1)^{2\lambda}\, d\theta_1 \,.\mathcal{G}_{m'}^\lambda(\cos\theta_2)(\sin\theta_2)^{2\lambda}\, d\theta_2 \right|$$

$$\leq \frac{2^{\frac{k_1+k_2}{2}}}{\sqrt{L_m^\lambda}\sqrt{L_{m'}^\lambda}} \frac{1}{2^{k_1+k_2}} \int_0^\pi \int_0^\pi \left|h\left(\frac{\cos\theta_1 + 2n-1}{2^{k_1}}, \frac{\cos\theta_2 + 2n'-1}{2^{k_2}}\right)\right| \left|\mathcal{G}_m^\lambda(\cos\theta_1)(\sin\theta_1)^{2\lambda}\right| d\theta_1 \,.\left|\mathcal{G}_{m'}^\lambda(\cos\theta_2)(\sin\theta_2)^{2\lambda}\right| d\theta_2$$

$$\leq \frac{\aleph_1}{\sqrt{L_m^\lambda}\sqrt{L_{m'}^\lambda}} \frac{1}{2^{\frac{k_1+k_2}{2}}} \int_0^\pi \left|\mathcal{G}_m^\lambda(\cos\theta_1)\right|(\sin\theta_1)^{2\lambda}\, d\theta_1 . \int_0^\pi . \left|\mathcal{G}_{m'}^\lambda(\cos\theta_2)\right|(\sin\theta_2)^{2\lambda}\, d\theta_2.$$

Now, by using the property of Gegenbauer polynomials, i.e.,

$$\left|\mathcal{G}_m^\lambda(\cos\theta)\right|(\sin\theta)^\lambda < \frac{2^{1-\lambda}\Gamma\left(m+\frac{3\lambda}{2}\right)}{\Gamma(\lambda)\Gamma\left(1+m+\frac{\lambda}{2}\right)}, \quad 0 \leq \theta \leq \pi,$$

it follows

$$\left|d_{n,m,n',m'}\right| < \frac{\aleph_1}{\sqrt{L_m^\lambda}\sqrt{L_{m'}^\lambda}} \frac{1}{2^{\frac{k_1+k_2}{2}}} \frac{2^{1-\lambda}\Gamma\left(m+\frac{3\lambda}{2}\right)}{\Gamma(\lambda)\Gamma\left(1+m+\frac{\lambda}{2}\right)} \frac{2^{1-\lambda}\Gamma\left(m'+\frac{3\lambda}{2}\right)}{\Gamma(\lambda)\Gamma\left(1+m'+\frac{\lambda}{2}\right)}$$

$$= \frac{\aleph_1\Pi}{2^{2\lambda-2}} \frac{\Gamma\left(m+\frac{3\lambda}{2}\right).\Gamma\left(m'+\frac{3\lambda}{2}\right)}{(\Gamma(\lambda))^2\Gamma\left(1+m+\frac{\lambda}{2}\right).\Gamma\left(1+m'+\frac{\lambda}{2}\right)}.$$

That is,

$$\left|d_{n,m,n',m'}\right| < \frac{\aleph_1\Pi}{2^{2\lambda-2}} \frac{\Gamma\left(m+\frac{3\lambda}{2}\right).\Gamma\left(m'+\frac{3\lambda}{2}\right)}{(\Gamma(\lambda))^2\Gamma\left(1+m+\frac{\lambda}{2}\right).\Gamma\left(1+m'+\frac{\lambda}{2}\right)}, \tag{4.11}$$

where

$$\Pi = \frac{1}{\sqrt{L_m^\lambda}\sqrt{L_{m'}^\lambda}} \frac{1}{2^{\frac{k_1+k_2}{2}}}.$$

Hence, the series $\sum_{n=1}^{2^{k_1-1}}\sum_{n'=1}^{2^{k_2-1}}\sum_{m=0}^{M-1}\sum_{m'=0}^{M'-1} d_{n,m,n',m'}$ is absolutely convergent, and it follows that the series $\sum_{n=1}^{2^{k_1-1}}\sum_{n'=1}^{2^{k_2-1}}\sum_{m=0}^{M-1}\sum_{m'=0}^{M'-1} d_{n,m,n',m'} \cdot \psi_{n,m,n',m'}(x,t)$ is uniformly convergent.

4.3.1.3 Error Analysis

In this part, the error analysis of GWM for solving the 2D nonlinear VIE has been derived.

Theorem 4.3.2

Let $\tilde{h}_{2^{k_1-1}M,2^{k_2-1}M'}(x,t)$ be the 2D GWs expansion of the real sufficiently smooth function $h(x,t) \in L^2(\Omega)$. Then, there exist real numbers $\Im_1, \Im_2$ and $\Im_3$ such that

$$\left\|h(x,t) - \tilde{h}_{2^{k_1-1}M,2^{k_2-1}M'}(x,t)\right\|_2 \le \frac{\Im_1}{2^{2^{k_1}M-1}\Gamma(2^{k_1-1}M+1)} + \frac{\Im_2}{2^{2^{k_2}M'-1}\Gamma(2^{k_2-1}M'+1)} + \frac{\Im_3}{2^{2^{k_1}M+2^{k_2}M'-2}\Gamma(2^{k_1-1}M+1)\Gamma(2^{k_2-1}M'+1)}. \tag{4.12}$$

Proof:

Assume that $h(x,t)$ is a sufficiently smooth function on $L^2(\Omega)$ and $\tilde{h}_{2^{k_1-1}M,2^{k_2-1}M'}(x,t)$ is the best approximation of h from $\Lambda_{2^{k_1-1}M,2^{k_2-1}M'}$, where

$$\Lambda_{2^{k_1-1}M,2^{k_2-1}M'} = \text{Span}\left\{\begin{matrix}\psi_{n,m,n',m'}(x,t) : n = 1,2,\ldots,2^{k_1-1}, m = 0,1,\ldots,M-1; \\ n' = 1,2,\ldots,2^{k_2-1}, m = 0,1,\ldots,M'-1\end{matrix}\right\}.$$

Then, for any $P_{2^{k_1-1}M,2^{k_2-1}M'}[h] \in \Lambda_{2^{k_1-1}M,2^{k_2-1}M'}$, the following inequality is held:

$$\left\|h(x,t) - \tilde{h}_{2^{k_1-1}M,2^{k_2-1}M'}(x,t)\right\|_2 \le \left\|h(x,t) - P_{2^{k_1-1}M,2^{k_2-1}M'}[h]\right\|_2, \tag{4.13}$$

where $P_{2^{k_1-1}M,2^{k_2-1}M'}[h]$ is the interpolating polynomial of h at the node points (x_p, t_q).

Here, $x_p, p = 1,2,\ldots,2^{k_1-1}M$ and $t_q, q = 1,2,\ldots,2^{k_2-1}M'$ are the roots of the $2^{k_1-1}M$-degree and $2^{k_2-1}M'$-degree shifted Chebyshev polynomial in [0,1], respectively.

Now,

$$h(x,t)-P_{2^{k_1-1}M,2^{k_2-1}M'}[h]=\frac{\partial^{2^{k_1-1}M}h(\xi_1,t)}{\partial x^{2^{k_1-1}M}}\frac{\prod_{p=1}^{2^{k_1-1}M}(x-x_p)}{\Gamma(2^{k_1-1}M+1)}+\frac{\partial^{2^{k_2-1}M'}\varphi(x,\xi_2)}{\partial t^{2^{k_2-1}M'}}$$
$$\frac{\prod_{q=1}^{2^{k_2-1}M'}(t-t_q)}{\Gamma(2^{k_2-1}M'+1)}+\frac{\partial^{2^{k_1-1}M+2^{k_2-1}M'}\varphi(\tilde{\xi}_1,\tilde{\xi}_2)}{\partial x^{2^{k_1-1}M}\partial t^{2^{k_2-1}M'}}\frac{\prod_{p=1}^{2^{k_1-1}M}(x-x_p)\prod_{q=1}^{2^{k_2-1}M'}(t-t_q)}{\Gamma(2^{k_1-1}M+1)\Gamma(2^{k_2-1}M'+1)}, \tag{4.14}$$

for some $\xi_1,\xi_2,\tilde{\xi}_1,\tilde{\xi}_2\in[0,1]$.
Now, utilizing the estimates for Chebyshev interpolating nodes on the interval Ω, it follows that [45]

$$\left|\prod_{p=1}^{2^{k_1-1}M}(x-x_p)\right|\le\frac{1}{2^{2^{k_1}M-1}}$$

and

$$\left|\prod_{q=1}^{2^{k_2-1}M'}(t-t_q)\right|\le\frac{1}{2^{2^{k_2}M'-1}}.$$

Therefore,

$$\left|h(x,t)-P_{2^{k_1-1}M,2^{k_2-1}M'}[h]\right|\le\sup_{t\in[0,1]}\left(\max_{\xi_1\in[0,1]}\left|\frac{\partial^{2^{k_1-1}M}h(\xi_1,t)}{\partial x^{2^{k_1-1}M}}\right|\right)\frac{1}{2^{2^{k_1}M-1}\Gamma(2^{k_1-1}M+1)}$$
$$+\sup_{x\in[0,1]}\left(\max_{\xi_2\in[0,1]}\left|\frac{\partial^{2^{k_2-1}M'}h(x,\xi_2)}{\partial t^{2^{k_2-1}M'}}\right|\right)\frac{1}{2^{2^{k_2}M'-1}\Gamma(2^{k_2-1}M'+1)}$$
$$+\sup_{(x,t)\in\Omega}\left(\max_{(\tilde{\xi}_1,\tilde{\xi}_2)\in\Omega}\left|\frac{\partial^{2^{k_1-1}M+2^{k_2-1}M'}h(\tilde{\xi}_1,\tilde{\xi}_2)}{\partial x^{2^{k_1-1}M}\partial t^{2^{k_2-1}M'}}\right|\right)$$
$$\frac{1}{2^{2^{k_1}M+2^{k_2}M'-2}\Gamma(2^{k_1-1}M+1)\Gamma(2^{k_2-1}M'+1)},$$

whence

$$\left|h(x,t)-P_{2^{k_1-1}M,2^{k_2-1}M'}[h]\right| \le \sup_{(x,t)\in\Omega}\left|\frac{\partial^{2^{k_1-1}M}h(x,t)}{\partial x^{2^{k_1-1}M}}\right|\frac{1}{2^{2^{k_1}M-1}\Gamma(2^{k_1-1}M+1)}$$
$$+\sup_{(x,t)\in\Omega}\left|\frac{\partial^{2^{k_2-1}M'}h(x,t)}{\partial t^{2^{k_2-1}M'}}\right|\frac{1}{2^{2^{k_2}M'-1}\Gamma(2^{k_2-1}M'+1)}$$
$$+\sup_{(x,t)\in\Omega}\left|\frac{\partial^{2^{k_1-1}M+2^{k_2-1}M'}h(x,t)}{\partial x^{2^{k_1-1}M}\partial t^{2^{k_2-1}M'}}\right| \tag{4.15}$$
$$\frac{1}{2^{2^{k_1}M+2^{k_2}M'-2}\Gamma(2^{k_1-1}M+1)\Gamma(2^{k_2-1}M'+1)}.$$

In particular, from Eq. (4.13),

$$\left\|h(x,t)-\tilde{h}_{2^{k_1-1}M,2^{k_2-1}M'}(x,t)\right\|_2^2 \le \left\|h(x,t)-P_{2^{k_1-1}M,2^{k_2-1}M'}[h]\right\|_2^2$$
$$=\int_0^1\int_0^1\left|h(x,t)-P_{2^{k_1-1}M,2^{k_2-1}M'}[h]\right|^2 dx\,dt. \tag{4.16}$$

Now, using Eqs. (4.15) and (4.16), it follows that

$$\left\|h(x,t)-\tilde{h}_{2^{k_1-1}M,2^{k_2-1}M'}(x,t)\right\|_2^2 \le \left[\frac{\Im_1}{2^{2^{k_1}M-1}\Gamma(2^{k_1-1}M+1)}+\frac{\Im_2}{2^{2^{k_2}M'-1}\Gamma(2^{k_2-1}M'+1)}\right.$$
$$\left.+\frac{\Im_3}{2^{2^{k_1}M+2^{k_2}M'-2}\Gamma(2^{k_1-1}M+1)\Gamma(2^{k_2-1}M'+1)}\right]^2, \tag{4.17}$$

assuming $\sup_{(x,t)\in\Omega}\left|\frac{\partial^{2^{k_1-1}M}h(x,t)}{\partial x^{2^{k_1-1}M}}\right| \le \Im_1$, $\sup_{(x,t)\in\Omega}\left|\frac{\partial^{2^{k_2-1}M'}h(x,t)}{\partial t^{2^{k_2-1}M'}}\right| \le \Im_2$ and $\sup_{(x,t)\in\Omega}\left|\frac{\partial^{2^{k_1-1}M+2^{k_2-1}M'}h(x,t)}{\partial x^{2^{k_1-1}M}\partial t^{2^{k_2-1}M'}}\right| \le \Im_3$.

Therefore,

$$\left\|h(x,t)-\tilde{h}_{2^{k_1-1}M,2^{k_2-1}M'}(x,t)\right\|_2 \le \frac{\Im_1}{2^{2^{k_1}M-1}\Gamma(2^{k_1-1}M+1)}+\frac{\Im_2}{2^{2^{k_2}M'-1}\Gamma(2^{k_2-1}M'+1)}$$
$$+\frac{\Im_3}{2^{2^{k_1}M+2^{k_2}M'-2}\Gamma(2^{k_1-1}M+1)\Gamma(2^{k_2-1}M'+1)}.$$

Hence, the theorem is proved. ■

Theorem 4.3.3

Let $\tilde{h}_{2^{k1-1}M,2^{k2-1}M'}(x,t) \approx \sum_{n=1}^{2^{k1-1}} \sum_{n'=1}^{2^{k2-1}} \sum_{m=0}^{M-1} \sum_{m'=0}^{M'-1} \tilde{d}_{n,m,n',m'} \cdot \psi_{n,m,n',m'}(x,t)$ is the best approximation of h defined on $L^2(\Omega)$ and $h^*_{2^{k1-1}M,2^{k2-1}M'}(x,t) \approx \sum_{n=1}^{2^{k1-1}} \sum_{n'=1}^{2^{k2-1}} \sum_{m=0}^{M-1} \sum_{m'=0}^{M'-1} d^*_{n,m,n',m'} \cdot \psi_{n,m,n',m'}(x,t)$ is the numerical approximation h defined on $L^2(\Omega)$. Then [46],

$$\left\| \tilde{h}_{2^{k1-1}M,2^{k2-1}M'}(x,t) - h^*_{2^{k1-1}M,2^{k2-1}M'}(x,t) \right\|_2 \leq \mathcal{M}$$

Proof:

$$\left\| \tilde{h}_{2^{k1-1}M,2^{k2-1}M'}(x,t) - h^*_{2^{k1-1}M,2^{k2-1}M'}(x,t) \right\|_2^2$$

$$\leq \int_0^1 \int_0^1 \left| \tilde{h}_{2^{k1-1}M,2^{k2-1}M'}(x,t) - h^*_{2^{k1-1}M,2^{k2-1}M'}(x,t) \right|^2 dx\, dt$$

$$\leq \int_0^1 \int_0^1 \left| \sum_{n=1}^{2^{k1-1}} \sum_{n'=1}^{2^{k2-1}} \sum_{m=0}^{M-1} \sum_{m'=0}^{M'-1} \tilde{d}_{n,m,n',m'} \cdot \psi_{n,m,n',m'}(x,t) - \sum_{n=1}^{2^{k1-1}} \sum_{n'=1}^{2^{k2-1}} \sum_{m=0}^{M-1} \sum_{m'=0}^{M'-1} d^*_{n,m,n',m'} \cdot \psi_{n,m,n',m'}(x,t) \right|^2 dxdt$$

$$\leq \int_0^1 \int_0^1 \left| \sum_{n=1}^{2^{k1-1}} \sum_{n'=1}^{2^{k2-1}} \sum_{m=0}^{M-1} \sum_{m'=0}^{M'-1} (\tilde{d}_{n,m,n',m'} - d^*_{n,m,n',m'}) \cdot \psi_{n,m,n',m'}(x,t) \right|^2 dxdt$$

$$\leq \sum_{n=1}^{2^{k1-1}} \sum_{n'=1}^{2^{k2-1}} \sum_{m=0}^{M-1} \sum_{m'=0}^{M'-1} \left| \tilde{d}_{n,m,n',m'} - d^*_{n,m,n',m'} \right|^2 \cdot \int_0^1 \int_0^1 \left| \psi_{n,m,n',m'}(x,t) \right|^2 dxdt$$

$$\leq \sum_{n=1}^{2^{k1-1}} \sum_{n'=1}^{2^{k2-1}} \sum_{m=0}^{M-1} \sum_{m'=0}^{M'-1} \left| \tilde{d}_{n,m,n',m'} - d^*_{n,m,n',m'} \right|^2 .$$

This implies

$$\left\|\tilde{h}_{2^{k_1-1}M,2^{k_2-1}M'}(x,t)-h^*_{2^{k_1-1}M,2^{k_2-1}M'}(x,t)\right\|_2 \le \left(\sum_{n=1}^{2^{k_1-1}}\sum_{n'=1}^{2^{k_2-1}}\sum_{m=0}^{M-1}\sum_{m'=0}^{M'-1}\left|\tilde{d}_{n,m,n',m'}-d^*_{n,m,n',m'}\right|^2\right)^{\frac{1}{2}} = \mathcal{M}, \tag{4.18}$$

where $\mathcal{M}=\left(\sum_{n=1}^{2^{k_1-1}}\sum_{n'=1}^{2^{k_2-1}}\sum_{m=0}^{M-1}\sum_{m'=0}^{M'-1}\left|\tilde{d}_{n,m,n',m'}-d^*_{n,m,n',m'}\right|^2\right)^{\frac{1}{2}}$.

Theorem 4.3.4

Let $h^*_{2^{k_1-1}M,2^{k_2-1}M'}(x,t)\approx\sum_{n=1}^{2^{k_1-1}}\sum_{n'=1}^{2^{k_2-1}}\sum_{m=0}^{M-1}\sum_{m'=0}^{M'-1}d^*_{n,m,n',m'}\cdot\psi_{n,m,n',m'}(x,t)$ be our numerical approximation defined on $L^2(\Omega)$. Then,

$$\left\|h(x,t)-h^*_{2^{k_1-1}M,2^{k_2-1}M'}(x,t)\right\|_2 \le \tilde{\mathcal{M}}+\mathcal{M},$$

where $\tilde{\mathcal{M}}=\dfrac{\Im_1}{2^{2^{k_1}M-1}\Gamma(2^{k_1-1}M+1)}+\dfrac{\Im_2}{2^{2^{k_2}M'-1}\Gamma(2^{k_2-1}M'+1)}$

$$+\frac{\Im_3}{2^{2^{k_1}M+2^{k_2}M'-2}\Gamma(2^{k_1-1}M+1)\Gamma(2^{k_2-1}M'+1)}.$$

Proof:

By using Theorems 4.3.2 and 4.3.3, it follows

$$\begin{aligned}\left\|h(x,t)-h^*_{2^{k_1-1}M,2^{k_2-1}M'}(x,t)\right\|_2 &\le \left\|\varphi(x,t)-\tilde{h}_{2^{k_1-1}M,2^{k_2-1}M'}(x,t)\right\|_2\\ &\quad+\left\|\tilde{h}_{2^{k_1-1}M,2^{k_2-1}M'}(x,t)-h^*_{2^{k_1-1}M,2^{k_2-1}M'}(x,t)\right\|_2\\ &\le \frac{\Im_1}{2^{2^{k_1}M-1}\Gamma(2^{k_1-1}M+1)}+\frac{\Im_2}{2^{2^{k_2}M'-1}\Gamma(2^{k_2-1}M'+1)}\\ &\quad+\frac{\Im_3}{2^{2^{k_1}M+2^{k_2}M'-2}\Gamma(2^{k_1-1}M+1)\Gamma(2^{k_2-1}M'+1)}++\mathcal{M}\\ &\le \tilde{\mathcal{M}}+\mathcal{M}.\end{aligned}$$

■

4.3.2 Function Approximation by CASWs

A continuous function $h(x,t)$ defined over $L^2(\Omega)$ may be expressed as follows:

$$h(x,t) \approx \sum_{n=0}^{2^{k_1}-1}\sum_{n'=0}^{2^{k_2}-1}\sum_{m=-M}^{M}\sum_{m'=-M'}^{M'} d_{n,m,n',m'}.\psi_{n,m,n',m'}(x,t) = \Psi^T(x).\Phi_2.\Psi(t), \quad (4.19)$$

where $\Psi(x)$ and $\Psi(t)$ are vectors of dimension $2^{k_1}(2M+1)$ and $2^{k_2}(2M'+1)$, respectively, given by

$$\Psi(x) = [\psi_{0,-M}(x),\ldots,\psi_{0,M}(x),\psi_{1,-M}(x),\ldots,\psi_{1,M}(x),\ldots,\psi_{2^{k_1}-1,-M}(x),\ldots,\psi_{2^{k_1}-1,M}(x)]^T, \quad (4.20)$$

$$\Psi(t) = [\psi_{0,-M'}(t),\ldots,\psi_{0,M'}(t),\psi_{1,-M'}(t),\ldots,\psi_{1,M'}(t),\ldots,\psi_{2^{k_2}-1,-M'}(t),\ldots,\psi_{2^{k_2}-1,M'}(t)]^T, \quad (4.21)$$

where $\psi_{n,m}(x),\ n = 0,1,\ldots,2^{k_1}-1,\ m = -M,-M+1,\ldots,M$ is the CASWs defined as follows:

$$\psi_{n,m}(x) = \begin{cases} 2^{\frac{k}{2}} CAS_m(2^k x - n), & \dfrac{n}{2^k} \le x < \dfrac{n+1}{2^k}, \\ 0, & \text{otherwise}. \end{cases} \quad (4.22)$$

Here, $\Phi_2 = [\hat{\varphi}_{p,q}]$ is the coefficient matrix of dimension $2^{k_1}(2M+1) \times 2^{k_2}(2M'+1)$ whose elements can be calculated as follows:

$$\hat{\varphi}_{p,q} = \langle \psi_{n,m}(x), \langle h(x,t), \psi_{n',m'}(t)\rangle\rangle,$$

where

$$p = n(2M+1) + m + M + 1,$$

$$q = n'(2M'+1) + m' + M' + 1,$$

for $n = 0,1,\ldots,2^{k_1}-1$, $n' = 0,1,\ldots,2^{k_2}-1$, $m = -M,-M+1,\ldots,M$, and $m' = -M',-M'+1,\ldots,M'$.

4.3.2.1 CASWM for Nonlinear Two-Dimensional VIEs

Consider the nonlinear VIEs given in Eq. (4.1). By using Eq. (4.19), we first approximate $u(x,t)$ and $h(x,t)$ as follows:

$$u(x,t) = \Psi^T(x) U \Psi(t) \quad (4.23)$$

and

$$h(x,t) = \Psi^T(\varsigma)\Phi_2\Psi(t). \tag{4.24}$$

From Eqs. (4.1), (4.23), and (4.24), we have

$$\Psi^T(x)U\Psi(t) = \Psi^T(x)\Phi_2\Psi(t) + \int_0^t\int_0^x \tilde{k}(x,t,y,z,\Psi^T(y)U\Psi(z))\,dy\,dz, \quad 0 \le x,t \le 1. \tag{4.25}$$

We now collocate Eq. (4.25) at $2^{k_1}(2M+1)\times 2^{k_2}(2M'+1)$ points (x_p,t_q) as follows:

$$\Psi^T(x_p)U\Psi(t_q) = \Psi^T(x_p)\Phi_2\Psi(t_q) + \int_0^{t_q}\int_0^{x_p} \tilde{k}(x_p,t_q,y,z,\Psi^T(y)U\Psi(z))\,dy\,dz, \tag{4.26}$$

where the collocation points x_p and t_q are

$$x_p = \frac{2p-1}{2(2M+1)2^{k_1}}, \quad p = 1,2,\ldots,(2M+1)2^{k_1},$$

$$t_q = \frac{2q-1}{2(2M'+1)2^{k_2}}, \quad q = 1,2,\ldots,(2M'+1)2^{k_2}.$$

The intervals $[0,t_q]$ and $[0,x_p]$ are each converted into interval $[-1,1]$ by the following transformations:

$$\tau = \frac{2}{t_q}z - 1, \quad \alpha = \frac{2}{x_p}y - 1,$$

to use the Gaussian quadrature formula for Eq. (4.26).

Let

$$\widehat{H}(x_p,t_q,\alpha,\tau) = \tilde{k}\left(x_p,t_q,\frac{x_p}{2}(\alpha+1),\frac{t_q}{2}(\tau+1),\Psi^T\left(\frac{x_p}{2}(\alpha+1)\right)U\Psi\left(\frac{t_q}{2}(\tau+1)\right)\right).$$

Equation (4.26) may then be approximately written as follows:

$$\Psi^T(x_p)U\Psi(t_q) \approx \Psi^T(x_p)\Phi_2\Psi(t_q) + \frac{x_p t_q}{4}\int_{-1}^{1}\int_{-1}^{1} \widehat{H}(x_p,t_q,\alpha,\tau)\,d\alpha\,d\tau.$$

From the Gaussian quadrature formula, we get

$$\Psi^T(x_p)U\Psi(t_q) \approx \Psi^T(x_p)\Phi_2\Psi(t_q) + \frac{x_p t_q}{4}\sum_{\tilde{p}=1}^{n_1}\sum_{\tilde{q}=1}^{n_2} \widehat{w}_{1\tilde{p}}\widehat{w}_{2\tilde{q}}\widehat{H}(x_p,t_q,\alpha_{\tilde{q}},\tau_{\tilde{p}}), \tag{4.27}$$

where $\tau_{\tilde{p}}$ and $\alpha_{\tilde{q}}$ are zeros of the Legendre polynomial of degrees n_1 and n_2, respectively, and $\widehat{w}_{1\tilde{p}}, \widehat{w}_{2\tilde{q}}$ are the corresponding weights. Equation (4.27) gives $2^{k_1}(2M+1)\times 2^{k_2}(2M'+1)$ nonlinear equations, which can be solved for the elements of U in Eq. (4.23) using Newton's iterative method.

4.3.2.2 Convergence Analysis

In this section, we presented that the CASWs expansion of a continuous function $h(x,t) \in L^2(\Omega)$ converges uniformly to h.

Theorem 4.3.5

Let $h(x,t)$ be a continuous function on $L^2(\Omega)$ and $|h(x,t)| < \wp$ can be expanded as an infinite sum of the CASWs and the series converges uniformly to $h(x,t)$, i.e.,

$$h(x,t) = \sum_{n=0}^{\infty}\sum_{n'=0}^{\infty}\sum_{m\in\mathbb{Z}}\sum_{m'\in\mathbb{Z}} d_{n,m,n',m'}.\psi_{n,m,n',m'}(x,t).$$

Proof:

The wavelet coefficients $d_{n,m,n',m'}$ can be calculated as follows:

$$\begin{aligned} d_{n,m,n',m'} &= <\psi_{n,m}(x), <h(x,t), \psi_{n',m'}(t)>> \\ &= \int_0^1 \left[\int_0^1 h(x,t).\psi_{n,m}(x)\,dx \right] .\psi_{n',m'}(t)dt \\ &= \int_{\frac{n'}{2^{k_2}}}^{\frac{n'+1}{2^{k_2}}} \left[\int_{\frac{n}{2^{k_1}}}^{\frac{n+1}{2^{k_1}}} h(x,t).2^{\frac{k_1}{2}}\,\mathrm{CAS}_m(2^{k_1}x-n)\,dx \right] .2^{\frac{k_2}{2}}\,\mathrm{CAS}_{m'}(2^{k_2}t-n')dt. \end{aligned} \tag{4.28}$$

By substituting $2^{k_1}x - n = s_1$ into Eq. (4.28), we get

$$= \frac{1}{2^{\frac{k_1}{2}}} \int_{\frac{n'}{2^{k_2}}}^{\frac{n'+1}{2^{k_2}}} \left[\int_0^1 h\left(\frac{s_1+n}{2^{k_1}}, t\right).\mathrm{CAS}_m(s_1)\,ds_1 \right] .2^{\frac{k_2}{2}}\,\mathrm{CAS}_{m'}(2^{k_2}t-n')dt.$$

Again substituting $2^{k_2}t - n' = s_2$, we get

$$= \frac{1}{2^{\frac{k_1}{2}}}.\frac{1}{2^{\frac{k_2}{2}}} \int_0^1 \left[\int_0^1 h\left(\frac{s_1+n}{2^{k_1}}, \frac{s_2+n'}{2^{k_2}}\right).\mathrm{CAS}_m(s_1)\,ds_1 \right] .\mathrm{CAS}_{m'}(s_2)ds_2.$$

This implies

$$\left|d_{n,m,n',m'}\right|^2 = \left|\frac{1}{2^{\frac{k_1}{2}}}\cdot\frac{1}{2^{\frac{k_2}{2}}}\int_0^1\left[\int_0^1 h\left(\frac{s_1+n}{2^{k_1}},\frac{s_2+n'}{2^{k_2}}\right).\mathrm{CAS}_m(s_1)\,ds_1\right].\mathrm{CAS}_{m'}(s_2)ds_2\right|^2$$

$$\leq \left(\frac{1}{2^{\frac{k_1}{2}}}\cdot\frac{1}{2^{\frac{k_2}{2}}}\right)^2\int_0^1\int_0^1\left|h\left(\frac{s_1+n}{2^{k_1}},\frac{s_2+n'}{2^{k_2}}\right)\right|^2.\left|\mathrm{CAS}_m(s_1)\right|^2.\left|\mathrm{CAS}_{m'}(s_2)\right|^2 ds_1\,ds_2$$

$$\leq \left(\frac{\wp}{2^{\frac{k_1}{2}}2^{\frac{k_2}{2}}}\right)^2\int_0^1\left|\mathrm{CAS}_m(s_1)\right|^2 ds_1.\int_0^1\left|\mathrm{CAS}_{m'}(s_2)\right|^2 ds_2.$$

From orthonormality of CASWs, we know that $\int_0^1\left|\mathrm{CAS}_m(s_1)\right|^2 ds_1 = 1$ and $\int_0^1\left|\mathrm{CAS}_{m'}(s_2)\right|^2 ds_2 = 1$. Since $n \leq 2^{k_1}, n' \leq 2^{k_2}$, we have

$$\left|d_{n,m,n',m'}\right| \leq \frac{\wp}{n^{\frac{1}{2}}n'^{\frac{1}{2}}}. \tag{4.29}$$

Hence, the series $\sum_{n=0}^{\infty}\sum_{n'=0}^{\infty}\sum_{m\in\mathbb{Z}}\sum_{m'\in\mathbb{Z}} d_{n,m,n',m'}$ is absolutely convergent, and it follows that the series $\sum_{n=0}^{\infty}\sum_{n'=0}^{\infty}\sum_{m\in\mathbb{Z}}\sum_{m'\in\mathbb{Z}} d_{n,m,n',m'}.\psi_{n,m,n',m'}(x,t)$ is uniformly convergent.

4.3.2.3 Error Analysis

In this part, the error analysis of CASWM for solving the 2D nonlinear VIE has been derived.

Theorem 4.3.6

Let $\widehat{h}_{2^{k_1}(2M+1),2^{k_2}(2M'+1)}(x,t)$ be the 2D CASWs expansion of the real sufficiently smooth function $h(x,t)\in L^2(\Omega)$. Then, there exist real numbers $\Im_1', \Im_2'$ and $\Im_3'$ such that

$$\left\|h(x,t)-\widehat{h}_{2^{k_1}(2M+1),2^{k_2}(2M'+1)}(x,t)\right\|_2 \leq \frac{\Im_1'}{2^{2^{k_1}M-1}\Gamma(2^{k_1}(2M+1)+1)}+\frac{\Im_2'}{2^{2^{k_2}M'-1}\Gamma(2^{k_2}(2M'+1)+1)}$$
$$+\frac{\Im_3'}{2^{2^{k_1}M+2^{k_2}M'-2}\Gamma(2^{k_1}(2M+1)+1)\Gamma(2^{k_2}(2M'+1)+1)}.$$

Proof:

The theorem can be proved in a similar manner as discussed in Theorem 4.3.2.

Theorem 4.3.7

Let $\widehat{h}_{2^{k_1}(2M+1),2^{k_2}(2M'+1)}(x,t) \approx \sum_{n=0}^{2^{k_1}-1}\sum_{n'=0}^{2^{k_2}-1}\sum_{m=-M}^{M}\sum_{m'=-M'}^{M'} \widehat{d}_{n,m,n',m'} \cdot \psi_{n,m,n',m'}(x,t)$ is the best approximation of h defined on $L^2(\Omega)$ and $h^*_{2^{k_1}(2M+1),2^{k_2}(2M'+1)}(x,t) \approx \sum_{n=0}^{2^{k_1}-1}\sum_{n'=0}^{2^{k_2}-1}\sum_{m=-M}^{M}\sum_{m'=-M'}^{M'} d^*_{n,m,n',m'} \cdot \psi_{n,m,n',m'}(x,t)$ is the numerical approximation h defined on $L^2(\Omega)$. Then,

$$\left\|\widehat{h}_{2^{k_1}(2M+1),2^{k_2}(2M'+1)}(x,t) - h^*_{2^{k_1}(2M+1),2^{k_2}(2M'+1)}(x,t)\right\|_2 \leq \mathcal{M}_1.$$

Proof:

The theorem can be proved in a similar manner as discussed in Theorem 4.3.3.

Theorem 4.3.8

Let $h^*_{2^{k_1}(2M+1),2^{k_2}(2M'+1)}(x,t) \approx \sum_{n=0}^{2^{k_1}-1}\sum_{n'=0}^{2^{k_2}-1}\sum_{m=-M}^{M}\sum_{m'=-M'}^{M'} d^*_{n,m,n',m'} \cdot \psi_{n,m,n',m'}(x,t)$ be our numerical approximation defined on $L^2(\Omega)$. Then,

$$\left\|h(x,t) - h^*_{2^{k_1}(2M+1),2^{k_2}(2M'+1)}(x,t)\right\|_2 \leq \tilde{\mathcal{M}}_1 + \mathcal{M}_1,$$

where $\tilde{\mathcal{M}}_1 = \dfrac{\Im_1'}{2^{2^{k_1}M-1}\Gamma(2^{k_1}(2M+1)+1)} + \dfrac{\Im_2'}{2^{2^{k_2}M'-1}\Gamma(2^{k_2}(2M'+1)+1)}$

$$+\frac{\Im_3'}{2^{2^{k_1}M+2^{k_2}M'-2}\Gamma(2^{k_1}(2M+1)+1)\Gamma(2^{k_2}(2M'+1)+1)}.$$

Proof:

The theorem can be proved in a similar manner as discussed in Theorem 4.3.4.

4.3.3 Numerical Experiments

This section deals with the performance of the proposed method for solving 2D nonlinear VIE. The accuracy of the approached technique can be demonstrated by using an absolute error function definition, i.e.,

$$\tilde{e}(x,t) = |u(x,t) - \tilde{u}(x,t)|,$$

TABLE 4.1
Comparison of Absolute Error for Example 4.3.1

$(x,t)=\left(\frac{1}{2^\alpha},\frac{1}{2^\alpha}\right)$	GWM		Haar Function [47]	
	$(k=1, M=8)$	$(k=1, M=16)$	$\tilde{m}=8$	$\tilde{m}=16$
$\alpha=1$	1.82698E-02	6.22613E-03	1.2E-01	6.2E-02
$\alpha=2$	5.65137E-04	4.47158E-03	1.2E-01	6.2E-02
$\alpha=3$	4.3066E-03	7.06562E-04	1.2E-01	6.2E-02
$\alpha=4$	1.41746E-02	2.94809E-03	2.2E-06	6.2E-02
$\alpha=5$	4.0627E-03	8.40636E-05	6.2E-02	6.9E-08
$\alpha=6$	1.08478E-02	6.38299E-03	9.4E-02	3.1E-02

TABLE 4.2
Comparison of Absolute Error for Example 4.3.2

$(x,t)=\left(\frac{1}{2^\alpha},\frac{1}{2^\alpha}\right)$	GWM	Haar Function [47]
	$(k=1, M=16)$	$\tilde{m}=16$
$\alpha=1$	3.84061E-03	6.5E-02
$\alpha=2$	4.82844E-03	3.4E-02
$\alpha=3$	1.44648E-03	1.8E-02
$\alpha=4$	3.2603E-03	1.0E-02
$\alpha=5$	5.95627E-04	6.5E-04
$\alpha=6$	6.05172E-03	2.1E-03

where $\tilde{u}(x,t)$ and $u(x,t)$ are the approximate and exact solutions, respectively.

Moreover, the comparison of absolute error with rationalized Haar functions method [47] is presented in Tables 4.1 and 4.2.

Furthermore, we compare the results obtained by presented methods with the exact solutions for examining the better accuracy. For a fixed value of t_l, the L^2 and L^∞ norms at the collocation points for GWM and CASWM are presented as follows:

$$L^2\text{error}=\left[\frac{1}{2^{k-1}M}\sum_{p=1}^{2^{k-1}M}(u(x_p,t_l)-u_m(x_p,t_l))^2\right]^{\frac{1}{2}},$$

$$L^\infty\text{error}=\underset{1\le p\le 2^{k-1}M}{\text{Max}}\left|u(x_p,t_l)-u_m(x_p,t_l)\right|,$$

and

$$L^2\text{error}=\left[\frac{1}{2^k(2M+1)}\sum_{p=1}^{2^k(2M+1)}(u(x_p,t_l)-u_m(x_p,t_l))^2\right]^{\frac{1}{2}},$$

$$L^{\infty}\text{error} = \underset{1 \le p \le 2^k (2M+1)}{\text{Max}} \left| u(x_p, t_l) - u_m(x_p, t_l) \right|,$$

respectively, where $u(x,t)$ is the exact solution and $u_m(x,t)$ is the approximate solution.

Example 4.3.1

Consider the following 2D VIE [48]:

$$u(x,t) = h(x,t) + \int_0^t \int_0^x (x + t - z - y) u^2(y,z)\, dy\, dz, \quad (x,t) \in \Omega, \tag{4.30}$$

where

$$h(x,t) = x + t - \frac{1}{12} xt(x^3 + 4x^2 t + 4xt^2 + t^3),$$

with the exact solution $u(x,t) = x + t$.

The values of errors for $k_1 = 1, k_2 = 1$, $M = 8, M' = 8$ and for different values of λ are shown in Table 4.3, and the comparisons of L^2 and L^{∞} error norms are shown in Table 4.4. For different values of λ, the numerical results and error analysis obtained from GWM are plotted in Figures 4.1–4.3. In addition, for different values of k and M, the numerical results and error analysis obtained from CASWM are plotted in Figures 4.4–4.6.

TABLE 4.3
Absolute Errors Obtained by GWs for Example 4.3.1 with $k_1 = 1, k_2 = 1$, $M = 8, M' = 8$

	Error Function $\tilde{e}(x,t)$				
(x,t)	$\lambda = 0.75$	$\lambda = 0.9$	$\lambda = 1.25$	$\lambda = 1.5$	$\lambda = 1.75$
(0.2,0.2)	1.19966E-2	1.95227E-2	3.6696E-2	4.77695E-2	5.72419E-2
(0.2.0.4)	2.78328E-3	4.89285E-3	1.04716E-2	1.46143E-2	1.86391E-2
(0.2,0.6)	1.07764E-2	1.76266E-2	3.40112E-2	4.56379E-2	5.70062E-2
(0.2,0.8)	8.33812E-3	1.62857E-2	4.14849E-2	6.37309E-2	8.82903E-2
(0.4,0.2)	2.78328E-3	4.89285E-3	1.04716E-2	1.46143E-2	1.86391E-2
(0.4,0.4)	7.44293E-3	1.22179E-2	2.40669E-2	3.30226E-2	4.23263E-2
(0.4,0.6)	9.54496E-5	8.98105E-4	4.68233E-3	8.87751E-3	1.42408E-2
(0.4,0.8)	2.94468E-4	1.32427E-3	9.54362E-3	1.86308E-2	3.00344E-2
(0.6,0.2)	1.07764E-2	1.76266E-2	3.40112E-2	4.56379E-2	5.70062E-2
(0.6,0.4)	9.54496E-5	8.98105E-4	4.68233E-3	8.87751E-3	1.42408E-2
(0.6,0.6)	5.2515E-3	7.38678E-3	9.68357E-3	9.12545E-3	6.73863E-3
(0.6,0.8)	5.45616E-3	1.06649E-2	2.79742E-2	4.44753E-2	6.44758E-2
(0.8,0.2)	8.33812E-3	1.62857E-2	4.14849E-2	6.37309E-2	8.82903E-2
(0.8,0.4)	2.94468E-4	1.32427E-3	9.54362E-3	1.86308E-2	3.00344E-2
(0.8,0.6)	5.45616E-3	1.06649E-2	2.79742E-2	4.44753E-2	6.44758E-2

TABLE 4.4
Comparison of L^2 and L^∞ Error Norms for Example 4.3.1

	GWM			CASWM		
t	L^2	L^∞	CPU Time (in sec.)	L^2	L^∞	CPU Time (in sec.)
0.2	1.1663E-2	2.04016E-2	337.749	1.36618E-2	2.09657E-2	5,940.91
0.4	7.88984E-3	1.42671E-2		1.49241E-2	3.50699E-2	
0.6	1.03182E-2	2.09812E-2		1.14453E-2	2.47597E-2	
0.8	1.18108E-2	2.49997E-2		1.78494E-2	3.64515E-2	

For $\lambda = 0.75$, we have

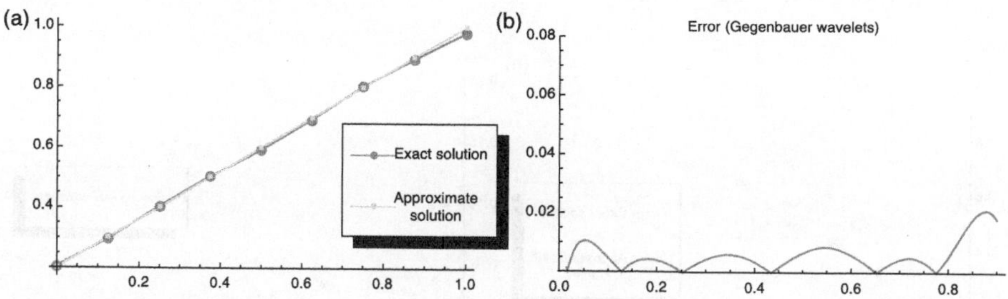

FIGURE 4.1 Numerical analysis and error analysis of GWs $\lambda = 0.75$ for Example 4.3.1.

For $\lambda = 0.9$, we have

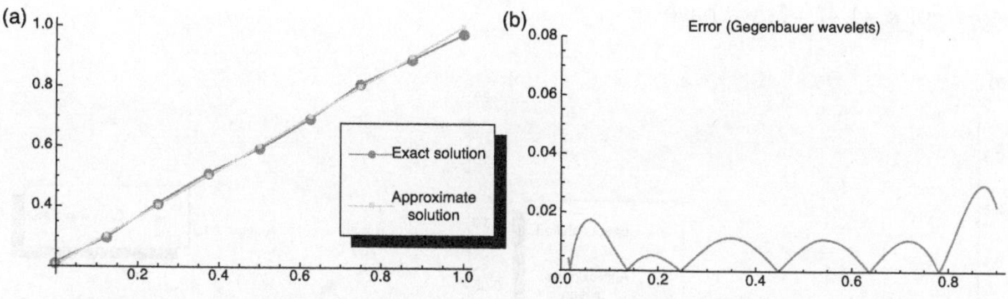

FIGURE 4.2 Numerical analysis and error analysis of GWs $\lambda = 0.9$ for Example 4.3.1.

For $\lambda = 1.25$, we have

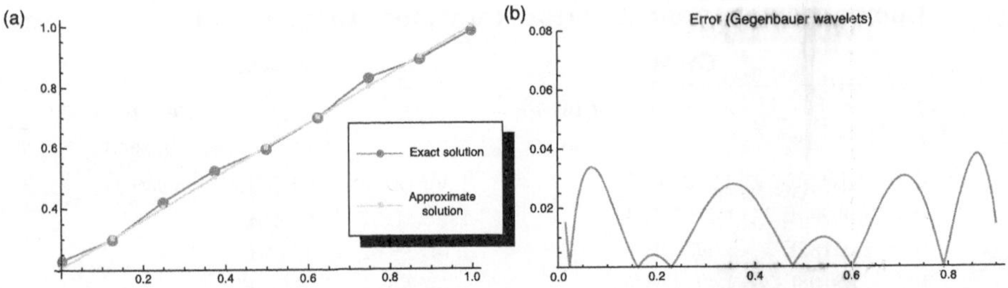

FIGURE 4.3 Numerical analysis and error analysis of GWs $\lambda = 1.25$ for Example 4.3.1.

Now for $k = 1, M = 2$, we have

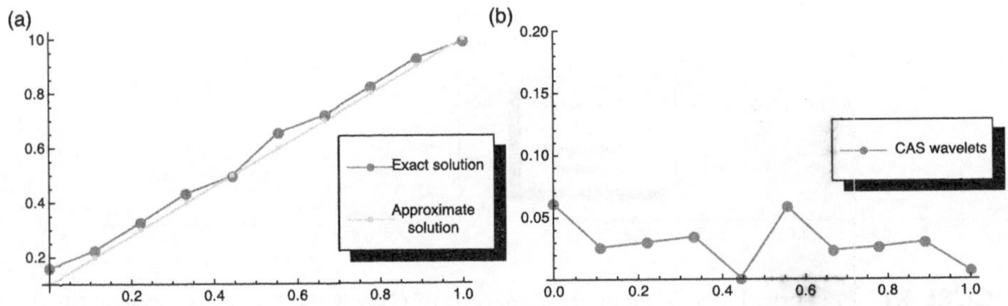

FIGURE 4.4 Numerical analysis and error analysis of CASWs $k = 1$, $M = 2$ for Example 4.3.1.

For $k = 1, M = 3$, we have

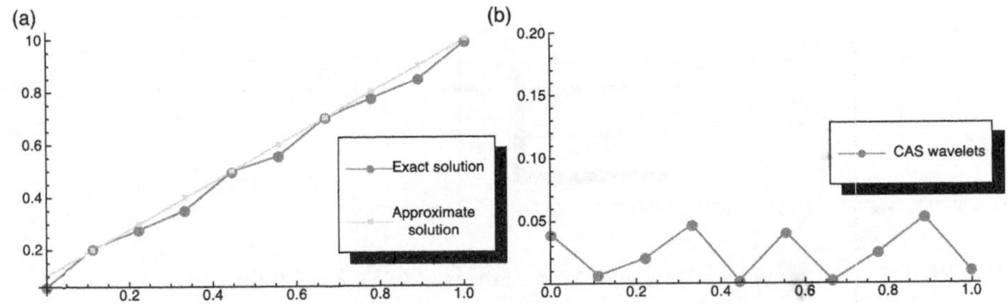

FIGURE 4.5 Numerical analysis and error analysis of CASWs $k = 1$, $M = 3$ for Example 4.3.1.

For $k = 1, M = 4$, we have

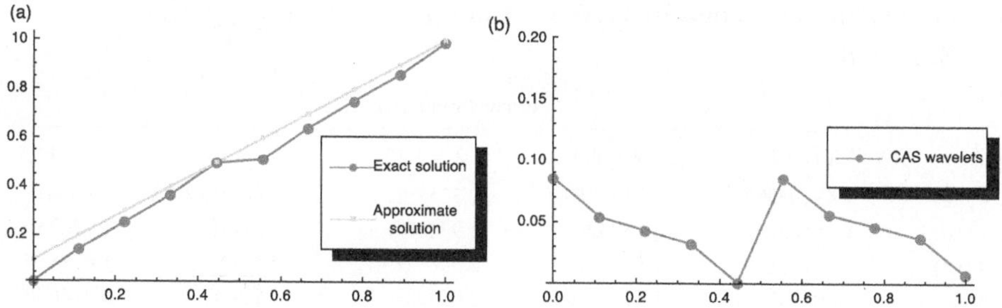

FIGURE 4.6 Numerical analysis and error analysis of CASWs $k = 1$, $M = 4$ for Example 4.3.1.

Example 4.3.2

Let us consider the 2D VIE [48]:

$$u(x,t) = h(x,t) + \int_0^t \int_0^x u^2(y,z)\,dy\,dz, \quad (x,t) \in \Omega, \tag{4.31}$$

where

$$\varphi(x,t) = x^2 + t^2 - \frac{1}{45}xt(9x^4 + 10x^2t^2 + 9t^4).$$

In this case, the exact solution is $u(x,t) = x^2 + t^2$.

The values of errors for $k_1 = 1, k_2 = 1, M = 8, M' = 8$ and for different values of λ are shown in Table 4.5 and the comparisons of L^2 and L^∞ error norms are shown in Table 4.6. For different values of λ, the numerical results and error analysis obtained from GWM are plotted in Figures 4.7–4.9. Moreover, for different values of k and M, the numerical results and error analysis obtained from CASWM are plotted in Figures 4.10–4.12.

TABLE 4.5
Absolute Errors Obtained by GWs for Example 4.3.2 with $k_1 = 1, k_2 = 1$, $M = 8, M' = 8$

	Error Function $\tilde{e}(x,t)$				
(x,t)	$\lambda = 0.75$	$\lambda = 0.9$	$\lambda = 1.25$	$\lambda = 1.5$	$\lambda = 1.75$
(0.2,0.2)	1.057515E-2	1.6827E-2	3.01448E-2	3.79615E-2	4.39879E-2
(0.2.0.4)	1.83565E-3	3.11511E-3	6.20617E-3	8.26322E-3	1.008E-2
(0.2,0.6)	1.00391E-2	1.61451E-2	3.00721E-2	3.94291E-2	4.81465E-2
(0.2,0.8)	7.61567E-3	1.46515E-2	3.65194E-2	5.55176E-2	7.6267E-2
(0.4,0.2)	1.83565E-2	3.11511E-3	6.20617E-3	8.26322E-3	1.008E-2

(Continued)

TABLE 4.5 (*Continued*)
Absolute Errors Obtained by GWs for Example 4.3.2 with $k_1 = 1, k_2 = 1$, $M = 8, M' = 8$

	Error Function $\tilde{e}(x,t)$				
(x,t)	$\lambda = 0.75$	$\lambda = 0.9$	$\lambda = 1.25$	$\lambda = 1.5$	$\lambda = 1.75$
(0.4,0.4)	7.49195E-3	1.21332E-2	2.32539E-2	3.13714E-2	3.96063E-2
(0.4,0.6)	1.36563E-4	8.18232E-4	3.93206E-3	7.31924E-3	1.15845E-2
(0.4,0.8)	2.12109E-3	1.87762E-3	2.54411E-3	8.57763E-3	1.67385E-2
(0.6,0.2)	1.00391E-2	1.61451E-2	3.00721E-2	3.94291E-2	4.81465E-2
(0.6,0.4)	1.36563E-4	8.18232E-4	3.93206E-3	7.31924E-3	1.15845E-2
(0.6,0.6)	6.23E-3	9.00871E-3	1.29866E-2	1.38537E-2	1.31797E-2
(0.6,0.8)	4.85011E-3	9.28701E-3	2.36554E-2	3.70865E-2	5.30958E-2
(0.8,0.2)	7.61567E-3	1.46515E-2	3.65194E-2	5.55176E-2	7.6267E-2
(0.8,0.4)	2.12109E-3	1.87762E-3	2.54411E-3	8.57763E-3	1.67385E-2
(0.8,0.6)	4.85011E-3	9.28701E-3	2.36554E-2	3.70865E-2	5.30958E-2

TABLE 4.6
Comparison of L^2 and L^∞ Error Norms for Example 4.3.2

	GWM			CASWM		
t	L^2	L^∞	CPU Time (in sec.)	L^2	L^∞	CPU Time (in sec.)
0.2	1.06692E-2	1.83697E-2	413.952	7.37205E-3	1.15663E-2	7,847.22
0.4	6.20569E-3	8.6434E-3		8.35135E-3	1.94478E-2	
0.6	9.03586E-3	1.76408E-2		7.80864E-3	9.99959E-3	
0.8	1.06146E-2	2.10834E-2		1.95741E-2	2.97759E-2	

For $\lambda = 0.75$, we have

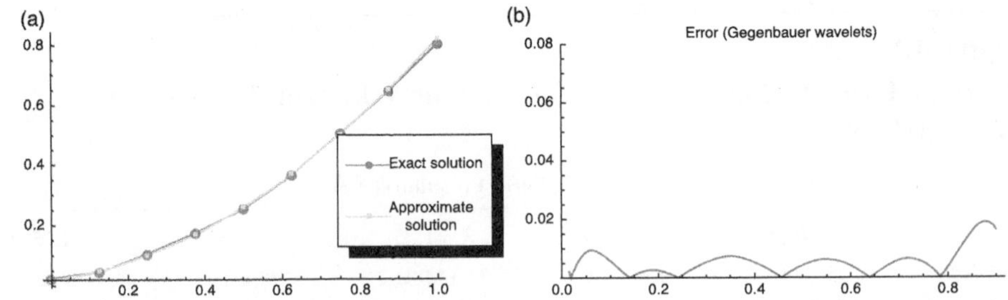

FIGURE 4.7 Numerical analysis and error analysis of GWs $\lambda = 0.75$ for Example 4.3.2.

For $\lambda = 0.9$, we have

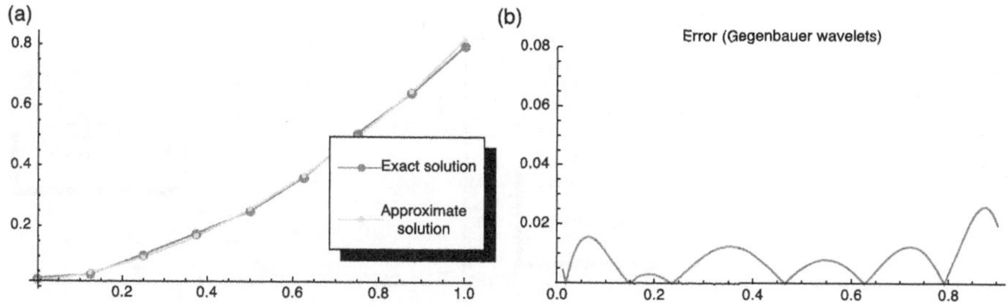

FIGURE 4.8 Numerical analysis and error analysis of GWs $\lambda = 0.9$ for Example 4.3.2.

For $\lambda = 1.25$, we have

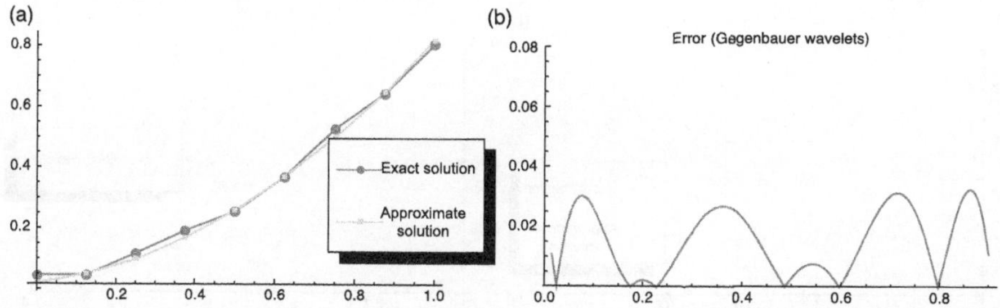

FIGURE 4.9 Numerical analysis and error analysis of GWs $\lambda = 1.25$ for Example 4.3.2.

Now for $k = 1, M = 2$, we have

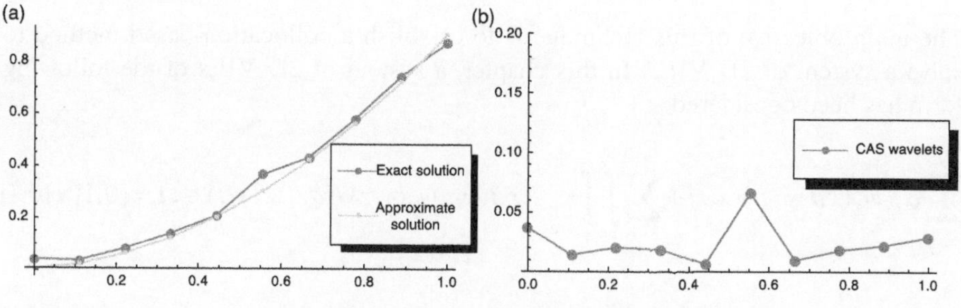

FIGURE 4.10 Numerical analysis and error analysis of CASWs $k = 1$, $M = 2$ for Example 4.3.2.

For $k=1, M=3$, we have

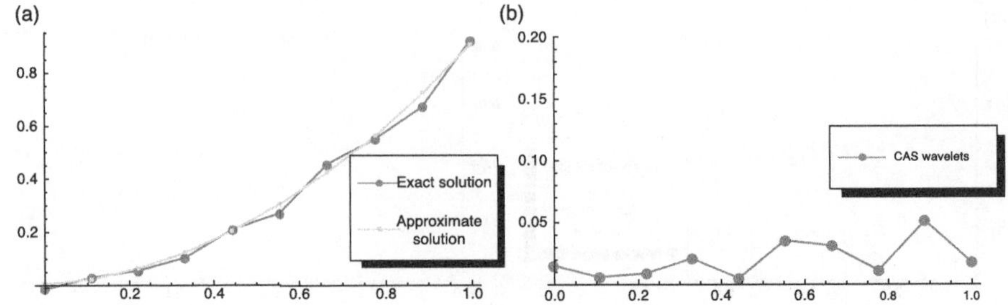

FIGURE 4.11 Numerical analysis and error analysis of CASWs $k = 1$, $M = 3$ for Example 4.3.2.

For $k=1, M=4$, we have

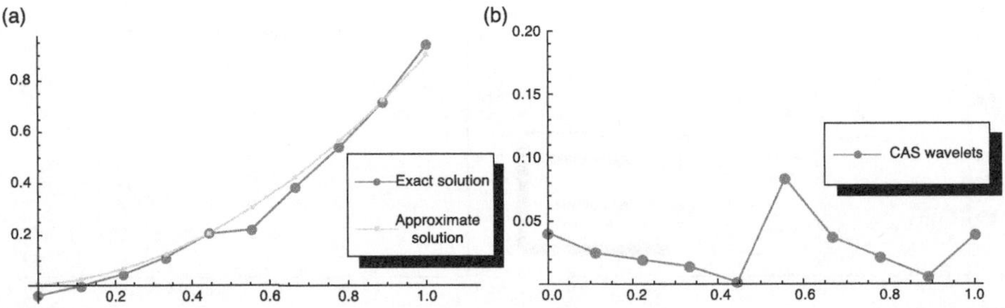

FIGURE 4.12 Numerical analysis and error analysis of CASWs $k = 1$, $M = 4$ for Example 4.3.2.

4.4 SYSTEM OF TWO-DIMENSIONAL VOLTERRA INTEGRAL EQUATIONS

The main objective of this technique is to establish a collocation-based method to solve a system of 2D VIEs. In this chapter, a system of 2D VIEs of the following form has been considered:

$$\sum_{q=1}^{l} \alpha_{p,q} u_q(x,t) = h_p(x,t) + \sum_{q=1}^{l} \int_0^t \int_0^x \kappa_{p,q}(x,t,y,z) u_q(y,z) dy dz, \quad (x,t) \in \Omega = [0,1] \times [0,1] \tag{4.32}$$

where $\kappa_{p,q}(x,t,y,z)$ and $h_p(x,t)$ for $p,q = 1,2,\ldots,l$ are assumed to be continuous functions defined on $\Omega \times \Omega$ and Ω, respectively; $\alpha_{p,q}$ are constants or linear functions; and $u_q(x,t)$, $p,q = 1,2,\ldots,l$ are unknown functions defined on Ω to be determined.

Equation (4.32) can also be written as follows:

$$\Lambda\tilde{\Phi} = H + \int_0^t \int_0^x \Pi\tilde{\Phi}\, dy\, dz, \tag{4.33}$$

where

$$\begin{cases} \tilde{\Phi}(x,t) = [u_1(x,t), u_2(x,t), \ldots, u_l(x,t)]^T, \\ H(x,t) = [h_1(x,t), h_2(x,t), \ldots, h_l(x,t)]^T, \\ \Pi(x,t,y,z) = \{\kappa_{p,q}(x,t,y,z)\},\ p,q = 1,2,\ldots,l, \\ \Lambda = \begin{bmatrix} \alpha_{1,1} & \alpha_{1,2} & . & . & \alpha_{1,l} \\ \alpha_{2,1} & \alpha_{2,2} & . & . & \alpha_{2,l} \\ . & . & . & . & . \\ . & . & . & . & . \\ \alpha_{l,1} & \alpha_{l,2} & . & . & \alpha_{l,l} \end{bmatrix}. \end{cases} \tag{4.34}$$

4.4.1 Function Approximation by JWs

A continuous function $u(x,t)$ defined over $L^2(\Omega)$ can be expanded as follows:

$$u(x,t) \approx \sum_{n=1}^{2^{k_1-1}} \sum_{n'=1}^{2^{k_2-1}} \sum_{m=0}^{M-1} \sum_{m'=0}^{M'-1} d_{n,m,n',m'}.\psi_{n,m,n',m'}(x,t) = \tilde{D}^T.\Psi(x,t), \tag{4.35}$$

where $\tilde{D}$ and $\Psi(x,t)$ are vectors of dimension $(2^{k_1-1}2^{k_2-1}MM' \times 1)$ given by

$$\tilde{D} = [d_{1,0,1,0}, \ldots, d_{1,0,1,M'-1}, \ldots, d_{1,0,2^{k_2-1},M'-1}, \ldots, d_{2^{k_1-1},M-1,2^{k_2-1},M'-1}]^T, \tag{4.36}$$

$$\Psi(x,t) = [\psi_{1,0,1,0}(x,t), \ldots, \psi_{1,0,1,M'-1}(x,t), \ldots, \psi_{1,0,2^{k_2-1},M'-1}(x,t), \ldots, \psi_{2^{k_1-1},M-1,2^{k_2-1},M'-1}(x,t)]^T, \tag{4.37}$$

where "$\psi_{n,m,n',m'}(x,t)$" is the 2D JWs defined as follows:

$$\psi_{n,m}(x) = \begin{cases} \sqrt{g_m^{(\delta,\eta)}}\, 2^{\frac{k-1}{2}}\, \wp_m^{(\delta,\eta)}(2^k x - 2n + 1), & \dfrac{(n-1)}{2^{k-1}} \le x < \dfrac{n}{2^{k-1}}, \\ 0, & \text{otherwise.} \end{cases} \tag{4.38}$$

The JWs coefficients $d_{n,m,n',m'}$ can be determined as follows:

$$d_{n,m,n',m'} =<< u(x,t), \psi_{n,m}(x) >, \psi_{n',m'}(t) >,$$

for $n = 1,2,\ldots,2^{k_1-1}$, $n' = 1,2,\ldots,2^{k_2-1}$, $m = 0,1,\ldots,M-1$, and $m' = 0,1,\ldots,M'-1$.

4.4.2 JWM for 2-D VIEs

Consider the system of 2D-VIE given in Eq. (4.32). To demonstrate the proposed scheme, the unknown functions $u_q(x,t),\ q = 1,2,\ldots,l$ can be approximated as follows:

$$u_q(x,t) = \Psi^T(x,t)\tilde{C}_q. \tag{4.39}$$

Now by using Eq. (4.39), Eq. (4.32) can be written as follows:

$$\sum_{q=1}^{l} \alpha_{p,q}\Psi^T(x,t)\tilde{C}_q = h_p(x,t) + \sum_{q=1}^{l}\int_0^t\int_0^x \kappa_{p,q}(x,t,y,z)\Psi^T(y,z)\tilde{C}_q dydz. \tag{4.40}$$

We now collocate Eq. (4.40) at $2^{k_1-1}M \times 2^{k_2-1}M'$ points $(\tilde{x}_i,\tilde{t}_j)$ as follows:

$$\sum_{q=1}^{l} \alpha_{p,q}\Psi^T(\tilde{x}_i,\tilde{t}_j)\tilde{C}_q = h_p(\tilde{x}_i,\tilde{t}_j) + \sum_{q=1}^{l}\int_0^{\tilde{t}_j}\int_0^{\tilde{x}_i} \kappa_{p,q}(\tilde{x}_i,\tilde{t}_j,y,z)\Psi^{JT}(y,z)\tilde{C}_q dydz, \tag{4.41}$$

where the collocation points $\tilde{x}_i$ and $\tilde{t}_j$ are

$$\tilde{x}_i = \frac{2i-1}{2.2^{k_1-1}.M}, \quad i = 1,2,\ldots,2^{k_1-1}M,$$

$$\tilde{t}_j = \frac{2j-1}{2.2^{k_2-1}.M'}, \quad j = 1,2,\ldots,2^{k_2-1}M'.$$

The intervals $[0,\tilde{t}_j]$ and $[0,\tilde{x}_i]$ are each converted into the interval $[-1,1]$ by the following transformations:

$$\gamma = \frac{2}{\tilde{t}_j}z - 1, \quad \beta = \frac{2}{\tilde{x}_i}y - 1,$$

to apply the Gaussian quadrature formula for Eq. (4.41).

Let

$$F(\tilde{x}_i,\tilde{t}_j,\beta,\gamma) = \kappa_{p,q}\left(\tilde{x}_i,\tilde{t}_j,\frac{(1+\beta)\tilde{x}_i}{2},\frac{(1+\gamma)\tilde{t}_j}{2}\right).$$

Equation (4.41) may then be approximately written as follows:

$$\sum_{q=1}^{l} \alpha_{p,q}\Psi^T(\tilde{x}_i,\tilde{t}_j)\tilde{C}_q \cong h_p(\tilde{x}_i,\tilde{t}_j) + \sum_{q=1}^{l}\left(\frac{\tilde{x}_i\tilde{t}_j}{4}\int_{-1}^{1}\int_{-1}^{1} F(\tilde{x}_i,\tilde{t}_j,\beta,\gamma)\Psi^T\left(\frac{(1+\beta)\tilde{x}_i}{2},\frac{(1+\gamma)\tilde{t}_j}{2}\right)\tilde{C}_q d\beta d\gamma\right).$$

From the Gaussian quadrature formula, we get

$$\sum_{q=1}^{l}\alpha_{p,q}\Psi^T(\tilde{x}_i,\tilde{t}_j)\tilde{C}_q \cong h_p(\tilde{x}_i,\tilde{t}_j)+\sum_{q=1}^{l}\left(\frac{\tilde{x}_i\tilde{t}_j}{4}\sum_{\bar{i}=1}^{n_1}\sum_{\bar{j}=1}^{n_2}\tilde{\omega}_{1\bar{i}}\tilde{\omega}_{2\bar{j}}F(\tilde{x}_i,\tilde{t}_j,\beta_{\bar{i}},\gamma_{\bar{j}}).\right.$$

$$\left.\Psi^T\left(\frac{(1+\beta_{\bar{i}})\tilde{x}_i}{2},\frac{(1+\gamma_{\bar{j}})\tilde{t}_j}{2}\right)\tilde{C}_q\right),$$

(4.42)

where $\beta_{\bar{i}}$ and $\gamma_{\bar{j}}$ are zeros of Legendre polynomials of degrees n_1 and n_2, respectively, and $\tilde{\omega}_{1\bar{i}},\tilde{\omega}_{2\bar{j}}$ are the corresponding weights. Equation (4.42) gives $2^{k_1-1}M\times 2^{k_2-1}M'$ system of algebraic equations which can be solved for the elements of $\tilde{C}_q$, $q=1,2,\ldots,l$ using Newton's method.

4.4.3 Convergence Analysis

In this section, we established that the expansion of any continuous function $u(x,t)\in L^2(\Omega)$, with respect to two-dimensional JWs, converges uniformly to u.

Theorem 4.4.1

Suppose that $u(x,t)\in L^2(\Omega)$ is a real-valued function and $|u''(x,t)|\le \aleph_2$, where $\aleph_2$ is a real and positive constant. Then, the JWs expansion of $u(x,t)$ converges uniformly to $u(x,t)$.

Proof:

The JWs coefficients can be determined as follows:

$$d_{n,m,n',m'} = \ll u(x,t),\psi_{n,m}(x)>,\psi_{n',m'}(t)>$$

$$=\int_0^1\int_0^1 u(x,t)\psi_{n,m}(x)\psi_{n',m'}(t)w(x)w(t)\,dx\,dt$$

$$=2^{\frac{(k_1+k_2)-2}{2}}\sqrt{g_m^{(\delta,\eta)}}\sqrt{g_{m'}^{(\delta,\eta)}}\int_{\frac{n'-1}{2^{k_2-1}}}^{\frac{n'}{2^{k_2-1}}}\left[\int_{\frac{n-1}{2^{k_1-1}}}^{\frac{n}{2^{k_1-1}}} u(x,t)\wp_m^{(\delta,\eta)}(2^k x-2n+1)w^{(\delta,\eta)}(2^k x-2n+1)\,dx\right]$$

$$.\wp_{m'}^{(\delta,\eta)}(2^{k_2}t-2n'+1)w^{(\delta,\eta)}(2^{k_2}t-2n'+1)\,dt.$$

Now, changing the variable $2^{k_1}x-2n+1=2\xi_1-1$, it follows

$$= 2^{\frac{(k_1+k_2)-2}{2}} \sqrt{g_m^{(\delta,\eta)}} \sqrt{g_{m'}^{(\delta,\eta)}} \frac{1}{2^{k_1-1}} \int_{\frac{n'-1}{2^{k_2-1}}}^{\frac{n'}{2^{k_2-1}}} \left[\int_0^1 u\left(\frac{\xi_1+n-1}{2^{k_1-1}}, t\right) \wp_m^{(\delta,\eta)}(2\xi_1-1) w^{(\delta,\eta)}(2\xi_1-1)\, d\xi_1 \right]$$

$$\cdot \wp_{m'}^{(\delta,\eta)}(2^{k_2}t-2n'+1) w^{(\delta,\eta)}(2^{k_2}t-2n'+1)\, dt.$$

Similarly, changing the variable $2^{k_2}t - 2n' + 1 = 2\xi_2 - 1$, it follows

$$= 2^{-\frac{(k_1+k_2)-2}{2}} \sqrt{g_m^{(\delta,\eta)}} \sqrt{g_{m'}^{(\delta,\eta)}} \int_0^1 \left[\int_0^1 u\left(\frac{\xi_1+n-1}{2^{k_1-1}}, \frac{\xi_2+n'-1}{2^{k_2-1}}\right) \wp_m^{(\delta,\eta)}(2\xi_1-1) w^{(\delta,\eta)}(2\xi_1-1)\, d\xi_1 \right]$$

$$\cdot \wp_{m'}^{(\delta,\eta)}(2\xi_2-1) w^{(\delta,\eta)}(2\xi_2-1)\, d\xi_2. \tag{4.43}$$

The Rodrigues formula of Jacobi polynomials results in [49]

$$w^{(\delta,\eta)}(\xi_1)\wp_m^{(\delta,\eta)}(\xi_1) = -\frac{1}{(\delta+\eta+m+1)} \frac{d}{d\xi_1}(w^{(\delta+1,\eta+1)}(\xi_1)\wp_{m-1}^{(\delta+1,\eta+1)}(\xi_1)) \tag{4.44}$$

and

$$w^{(\delta,\eta)}(\xi_2)\wp_{m'}^{(\delta,\eta)}(\xi_2) = -\frac{1}{(\delta+\eta+m'+1)} \frac{d}{d\xi_2}(w^{(\delta+1,\eta+1)}(\xi_2)\wp_{m'-1}^{(\delta+1,\eta+1)}(\xi_2)). \tag{4.45}$$

Applying the integration by parts technique for Eq. (4.43) and using Eq. (4.44), it follows

$$= 2^{-\frac{(k_1+k_2)-2}{2}} \sqrt{g_m^{(\delta,\eta)}} \sqrt{g_{m'}^{(\delta,\eta)}} \frac{(-1)}{2(\delta+\eta+m+1)} \int_0^1 \left[\int_0^1 u'\left(\frac{\xi_1+n-1}{2^{k_1-1}}, \frac{\xi_2+n'-1}{2^{k_2-1}}\right) \wp_{m-1}^{(\delta+1,\eta+1)}(2\xi_1-1) \right.$$

$$\left. w^{(\delta+1,\eta+1)}(2\xi_1-1)\, d\xi_1 \right] \cdot \wp_{m'}^{(\delta,\eta)}(2\xi_2-1) w^{(\delta,\eta)}(2\xi_2-1)\, d\xi_2. \tag{4.46}$$

Again, applying the integration by parts technique for Eq. (4.46) and using Eq. (4.45), it follows

$$= 2^{\frac{2-(k_1+k_2)}{2}} \sqrt{g_m^{(\delta,\eta)}} \sqrt{g_{m'}^{(\delta,\eta)}} \frac{1}{4(\delta+\eta+m+1)(\delta+\eta+m'+1)} \int_0^1 \left[\int_0^1 u''\left(\frac{\xi_1+n-1}{2^{k_1-1}}, \frac{\xi_2+n'-1}{2^{k_2-1}}\right) \right.$$

$$\left. \wp_{m-1}^{(\delta+1,\eta+1)}(2\xi_1-1)\ w^{(\delta+1,\eta+1)}(2\xi_1-1)\, d\xi_1 \right] \cdot \wp_{m'-1}^{(\delta+1,\eta+1)}(2\xi_2-1) w^{(\delta+1,\eta+1)}(2\xi_2-1)\, d\xi_2. \tag{4.47}$$

By using Hölder's inequality, Eq. (4.47) can be written as follows:

$$\left|d_{n,m,n',m'}\right| \le \frac{2^{-\frac{2+(k_1+k_2)}{2}}\sqrt{g_m^{(\delta,\eta)}}\sqrt{g_{m'}^{(\delta,\eta)}}\aleph_2}{(\delta+\eta+m+1)(\delta+\eta+m'+1)}\left(\int_0^1 (\wp_{m-1}^{(\delta+1,\eta+1)}(2\xi_1-1)w^{(\delta+1,\eta+1)}(2\xi_1-1))^2\,d\xi_1\right)^{\frac{1}{2}}$$
$$\left(\int_0^1 (\wp_{m'-1}^{(\delta+1,\eta+1)}(2\xi_2-1)w^{(\delta+1,\eta+1)}(2\xi_2-1))^2\,d\xi_2\right)^{\frac{1}{2}}. \tag{4.48}$$

Again, changing the variable $2\xi_1-1=s_1$ and $2\xi_2-1=s_2$, it follows

$$\left|d_{n,m,n',m'}\right| \le \frac{2^{-\frac{6+(k_1+k_2)}{2}}\sqrt{g_m^{(\delta,\eta)}}\sqrt{g_{m'}^{(\delta,\eta)}}\aleph_2}{(\delta+\eta+m+1)(\delta+\eta+m'+1)}\left(\int_{-1}^1 (\wp_{m-1}^{(\delta+1,\eta+1)}(s_1)w^{(\delta+1,\eta+1)}(s_1))^2\,ds_1\right)^{\frac{1}{2}}$$
$$\cdot\left(\int_{-1}^1 (\wp_{m'-1}^{(\delta+1,\eta+1)}(s_2)w^{(\delta+1,\eta+1)}(s_2))^2\,ds_2\right)^{\frac{1}{2}}. \tag{4.49}$$

Now putting,

$$r_1(s_1)=w^{(\delta+1,\eta+1)}(s_1),\quad z_1(s_1)=w^{(\delta+1,\eta+1)}(s_1)(\wp_{m-1}^{(\delta+1,\eta+1)}(s_1))^2$$

and

$$r_2(s_2)=w^{(\delta+1,\eta+1)}(s_2),\quad z_2(s_2)=w^{(\delta+1,\eta+1)}(s_2)(\wp_{m-1}^{(\delta+1,\eta+1)}(s_2))^2.$$

From Eq. (4.49) and the fact $r_1(s_1), r_2(s_2), z_1(s_1), z_2(s_2)>0$ on $[-1,1]$, it follows

$$\left|d_{n,m,n',m'}\right| \le \frac{2^{-\frac{6+(k_1+k_2)}{2}}\sqrt{g_m^{(\delta,\eta)}}\sqrt{g_{m'}^{(\delta,\eta)}}\aleph_2}{(\delta+\eta+m+1)(\delta+\eta+m'+1)}\left(\left(\max_{-1\le x_1\le 1} r_1(s_1)\right)^{\frac{1}{2}}\left(\int_{-1}^1 z_1(s_1)\,ds_1\right)^{\frac{1}{2}}\right)$$
$$\cdot\left(\left(\max_{-1\le x_2\le 1} r_2(s_2)\right)^{\frac{1}{2}}\left(\int_{-1}^1 z_2(s_2)\,ds_2\right)^{\frac{1}{2}}\right). \tag{4.50}$$

A simple computation shows that $s_1^*=\dfrac{\eta-\delta}{\delta+\eta+2}$ and $s_2^*=\dfrac{\eta-\delta}{\delta+\eta+2}$ maximize $r_1(s_1)$ and $r_2(s_2)$, respectively, on $[-1,1]$. This maximum value is

$$\begin{cases} r_1(s_1^*) = \max\limits_{-1\le x_1\le 1} r_1(s_1) = \left(\dfrac{2\delta+2}{\delta+\eta+2}\right)^{\delta+1}\left(\dfrac{2\eta+2}{\delta+\eta+2}\right)^{\eta+1}, \\ r_2(s_2^*) = \max\limits_{-1\le x_2\le 1} r_2(s_2) = \left(\dfrac{2\delta+2}{\delta+\eta+2}\right)^{\delta+1}\left(\dfrac{2\eta+2}{\delta+\eta+2}\right)^{\eta+1}. \end{cases} \tag{4.51}$$

From Eq. (4.50), it is clear that

$$\begin{cases} \displaystyle\int_{-1}^{1} z_1(s_1)\,ds_1 = \frac{1}{g_m^{(\delta,\eta)}}, \\ \displaystyle\int_{-1}^{1} z_2(s_2)\,ds_2 = \frac{1}{g_{m'}^{(\delta,\eta)}}. \end{cases} \tag{4.52}$$

Now from Eqs. (4.50)–(4.52), it follows that

$$\left|d_{n,m,n',m'}\right| \le \frac{\aleph_2.\left(\dfrac{2\delta+2}{\delta+\eta+2}\right)^{\delta+1}\left(\dfrac{2\eta+2}{\delta+\eta+2}\right)^{\eta+1}}{2^{\frac{6+(k_1+k_2)}{2}}.(\delta+\eta+m+1)(\delta+\eta+m'+1)}\sqrt{\frac{g_m^{(\delta,\eta)}}{g_{m-1}^{(\delta+1,\eta+1)}}}\sqrt{\frac{g_{m'}^{(\delta,\eta)}}{g_{m'-1}^{(\delta+1,\eta+1)}}}$$

$$= \frac{\aleph_2.\left(\dfrac{2\delta+2}{\delta+\eta+2}\right)^{\delta+1}\left(\dfrac{2\eta+2}{\delta+\eta+2}\right)^{\eta+1}}{2^{\frac{6+(k_1+k_2)}{2}}.(\delta+\eta+m+1)(\delta+\eta+m'+1)}\sqrt{\frac{4.m.m'}{(m+\delta+\eta+1)(m'+\delta+\eta+1)}}$$

$$= \frac{\aleph_2.\left(\dfrac{2\delta+2}{\delta+\eta+2}\right)^{\delta+1}\left(\dfrac{2\eta+2}{\delta+\eta+2}\right)^{\eta+1}\sqrt{m.m'}}{2^{\frac{4+(k_1+k_2)}{2}}.(\delta+\eta+m+1)^{\frac{3}{2}}(\delta+\eta+m'+1)^{\frac{3}{2}}}.$$

Hence, the series $\displaystyle\sum_{n=1}^{2^{k_1-1}}\sum_{n'=1}^{2^{k_2-1}}\sum_{m=0}^{M-1}\sum_{m'=0}^{M'-1} d_{n,m,n',m'}$ is absolutely convergent, and it follows that the series $\displaystyle\sum_{n=1}^{2^{k_1-1}}\sum_{n'=1}^{2^{k_2-1}}\sum_{m=0}^{M-1}\sum_{m'=0}^{M'-1} d_{n,m,n',m'}.\psi_{n,m,n',m'}(x,t)$ is uniformly convergent.

4.4.4 Error Analysis

In this part, the error analysis of JWM for solving the system of 2D VIEs has been derived.

Theorem 4.4.2

Let $\tilde{u}_{2^{k_1-1}M,2^{k_2-1}M'}(x,t)$ be the 2D JWs expansion of the real sufficiently smooth function $u(x,t)\in L^2(\Omega)$. Then, there exist real numbers $\Im_1,\Im_2$ and $\Im_3$ such that

$$\left\|u(x,t)-\tilde{u}_{2^{k_1-1}M,2^{k_2-1}M'}(x,t)\right\|_2 \le \frac{\Im_1}{2^{2^{k_1}M-1}\Gamma(2^{k_1-1}M+1)}+\frac{\Im_2}{2^{2^{k_2}M'-1}\Gamma(2^{k_2-1}M'+1)} + \frac{\Im_3}{2^{2^{k_1}M+2^{k_2}M'-2}\Gamma(2^{k_1-1}M+1)\Gamma(2^{k_2-1}M'+1)}. \tag{4.53}$$

Proof:

Assume that $u(x,t)$ is a sufficiently smooth function on $L^2(\Omega)$ and $\tilde{\varphi}_{2^{k_1-1}M,2^{k_2-1}M'}(x,t)$ is the best approximation of u from $\Lambda_{2^{k_1-1}M,2^{k_2-1}M'}$, where

$$\Lambda_{2^{k_1-1}M,2^{k_2-1}M'}=\text{Span}\left\{\begin{matrix}\psi_{n,m,n',m'}(x,t):n=1,2,...,2^{k_1-1},m=0,1,...,M-1;\\ n'=1,2,...,2^{k_2-1},m=0,1,...,M'-1\end{matrix}\right\}.$$

Then, for any $P_{2^{k_1-1}M,2^{k_2-1}M'}[u]\in\Lambda_{2^{k_1-1}M,2^{k_2-1}M'}$, the following inequality is held:

$$\left\|u(x,t)-\tilde{u}_{2^{k_1-1}M,2^{k_2-1}M'}(x,t)\right\|_2 \le \left\|u(x,t)-P_{2^{k_1-1}M,2^{k_2-1}M'}[u]\right\|_2, \tag{4.54}$$

where $P_{2^{k_1-1}M,2^{k_2-1}M'}[u]$ is the interpolating polynomial of u at the node points (x_p,t_q).

Here, $x_p, p=1,2,...,2^{k_1-1}M$ and $t_q,q=1,2,...,2^{k_2-1}M'$ are the roots of the $2^{k_1-1}M$-degree and $2^{k_1-1}M$-degree shifted Chebyshev polynomial in [0,1], respectively.

Now,

$$u(x,t)-P_{2^{k_1-1}M,2^{k_2-1}M'}[u]=\frac{\partial^{2^{k_1-1}M}u(\zeta_1,t)}{\partial x^{2^{k_1-1}M}}\frac{\prod_{p=1}^{2^{k_1-1}M}(x-x_p)}{\Gamma(2^{k_1-1}M+1)}+\frac{\partial^{2^{k_2-1}M'}u(x,\zeta_2)}{\partial t^{2^{k_2-1}M'}}\frac{\prod_{q=1}^{2^{k_2-1}M'}(t-t_q)}{\Gamma(2^{k_2-1}M'+1)} + \frac{\partial^{2^{k_1-1}M+2^{k_2-1}M'}\varphi(\tilde{\zeta}_1,\tilde{\zeta}_2)}{\partial x^{2^{k_1-1}M}\partial t^{2^{k_2-1}M'}}\frac{\prod_{p=1}^{2^{k_1-1}M}(x-x_p)\prod_{q=1}^{2^{k_2-1}M'}(t-t_q)}{\Gamma(2^{k_1-1}M+1)\Gamma(2^{k_2-1}M'+1)}, \tag{4.55}$$

for some $\zeta_1,\zeta_2,\tilde{\zeta}_1,\tilde{\zeta}_2\in[0,1]$.

Now, utilizing the estimates for Chebyshev interpolating nodes on the interval Ω, it follows that [45]

$$\left|\prod_{p=1}^{2^{k_1-1}M}(x-x_p)\right| \le \frac{1}{2^{2^{k_1}M-1}}$$

and

$$\left|\prod_{q=1}^{2^{k_2-1}M'}(t-t_q)\right| \le \frac{1}{2^{2^{k_2}M'-1}}.$$

Therefore,

$$\begin{aligned}\left|u(x,t)-P_{2^{k_1-1}M,2^{k_2-1}M'}[u]\right| &\le \sup_{t\in[0,1]}\left(\max_{\zeta_1\in[0,1]}\left|\frac{\partial^{2^{k_1-1}M}u(\zeta_1,t)}{\partial x^{2^{k_1-1}M}}\right|\right)\frac{1}{2^{2^{k_1}M-1}\Gamma(2^{k_1-1}M+1)}\\ &+\sup_{x\in[0,1]}\left(\max_{\zeta_2\in[0,1]}\left|\frac{\partial^{2^{k_2-1}M'}u(x,\zeta_2)}{\partial t^{2^{k_2-1}M'}}\right|\right)\frac{1}{2^{2^{k_2}M'-1}\Gamma(2^{k_2-1}M'+1)}\\ &+\sup_{(x,t)\in\Omega}\left(\max_{(\tilde{\zeta}_1,\tilde{\zeta}_2)\in\Omega}\left|\frac{\partial^{2^{k_1-1}M+2^{k_2-1}M'}u(\tilde{\zeta}_1,\tilde{\zeta}_2)}{\partial x^{2^{k_1-1}M}\partial t^{2^{k_2-1}M'}}\right|\right)\\ &\frac{1}{2^{2^{k_1}M+2^{k_2}M'-2}\Gamma(2^{k_1-1}M+1)\Gamma(2^{k_2-1}M'+1)},\end{aligned}$$

whence

$$\begin{aligned}\left|u(x,t)-P_{2^{k_1-1}M,2^{k_2-1}M'}[u]\right| &\le \sup_{(x,t)\in\Omega}\left|\frac{\partial^{2^{k_1-1}M}u(x,t)}{\partial x^{2^{k_1-1}M}}\right|\frac{1}{2^{2^{k_1}M-1}\Gamma(2^{k_1-1}M+1)}\\ &+\sup_{(x,t)\in\Omega}\left|\frac{\partial^{2^{k_2-1}M'}u(x,t)}{\partial t^{2^{k_2-1}M'}}\right|\frac{1}{2^{2^{k_2}M'-1}\Gamma(2^{k_2-1}M'+1)}\\ &+\sup_{(x,t)\in\Omega}\left|\frac{\partial^{2^{k_1-1}M+2^{k_2-1}M'}u(x,t)}{\partial x^{2^{k_1-1}M}\partial t^{2^{k_2-1}M'}}\right|\\ &\frac{1}{2^{2^{k_1}M+2^{k_2}M'-2}\Gamma(2^{k_1-1}M+1)\Gamma(2^{k_2-1}M'+1)}.\end{aligned} \tag{4.56}$$

In particular, from Eq. (4.54),

$$\left\|u(x,t)-\tilde{u}_{2^{k_1-1}M,2^{k_2-1}M'}(x,t)\right\|_2^2 \le \left\|u(x,t)-P_{2^{k_1-1}M,2^{k_2-1}M'}[u]\right\|_2^2$$
$$=\int_0^1\int_0^1\left|u(x,t)-P_{2^{k_1-1}M,2^{k_2-1}M'}[u]\right|^2 dx\,dt. \tag{4.57}$$

Now, using Eqs. (4.56) and (4.57), it follows that

$$\left\|u(x,t)-\tilde{u}_{2^{k_1-1}M,2^{k_2-1}M'}(x,t)\right\|_2^2 \le \left[\frac{\Im_1}{2^{2^{k_1}M-1}\Gamma(2^{k_1-1}M+1)}+\frac{\Im_2}{2^{2^{k_2}M'-1}\Gamma(2^{k_2-1}M'+1)}\right.$$
$$\left.+\frac{\Im_3}{2^{2^{k_1}M+2^{k_2}M'-2}\Gamma(2^{k_1-1}M+1)\Gamma(2^{k_2-1}M'+1)}\right]^2, \tag{4.58}$$

assuming $\sup_{(x,t)\in\Omega}\left|\dfrac{\partial^{2^{k_1-1}M}u(x,t)}{\partial x^{2^{k_1-1}M}}\right|\le\Im_1$, $\sup_{(x,t)\in\Omega}\left|\dfrac{\partial^{2^{k_2-1}M'}u(x,t)}{\partial t^{2^{k_2-1}M'}}\right|\le\Im_2$, and $\sup_{(x,t)\in\Omega}\left|\dfrac{\partial^{2^{k_1-1}M+2^{k_2-1}M'}u(x,t)}{\partial x^{2^{k_1-1}M}\partial t^{2^{k_2-1}M'}}\right|\le\Im_3$.

Therefore,

$$\left\|u(x,t)-\tilde{u}_{2^{k_1-1}M,2^{k_2-1}M'}(x,t)\right\|_2 \le \frac{\Im_1}{2^{2^{k_1}M-1}\Gamma(2^{k_1-1}M+1)}+\frac{\Im_2}{2^{2^{k_2}M'-1}\Gamma(2^{k_2-1}M'+1)}$$
$$+\frac{\Im_3}{2^{2^{k_1}M+2^{k_2}M'-2}\Gamma(2^{k_1-1}M+1)\Gamma(2^{k_2-1}M'+1)}.$$

Hence, the theorem is proved. ■

Theorem 4.4.3

Let $\tilde{u}_{2^{k_1-1}M,2^{k_2-1}M'}(x,t)\approx\sum_{n=1}^{2^{k_1-1}}\sum_{n'=1}^{2^{k_2-1}}\sum_{m=0}^{M-1}\sum_{m'=0}^{M'-1}\tilde{d}_{n,m,n',m'}\cdot\psi_{n,m,n',m'}(x,t)$ is the best approximation of u defined on $L^2(\Omega)$ and $u^*_{2^{k_1-1}M,2^{k_2-1}M'}(x,t)\approx\sum_{n=1}^{2^{k_1-1}}\sum_{n'=1}^{2^{k_2-1}}\sum_{m=0}^{M-1}\sum_{m'=0}^{M'-1}d^*_{n,m,n',m'}\cdot\psi_{n,m,n',m'}(x,t)$ is the numerical approximation u defined on $L^2(\Omega)$. Then [46],

$$\left\|\tilde{\varphi}_{2^{k_1-1}M,2^{k_2-1}M'}(x,t)-\varphi^*_{2^{k_1-1}M,2^{k_2-1}M'}(x,t)\right\|_2 \le \mathcal{M}.$$

Proof:

$$\left\|\tilde{u}_{2^{k_1-1}M,2^{k_2-1}M'}(x,t)-u^*_{2^{k_1-1}M,2^{k_2-1}M'}(x,t)\right\|_2^2$$

$$\le \int_0^1\int_0^1\left|\tilde{u}_{2^{k_1-1}M,2^{k_2-1}M'}(x,t)-u^*_{2^{k_1-1}M,2^{k_2-1}M'}(x,t)\right|^2 dx\,dt$$

$$\le \int_0^1\int_0^1\left|\sum_{n=1}^{2^{k_1-1}}\sum_{n'=1}^{2^{k_2-1}}\sum_{m=0}^{M-1}\sum_{m'=0}^{M'-1}\tilde{d}_{n,m,n',m'}\cdot\psi_{n,m,n',m'}(x,t)-\sum_{n=1}^{2^{k_1-1}}\sum_{n'=1}^{2^{k_2-1}}\sum_{m=0}^{M-1}\sum_{m'=0}^{M'-1}d^*_{n,m,n',m'}\cdot\psi_{n,m,n',m'}(x,t)\right|^2 dxdt$$

$$\le \int_0^1\int_0^1\left|\sum_{n=1}^{2^{k_1-1}}\sum_{n'=1}^{2^{k_2-1}}\sum_{m=0}^{M-1}\sum_{m'=0}^{M'-1}(\tilde{d}_{n,m,n',m'}-d^*_{n,m,n',m'}).\psi_{n,m,n',m'}(x,t)\right|^2 dxdt$$

$$\le \sum_{n=1}^{2^{k_1-1}}\sum_{n'=1}^{2^{k_2-1}}\sum_{m=0}^{M-1}\sum_{m'=0}^{M'-1}\left|\tilde{d}_{n,m,n',m'}-d^*_{n,m,n',m'}\right|^2\cdot\int_0^1\int_0^1\left|\psi_{n,m,n',m'}(x,t)\right|^2 dxdt$$

$$\le \sum_{n=1}^{2^{k_1-1}}\sum_{n'=1}^{2^{k_2-1}}\sum_{m=0}^{M-1}\sum_{m'=0}^{M'-1}\left|\tilde{d}_{n,m,n',m'}-d^*_{n,m,n',m'}\right|^2.$$

This implies

$$\left\|\tilde{u}_{2^{k_1-1}M,2^{k_2-1}M'}(x,t)-u^*_{2^{k_1-1}M,2^{k_2-1}M'}(x,t)\right\|_2 \le \left(\sum_{n=1}^{2^{k_1-1}}\sum_{n'=1}^{2^{k_2-1}}\sum_{m=0}^{M-1}\sum_{m'=0}^{M'-1}\left|\tilde{d}_{n,m,n',m'}-d^*_{n,m,n',m'}\right|^2\right)^{\frac{1}{2}} = \mathcal{M}, \tag{4.59}$$

where $\mathcal{K} = \left(\sum_{n=1}^{2^{k_1-1}}\sum_{n'=1}^{2^{k_2-1}}\sum_{m=0}^{M-1}\sum_{m'=0}^{M'-1}\left|\tilde{d}_{n,m,n',m'}-d^*_{n,m,n',m'}\right|^2\right)^{\frac{1}{2}}$.

Theorem 4.4.4

Let $u^{*}_{2^{k_1-1}M,2^{k_2-1}M'}(x,t) \approx \sum_{n=1}^{2^{k_1-1}} \sum_{n'=1}^{2^{k_2-1}} \sum_{m=0}^{M-1} \sum_{m'=0}^{M'-1} d^{*}_{n,m,n',m'} \cdot \psi_{n,m,n',m'}(x,t)$ be our numerical approximation defined on $L^2(\Omega)$. Then,

$$\left\| u(x,t) - u^{*}_{2^{k_1-1}M,2^{k_2-1}M'}(x,t) \right\|_2 \le \tilde{\mathcal{M}} + \mathcal{M},$$

where
$$\tilde{\mathcal{M}} = \frac{\Im_1}{2^{2^{k_1}M-1}\Gamma(2^{k_1-1}M+1)} + \frac{\Im_2}{2^{2^{k_2}M'-1}\Gamma(2^{k_2-1}M'+1)} + \frac{\Im_3}{2^{2^{k_1}M+2^{k_2}M'-2}\Gamma(2^{k_1-1}M+1)\Gamma(2^{k_2-1}M'+1)}.$$

Proof:

By using Theorems 4.4.2 and 4.4.3, it follows

$$\begin{aligned}
\left\| u(x,t) - u^{*}_{2^{k_1-1}M,2^{k_2-1}M'}(x,t) \right\|_2 &\le \left\| u(x,t) - \tilde{u}_{2^{k_1-1}M,2^{k_2-1}M'}(x,t) \right\|_2 \\
&\quad + \left\| \tilde{u}_{2^{k_1-1}M,2^{k_2-1}M'}(x,t) - u^{*}_{2^{k_1-1}M,2^{k_2-1}M'}(x,t) \right\|_2 \\
&\le \frac{\Im_1}{2^{2^{k_1}M-1}\Gamma(2^{k_1-1}M+1)} + \frac{\Im_2}{2^{2^{k_2}M'-1}\Gamma(2^{k_2-1}M'+1)} \\
&\quad + \frac{\Im_3}{2^{2^{k_1}M+2^{k_2}M'-2}\Gamma(2^{k_1-1}M+1)\Gamma(2^{k_2-1}M'+1)} + \mathcal{M} \\
&\le \tilde{\mathcal{M}} + \mathcal{M}.
\end{aligned}$$

4.4.5 Numerical Experiments

In this part, some numerical examples have been presented to establish the efficiency and applicability of the approached scheme. The accuracy of the proposed technique can be shown by using absolute error function definition, i.e.,

$$e_i(x,t) = \left| u_i(x,t) - \tilde{u}_i(x,t) \right|,$$

where $u_i(x,t)$, $i = 1,2,\ldots,l$ are exact solutions and $\tilde{u}_i(x,t)$, $i = 1,2,\ldots,l$ are approximate solutions. For different values of k and M, L^∞ and L^2 error norms for the systems of 2D-VIE are presented in Tables 4.7–4.10, respectively, to show the efficiency of the proposed method. For a fixed value of t, the L^∞ and L^2 error norms at the collocation points for JWs and GWs are given as follows:

$$L_\infty \text{error} = \max_{1 \le p \le 2^{k-1}M} \left| u(x_p,t) - u_m(x_p,t) \right|$$

and

$$L_2 \text{ error} = \left[\frac{1}{2^{k-1}M} \sum_{p=1}^{2^{k-1}M} (u(x_p,t) - u_m(x_p,t))^2 \right]^{\frac{1}{2}},$$

respectively, where $u(x_p,t)$ and $u_m(x_p,t)$ are exact and approximate solutions, respectively.

TABLE 4.7
L^∞ and L^2 Error Norms for JWM and GWM of Example 4.4.1 with $k = 2$, $M = 4$

	JWM ($\delta = \eta = 0.5$)		GWM ($\lambda = 0.9$)	
t	L^∞	L^2	L^∞	L^2
0.1	(4.56424E-3, 7.75679E-4)	(2.83255E-3, 3.92649E-4)	(3.42263E-3, 6.19655E-4)	(2.13053E-3, 3.13653E-4)
0.3	(8.12487E-3, 3.00458E-4)	(3.099E-3, 1.50208E-4)	(6.67637E-3, 2.33123E-4)	(2.5637E-3, 1.16425E-4)
0.7	(2.20027E-2, 1.55611E-3)	(8.10101E-3, 6.85411E-4)	(1.8577E-2, 1.16867E-3)	(6.82982E-3, 5.12451E-4)
0.9	(2.58081E-2, 1.87281E-2)	(9.83911E-3, 1.00637E-2)	(2.31435E-2, 1.4652E-2)	(8.82258E-3, 7.8971E-3)

TABLE 4.8
L^∞ and L^2 Error Norms for JWM and GWM of Example 4.4.1 with $k = 2$, $M = 6$

	JWM ($\delta = \eta = 0.5$)		GWM ($\lambda = 0.9$)	
t	L^∞	L^2	L^∞	L^2
0.1	(8.61934E-4, 1.77165E-4)	(5.68437E-4, 8.6078E-5)	(6.17452E-4, 1.35319E-4)	(4.03668E-4, 6.57431E-5)
0.3	(3.05216E-3, 5.12676E-5)	(1.0252E-3, 2.42256E-5)	(2.44351E-3, 3.81595E-5)	(8.19739E-4, 1.80086E-5)
0.7	(6.52461E-3, 4.72092E-4)	(1.91683E-3, 1.87747E-4)	(5.22849E-3, 3.53651E-4)	(1.53247E-3, 1.40828E-4)
0.9	(8.17518E-3, 4.26283E-3)	(3.86784E-3, 2.27705E-3)	(6.42574E-3, 3.24737E-3)	(3.03901E-3, 1.70949E-3)

TABLE 4.9

L^∞ and L^2 Error Norms for JWM and GWM of Example 4.4.2 with $k = 1$, $M = 7$

	JWM ($\delta = \eta = 0.3$)		GWM ($\lambda = 0.75$)	
t	L^∞	L^2	L^∞	L^2
0.1	(2.35264E-2, 3.90114E-2)	(1.39389E-2, 2.04926E-2)	(1.89818E-2, 3.1605E-2)	(1.13627E-2, 1.65016E-2)
0.3	(3.05059E-2, 2.82887E-2)	(1.29798E-2, 1.41848E-2)	(2.46268E-2, 2.29955E-2)	(1.05138E-2, 1.14854E-2)
0.7	(1.00055E-2, 1.0294E-2)	(6.6539E-3, 4.88057E-3)	(8.06567E-3, 8.38122E-3)	(5.48105E-3, 3.94179E-3)
0.9	(9.81188E-3, 1.02496E-2)	(5.38956E-3, 6.34781E-3)	(7.6737E-3, 8.16253E-3)	(4.21283E-3, 5.08466E-3)

TABLE 4.10

L^∞ and L^2 Error Norms for JWM and GWM of Example 4.4.2 with $k = 1$, $M = 8$

	JWM ($\delta = \eta = 0.3$)		GWM ($\lambda = 0.75$)	
t	L^∞	L^2	L^∞	L^2
0.1	(6.18572E-2, 4.4584E-2)	(2.44488E-2, 1.84626E-2)	(5.04561E-2, 3.63594E-2)	(1.99577E-2, 1.4993E-2)
0.3	(4.74167E-2, 3.46118E-2)	(1.72572E-2, 1.44682E-5)	(3.87487E-2, 2.82058E-2)	(1.41118E-2, 1.17418E-2)
0.7	(1.77805E-2, 8.03049E-3)	(8.74574E-3, 3.98389E-3)	(1.45494E-2, 6.53869E-3)	(7.18682E-3, 3.23667E-3)
0.9	(1.16207E-2, 3.66203E-3)	(9.79889E-3, 1.41099E-3)	(9.4402E-3, 3.00483E-3)	(7.93856E-3, 1.17449E-3)

Example 4.4.1

Let us consider a system of 2D-VIEs [50], where

$$\Lambda = \begin{bmatrix} 1 & -1 \\ 0 & -1 \end{bmatrix}, \quad \Pi(x,t,y,z) = \begin{bmatrix} 2x & e^{-z} \\ y^2 z^3 & zt \end{bmatrix},$$

$$h_1(x,t) = e^{-t}\left(\frac{1}{3}x^3t^3 + x^3t^2 + 2x^3t + 2x^3\right) + \frac{e^t}{4}(x - x^3) - \frac{7}{4}x^3 - x^2t^3,$$

$$h_2(x,t) = -\frac{1}{16}x^4(e^t(t((t-3)t+6)-6)+6) - \frac{1}{15}x^3t^6 - x^2t^3,$$

where $u_1(x,t) = \frac{1}{4}xe^t$ and $u_2(x,t) = x^2t^3$ are exact solutions. For different values of k and M, the values of absolute errors for JWM and GWM are shown in Tables 4.11–4.14, respectively. For $k = 2$, $M = 4$, the numerical analysis graphs for both JWM and GWM are plotted in Figures 4.13 and 4.14, respectively, and for $k = 2$, $M = 6$, the numerical analysis graphs for both JWs and GWs are plotted in Figures 4.15 and 4.16, respectively. Table 4.15 shows the comparison of absolute errors obtained by JWM, GWM, and Legendre wavelet method (LWM) [50] for $k = 2$, $M = 4$.

For $k = 2, M = 4$, Figures 4.13 and 4.14 are obtained by JWM and GWM, respectively.

For $k = 2, M = 6$, Figure 4.15 represents for JWM and Figure 4.16 represents for GWM.

TABLE 4.11
Numerical Results of the Absolute Error Functions $(e_1(x,t), e_2(x,t))$ by JWM for Example 4.4.1 with $k = 2$, $M = 4$

	Error Functions $(e_1(x,t), e_2(x,t))$	
(x,t)	$\delta = \eta = 0.3$	$\delta = \eta = 0.4$
(0.0625, 0.0625)	(4.63834E-7, 7.76714E-13)	(6.02098E-7, 3.37728E-12)
(0.0625, 0.1875)	(7.42954E-7, 1.00854E-9)	(9.7199E-7, 1.26732E-9)
(0.0625, 0.3125)	(1.29833E-6, 2.09987E-8)	(1.69819E-6, 2.56466E-8)
(0.0625, 0.4375)	(2.00825E-6, 1.58657E-7)	(2.60338E-6, 1.94945E-7)
(0.1875, 0.0625)	(8.83622E-6, 1.64867E-10)	(1.14752E-5, 2.07994E-10)
(0.1875, 0.1875)	(6.11139E-6, 1.85141E-9)	(7.68539E-6, 2.6776E-9)
(0.1875, 0.3125)	(6.35706E-6, 8.47428E-9)	(8.02002E-6, 8.89726E-9)
(0.1875, 0.4375)	(2.92731E-6, 1.03037E-8)	(3.16932E-6, 3.47885E-8)
(0.3125, 0.0625)	(4.26919E-5, 7.94262E-10)	(5.56273E-5, 1.00189E-9)
(0.3125, 0.1875)	(3.43121E-5, 8.05581E-9)	(4.42474E-5, 1.11373E-8)
(0.3125, 0.3125)	(4.01349E-5, 4.66155E-8)	(5.26534E-5, 6.17322E-8)
(0.3125, 0.4375)	(2.95518E-5, 6.01883E-8)	(3.80972E-5, 9.98092E-8)
(0.4375, 0.0625)	(1.11794E-4, 2.18524E-9)	(1.45391E-4, 2.76341E-9)
(0.4375, 0.1875)	(7.53266E-5, 3.28689E-8)	(9.53384E-5, 4.54907E-8)
(0.4375, 0.3125)	(7.56984E-5, 1.22856E-8)	(9.67563E-5, 1.42462E-8)
(0.4375, 0.4375)	(2.80474E-5, 9.81816E-7)	(3.09307E-5, 1.37236E-6)

TABLE 4.12
Numerical Results of the Absolute Error Functions $(e_1(x,t), e_2(x,t))$ by JWM for Example 4.4.1 with $k = 2$, $M = 6$

	Error Functions $(e_1(x,t), e_2(x,t))$	
(x,t)	$\delta = \eta = 0.3$	$\delta = \eta = 0.4$
(0.0625, 0.0625)	(2.4072E-4, 7.76661E-7)	(3.39711E-4, 1.09287E-6)
(0.0625, 0.1875)	(1.18549E-4, 1.90205E-6)	(1.66389E-4, 2.59647E-6)
(0.0625, 0.3125)	(1.34717E-4, 9.41741E-6)	(1.89104E-4, 1.29159E-5)
(0.0625, 0.4375)	(3.50474E-5, 3.03924E-5)	(4.947E-4, 4.20093E-5)
(0.1875, 0.0625)	(3.18442E-4, 5.777E-6)	(4.47238E-4, 7.91566E-6)
(0.1875, 0.1875)	(9.84598E-5, 2.24418E-6)	(1.42116E-4, 3.20479E-6)
(0.1875, 0.3125)	(1.10695E-4, 5.32828E-6)	(1.5991E-4, 7.634E-6)
(0.1875, 0.4375)	(4.63139E-4, 2.30487E-6)	(6.50211E-4, 3.18979E-5)
(0.3125, 0.0625)	(5.16603E-4, 1.60882E-5)	(7.26896E-4, 2.20558E-5)
(0.3125, 0.1875)	(1.76899E-4, 5.52508E-6)	(2.53627E-4, 7,89532E-6)
(0.3125, 0.3125)	(1.95771E-4, 1.1517E-5)	(2.81329E-4, 1.653E-5)
(0.3125, 0.4375)	(7.62555E-4, 7.3079E-5)	(1.07141E-3, 1.01572E-4)
(0.4375, 0.0625)	(1.62502E-3, 3.23164E-5)	(2.29944E-3, 4.44997E-5)
(0.4375, 0.1875)	(7.73487E-4, 1.80942E-6)	(1.09078E-3, 2.26822E-6)
(0.4375, 0.3125)	(8.90916E-4, 3.61957E-5)	(1.255E-3, 5.02732E-5)
(0.4375, 0.4375)	(2.40697E-3, 3.06464E-4)	(3.40144E-3, 4.29884E-4)

TABLE 4.13
Numerical Results of the Absolute Error Functions $(e_1(x,t), e_2(x,t))$ by GWM for Example 4.4.1 with $k = 2$, $M = 4$

	Error Functions $(e_1(x,t), e_2(x,t))$	
(x,t)	$\lambda = 0.75$	$\lambda = 1.25$
(0.0625, 0.0625)	(3.91888E-7, 2.77707E-14)	(1.03379E-6, 1.364E-11)
(0.0625, 0.1875)	(6.2506E-7, 8.66617E-10)	(1.70568E-6, 1.96759E-9)
(0.0625, 0.3125)	(1.0927E-6, 1.82993E-8)	(2.9914E-6, 3.59679E-8)
(0.0625, 0.4375)	(1.69818E-6, 1.37856E-7)	(4.46492E-6, 2.79713E-7)
(0.1875, 0.0625)	(7.46158E-6, 1.41435E-10)	(1.96482E-5, 3.28976E-10)
(0.1875, 0.1875)	(5.23666E-6, 1.47607E-9)	(1.13375E-5, 6.32806E-9)
(0.1875, 0.3125)	(5.43848E-6, 7.87548E-9)	(1.19725E-5, 4.23725E-9)
(0.1875, 0.4375)	(2.66591E-6, 1.61634E-9)	(9.70641E-7, 1.8328E-7)
(0.3125, 0.0625)	(3.59912E-5, 6.80889E-10)	(9.64133E-5, 1.56392E-9)
(0.3125, 0.1875)	(2.90615E-5, 6.58523E-9)	(7.31566E-5, 2.32584E-8)
(0.3125, 0.3125)	(3.37105E-5, 3.90524E-8)	(9.30046E-5, 1.17113E-7)
(0.3125, 0.4375)	(2.50264E-5, 4.36313E-8)	(6.2557E-5, 2.92775E-7)
(0.4375, 0.0625)	(9.43354E-5, 1.87078E-9)	(2.50274E-4, 4.34463E-9)
(0.4375, 0.1875)	(6.43582E-5, 2.68671E-8)	(1.45209E-4, 9.61189E-8)
(0.4375, 0.3125)	(6.43714E-5, 1.10036E-8)	(1.53477E-4, 1.84929E-8)
(0.4375, 0.4375)	(2.53595E-5, 7.9816E-7)	(1.38865E-5, 2.97575E-6)

TABLE 4.14
Numerical Results of the Absolute Error Functions $(e_1(x,t), e_2(x,t))$ by GWM for Example 4.4.1 with $k = 2$, $M = 6$

	Error Functions $(e_1(x,t), e_2(x,t))$	
(x,t)	$\lambda = 0.75$	$\lambda = 1.25$
(0.0625, 0.0625)	(1.9483E-4, 6.29784E-7)	(7.65674E-4, 2.45661E-6)
(0.0625, 0.1875)	(9.61982E-5, 1.56517E-6)	(3.6689E-4, 5.19934E-6)
(0.0625, 0.3125)	(1.09311E-4, 7.73245E-6)	(4.17121E-4, 2.63836E-5)
(0.0625, 0.4375)	(2.83628E-4, 2.48619E-5)	(1.11569E-3, 8.85245E-5)
(0.1875, 0.0625)	(2.58314E-4, 4.74567E-6)	(9.87909E-4, 1.61209E-5)
(0.1875, 0.1875)	(7.88062E-5, 1.80582E-6)	(3.47307E-4, 7.54307E-6)
(0.1875, 0.3125)	(8.8558E-5, 4.28038E-6)	(3.91654E-4, 1.81696E-5)
(0.1875, 0.4375)	(3.7577E-4, 1.88355E-5)	(1.43487E-3, 6.69314E-5)
(0.3125, 0.0625)	(4.18639E-4, 1.32129E-5)	(1.61408E-3, 4.50247E-5)
(0.3125, 0.1875)	(1.42096E-4, 4.4444E-6)	(6.07968E-4, 1.86307E-5)
(0.3125, 0.3125)	(1.57058E-4, 9.24375E-6)	(6.78511E-4, 3.95817E-5)
(0.3125, 0.4375)	(6.1845E-4, 5.95977E-5)	(2.37031E-3, 2.16826E-4)
(0.4375, 0.0625)	(1.31333E-3, 2.64858E-5)	(5.22208E-3, 9.25393E-5)
(0.4375, 0.1875)	(6.2608E-4, 1.54404E-6)	(2.43874E-3, 2.66496E-6)
(0.4375, 0.3125)	(7.21569E-4, 2.9524E-5)	(2.79796E-3, 1.06665E-4)
(0.4375, 0.4375)	(1.94672E-3, 2.48818E-4)	(7.69934E-3, 9.50595E-4)

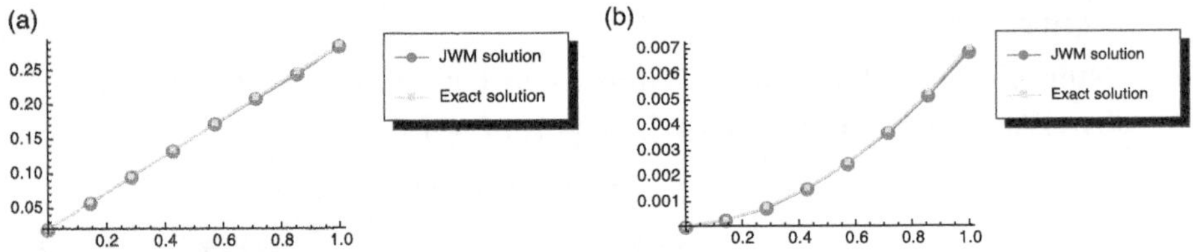

FIGURE 4.13 (a) Comparison of JWM solution with exact solution for function u_1 with $k = 2$, $M = 4$ for Example 4.4.1. (b) Comparison of JWM solution with the exact solution for function u_2 for Example 4.4.1.

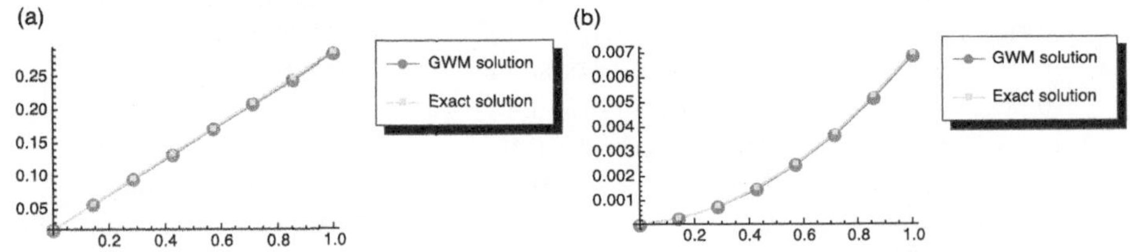

FIGURE 4.14 (a) Comparison of GWM solution with exact solution for function u_1 with $k = 2$, $M = 4$ for Example 4.4.1. (b) Comparison of GWM solution with the exact solution for function u_2 with $k = 2$, $M = 6$ for Example 4.4.1.

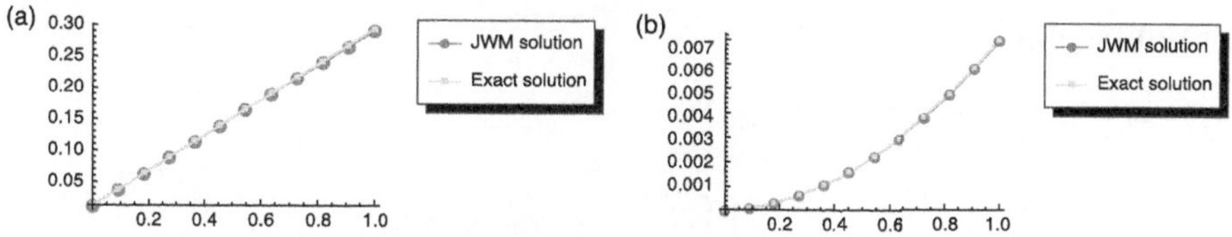

FIGURE 4.15 (a) Comparison of JWM solution with the exact solution for function u_1 with $k = 2, M = 6$ for Example 4.4.1. (b) Comparison of JWM solution with the exact solution for function u_2 with $k = 2, M = 6$ for Example 4.4.1.

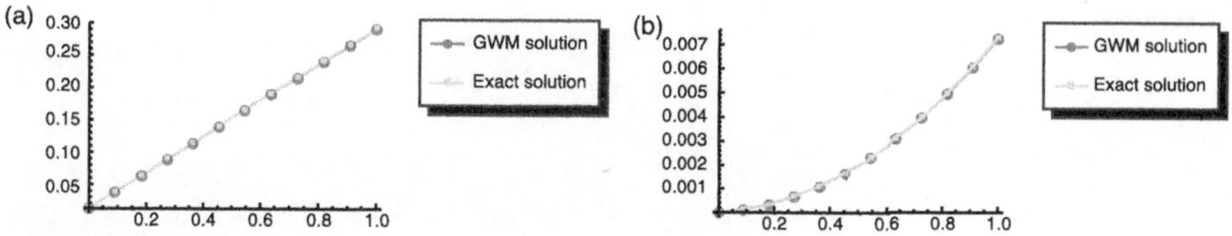

FIGURE 4.16 (a) Comparison of GWM solution with the exact solution for function u_1 with $k = 2, M = 6$ for Example 4.4.1. (b) Comparison of GWM solution with the exact solution for function u_2 with $k = 2, M = 6$ for Example 4.4.1.

TABLE 4.15
Comparison of Absolute Error Functions $(e_1(x,t), e_2(x,t))$ for Example 4.4.1 with $k = 2$, $M = 4$

	Error Functions $(e_1(x,t), e_2(x,t))$		
(x,t)	JWM ($\delta = \eta = 0.3$)	GWM ($\lambda = 0.75$)	LWM [50]
(1/8, 1/8)	(7.99321E-4, 1.12971E-5)	(6.4931E-4, 9.28167E-6)	(2.43863E-4, 2.43736E-4)
(1/8, 3/8)	(1.02674E-3, 7.64724E-5)	(8.33942E-4, 6.29523E-5)	(7.31668E-4, 7.28113E-4)
(1/8, 5/8)	(1.34473E-3, 3.25612E-4)	(1.09248E-3, 2.6815E-4)	(1.22551E-3, 1.20748E-3)
(1/8, 7/8)	(1.64807E-3, 8.51195E-4)	(1.33792E-3, 7.01391E-4)	(1.72608E-3, 1.67253E-3)
(3/8, 1/8)	(2.34327E-3, 8.44041E-5)	(1.9017E-3, 6.9743E-5)	(2.19409E-3, 2.17918E-3)
(3/8, 3/8)	(3.01933E-3, 3.17128E-4)	(2.4508E-3, 2.59892E-4)	(6.6283E-3, 6.44777E-3)
(3/8, 5/8)	(4.29911E-3, 1.20711E-3)	(3.50753E-3, 9.86505E-4)	(1.12383E-2, 1.05111E-2)
(3/8, 7/8)	(3.89224E-3, 3.02622E-3)	(3.10969E-3, 2.47487E-3)	(1.59781E-2, 1.41138E-2)
(5/8, 1/8)	(4.18615E-3, 2.30081E-4)	(3.4143E-3, 1.90226E-4)	(6.13283E-3, 6.02074E-3)
(5/8, 3/8)	(6.5255E-3, 7.91006E-4)	(5.37399E-3, 6.47432E-4)	(1.88167E-2, 1.76147E-2)
(5/8, 5/8)	(1.15777E-2, 2.80238E-3)	(9.6408E-3, 2.28072E-3)	(3.24337E-2, 2.80648E-2)
(5/8, 7/8)	(7.68445E-3, 6.94721E-3)	(6.2795E-3, 5.65198E-3)	(4.59036E-2, 3.57421E-2)
(7/8, 1/8)	(3.47176E-3, 4.50576E-4)	(2.728E-3, 3.72578E-4)	(1.21687E-2, 1.17286E-2)
(7/8, 3/8)	(1.38079E-3, 1.54138E-3)	(8.52654E-4, 1.26259E-3)	(3.82318E-2, 3.38868E-2)
(7/8, 5/8)	(3.17035E-3, 5.67497E-3)	(2.32222E-3, 4.62612E-3)	(6.72218E-2, 5.2358E-2)
(7/8, 7/8)	(1.88376E-2, 1.64098E-2)	(1.69669E-2, 1.35082E-2)	(9.3163E-2, 6.1493E-2)

Example 4.4.2

Consider a system of 2D-VIE [51]:

$$\begin{cases} u_1(x,t)+(x+t)u_2(x,t)=h_1(x,t)+\int_0^t\int_0^x (y-z-x-t)u_1(y,z)\,dz\,dy, \\ (x+t)u_1(x,t)+(x+t)^2u_2(x,t)=h_2(x,t)+\int_0^t\int_0^x (-u_1(y,z)-5e^{(x-t+y+z)}u_2(y,z))\,dz\,dy, \end{cases} \tag{4.60}$$

where

$$h_1(x,t)=\frac{x^4t}{12}+\frac{x^3t^2}{2}-\frac{x^2t^2}{4}+\frac{x^2t}{2}+x^2-\frac{5xt^3}{6}+\frac{3xt^2}{2}+e^{-(x+t)}(x+t)-t+1,$$

$$h_2(x,t)=\frac{x^3t}{3}+(x^2-t+1)(x+t)-\frac{xt^2}{2}+5xte^{x-t}+xt+e^{-(x+t)}(x+t)^2,$$

and $u_1(x,t)=x^2-t+1$, $u_2(x,t)=e^{-(x+t)}$ are exact solutions. For different values of k and M, the values of absolute errors for JWM and GWM are shown in Tables 4.16–4.19,

TABLE 4.16
Numerical Results of the Absolute Error Functions $(e_1(x,t),e_2(x,t))$ by JWM for Example 4.4.2 with $k=1$, $M=7$

	Error Functions $(e_1(x,t),e_2(x,t))$	
(x,t)	$\delta=\eta=0.3$	$\delta=\eta=0.4$
(0.0625, 0.0625)	(2.29888E-2, 7.29633E-2)	(3.21708E-2, 1.02591E-1)
(0.0625, 0.1875)	(7.32095E-3, 1.96152E-2)	(9.99503E-3, 2.72425E-2)
(0.0625, 0.3125)	(1.60335E-2, 2.7122E-2)	(2.28274E-2, 3.85469E-2)
(0.0625, 0.4375)	(1.02328E-2, 1.71784E-2)	(1.44906E-2, 2.43379E-2)
(0.1875, 0.0625)	(9.07813E-3, 1.73735E-2)	(1.24026E-2, 2.40732E-2)
(0.1875, 0.1875)	(1.35509E-2, 2.1383E-2)	(1.89902E-2, 2.9721E-2)
(0.1875, 0.3125)	(6.83687E-3, 9.94233E-3)	(9.17915E-3, 1.33904E-2)
(0.1875, 0.4375)	(1.14332E-2, 1.08967E-2)	(1.57044E-2, 1.49017E-2)
(0.3125, 0.0625)	(1.77455E-2, 5.13819E-3)	(2.50813E-2, 3.65475E-2)
(0.3125, 0.1875)	(6.9142E-3, 5.13819E-3)	(9.36602E-3, 1.11372E-2)
(0.3125, 0.3125)	(1.73604E-4, 5.13819E-3)	(1.02346E-3, 7.06894E-4)
(0.3125, 0.4375)	(4.74702E-3, 5.13819E-3)	(6.05819E-3, 2.8963E-3)
(0.4375, 0.0625)	(1.50307E-2, 5.13819E-3)	(2.10316E-2, 2.55408E-2)
(0.4375, 0.1875)	(1.09227E-2, 5.13819E-3)	(1.5104E-2, 1.35222E-2)
(0.4375, 0.3125)	(3.68304E-3, 5.13819E-3)	(4.56329E-3, 2.91594E-3)
(0.4375, 0.4375)	(7.718E-3, 5.13819E-3)	(1.03672E-2, 4.82259E-3)

TABLE 4.17
Numerical Results of the Absolute Error Functions $(e_1(x,t), e_2(x,t))$ by JWM for Example 4.4.2 with $k = 1$, $M = 8$

	Error Functions $(e_1(x,t), e_2(x,t))$	
(x,t)	$\delta = \eta = 0.3$	$\delta = \eta = 0.4$
(0.0625, 0.0625)	(6.50751E-3, 5.27486E-2)	(9.2881E-3, 7.51992E-2)
(0.0625, 0.1875)	(3.36576E-3, 1.50853E-2)	(4.86625E-3, 2.15765E-2)
(0.0625, 0.3125)	(6.12983E-3, 1.89552E-2)	(9.04065E-3, 2.75176E-2)
(0.0625, 0.4375)	(3.75175E-3, 1.09602E-2)	(5.68699E-3, 1.58955E-2)
(0.1875, 0.0625)	(3.08479E-3, 1.33923E-2)	(4.43871E-3, 1.91286E-2)
(0.1875, 0.1875)	(6.1343E-3, 1.46308E-2)	(8.44151E-3, 2.02735E-2)
(0.1875, 0.3125)	(4.98977E-3, 7.4628E-3)	(6.5982E-3, 9.9209E-3)
(0.1875, 0.4375)	(7.65419E-3, 9.12373E-3)	(1.03588E-2, 1.25079E-2)
(0.3125, 0.0625)	(6.36086E-3, 1.84029E-2)	(9.26828E-3, 2.66006E-3)
(0.3125, 0.1875)	(4.10517E-3, 6.1865E-3)	(5.40261E-3, 8.17207E-3)
(0.3125, 0.3125)	(2.25519E-3, 8.38755E-4)	(2.55624E-3, 4.61404E-4)
(0.3125, 0.4375)	(4.86545E-3, 3.20176E-3)	(6.29877E-3, 4.08714E-3)
(0.4375, 0.0625)	(5.06983E-3, 1.19764E-2)	(7.36978E-3, 1.71472E-2)
(0.4375, 0.1875)	(6.63051E-3, 8.29719E-3)	(9.03352E-3, 1.14398E-2)
(0.4375, 0.3125)	(4.64668E-3, 3.21368E-3)	(6.03724E-3, 4.13135E-3)
(0.4375, 0.4375)	(6.89064E-3, 4.49103E-3)	(9.27289E-3, 6.14115E-3)

TABLE 4.18
Numerical Results of the Absolute Error Functions $(e_1(x,t), e_2(x,t))$ by GWM for Example 4.4.2 with $k = 1$, $M = 7$

	Error Functions $(e_1(x,t), e_2(x,t))$	
(x,t)	$\lambda = 0.75$	$\lambda = 0.9$
(0.0625, 0.0625)	(1.8681E-2, 5.91318E-2)	(3.21708E-2, 1.02591E-1)
(0.0625, 0.1875)	(6.01147E-3, 1.59693E-2)	(9.99503E-3, 2.72425E-2)
(0.0625, 0.3125)	(1.29107E-2, 2.18528E-3)	(2.28274E-2, 3.85469E-2)
(0.0625, 0.4375)	(8.25304E-3, 1.38572E-2)	(1.44906E-2, 2.43379E-2)
(0.1875, 0.0625)	(7.45594E-3, 1.41601E-2)	(1.24026E-2, 2.40732E-2)
(0.1875, 0.1875)	(1.10137E-2, 1.74548E-2)	(1.89902E-2, 2.9721E-2)
(0.1875, 0.3125)	(5.67844E-3, 8.24546E-3)	(9.17915E-3, 1.33904E-2)
(0.1875, 0.4375)	(9.38831E-3, 8.96664E-3)	(1.57044E-2, 1.49017E-2)
(0.3125, 0.0625)	(1.4346E-2, 2.07926E-2)	(2.50813E-2, 3.65475E-2)
(0.3125, 0.1875)	(5.71742E-3, 6.90503E-3)	(9.36602E-3, 1.11372E-2)
(0.3125, 0.3125)	(8.98822E-5, 2.44555E-4)	(1.02346E-3, 7.06894E-4)
(0.3125, 0.4375)	(4.03456E-3, 2.10575E-3)	(6.05819E-3, 2.8963E-3)
(0.4375, 0.0625)	(1.22068E-2, 1.46948E-2)	(2.10316E-2, 2.55408E-2)
(0.4375, 0.1875)	(8.9384E-3, 8.08564E-3)	(1.5104E-2, 1.35222E-2)
(0.4375, 0.3125)	(3.17069E-3, 2.09271E-3)	(4.56329E-3, 2.91594E-3)
(0.4375, 0.4375)	(6.40481E-3, 3.03759E-3)	(1.03672E-2, 4.82259E-3)

TABLE 4.19
Numerical Results of the Absolute Error Functions $(e_1(x,t), e_2(x,t))$ by GWM for Example 4.4.2 with $k=1$, $M=8$

	Error Functions $(e_1(x,t), e_2(x,t))$	
(x,t)	$\lambda = 0.75$	$\lambda = 0.9$
(0.0625, 0.0625)	(5.23251E-3, 4.24412E-2)	(9.2881E-3, 7.51992E-2)
(0.0625, 0.1875)	(2.68148E-3, 1.20939E-2)	(4.86625E-3, 2.15765E-2)
(0.0625, 0.3125)	(4.83896E-3, 1.50997E-2)	(9.04065E-3, 2.75176E-2)
(0.0625, 0.4375)	(2.90975E-3, 8.72722E-3)	(5.68699E-3, 1.58955E-2)
(0.1875, 0.0625)	(2.46427E-3, 1.07441E-2)	(4.43871E-3, 1.91286E-2)
(0.1875, 0.1875)	(5.03456E-3, 1.19661E-2)	(8.44151E-3, 2.02735E-2)
(0.1875, 0.3125)	(4.1744E-3, 6.22756E-3)	(6.5982E-3, 9.9209E-3)
(0.1875, 0.4375)	(6.33175E-3, 7.49926E-3)	(1.03588E-2, 1.25079E-2)
(0.3125, 0.0625)	(5.05729E-3, 1.46975E-2)	(9.26828E-3, 2.66006E-2)
(0.3125, 0.1875)	(3.4418E-3, 5.17772E-3)	(5.40261E-3, 8.17207E-3)
(0.3125, 0.3125)	(2.01081E-3, 8.90848E-4)	(2.55624E-3, 4.61404E-4)
(0.3125, 0.4375)	(4.10725E-3, 2.71875E-3)	(6.29877E-3, 4.08714E-3)
(0.4375, 0.0625)	(4.03199E-3, 9.60722E-3)	(7.36978E-3, 1.71472E-2)
(0.4375, 0.1875)	(5.46657E-3, 6.80017E-3)	(9.03352E-3, 1.14398E-2)
(0.4375, 0.3125)	(3.91581E-3, 2.72016E-3)	(6.03724E-3, 4.13135E-3)
(0.4375, 0.4375)	(5.71185E-5, 3.69304E-3)	(9.27289E-3, 6.14115E-3)

respectively. For $k=1$, $M=7$, the numerical analysis graphs for both JWM and GWM are plotted in Figures 4.17 and 4.18, respectively, and for $k=1$, $M=8$, the numerical analysis graphs for both JWs and GWs are plotted in Figures 4.19 and 4.20, respectively. Table 4.20 shows the comparison of absolute errors obtained by JWM, GWM, and LWM [50] for $k=1$, $M=2$.

For $k=1$, $M=7$, Figures 4.17 and 4.18 are obtained by JWM and GWM, respectively.

For $k=1$, $M=8$, Figures 4.19 and 4.20 are obtained by JWM and GWM, respectively.

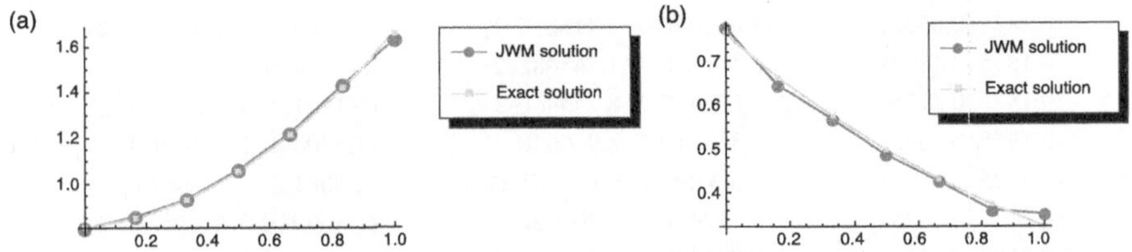

FIGURE 4.17 (a) Comparison of JWM solution with the exact solution for function u_1 with $k=2$, $M=7$ for Example 4.4.2. (b) Comparison of JWM solution with the exact solution for function u_2 with $k=2$, $M=7$ for Example 4.4.2.

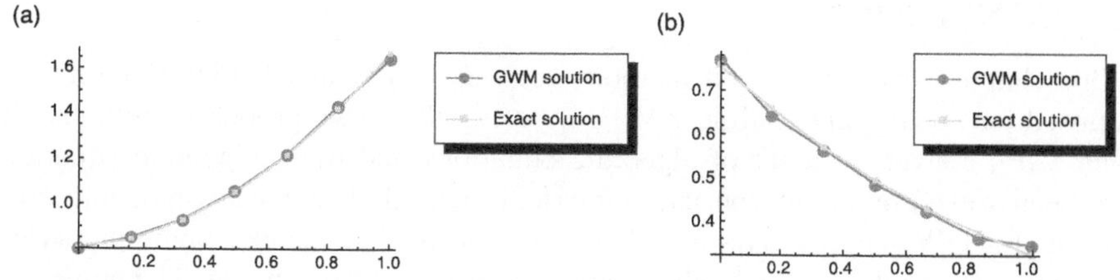

FIGURE 4.18 (a) Comparison of JWM solution with the exact solution for function u_1 with $k = 2, M = 7$ for Example 4.4.2. (b) Comparison of JWM solution with the exact solution for function u_2 with $k = 2, M = 7$ for Example 4.4.2.

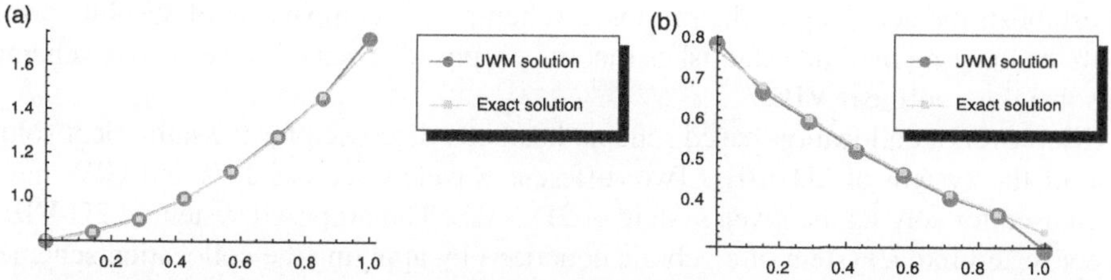

FIGURE 4.19 (a) Comparison of JWM solution with the exact solution for function u_1 with $k = 2, M = 8$ for Example 4.4.2. (b) Comparison of JWM solution with the exact solution for function u_2 with $k = 2, M = 8$ for Example 4.4.2.

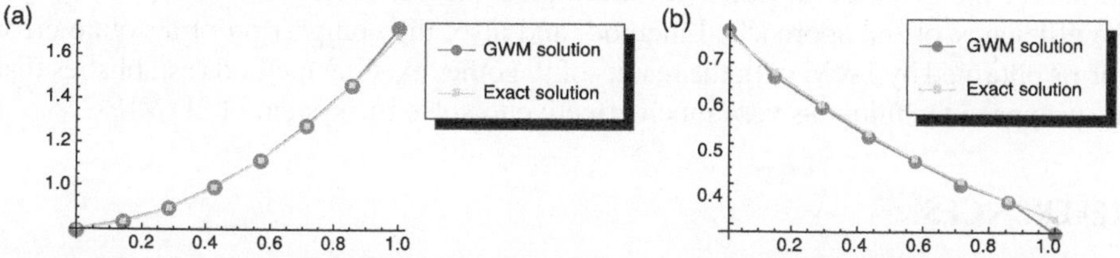

FIGURE 4.20 (a) Comparison of JWM solution with the exact solution for function u_1 with $k = 2, M = 8$ for Example 4.4.2. (b) Comparison of JWM solution with the exact solution for function u_2 with $k = 2, M = 8$ for Example 4.4.2.

TABLE 4.20

Comparison of Absolute Error Functions $(e_1(x,t), e_2(x,t))$ for Example 4.4.2 with $k = 1, M = 2$

	Error Functions $(e_1(x,t), e_2(x,t))$		
(x,t)	JWM ($\delta = \eta = 0.1$)	GWM ($\lambda = 0.2$)	LWM [50]
(1/4, 1/4)	(2.82854E-2, 7.1278E-2)	(3.56778E-2, 8.7972E-2)	(1.95618E-1, 2.70472E-1)
(1/4, 3/4)	(8.68884E-2, 6.06471E-2)	(4.60131E-2, 3.52669E-2)	(2.30878E-1, 1.36598E-1)
(3/4, 1/4)	(6.88689E-2, 5.53532E-2)	(6.29105E-2, 3.32941E-2)	(8.64944E-2, 4.07469E-2)
(3/4, 3/4)	(2.44284E-2, 1.67656E-2)	(9.20728E-2, 4.56812E-2)	(7.05652E-2, 9.21723E-2)

4.5 CONCLUSION

In this chapter, two collocation methods such as GWM and CASWM have been proposed for solving 2D nonlinear VIE. By using these two proposed methods, the given VIEs convert to a set of algebraic equations and those algebraic equations have been solved by an appropriate numerical method. The convergence and error analyses for GWs and CASWs are also presented in this chapter. Some numerical examples are included to establish the applicability of the technique. The numerical results show that the efficiency of the obtained solutions is satisfactory. Furthermore, the comparison of L^2 and L^∞ errors presented in Tables 4.4 and 4.6 justifies the plausibility for the accuracy of the proposed collocation methods. Moreover, the graphical representations of absolute errors for GWM and CASWM are plotted in this chapter to establish the accuracy of the proposed scheme. The comparison of absolute error with the existing method establishes that the proposed scheme is an effective scheme to solve the nonlinear VIEs.

Moreover, a collocation-based scheme has been discussed for the numerical solution of the system of 2D-VIEs. Two different wavelets such as JWs and GWs have been used for solving the given system of 2D-VIEs. The proposed system of 2D-VIEs is converted into a system of algebraic equations by applying the collocation scheme, and by solving the algebraic system, the desirable accurate approximate solutions have been obtained. Moreover, the convergence analysis has been discussed with respect to 2D JWs. Furthermore, some numerical problems are solved which establish the applicability of the proposed method. Moreover, the comparison of L^∞ and L^2 error norms for the given 2D wavelets has been presented in Tables 4.13 and 4.15 to show the efficiency of the approached method, and also, the comparison of absolute error norms obtained by JWM with the results of the other existing methods establishes that the proposed technique is very much effective to solve the system of 2D VIEs.

REFERENCES

1. Wazwaz A.M., 2011, *Linear and Nonlinear Integral Equations Methods and Applications*, Higher Education Press, Beijing and Springer-Verlag, Berlin.
2. Atkinson K.E., 1997, "The numerical solution of boundary integral equations", *Institute of Mathematics and Its Applications Conference Series*, **63**, pp. 223–260.
3. Polyanin A.D., and Manzhirov A.V., 1998, *Handbook of Integral Equations*, CRC press, Boca Raton, FL.
4. Baker C., 1978, *The Numerical Treatment of Integral Equations*, Clarendon Press, Oxford.
5. Bartoshevich M.A., 1975, "A heat-conduction problem", *Journal of Engineering Physics*, **28**(2), pp. 240–244.
6. Baratella P., 2009, "A Nyström interpolant for some weakly singular linear Volterra integral equations", *Journal of Computational and Applied Mathematics*, **231**(2), pp. 725–734.
7. Ding H.J., Wang H.M., and Chen W.Q., 2003, "Analytical solution for the electroelastic dynamics of a nonhomogeneous spherically isotropic piezoelectric hollow sphere", *Archive of Applied Mechanics*, **73**(1), pp. 49–62.
8. Farengo R., Lee Y.C., and Guzdar P.N., 1983, "An electromagnetic integral equation: application to microtearing modes", *The Physics of Fluids*, **26**(12), pp. 3515–3523.

9. Kit G.S., and Maksymuk A.V., 1998, "The method of Volterra integral equations in contact problems for thin-walled structural elements", *Journal of Mathematical Sciences*, **90**(1), pp. 1863–1867.
10. Galdi G.P., Pileckas K., and Silvestre, A.L., 2007, "On the unsteady Poiseuille flow in a pipe", *Zeitschrift für angewandte Mathematik und Physik*, **58**(6), pp. 994–1007.
11. Sahu P.K., and Saha Ray S., 2016, "Legendre spectral collocation method for the solution of the model describing biological species living together", *Journal of Computational and Applied Mathematics*, **296**, pp. 47–55.
12. Sahu P.K., and Saha Ray S., 2016, "Comparative experiment on the numerical solutions of Hammerstein integral equation arising from chemical phenomenon", *Journal of Computational and Applied Mathematics*, **291**, pp. 402–409.
13. Atkinson K.E., 1996, The numerical solution of integral equations of the second kind. *Cambridge Monographs on Applied and Computational Mathematics*, **1996**, p. 572.
14. Delves L.M., and Mohamed J.L., 1985, *Computational Methods for Integral Equations*, Cambridge University Press, Cambridge.
15. Jerri A., 1999, *Introduction to Integral Equations with Applications*, John Wiley & Sons, Hoboken, NJ.
16. Kythe P., and Puri P., 2011, *Computational Methods for Linear Integral Equations*, Springer Science & Business Media, Berlin.
17. Singh P., 1976, "A note on the solution of two-dimensional Volterra integral equations by splines", *Indian Journal of Mathematics*, **18**, pp. 61–64.
18. Brunner H., and Kauthen J.P., 1989, "The numerical solution of two-dimensional Volterra integral equations by collocation and iterated collocation", *IMA Journal of Numerical Analysis*, **9**(1), pp. 47–59.
19. Guoqiang H., and Liqing Z., 1994, "Asymptotic error expansion of two-dimensional Volterra integral equation by iterated collocation", *Applied Mathematics and Computation*, **61**(2–3), pp. 269–285.
20. McKee S., Tang T., and Diogo T., 2000, "An Euler-type method for two-dimensional Volterra integral equations of the first kind", *IMA Journal of Numerical Analysis*, **20**(3), pp. 423–440.
21. Mirzaee F., and Alipour S., 2020, "A hybrid approach of nonlinear partial mixed integro-differential equations of fractional order", *Iranian Journal of Science and Technology, Transactions A: Science*, **44**(3), pp. 725–737.
22. Singh I., and Kumar S., 2016, "Haar wavelet method for some nonlinear Volterra integral equations of the first kind", *Journal of Computational and Applied Mathematics*, **292**, pp. 541–552.
23. Assari P., and Dehghan M., 2019, "A meshless local discrete Galerkin (MLDG) scheme for numerically solving two-dimensional nonlinear Volterra integral equations", *Applied Mathematics and Computation*, **350**, pp. 249–265.
24. Assari P., and Dehghan, M., 2018, "A meshless local discrete collocation (MLDC) scheme for solving 2-dimensional singular integral equations with logarithmic kernels", *International Journal of Numerical Modelling: Electronic Networks, Devices and Fields*, **31**(3), p. e2311.
25. Sahu P.K., and Saha Ray S., 2015, "Legendre wavelets operational method for the numerical solutions of nonlinear Volterra integro-differential equations system", *Applied Mathematics and Computation*, **256**, pp. 715–723.
26. Assari P., 2019, "On the numerical solution of two-dimensional integral equations using a meshless local discrete Galerkin scheme with error analysis", *Engineering with Computers*, **35**(3), pp. 893–916.
27. Khajehnasiri A.A., Ezzati R., and Kermani M.A., 2021, "Solving fractional two-dimensional nonlinear partial volterra integral equation by using Bernoulli Wavelet", *Iranian Journal of Science and Technology, Transactions A: Science*, **45**(3), pp. 983–995.

28. Maleknejad K., Mohammadikia H., and Rashidinia J., 2019, "Analysis and solution of a class of nonlinear two-dimensional Volterra–Fredholm integral equations via hybrid of radial basis functions", *Iranian Journal of Science and Technology, Transactions A: Science*, **43**(5), pp. 2253–2260.
29. Assari P., and Dehghan M., 2017, "The numerical solution of two-dimensional logarithmic integral equations on normal domains using radial basis functions with polynomial precision", *Engineering with Computers*, **33**(4), pp. 853–870.
30. Assari P., and Dehghan M., 2018, "The approximate solution of nonlinear Volterra integral equations of the second kind using radial basis functions", *Applied Numerical Mathematics*, **131**, pp. 140–157.
31. Maleknejad K., and Mirzaee F., 2003, "Numerical solution of linear Fredholm integral equations system by rationalized Haar functions method", *International Journal of Computer Mathematics*, **80**(11), pp. 1397–1405.
32. Biazar J., Babolian E., and Islam R., 2003, "Solution of a system of Volterra integral equations of the first kind by Adomian method", *Applied Mathematics and Computation*, **139**(2–3), pp. 249–258.
33. Yusufoğlu E., 2007, "An efficient algorithm for solving integro-differential equations system", *Applied Mathematics and Computation*, **192**(1), pp. 51–55.
34. Arikoglu A., and Ozkol I., 2008, "Solutions of integral and integro-differential equation systems by using differential transform method", *Computers & Mathematics with Applications*, **56**(9), pp. 2411–2417.
35. Sorkun H.H., and Yalçinbaş S., 2010, "Approximate solutions of linear Volterra integral equation systems with variable coefficients", *Applied Mathematical Modelling*, **34**(11), pp. 3451–3464.
36. Maleknejad K., and Kajani M.T., 2004, "Solving linear integro-differential equation system by Galerkin methods with hybrid functions", *Applied Mathematics and Computation*, **159**(3), pp. 603–612.
37. Maleknejad K., Safdari H., and Nouri M., 2011, "Numerical solution of an integral equations system of the first kind by using an operational matrix with block pulse functions", *International Journal of Systems Science*, **42**(1), pp. 195–199.
38. Jafarian A., Nia S.A.M., Golmankhaneh A.K., and Baleanu D., 2013, "Numerical solution of linear integral equations system using the Bernstein collocation method", *Advances in Difference Equations*, **2013**(1), p. 123.
39. Hasan T.I., Salleh S., and Sulaiman N.A., 2015, "Solving a system of Volterra-Fredholm integral equations of the second kind via fixed point method", *AIP Conference Proceedings*, **1691**(1), p. 040006.
40. Babolian E., Biazar J., and Vahidi A.R., 2004, "On the decomposition method for system of linear equations and system of linear Volterra integral equations", *Applied Mathematics and Computation*, **147**(1), pp. 19–27.
41. Javidi M., 2009, "Modified homotopy perturbation method for solving system of linear Fredholm integral equations", *Mathematical and Computer Modelling*, **50**(1–2), pp. 159–165.
42. Golbabai A., Mammadov M., and Seifollahi S., 2009, "Solving a system of nonlinear integral equations by an RBF network", *Computers & Mathematics with Applications*, **57**(10), pp. 1651–1658.
43. Shidfar A., and Molabahrami A., 2011, "Solving a system of integral equations by an analytic method", *Mathematical and Computer Modelling*, **54**(1–2), pp. 828–835.
44. Banifatemi E., Razzaghi M., and Yousefi S., 2007, "Two-dimensional Legendre Wavelets Method for the mixed Volterra-Fredholm integral equations", *Journal of Vibration and Control*, **13**(11), pp. 1667–1675.
45. Narumi S., 1920, "Some formulas in the theory of interpolation of many independent variables", *Tohoku Mathematical Journal, First Series*, **18**, pp. 309–321.

46. Rahimkhani P., and Ordokhani Y., 2018, "Numerical solution a class of 2D fractional optimal control problems by using 2D Müntz-Legendre wavelets", *Optimal Control Applications and Methods*, **39**(6), pp. 1916–1934.
47. Babolian E., Bazm S., and Lima, P., 2011, "Numerical solution of nonlinear two-dimensional integral equations using rationalized Haar functions", *Communications in Nonlinear Science and Numerical Simulation*, **16**(3), pp. 1164–1175.
48. Saberi Najafi J., Navid Samadi O., and Tohidi E., 2011, "Numerical solution of two-dimensional Volterra integral equations by spectral Galerkin method", *Journal of Applied Mathematics and Bioinformatics*, **1**(2), pp. 159–174.
49. Van Assche W., and Coussement E., 2001, "Some classical multiple orthogonal polynomials", *Journal of Computational and Applied Mathematics*, **127**, pp. 317–347.
50. Karimi A., Maleknejad K., and Ezzati R., 2020, "Numerical solutions of system of two-dimensional Volterra integral equations via Legendre wavelets and convergence", *Applied Numerical Mathematics*, **156**, pp. 228–241.
51. Bulatov M.V., and Lima P.M., 2011, "Two-dimensional integral–algebraic systems: analysis and computational methods", *Journal of Computational and Applied Mathematics*, **236**(2), pp. 132–140.

5 Numerical Solutions of Linear and Nonlinear Volterra Integro-Differential Equations

5.1 INTRODUCTION

The real-world physical problems are designed as integral, differential, and integro-differential equations (IDEs). The studies of IDEs have emerged very rapidly. In recent years, a lot of considerable effort has been devoted by many physicists and mathematicians to find the numerical solution of the IDEs. The application of IDEs arises in many areas of scientific problems such as physics, engineering, and mechanics [1,2]. IDEs are applied in many problems of stereology [3], heat conductions [4], heat flow [5], control system theory [6], and viscoelasticity [7], and also, these equations arise in many physical processes, such as nano-hydrodynamics [8], wind ripple in the desert [9], glass-forming process [10], drop-wise condensation [11], and biological model [12]. The field of increasing mathematical interest over the past decade is the study of the Volterra integro-differential equation (VIDE).

In recent years, many researchers have found wavelets to be powerful tools to utilize in different numerical methods. The theory of wavelets is employed mainly in the area of mathematics and engineering. The application of wavelet analysis has been found in various fields such as signal analysis, image analysis, and communication systems [13].

The concept of linear and non-linear VIDEs provides important mathematical tools for many real-world phenomena in science and engineering. Therefore, it is very important during investigations which are related to sciences and engineering to have information about the qualitative properties of solutions of linear and non-linear VIDEs without solving them. The analytical solutions of nonlinear IDEs are difficult to find, and exact solutions are rather difficult to obtain. IDEs arise in various applications such as initial value problems or boundary value problems which can be converted into Volterra integral equation (VIE) and Fredholm integral equation (FIE), respectively, and vice versa. Different methods like the Haar wavelet method [14], Adomian decomposition method [15], Taylor polynomial solution method [16], etc. are used to solve the IDEs. Moreover, several techniques have been introduced for the solution of VIDEs such as the Tau method [17], hybrid methods

DOI: 10.1201/9781003453451-5

[18], collocation method [19], variational iteration method [20], CAS wavelet method [21], and Legendre wavelet method [22].

Several studies have been done by many researchers for the numerical solutions of IDEs. In an earlier study [23], Venkatesh et al. have applied the Legendre wavelet method to solve a class of nonlinear IDEs. Biazer et al. [24] have introduced the Homotopy perturbation method for the solution of system of nonlinear VIDEs. Nemati et al. [25] applied Legendre polynomials to solve two-dimensional (2D) VIEs. Babolin et al. [26] introduced differential transform and homotopy perturbation methods to find the approximate solution of 2D linear and nonlinear VIEs. Sahu and Saha Ray [27] introduced the Legendre spectral collocation method for the numerical solution of Fredholm IDEs. Maleknejad et al. [28] have solved nonlinear Volterra–Fredholm integral equations by applying the triangular functions method. The Tau method has been introduced for the numerical solution of two-dimensional (2D) linear VIDEs [29]. Nasser et al. [30] analysed the solution of nonlinear 2D VIDEs by employing the block pulse function. Sahu and Saha Ray [31] have applied the Bernoulli wavelet method for solving the weakly singular VIDEs. Das et al. [32–35] have proposed homotopy perturbation technique and also established the convergence analysis for fractional-order IDEs. The error analysis of the adaptive approach may be found in previous works [36–39].

The stability of the approximate solutions is not straightforward for PDEs, and therefore, extra conditions are needed on the discrete scheme under which the matrices will be invertible. Several stability properties are handled previously [40–43], where a details analysis on combination has been demonstrated.

5.2 OUTLINE OF THE PRESENT STUDY

In this chapter, an effective approach to obtain the approximate solutions of linear and nonlinear 2D VIDEs has been proposed. The operational matrices of integration, differentiation, and product based on 2D Bernoulli wavelets (BWs) and Legendre wavelets (LWs) have been constructed. By utilizing the properties and matrices of wavelets along with the collocation point, the 2D linear and nonlinear VIDEs are reduced into the system of linear and nonlinear algebraic equations, respectively. The convergence and error analyses have been extensively studied by the help of 2D wavelets approximation. Some illustrative examples are examined to clarify the accuracy and effectiveness of the proposed scheme.

5.3 LINEAR AND NONLINEAR VOLTERRA INTEGRO-DIFFERENTIAL EQUATIONS

A general form of the 2D VIDE of the second kind can be written as follows:

$$u_{xx}(x,t)+u_{xt}(x,t)+u_{tt}(x,t)+u(x,t)=f(x,t)+\int_0^x\int_0^t k(x,t,y,z)H(u(y,z))dydz,$$

$$(x,t)\in[0,1]\times[0,1], \tag{5.1}$$

with the given Dirichlet boundary conditions, where $u(x,t)$ is an unknown function which should be determined; $f(x,t)$ and $k(x,t,y,z)$ are known analytic functions. In this chapter, the term $H(u(y,z))$ is considered as $H(u(y,z))=u(y,z)$ and $H(u(y,z))=u^{\lambda}(y,z)$, $\lambda>1$ for linear and nonlinear cases, respectively.

5.3.1 Function Approximation by BWs

Let us consider a continuous function $f(x,t)\in L^2([0,1)\times[0,1))$, and then, it can be approximated as follows:

$$f(x,t)=\sum_{n=1}^{2^{k_1-1}}\sum_{n'=1}^{2^{k_2-1}}\sum_{m=0}^{M_1-1}\sum_{m'=0}^{M_2-1}\tilde{f}_{n,m,n',m'}\cdot\tilde{\psi}_{n,m,n_1,m_1}(x,t)=\tilde{F}^T.\Psi(x,t), \tag{5.2}$$

where $\tilde{F}$ and $\Psi(x,t)$ are matrices of dimension $(2^{k_1-1}2^{k_2-1}M_1M_2\times 1)$ given by

$$\tilde{F}=[\tilde{f}_{1,0,1,0},\ldots,\tilde{f}_{1,0,1,M_2-1},\ldots,\tilde{f}_{1,0,2^{k_2-1},M_2-1},\ldots,\tilde{f}_{2^{k_1-1},M_1-1,2^{k_2-1},M_2-1}]^T, \tag{5.3}$$

$$\begin{aligned}\Psi(x,t)=[&\psi_{1,0,1,0}(x,t),\ldots,\psi_{1,0,1,M_2-1}(x,t),\ldots,\psi_{1,0,2^{k_2-1},M_2-1}(x,t),\ldots,\\&\psi_{2^{k_1-1},M_1-1,2^{k_2-1},M_2-1}(x,t)]^T,\end{aligned} \tag{5.4}$$

where $\psi_{n,m,n',m'}(x,t)$, $n=1,2,\ldots,2^{k_1-1}$, $n'=1,2,\ldots,2^{k_2-1}$, $m=0,1,2,\ldots,M_1-1$, and $m'=0,1,2,\ldots,M_2-1$ is the 2D BWs defined over the domain $[0,1)\times[0,1)$ as follows:

$$\psi_{n,m,n',m'}(x,t)=\begin{cases}2^{\frac{k_1+k_2}{2}}\tilde{\beta}_m(2^{k_1-1}x-n+1).\tilde{\beta}_{m'} & \text{for } \dfrac{n-1}{2^{k_1-1}}\le x<\dfrac{n}{2^{k_1-1}},\\ (2^{k_2-1}t-n'+1), & \dfrac{n'-1}{2^{k_2-1}}\le t<\dfrac{n'}{2^{k_2-1}},\\ 0, & \text{otherwise.}\end{cases} \tag{5.5}$$

The coefficients $\tilde{f}_{n,m,n',m'}$ of $2^{k_1-1}2^{k_2-1}M_1M_2$-dimensional vector $\tilde{F}$ are defined as follows:

$$\tilde{f}_{n,m,n',m'}=\frac{<\psi_{n,m}(x),<f(x,t),\psi'_{n',m'}(t)>>}{<\psi_{n,m}(x),\psi_{n,m}(x)>.<\psi'_{n',m'}(t),\psi'_{n',m'}(t)>},$$

where

$$<\psi_{n,m}(x),\psi_{n,m}(x)>=\int_{\frac{n-1}{2^{k_1-1}}}^{\frac{n}{2^{k_1-1}}}\psi_{n,m}(x).\psi_{n,m}(x)\,dx=1,$$

$$< \psi'_{n',m'}(t), \psi'_{n',m'}(t) > = \int_{\frac{n'-1}{2^{k_2-1}}}^{\frac{n'}{2^{k_2-1}}} \psi'_{n',m'}(t).\psi'_{n',m'}(t)\,dt = 1.$$

We have

$$\tilde{f}_{n,m,n',m'} = \int_{\frac{n-1}{2^{k_1-1}}}^{\frac{n}{2^{k_1-1}}} \psi_{n,m}(x) \int_{\frac{n'-1}{2^{k_2-1}}}^{\frac{n'}{2^{k_2-1}}} f(x,t).\psi'_{n',m'}(t)\,dx\,dt. \tag{5.6}$$

The kernel function $k(x,t,y,z)$ can be approximated as follows:

$$k(x,t,y,z) = \psi^T(x,t)\tilde{K}\psi(y,z),$$

where $\tilde{K}$ is the coefficient matrix of dimension $(2^{k_1-1}2^{k_2-1}M_1M_2 \times 2^{k_1-1}2^{k_2-1}M_1M_2)$ and its components are defined as follows:

$$\tilde{K}_{i,j} = < \psi_{n,m}(x), < \psi'_{n',m'}(t), < \psi_{n_1,m_1}(y), < k(x,t,y,z), \psi'_{n'_1,m'_1}(z) >>>>,$$

in which

$$i = (n.M_1.2^{k_1-1}.M_2) - ((M_1-(m+1)).2^{k_1-1}.M_2) - ((2^{k_1-1}-n').M_2) - (M_2-(m'+1)),$$

$$j = (n_1.M_1.2^{k_1-1}.M_2) - ((M_1-(m_1+1)).2^{k_1-1}.M_2) - ((2^{k_1-1}-n'_1).M_2) - (M_2-(m'_1+1)),$$

where $n, n' = 1,2,\ldots,2^{k_1-1}$, $n_1, n'_1 = 1,2,\ldots,2^{k_2-1}$, $m, m' = 0,1,\ldots,M_1-1$, and $m_1, m'_1 = 0,1,\ldots,M_2-1$.

5.4 TWO-DIMENSIONAL BLOCK-PULSE FUNCTION

The $2^{k_1-1}2^{k_2-1}M_1M_2$ - set of two-dimensional Block-Pulse functions (2D-BPFs) is defined on $L^2([0,1)\times[0,1))$ as follows:

$$b_{p,r}(x,t) = \begin{cases} 1, & \frac{p-1}{2^{k_1-1}} \le x < \frac{p}{2^{k_1-1}}, \frac{r-1}{2^{k_2-1}} \le t < \frac{r}{2^{k_2-1}}, \\ 0, & \text{otherwise}, \end{cases} \tag{5.7}$$

where $p = 1,2,\ldots,2^{k_1-1}$, $r = 1,2,\ldots,2^{k_2-1}$.

The 2D-BPFs are disjoined with each other, i.e.,

$$b_{p,r}(x,t)b_{q,s}(x,t) = \begin{cases} b_{p,r}(x,t), & p = r, q = s, \\ 0, & \text{otherwise}. \end{cases} \tag{5.8}$$

The 2D-BPFs are also orthogonal with each other [44].

The 2D BWs may be expanded into $2^{k_1-1}2^{k_2-1}M_1M_2$ - set of 2D-BPFs as

$$\Psi(x,t) = \tilde{Q}B(x,t), \tag{5.9}$$

where

$$B(x,t) = [b_{1,1}(x,t),\ldots,b_{1,2^{k_2-1}}(x,t),\ldots,b_{2^{k_1-1},1}(x,t),\ldots,b_{2^{k_1-1},2^{k_2-1}}(x,t)]^T,$$

and

$$\tilde{Q} = \begin{bmatrix} \Psi(x_1,t_1) & \Psi(x_1,t_2) & . & . & . & \Psi(x_1,t_{2^{k_2-1}}) \\ \Psi(x_2,t_1) & \Psi(x_2,t_2) & . & . & . & \Psi(x_2,t_{2^{k_2-1}}) \\ . & . & . & . & . & . \\ . & . & . & . & . & . \\ . & . & . & . & . & . \\ \Psi(x_{2^{k_1-1}},t_1) & . & . & . & . & \Psi(x_{2^{k_1-1}},t_{2^{k_2-1}}) \end{bmatrix}.$$

The following definitions are consequences of relations of BPFs [45].

Definition 5.4.1

For $2^{k_1-1}2^{k_2-1}M_1M_2$-vector W, we have

$$\Psi(x,t)\Psi^T(x,t)W = \tilde{W}\Psi(x,t),$$

in which $\tilde{W}$ is a $(2^{k_1-1}M_1 \times 2^{k_2-1}M_2)$ matrix of the form

$$\tilde{W} = \tilde{Q}\overline{W}\tilde{Q}^{-1},$$

where $\overline{W} = \mathrm{diag}(\tilde{Q}^T W)$.

Definition 5.4.2

Let A be an arbitrary $(2^{k_1-1}M_1 \times 2^{k_2-1}M_2)$-dimensional matrix. Then, we have

$$\Psi^T(x,t)A\Psi(x,t) = \tilde{A}\Psi(x,t),$$

where $\tilde{A} = \mathrm{diag}(\tilde{Q}^T A\tilde{Q}).\tilde{Q}^{-1}$.

5.5 OPERATIONAL MATRICES OF INTEGRATION AND DIFFERENTIATION

In this section, we derive the operational matrices (OMs) of integration and differentiation with respect to 2D wavelets.

5.5.1 OM of Integration

The integration of 2D BWs $\Psi(x,t)$ with respect to x defined in Eq. (5.4) is given by [46]

$$\int_0^x \Psi(\alpha,t)\,d\alpha \cong \tilde{P}_x \Psi(x,t), \tag{5.10}$$

where $\tilde{P}_x$ is a $(2^{k_1-1}2^{k_2-1}M_1M_2 \times 2^{k_1-1}2^{k_2-1}M_1M_2)$-dimensional matrix.

Similarly, OM of integration for the variable t can be derived as follows [46]:

$$\int_0^t \Psi(x,\beta)\,d\beta \cong \tilde{P}_t' \Psi(x,t), \tag{5.11}$$

where $\tilde{P}_t'$ is a $(2^{k_1-1}2^{k_2-1}M_1M_2 \times 2^{k_1-1}2^{k_2-1}M_1M_2)$-dimensional matrix.

5.5.2 OM of Differentiation

The OM of differentiation of $\Psi(x,t)$ with respect to x defined in Eq. (5.4) is derived as follows [46]:

$$\frac{\partial}{\partial x}\Psi(x,t) \cong \tilde{D}_x \Psi(x,t), \tag{5.12}$$

where $\tilde{D}_x$ is a $(2^{k_1-1}2^{k_2-1}M_1M_2 \times 2^{k_1-1}2^{k_2-1}M_1M_2)$-dimensional matrix.

Similarly, OM for differentiation with respect to t can be computed as follows:

$$\frac{\partial}{\partial t}\Psi(x,t) \cong \tilde{D}_t' \Psi(x,t), \tag{5.13}$$

where $\tilde{D}_t'$ is a $(2^{k_1-1}2^{k_2-1}M_1M_2 \times 2^{k_1-1}2^{k_2-1}M_1M_2)$-dimensional matrix.

5.6 BERNOULLI WAVELET METHOD FOR LINEAR AND NONLINEAR TWO-DIMENSIONAL VIDES

In our scheme, the linear and nonlinear VIDEs have been solved by employing the OM approach derived by 2D wavelets.

5.6.1 Linear Case

For the linear VIDEs, let $H(u(y,z)) = u(y,z)$ in Eq. (5.1). Therefore, we have

$$u_{xx}(x,t)+u_{xt}(x,t)+u_{tt}(x,t)+u(x,t) = f(x,t)+\int_0^x\int_0^t k(x,t,y,z)u(y,z)\,dydz, \tag{5.14}$$

$$(x,t)\in[0,1]\times[0,1].$$

To discuss the proposed scheme for solving Eq. (5.14), we use the following approximations:

$$k(x,t,y,z) \approx \Psi^T(x_1,x_2)\tilde{K}\Psi(y,z) = \Psi^T(y,z)\tilde{K}^T\Psi(x,t), \tag{5.15}$$

$$f(x,t) \approx \Psi^T(x,t)\tilde{F} = \tilde{F}^T\Psi(x,t), \tag{5.16}$$

$$u_{xt}(x,t) \approx \Psi^T(x,t)\tilde{C} = \tilde{C}^T\Psi(x,t), \tag{5.17}$$

$$u(x,0) \approx \Psi^T(x,t)U = U^T\Psi(x,t), \tag{5.18}$$

$$u_t(0,t) \approx \Psi^T(x,t)V = V^T\Psi(x,t), \tag{5.19}$$

$$u_t(x,0) \approx \Psi^T(x,t)Y = Y^T\Psi(x,t), \tag{5.20}$$

$$u_x(0,t) \approx \Psi^T(x,t)Z = Z^T\Psi(x,t). \tag{5.21}$$

By integrating Eq. (5.17) with respect to x and by using Eq. (5.10), we have

$$\begin{aligned} u_t(x,t) &= u_t(0,t) + \tilde{C}^T\tilde{P}_x\Psi(x,t) \\ &= V^T\Psi(x,t) + \tilde{C}^T\tilde{P}_x\Psi(x,t) \\ &= (V^T + \tilde{C}^T\tilde{P}_x)\Psi(x,t) \\ &= \Psi^T(x,t)C_1, \end{aligned} \tag{5.22}$$

where $C_1 = V + \tilde{P}_x^T\tilde{C}$.

Again integrating Eq. (5.22) with respect to t and by using Eq. (5.11), we have

$$\begin{aligned} u(x,t) &= u(x,0) + {C_1}^T\tilde{P}_t'\Psi(x,t) \\ &= U^T\Psi(x,t) + {C_1}^T\tilde{P}_t'\Psi(x,t) \\ &= (U^T + {C_1}^T\tilde{P}_t')\Psi(x,t) \\ &= \Psi^T(x,t)C_2, \end{aligned} \tag{5.23}$$

where $C_2 = U + \tilde{P}_t'^T C_1$.

Let

$$u_{tt}(x,t) \approx \Psi^T(x,t)\tilde{D}_1 = \tilde{D}_1^T\Psi(x,t). \tag{5.24}$$

Now integrating Eq. (5.24) with respect to t and by using Eq. (5.11), we have

$$\begin{aligned} u_t(x,t) &= u_t(x,0) + \tilde{D}_1^T \tilde{P}_t' \Psi(x,t) \\ &= Y^T \Psi(x,t) + \tilde{D}_1^T \tilde{P}_t' \Psi(x,t) \\ &= (Y^T + \tilde{D}_1^T \tilde{P}_t') \Psi(x,t). \end{aligned} \tag{5.25}$$

From Eqs. (5.22) and (5.25), we get

$$\tilde{D}_1 = (\tilde{P}_t'^T)^{-1}(C_1 - Y). \tag{5.26}$$

Let

$$u_{xx}(x,t) \approx \Psi^T(x,t)\tilde{D}_2 = \tilde{D}_2^T \Psi(x,t). \tag{5.27}$$

Now integrating Eq. (5.27) with respect to x and by using Eq. (5.10), we have

$$\begin{aligned} u_x(x,t) &= u_x(0,t) + \tilde{D}_2^T \tilde{P}_x \Psi(x,t) \\ &= Z^T \Psi(x,t) + \tilde{D}_2^T \tilde{P}_x \Psi(x,t) \\ &= (Z^T + \tilde{D}_2^T \tilde{P}_x) \Psi(x,t). \end{aligned} \tag{5.28}$$

Differentiating Eq. (5.23) with respect to x and by using Eq. (5.12), we have

$$u_x(x,t) = C_2^T \tilde{D}_x \Psi(x,t). \tag{5.29}$$

From Eqs. (5.28) and (5.29), we get

$$D_2 = (\tilde{P}_x^T)^{-1}(\tilde{D}_x^T C_2 - Z). \tag{5.30}$$

Now the integral part of Eq. (5.14) can be derived as follows:

$$\begin{aligned} &\int_0^x \int_0^t k(x,t,y,z) u(y,z)\, dy\, dz \\ &= \int_0^x \int_0^t \Psi^T(x,t) \tilde{K} \Psi(y,z) \Psi^T(y,z) C_2\, dy\, dz \\ &= \Psi^T(x,t) \tilde{K} \int_0^x \int_0^t \Psi(y,z) \Psi^T(y,z) C_2\, dy\, dz \\ &= \Psi^T(x,t) \tilde{K} \int_0^x \int_0^t \tilde{C}_2 \Psi(y,z)\, dy\, dz \\ &= \Psi^T(x,t) \tilde{K} \tilde{C}_2 (P \otimes P) \Psi(x,t) \\ &= \Psi^T(x,t) C_3 \Psi(x,t), \end{aligned}$$

where $C_3 = \tilde{K}\tilde{C}_2(P \otimes P)$.

$$= \tilde{C}_3 \Psi(x,t). \tag{5.31}$$

Now by substituting Eqs. (5.15)–(5.31) into Eq. (5.14), we have

$$\Psi^T(x,t)\tilde{D}_2 + \Psi^T(x,t)\tilde{C} + \Psi^T(x,t)\tilde{D}_1 + \Psi^T(x,t)C_2 = \Psi^T(x,t)\tilde{F} + \Psi^T(x,t)\tilde{C}_3. \tag{5.32}$$

Collocating Eq. (5.32) at $2^{k_1-1}M_1 \times 2^{k_2-1}M_2$ points (x_i, t_j) as

$$\Psi^T(x_i,t_j)\tilde{D}_2 + \Psi^T(x_i,t_j)\tilde{C} + \Psi^T(x_i,t_j)\tilde{D}_1 + \Psi^T(x_i,t_j)C_2 = \Psi^T(x_i,t_j)\tilde{F} + \Psi^T(x_i,t_j)\tilde{C}_3, \tag{5.33}$$

where the collocation points x_i and t_j are given as follows:

$$x_i = \frac{2i-1}{2.2^{k_1-1}.M_1}, \quad i = 1,2,\ldots,2^{k_1-1}.M_1,$$

$$t_j = \frac{2j-1}{2.2^{k_2-1}.M_2}, \quad j = 1,2,\ldots,2^{k_2-1}.M_2.$$

Now by substituting the collocation points in Eq. (5.33), a system of $2^{k_1-1}2^{k_2-1}M_1M_2$-linear algebraic equations are formed. Furthermore, the value of $\tilde{C}$ is calculated by solving the above system. Then, by using the value of $\tilde{C}$, the value of C_2 can be obtained.

Therefore, by plugging the value of C_2 in Eq. (5.23), the approximate solution of Eq. (5.14) can be calculated numerically.

5.6.2 Nonlinear Case

For the nonlinear VIDEs, let $H(u(y,z)) = u^\lambda(y,z)$, $\lambda > 1$ in Eq. (5.1). Therefore, we have

$$u_{xx}(x,t) + u_{xt}(x,t) + u_{tt}(x,t) + u(x,t) = f(x,t) + \int_0^x \int_0^t k(x,t,y,z)u^\lambda(y,z)dydz, \tag{5.34}$$
$$(x,t) \in [0,1] \times [0,1].$$

Now in the case of $\lambda = 2$, we have

$$\begin{aligned}[u(x,t)]^2 &= (C_2^T\Psi(x,t)).(C_2^T\Psi(x,t)) \\ &= C_2^T\Psi(x,t)\Psi^T(x,t)C_2 \\ &= C_2^T\tilde{C}_2\Psi(x,t) \\ &= C_4^T\Psi(x,t),\end{aligned} \tag{5.35}$$

where $C_4^T = C_2^T \tilde{C}_2$ and $\tilde{C}_2 = \int_0^1 \int_0^1 \Psi(x,t)\Psi^T(x,t)C_2\Psi^T(x,t)\,dx\,dt$.

Now, the integral part of Eq. (5.34) can be derived as follows:

$$\int_0^x \int_0^t k(x,t,y,z)u^2(y,z)\,dy\,dz$$

$$= \int_0^x \int_0^t \Psi^T(x,t)\tilde{K}\Psi(y,z)\Psi^T(y,z)C_4\,dy\,dz$$

$$= \Psi^T(x,t)\tilde{K}\int_0^x \int_0^t \Psi(y,z)\Psi^T(y,z)C_4\,dy\,dz$$

$$= \Psi^T(x,t)\tilde{K}\int_0^x \int_0^t \tilde{C}_4\Psi(y,z)\,dy\,dz$$

$$= \Psi^T(x,t)\tilde{K}\tilde{C}_4(P\otimes P)\Psi(x,t)$$

$$= \Psi^T(x,t)C_5\Psi(x,t),$$

where $C_5 = \tilde{K}\tilde{C}_4(P\otimes P)$.

$$= \tilde{C}_5\Psi(x,t). \tag{5.36}$$

Now by substituting Eqs. (5.15)–(5.30) and Eq. (5.36) into Eq. (5.34), we have

$$\Psi^T(x,t)\tilde{D}_2 + \Psi^T(x,t)\tilde{C} + \Psi^T(x,t)\tilde{D}_1 + \Psi^T(x,t)C_2 = \Psi^T(x,t)\tilde{F} + \Psi^T(x,t)\tilde{C}_5. \tag{5.37}$$

Collocating Eq. (5.37) at $2^{k_1-1}M_1 \times 2^{k_2-1}M_2$ points (x_i,t_j) as

$$\Psi^T(x_i,t_j)\tilde{D}_2 + \Psi^T(x_i,t_j)\tilde{C} + \Psi^T(x_i,t_j)\tilde{D}_1 + \Psi^T(x_i,t_j)C_2 = \Psi^T(x_i,t_j)\tilde{F} + \Psi^T(x_i,t_j)\tilde{C}_5, \tag{5.38}$$

where the collocation points x_i and t_j are given as follows:

$$x_i = \frac{2i-1}{2.2^{k_1-1}.M_1}, \quad i = 1,2,\ldots,2^{k_1-1}.M_1,$$

$$t_j = \frac{2j-1}{2.2^{k_2-1}.M_2}, \quad j = 1,2,\ldots,2^{k_2-1}.M_2.$$

Now by substituting the collocation points in Eq. (5.38), a system of $2^{k_1-1}2^{k_2-1}M_1M_2$-nonlinear algebraic equations are formed. Furthermore, the value of $\tilde{C}$ is calculated by solving the above system. Then, by using the value of $\tilde{C}$, the value of C_2 can be obtained.

Therefore, by plugging the value of C_2 in Eq. (5.23), the approximate solution of Eq. (5.34) can be calculated numerically.

5.7 CONVERGENCE ANALYSIS

In the present analysis, we established that the expansion of any continuous function $u(x,t) \in L^2(\Omega)$, $\Omega = [0,1) \times [0,1)$ with respect to two-dimensional BWs, as given in Eq. (5.2), converges uniformly to u.

Theorem 5.7.1

Consider $u(x,t)$ be a continuous function in $L^2(\Omega)$ and $|u(x,t)| \le \tilde{\Upsilon}$, where $\tilde{\Upsilon}$ is a real and positive constant. Then, the BWs expansion of $u(x,t)$ converges uniformly to $u(x,t)$ and also

$$\left|\tilde{f}_{n,m,n',m'}\right| < \tilde{\Upsilon}\frac{\Pi_1\Pi_2}{2^{\frac{k_1-1}{2}}2^{\frac{k_2-1}{2}}}\frac{16m!}{(2\pi)^{m+1}}\frac{16m'!}{(2\pi)^{m'+1}},$$

where $\Pi_1 = \dfrac{1}{\sqrt{\dfrac{(-1)^{m-1}(m!)^2}{(2m)!}\alpha_{2m}}}$ and $\Pi_2 = \dfrac{1}{\sqrt{\dfrac{(-1)^{m'-1}(m'!)^2}{(2m')!}\alpha_{2m'}}}$.

Proof:

The BWs coefficients can be determined as follows:

$$\tilde{f}_{n,m,n',m'} = \langle\langle u(x,t)\psi_{n,m}(x)\rangle, \psi_{n',m'}(t)\rangle$$

$$= \int_0^1\int_0^1 u(x,t)\psi_{n,m}(x)\psi_{n',m'}(t)\,dx\,dt$$

$$= 2^{\frac{k_1-1}{2}}2^{\frac{k_2-1}{2}}\Pi_1\Pi_2\int_{\frac{n'-1}{2^{k_2-1}}}^{\frac{n'}{2^{k_2-1}}}\left[\int_{\frac{n-1}{2^{k_1-1}}}^{\frac{n}{2^{k_1-1}}} u(x,t)\tilde{\mathrm{B}}_m(2^{k_1-1}x-n+1)\,dx\right]\tilde{\mathrm{B}}_{m'}(2^{k_2-1}t-n'+1)dt.$$

Now, changing the variable $2^{k_1-1}x-n+1=\xi_1$, it follows

$$= 2^{\frac{k_1-1}{2}}2^{\frac{k_2-1}{2}}\Pi_1\Pi_2\frac{1}{2^{k_1-1}}\int_{\frac{n'-1}{2^{k_2-1}}}^{\frac{n'}{2^{k_2-1}}}\left[\int_0^1 u\left(\frac{\xi_1+n-1}{2^{k_1-1}},t\right)\tilde{\mathrm{B}}_m(\xi_1)\,d\xi_1\right]\tilde{\mathrm{B}}_{m'}(2^{k_2-1}t-n'+1)dt.$$

Similarly, changing the variable $2^{k_2-1}t-n'+1=\xi_2$, it follows

$$=2^{\frac{k_1-1}{2}}2^{\frac{k_2-1}{2}}\Pi_1\Pi_2\frac{1}{2^{k_1-1}2^{k_2-1}}\int_0^1\left[\int_0^1 u(\frac{\xi_1+n-1}{2^{k_1-1}},\frac{\xi_2+n'-1}{2^{k_2-1}})\tilde{\mathbf{B}}_m(\xi_1)\,d\xi_1\right]\tilde{\mathbf{B}}_{m'}(\xi_2)d\xi_2$$

$$=\frac{\Pi_1\Pi_2}{2^{\frac{k_1-1}{2}}2^{\frac{k_2-1}{2}}}\int_0^1\left[\int_0^1 u(\frac{\xi_1+n-1}{2^{k_1-1}},\frac{\xi_2+n'-1}{2^{k_2-1}})\tilde{\mathbf{B}}_m(\xi_1)\,d\xi_1\right]\tilde{\mathbf{B}}_{m'}(\xi_2)d\xi_2.$$

This implies

$$\left|\tilde{f}_{n,m,n',m'}\right|=\left|\frac{\Pi_1\Pi_2}{2^{\frac{k_1-1}{2}}2^{\frac{k_2-1}{2}}}\int_0^1\left[\int_0^1 u(\frac{\xi_1+n-1}{2^{k_1-1}},\frac{\xi_2+n'-1}{2^{k_2-1}})\tilde{\mathbf{B}}_m(\xi_1)\,d\xi_1\right]\tilde{\mathbf{B}}_{m'}(\xi_2)d\xi_2\right|$$

$$\leq\frac{\Pi_1\Pi_2}{2^{\frac{k_1-1}{2}}2^{\frac{k_2-1}{2}}}\int_0^1\int_0^1\left|u(\frac{\xi_1+n-1}{2^{k_1-1}},\frac{\xi_2+n'-1}{2^{k_2-1}})\right|\left|\tilde{\mathbf{B}}_m(\xi_1)\right|d\xi_1\left|\tilde{\mathbf{B}}_{m'}(\xi_2)\right|d\xi_2$$

$$\leq\tilde{\Upsilon}\frac{\Pi_1\Pi_2}{2^{\frac{k_1-1}{2}}2^{\frac{k_2-1}{2}}}\left(\int_0^1\left|\tilde{\mathbf{B}}_m(\xi_1)\right|d\xi_1\right)\left(\int_0^1\left|\tilde{\mathbf{B}}_{m'}(\xi_2)\right|d\xi_2\right)$$

$$\leq\tilde{\Upsilon}\frac{\Pi_1\Pi_2}{2^{\frac{k_1-1}{2}}2^{\frac{k_2-1}{2}}}\frac{16m!}{(2\pi)^{m+1}}\frac{16m'!}{(2\pi)^{m'+1}},\ \text{using the property of Bernoulli polynomials.}\tag{5.39}$$

Hence, the series $\sum_{n=1}^{2^{k_1-1}}\sum_{n'=1}^{2^{k_2-1}}\sum_{m=0}^{M_1-1}\sum_{m'=0}^{M_2-1}\tilde{f}_{n,m,n',m'}$ is absolutely convergent, and it follows that the series $\sum_{n=1}^{2^{k_1-1}}\sum_{n'=1}^{2^{k_2-1}}\sum_{m=0}^{M_1-1}\sum_{m'=0}^{M_2-1}\tilde{f}_{n,m,n',m'}\cdot\psi_{n,m,n',m'}(x,t)$ is uniformly convergent.

5.8 ERROR ANALYSIS

In this part, the error analysis of the Bernoulli wavelet method (BWM) for solving the 2D VIDE has been derived.

Theorem 5.8.1

Let $\tilde{u}_{2^{k_1-1}M_1,2^{k_2-1}M_2}(x,t)$ be the 2D BWs expansion of the real sufficiently smooth function $u(x,t)\in C^\infty(\Omega)\cap L^2(\Omega)$. Then, there exist real numbers $\Re_1,\Re_2$ and $\Re_3$ such that

$$\left\|u(x,t)-\tilde{u}_{2^{k_1-1}M_1,2^{k_2-1}M_2}(x,t)\right\|_2 \le \frac{\Re_1}{2^{2^{k_1}M_1-1}\Gamma(2^{k_1-1}M_1+1)}+\frac{\Re_2}{2^{2^{k_2}M_2-1}\Gamma(2^{k_2-1}M_2+1)}$$
$$+\frac{\Re_3}{2^{2^{k_1}M_1+2^{k_2}M_2-2}\Gamma(2^{k_1-1}M_1+1)\Gamma(2^{k_2-1}M_2+1)}. \tag{5.40}$$

Proof:

Assume that $u(x,t)$ is a sufficiently smooth function on $C^{\infty}(\Omega)\cap L^2(\Omega)$, and $\tilde{u}_{2^{k_1-1}M_1,2^{k_2-1}M_2}(x,t)$ is the best approximation of u from $\Theta_{2^{k_1-1}M_1,2^{k_2-1}M_2}$, where $\Theta_{2^{k_1-1}M_1,2^{k_2-1}M_2}=\text{Span}\{\psi_{n,m,n',m'}(x,t): n=1,2,\ldots,2^{k_1-1}, m=0,1,\ldots,M_1-1; n'=1,2,\ldots,2^{k_2-1}, m'=0,1,\ldots,M_2-1\}$.

Then, for any $P_{2^{k_1-1}M_1,2^{k_2-1}M_2}[u]\in\Theta_{2^{k_1-1}M_1,2^{k_2-1}M_2}$, the following inequality is held:

$$\left\|u(x,t)-\tilde{u}_{2^{k_1-1}M_1,2^{k_2-1}M_2}(x,t)\right\|_2 \le \left\|u(x,t)-P_{2^{k_1-1}M_1,2^{k_2-1}M_2}[u]\right\|_2, \tag{5.41}$$

where $P_{2^{k_1-1}M_1,2^{k_2-1}M_2}[u]$ is the interpolating polynomial of u at the node points (x_p,t_q).

Here, $x_p, p=1,2,\ldots,2^{k_1-1}M_1$ and $t_q, q=1,2,\ldots,2^{k_2-1}M_2$ are the roots of the $2^{k_1-1}M_1$-degree and $2^{k_2-1}M_2$-degree shifted Chebyshev polynomial in $[0,1]$, respectively.

Now,

$$u(x,t)-P_{2^{k_1-1}M_1,2^{k_2-1}M_2}[u]=\frac{\partial^{2^{k_1-1}M_1}u(\zeta_1,t)}{\partial x^{2^{k_1-1}M_1}}\frac{\prod_{p=1}^{2^{k_1-1}M_1}(x-x_p)}{\Gamma(2^{k_1-1}M_1+1)}$$
$$+\frac{\partial^{2^{k_2-1}M_2}u(x,\zeta_2)}{\partial t^{2^{k_2-1}M_2}}\frac{\prod_{q=1}^{2^{k_2-1}M_2}(t-t_q)}{\Gamma(2^{k_2-1}M_2+1)} \tag{5.42}$$
$$+\frac{\partial^{2^{k_1-1}M_1+2^{k_2-1}M_2}u(\tilde{\zeta}_1,\tilde{\zeta}_2)}{\partial x^{2^{k_1-1}M_1}\partial t^{2^{k_2-1}M_2}}\frac{\prod_{p=1}^{2^{k_1-1}M_1}(x-x_p)\prod_{q=1}^{2^{k_2-1}M_2}(t-t_q)}{\Gamma(2^{k_1-1}M_1+1)\Gamma(2^{k_2-1}M_2+1)},$$

for some $\zeta_1,\zeta_2,\tilde{\zeta}_1,\tilde{\zeta}_2\in[0,1]$.

Now, utilizing the estimates for Chebyshev interpolating nodes on the interval Ω, it follows that [47]

$$\left|\prod_{p=1}^{2^{k_1-1}M_1}(x-x_p)\right|\le\frac{1}{2^{2^{k_1}M_1-1}}$$

and

$$\left|\prod_{q=1}^{2^{k_2-1}M_2}(t-t_q)\right| \le \frac{1}{2^{2^{k_2}M_2-1}}.$$

Therefore,

$$\begin{aligned}\left|u(x,t)-P_{2^{k_1-1}M_1,2^{k_2-1}M_2}[u]\right| \le{} & \sup_{t\in[0,1]}\left(\max_{\zeta_1\in[0,1]}\left|\frac{\partial^{2^{k_1-1}M_1}u(\zeta_1,t)}{\partial x^{2^{k_1-1}M_1}}\right|\right)\frac{1}{2^{2^{k_1}M_1-1}\Gamma(2^{k_1-1}M_1+1)}\\ &+\sup_{x\in[0,1]}\left(\max_{\zeta_2\in[0,1]}\left|\frac{\partial^{2^{k_2-1}M_2}u(x,\zeta_2)}{\partial t^{2^{k_2-1}M_2}}\right|\right)\frac{1}{2^{2^{k_2}M_2-1}\Gamma(2^{k_2-1}M_2+1)}\\ &+\sup_{(x,t)\in\Omega}\left(\max_{(\tilde{\zeta}_1,\tilde{\zeta}_2)\in\Omega}\left|\frac{\partial^{2^{k_1-1}M_1+2^{k_2-1}M_2}u(\tilde{\zeta}_1,\tilde{\zeta}_2)}{\partial x^{2^{k_1-1}M_1}\partial t^{2^{k_2-1}M_2}}\right|\right)\\ &\frac{1}{2^{2^{k_1}M_1+2^{k_2}M_2-2}\Gamma(2^{k_1-1}M_1+1)\Gamma(2^{k_2-1}M_2+1)},\end{aligned}$$

whence

$$\begin{aligned}\left|u(x,t)-P_{2^{k_1-1}M_1,2^{k_2-1}M_2}[u]\right| \le{} & \sup_{(x,t)\in\Omega}\left|\frac{\partial^{2^{k_1-1}M_1}u(x,t)}{\partial x^{2^{k_1-1}M_1}}\right|\frac{1}{2^{2^{k_1}M_1-1}\Gamma(2^{k_1-1}M_1+1)}\\ &+\sup_{(x,t)\in\Omega}\left|\frac{\partial^{2^{k_2-1}M_2}u(x,t)}{\partial t^{2^{k_2-1}M_2}}\right|\frac{1}{2^{2^{k_2}M_2-1}\Gamma(2^{k_2-1}M_2+1)}\\ &+\sup_{(x,t)\in\Omega}\left|\frac{\partial^{2^{k_1-1}M_1+2^{k_2-1}M_2}u(x,t)}{\partial x^{2^{k_1-1}M_1}\partial t^{2^{k_2-1}M_2}}\right|\\ &\frac{1}{2^{2^{k_1}M_1+2^{k_2}M_2-2}\Gamma(2^{k_1-1}M_1+1)\Gamma(2^{k_2-1}M_2+1)}.\end{aligned} \tag{5.43}$$

In particular, from Eq. (5.41),

$$\begin{aligned}\left\|u(x,t)-\tilde{u}_{2^{k_1-1}M_1,2^{k_2-1}M_2}(x,t)\right\|_2^2 &\le \left\|u(x,t)-P_{2^{k_1-1}M_1,2^{k_2-1}M_2}[u]\right\|_2^2\\ &=\int_0^1\int_0^1\left|u(x,t)-P_{2^{k_1-1}M_1,2^{k_2-1}M_2}[u]\right|^2 dx\,dt.\end{aligned} \tag{5.44}$$

Now, using Eqs. (5.43) and (5.44), it follows that

$$\left\|u(x,t)-\tilde{u}_{2^{k_1-1}M_1,2^{k_2-1}M_2}(x,t)\right\|_2^2 \leq \left[\frac{\Re_1}{2^{2^{k_1}M_1-1}\Gamma(2^{k_1-1}M_1+1)}+\frac{\Re_2}{2^{2^{k_2}M_2-1}\Gamma(2^{k_2-1}M_2+1)} + \frac{\Re_3}{2^{2^{k_1}M_1+2^{k_2}M_2-2}\Gamma(2^{k_1-1}M_1+1)\Gamma(2^{k_2-1}M_2+1)}\right]^2. \tag{5.45}$$

Assuming $\sup_{(x,t)\in\Omega}\left|\frac{\partial^{2^{k_1-1}M_1}u(x,t)}{\partial x^{2^{k_1-1}M_1}}\right| \leq \Re_1$, $\sup_{(x,t)\in\Omega}\left|\frac{\partial^{2^{k_2-1}M_2}u(x,t)}{\partial t^{2^{k_2-1}M_2}}\right| \leq \Re_2$, and $\sup_{(x,t)\in\Omega}\left|\frac{\partial^{2^{k_1-1}M_1+2^{k_2-1}M_2}u(x,t)}{\partial x^{2^{k_1-1}M_1}\partial t^{2^{k_2-1}M_2}}\right| \leq \Re_3$.

Therefore,

$$\left\|u(x,t)-\tilde{u}_{2^{k_1-1}M_1,2^{k_2-1}M_2}(x,t)\right\|_2 \leq \frac{\Re_1}{2^{2^{k_1}M_1-1}\Gamma(2^{k_1-1}M_1+1)}+\frac{\Re_2}{2^{2^{k_2}M_2-1}\Gamma(2^{k_2-1}M_2+1)} + \frac{\Re_3}{2^{2^{k_1}M_1+2^{k_2}M_2-2}\Gamma(2^{k_1-1}M_1+1)\Gamma(2^{k_2-1}M_2+1)}.$$

Hence, the theorem is proved. ■

Theorem 5.8.2

Let $\tilde{u}_{2^{k_1-1}M_1,2^{k_2-1}M_2}(x,t) \approx \sum_{n=1}^{2^{k_1-1}}\sum_{n'=1}^{2^{k_2-1}}\sum_{m=0}^{M_1-1}\sum_{m'=0}^{M_2-1}\tilde{f}_{n,m,n',m'}\cdot\psi_{n,m,n',m'}(x,t)$ is the best approximation of u defined on $L^2(\Omega)$, and $\overset{*}{u}_{2^{k_1-1}M_1,2^{k_2-1}M_2}(x,t) \approx \sum_{n=1}^{2^{k_1-1}}\sum_{n'=1}^{2^{k_2-1}}\sum_{m=0}^{M_1-1}\sum_{m'=0}^{M_2-1}\tilde{f}^{*}_{n,m,n',m'}\cdot\psi_{n,m,n',m'}(x,t)$is the numerical approximation u defined on $L^2(\Omega)$. Then [48],

$$\left\|\tilde{u}_{2^{k_1-1}M_1,2^{k_2-1}M_2}(x,t)-\overset{*}{u}_{2^{k_1-1}M_1,2^{k_2-1}M_2}(x,t)\right\|_2 \leq \mathcal{M}.$$

Proof:

$$\left\| \tilde{u}_{2^{k_1-1}M_1,2^{k_2-1}M_2}(x,t) - u^*_{2^{k_1-1}M_1,2^{k_2-1}M_2}(x,t) \right\|_2^2$$

$$\leq \int_0^1\int_0^1 \left| \tilde{u}_{2^{k_1-1}M_1,2^{k_2-1}M_2}(x,t) - u^*_{2^{k_1-1}M_1,2^{k_2-1}M_2}(x,t) \right|^2 dx\,dt$$

$$\leq \int_0^1\int_0^1 \left| \sum_{n=1}^{2^{k_1-1}}\sum_{n'=1}^{2^{k_2-1}}\sum_{m=0}^{M_1-1}\sum_{m'=0}^{M_2-1} \tilde{f}_{n,m,n',m'}.\psi_{n,m,n',m'}(x,t) - \sum_{n=1}^{2^{k_1-1}}\sum_{n'=1}^{2^{k_2-1}}\sum_{m=0}^{M_1-1}\sum_{m'=0}^{M_2-1} \tilde{f}^*_{n,m,n',m'}.\psi_{n,m,n',m'}(x,t) \right|^2 dxdt$$

$$\leq \int_0^1\int_0^1 \left| \sum_{n=1}^{2^{k_1-1}}\sum_{n'=1}^{2^{k_2-1}}\sum_{m=0}^{M_1-1}\sum_{m'=0}^{M_2-1} (\tilde{f}_{n,m,n',m'} - \tilde{f}^*_{n,m,n',m'}).\psi_{n,m,n',m'}(x,t) \right|^2 dxdt$$

$$\leq \sum_{n=1}^{2^{k_1-1}}\sum_{n'=1}^{2^{k_2-1}}\sum_{m=0}^{M_1-1}\sum_{m'=0}^{M_2-1} \left| \tilde{f}_{n,m,n',m'} - \tilde{f}^*_{n,m,n',m'} \right|^2 . \int_0^1\int_0^1 \left| \psi_{n,m,n',m'}(x,t) \right|^2 dxdt$$

$$\leq \sum_{n=1}^{2^{k_1-1}}\sum_{n'=1}^{2^{k_2-1}}\sum_{m=0}^{M_1-1}\sum_{m'=0}^{M_2-1} \left| \tilde{f}_{n,m,n',m'} - \tilde{f}^*_{n,m,n',m'} \right|^2 .$$

This implies

$$\left\| \tilde{u}_{2^{k_1-1}M_1,2^{k_2-1}M_2}(x,t) - u^*_{2^{k_1-1}M_1,2^{k_2-1}M_2}(x,t) \right\|_2 \leq \left(\sum_{n=1}^{2^{k_1-1}}\sum_{n'=1}^{2^{k_2-1}}\sum_{m=0}^{M_1-1}\sum_{m'=0}^{M_2-1} \left| \tilde{f}_{n,m,n',m'} - \tilde{f}^*_{n,m,n',m'} \right|^2 \right)^{\frac{1}{2}}$$

$$= \boldsymbol{\mathcal{M}}, \tag{5.46}$$

where $\boldsymbol{\mathcal{M}} = \left(\sum_{n=1}^{2^{k_1-1}}\sum_{n'=1}^{2^{k_2-1}}\sum_{m=0}^{M_1-1}\sum_{m'=0}^{M_2-1} \left| \tilde{f}_{n,m,n',m'} - \tilde{f}^*_{n,m,n',m'} \right|^2 \right)^{\frac{1}{2}}$.

Theorem 5.8.3

Let $\overset{*}{u}_{2^{k_1-1}M_1,2^{k_2-1}M_2}(x,t) \approx \sum_{n=1}^{2^{k_1-1}} \sum_{n'=1}^{2^{k_2-1}} \sum_{m=0}^{M_1-1} \sum_{m'=0}^{M_2-1} \tilde{f}^{*}_{n,m,n',m'} \cdot \psi_{n,m,n',m'}(x,t)$ be our numerical approximation defined on $L^2(\Omega)$. Then,

$$\left\| u(x,t) - \overset{*}{u}_{2^{k_1-1}M_1,2^{k_2-1}M_2}(x,t) \right\|_2 \leq \tilde{\mathcal{M}} + \mathcal{M},$$

where $\tilde{\mathcal{M}} = \dfrac{\Re_1}{2^{2^{k_1}M_1-1}\Gamma(2^{k_1-1}M_1+1)} + \dfrac{\Re_2}{2^{2^{k_2}M_2-1}\Gamma(2^{k_2-1}M_2+1)}$

$$+ \frac{\Re_3}{2^{2^{k_1}M_1+2^{k_2}M_2-2}\Gamma(2^{k_1-1}M_1+1)\Gamma(2^{k_2-1}M_2+1)}.$$

Proof:

By using Theorem 5.8.1 and Theorem 5.8.2, it follows

$$\begin{aligned}
\left\| u(x,t) - \overset{*}{u}_{2^{k_1-1}M_1,2^{k_2-1}M_2}(x,t) \right\|_2 &\leq \left\| u(x,t) - \tilde{u}_{2^{k_1-1}M_1,2^{k_2-1}M_2}(x,t) \right\|_2 + \\
&\quad \left\| \tilde{u}_{2^{k_1-1}M_1,2^{k_2-1}M_2}(x,t) - \overset{*}{u}_{2^{k_1-1}M_1,2^{k_2-1}M_2}(x,t) \right\|_2 \\
&\leq \frac{\Re_1}{2^{2^{k_1}M_1-1}\Gamma(2^{k_1-1}M_1+1)} + \frac{\Re_2}{2^{2^{k_2}M_2-1}\Gamma(2^{k_2-1}M_2+1)} \\
&\quad + \frac{\Re_3}{2^{2^{k_1}M_1+2^{k_2}M_2-2}\Gamma(2^{k_1-1}M_1+1)\Gamma(2^{k_2-1}M_2+1)} + \mathcal{M} \\
&\leq \tilde{\mathcal{M}} + \mathcal{M}.
\end{aligned} \tag{5.47}$$

5.9 NUMERICAL EXPERIMENTS

In this section, the accuracy of the approached scheme has been examined by choosing some examples of linear and nonlinear VIDEs. In addition, the efficiency of the approached method is measured with the aid of the absolute error defined by

$$\tilde{e}(x,t) = |u(x,t) - \tilde{u}(x,t)|,$$

where $\tilde{u}(x,t)$ is the approximate solution and $u(x,t)$ is the exact solution.

Furthermore, the absolute error values of $\tilde{e}(x,t)$ obtained by BWM and Legendre wavelet method (LWM) are presented in Tables 5.1, 5.2, 5.4, 5.5, 5.7, 5.10, and 5.11, respectively. Moreover, the L^2 and L^∞ error values for the of 2D VIDEs have been presented in Tables 5.3, 5.6, 5.8, and 5.12, respectively.

For a fixed value of x, the L^2 and L^∞-error norms for BWs and LWs are defined as follows:

$$L_2 \text{ error} = \left[\frac{1}{2^{k-1}M}\sum_{i=1}^{2^{k-1}M}\left(u(x,t_i)-u_m(x,t_i)\right)^2\right]^{\frac{1}{2}}$$

and

$$L_\infty \text{ error} = \underset{1\le i\le 2^{k-1}M}{\text{Max}} \left|u(x,t_i)-u_m(x,t_i)\right|,$$

respectively, where $u(x,t_i)$ and $u_m(x,t_i)$ are the exact and approximate solutions, respectively, and $t_i, i = 1,2,\ldots,2^{k-1}M$ are the collocation points. Moreover, the time complexity analysis of numerical experiments is presented in Table 5.13.

Example 5.9.1

Consider the following linear 2D VIDE:

$$u_{xt}(x,t)+u_t(x,t)+\int_0^x\int_0^t u(y,z)\,dy\,dz = f(x,t), \tag{5.48}$$

where

$$f(x,t) = 2t+\frac{t^3x}{3}+\frac{x^3t}{3},$$

with the boundary conditions as follows:

$$u(x,0) = x^2,\; u_t(0,t) = 2t.$$

The exact solution is $u(x,t) = x^2+t^2$.

For $k_1 = 2, k_2 = 2, M_1 = 3, M_2 = 3$, the comparison of approximate solution with exact solution and error analysis is depicted in Figures 5.1, 5.3 and 5.2, 5.4 for BWM and LWM, respectively, and for $k_1 = 3, k_2 = 3, M_1 = 3, M_2 = 3$, the comparison of approximate solution with exact solution and error analysis is depicted in Figures 5.5, 5.7 and 5.6, 5.8 for BWM and LWM, respectively.

TABLE 5.1
Absolute Error Values for Example 5.9.1 with $k = 2, M = 3$

(x,t)	$\tilde{e}(x,t)$ BWM	$\tilde{e}(x,t)$ LWM
(0.01,0.01)	5.71917E-18	5.71917E-18
(0.02,0.02)	4.65123E-17	1.87567E-17
(0.03,0.03)	3.16587E-17	3.16587E-17
(0.04,0.04)	5.63785E-18	5.63785E-18
(0.05,0.05)	5.20417E-17	5.20417E-17
(0.06,0.06)	4.51028E-17	4.51028E-17
(0.07,0.07)	5.37764E-17	3.98986E-17
(0.08,0.08)	5.20417E-18	5.20417E-18
(0.09,0.09)	6.93889E-18	6.93889E-18

TABLE 5.2
Absolute Error Values for Example 5.9.1 with $k = 3, M = 3$

(x,t)	$\tilde{e}(x,t)$ BWM	$\tilde{e}(x,t)$ LWM
(0.01,0.01)	1.50975E-17	8.15862E-18
(0.02,0.02)	1.18178E-17	1.52873E-17
(0.03,0.03)	1.0842E-17	1.0842E-17
(0.04,0.04)	1.25767E-17	1.25767E-17
(0.05,0.05)	1.04083E-17	1.04083E-17
(0.06,0.06)	1.73472E-17	1.73472E-17
(0.07,0.07)	1.21431E-17	1.21431E-17
(0.08,0.08)	1.04083E-18	8.67362E-18
(0.09,0.09)	6.93889E-18	1.38778E-17

TABLE 5.3
Comparison of L^2 and L^∞ Error Values for Example 5.9.1

	BWM				LWM			
	$k=2, M=3$		$k=3, M=3$		$k=2, M=3$		$k=3, M=3$	
t	L^2	L^∞	L^2	L^∞	L^2	L^∞	L^2	L^∞
0.2	1.72312E-16	2.77556E-16	4.72321E-17	1.11022E-16	1.72312E-16	2.77556E-16	3.73671E-17	1.11022E-16
0.4	1.87564E-16	3.33067E-16	5.00371E-17	1.11022E-16	1.87564E-16	3.33067E-16	5.25405E-17	1.11022E-16
0.6	1.46869E-16	2.22045E-16	8.32667E-17	2.22045E-16	1.06296E-16	2.22045E-16	8.32667E-17	2.22045E-16
0.8	1.28198E-16	2.22045E-16	1.35974E-16	2.22045E-16	9.06493E-17	2.22045E-16	1.19918E-16	2.22045E-16

For $k = 2, M = 3$, Figures 5.1, 5.3 and 5.2, 5.4 are obtained by BWM and LWM, respectively.

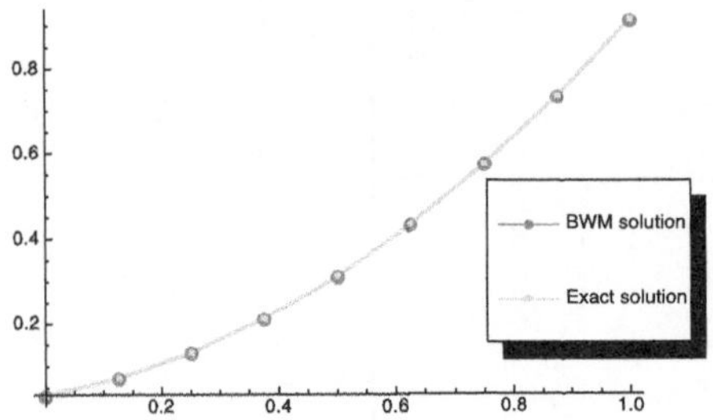

FIGURE 5.1 Comparison of BWM solution with exact solution for function $u(x,t)$.

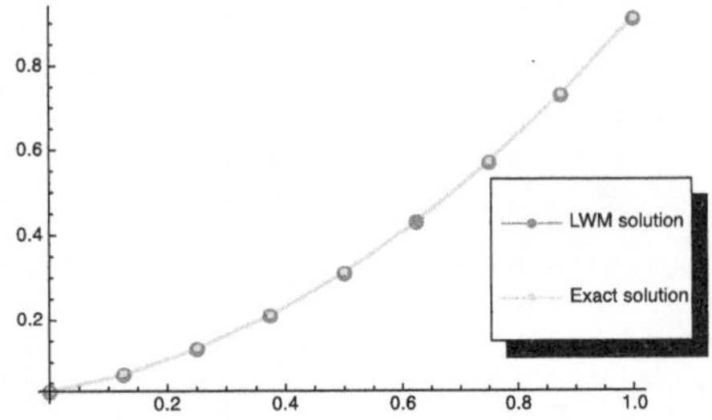

FIGURE 5.2 Comparison of LWM solution with exact solution for function $u(x,t)$.

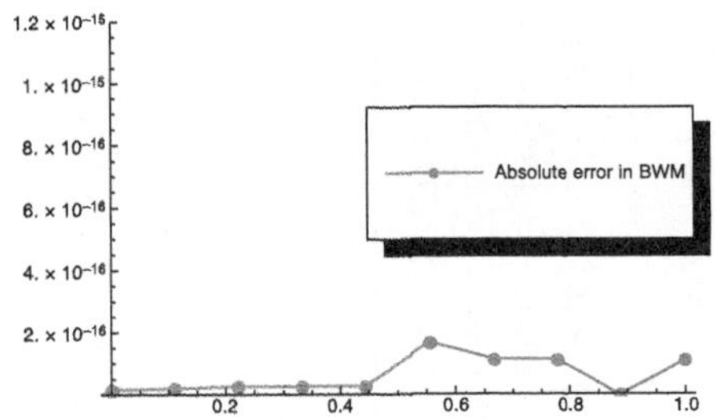

FIGURE 5.3 Error distribution for function $\tilde{e}(x,t)$ in BWM.

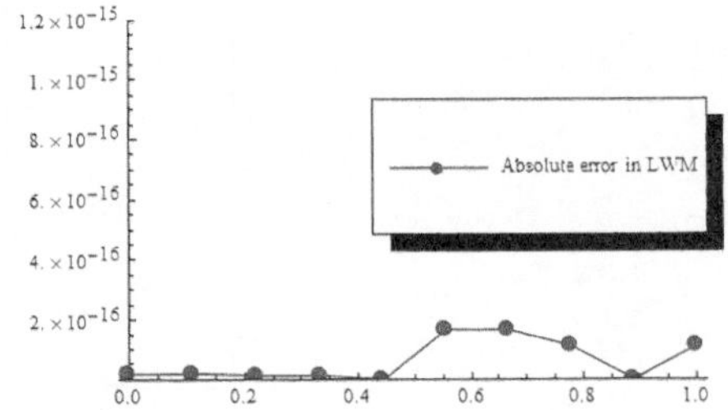

FIGURE 5.4 Error distribution for function $\tilde{e}(x,t)$ in LWM.

For $k = 3$, $M = 3$, Figures 5.5, 5.6 and 5.7, 5.8 are obtained by BWM and LWM, respectively.

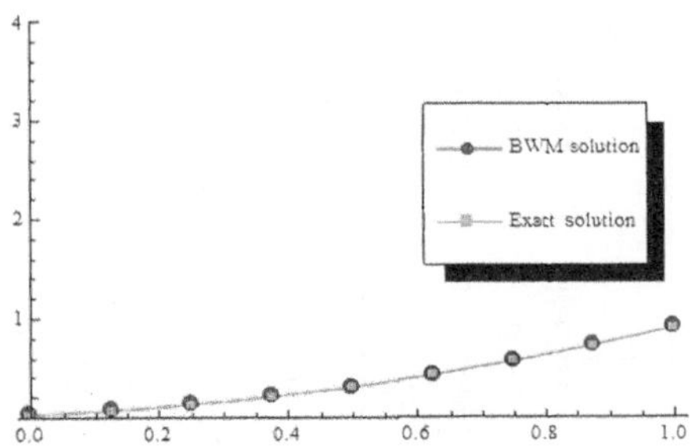

FIGURE 5.5 Comparison of BWM solution with exact solution for function $u(x,t)$.

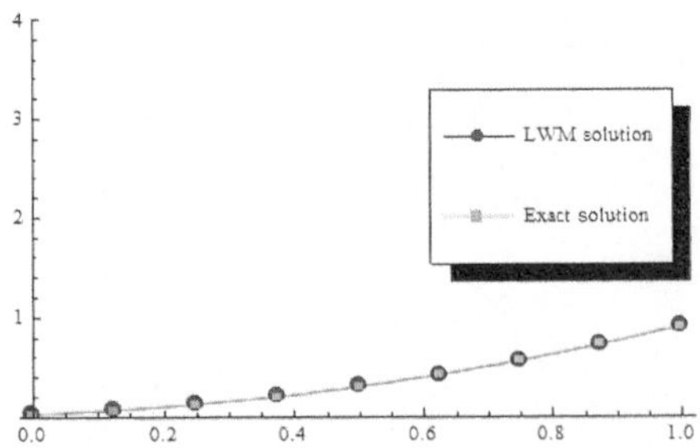

FIGURE 5.6 Comparison of LWM solution with exact solution for function $u(x,t)$.

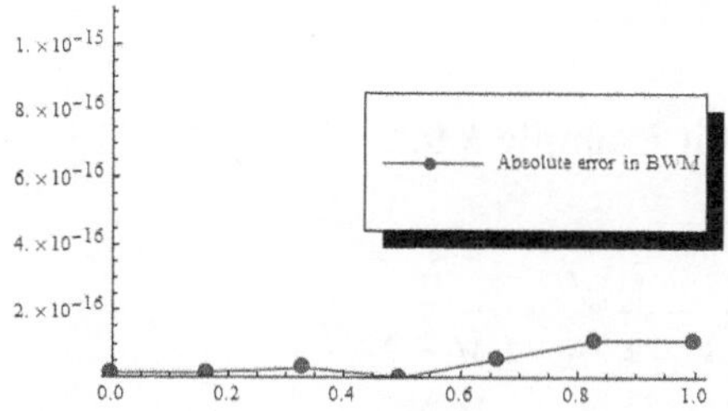

FIGURE 5.7 Error distribution for function $\tilde{e}(x,t)$ in BWM.

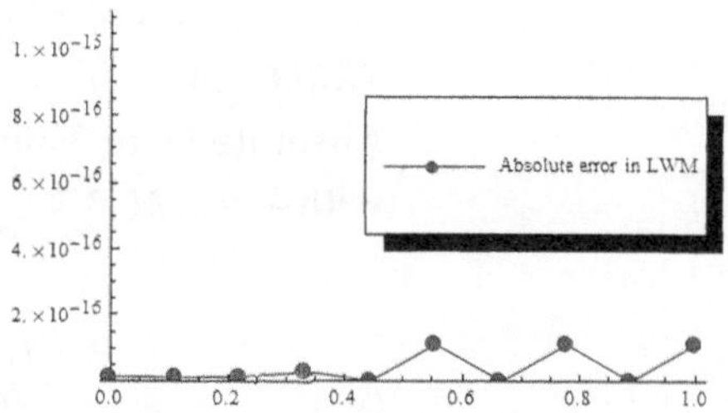

FIGURE 5.8 Error distribution for function $\tilde{e}(x,t)$ in LWM.

Example 5.9.2

Consider the nonlinear 2D VIDE:

$$u_{xt}(x,t)+u_t(x,t)+\int_0^x\int_0^t u^3(y,z)\,dy\,dz=f(x,t), \tag{5.49}$$

where

$$f(x,t)=2t+\frac{x^7t}{7}+\frac{x^3t^5}{5}+\frac{t^3x^5}{5}+\frac{t^7x}{7},$$

with the boundary conditions as follows:

$$u(x,0)=x^2,\ u_t(0,t)=2t.$$

The exact solution is $u(x,t)=x^2+t^2$.

For $k_1=2,k_2=2,M_1=2,M_2=2$, the comparison of approximate solution with exact solution and error analysis is depicted in Figures 5.9, 5.11 and 5.10, 5.12 for BWM and LWM, respectively, and for $k_1=2,k_2=2,M_1=3,M_2=3$, the comparison of approximate solution with exact solution and error analysis is depicted in Figures 5.13, 5.15 and 5.14, 5.16 for BWM and LWM, respectively.

TABLE 5.4
Absolute Error Values for Example 5.9.2 with $k = 2, M = 2$

	$\tilde{e}(x,t)$	
	$k_1 = 2, k_2 = 2, M_1 = 2, M_2 = 2$	
(x,t)	BWM	LWM
(0.125,0.125)	1.04046E-2	1.04046E-2
(0.125,0.375)	1.03852E-2	1.03852E-2
(0.125,0.625)	1.03797E-2	1.03797E-2
(0.125,0.875)	1.03834E-2	1.03834E-2
(0.375,0.125)	1.03814E-2	1.03814E-2
(0.375,0.375)	1.03549E-2	1.03549E-2
(0.375,0.625)	1.03696E-2	1.03696E-2
(0.375,0.875)	1.03732E-2	1.03732E-2
(0.625,0.125)	1.03062E-2	1.03062E-2
(0.625,0.375)	1.01428E-2	1.01428E-2
(0.625,0.625)	1.01166E-2	1.01166E-2
(0.625,0.875)	1.01943E-2	1.01943E-2
(0.875,0.125)	1.03127E-2	1.03127E-2
(0.875,0.375)	1.01933E-2	1.01933E-2
(0.875,0.625)	1.02071E-2	1.02071E-2
(0.875,0.875)	1.02302E-2	1.02302E-2

TABLE 5.5
Absolute Error Values for Example 5.9.2 with $k = 2, M = 3$

	$\tilde{e}(x,t)$	
(x,t)	BWM	LWM
(0.01,0.01)	5.8514E-7	5.8514E-7
(0.02,0.02)	9.93551E-7	9.93551E-7
(0.03,0.03)	1.47438E-6	1.47438E-6
(0.04,0.04)	2.02061E-6	2.02061E-6
(0.05,0.05)	2.62518E-6	2.62518E-6
(0.06,0.06)	3.28105E-6	3.28105E-6
(0.07,0.07)	3.98115E-6	3.98115E-6
(0.08,0.08)	4.71839E-6	4.71839E-6
(0.09,0.09)	5.48569E-6	5.48569E-6

TABLE 5.6
Comparison of L^2 and L^∞ Error Values for Example 5.9.2

	BWM				LWM			
	$k = 2, M = 2$		$k = 2, M = 3$		$k = 2, M = 2$		$k = 2, M = 3$	
t	L^2	L^∞	L^2	L^∞	L^2	L^∞	L^2	L^∞
0.2	2.34517E-2	2.35238E-2	2.9197E-5	5.8981E-5	2.34517E-2	2.35238E-2	2.9197E-5	5.8981E-5
0.4	3.38758E-3	3.50827E-3	5.82488E-5	1.19338E-4	3.38758E-3	3.50827E-3	5.82488E-5	1.19338E-4
0.6	3.39248E-3	3.50433E-3	7.75009E-5	1.61386E-4	3.39248E-3	3.50433E-3	7.75009E-5	1.61386E-4
0.8	2.34123E-2	2.35073E-2	8.53284E-5	1.73825E-4	2.34123E-2	2.35073E-2	8.53284E-5	1.73825E-4

For $k = 2$, $M = 2$, Figures 5.9, 5.11 and 5.10, 5.12 are obtained by BWM and LWM, respectively.

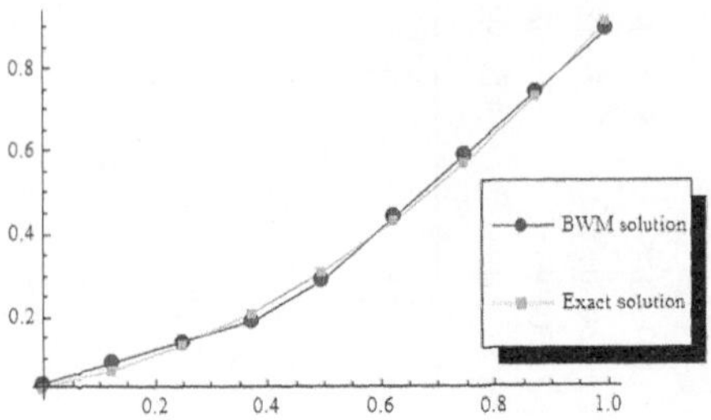

FIGURE 5.9 Comparison of BWM solution with exact solution for function $u(x,t)$.

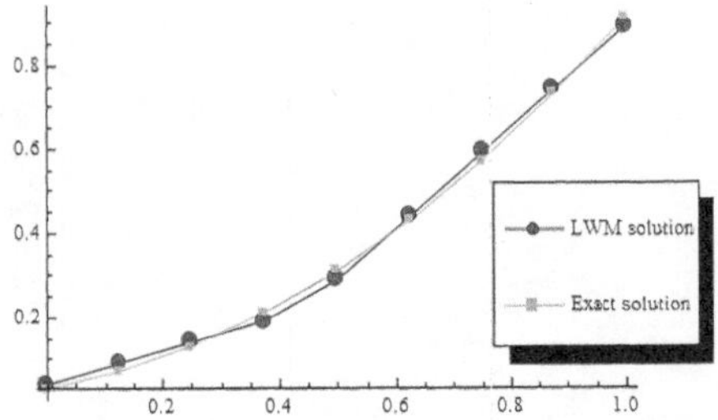

FIGURE 5.10 Comparison of LWM solution with exact solution for function $u(x,t)$.

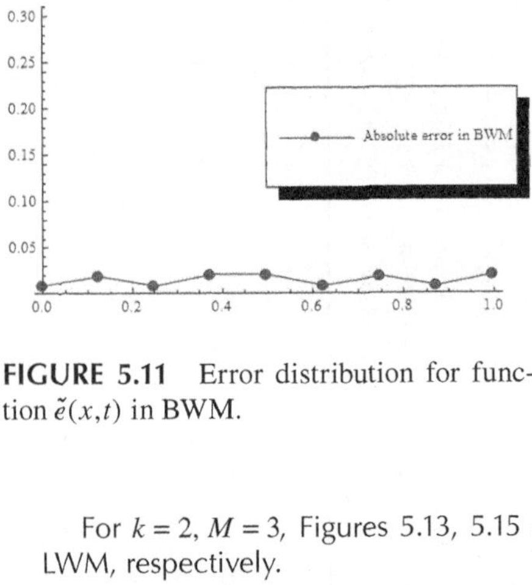

FIGURE 5.11 Error distribution for function $\tilde{e}(x,t)$ in BWM.

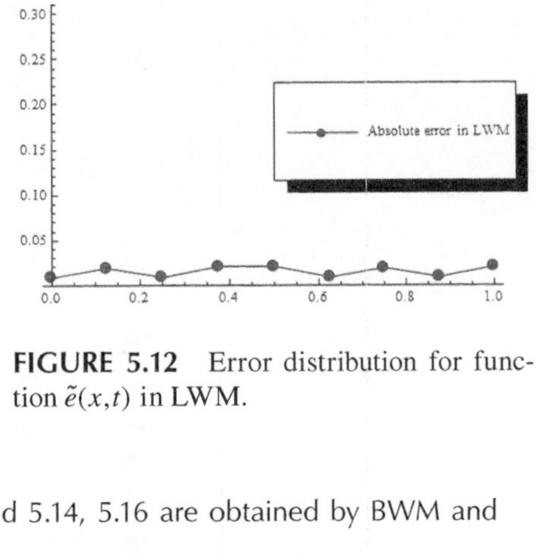

FIGURE 5.12 Error distribution for function $\tilde{e}(x,t)$ in LWM.

For $k = 2$, $M = 3$, Figures 5.13, 5.15 and 5.14, 5.16 are obtained by BWM and LWM, respectively.

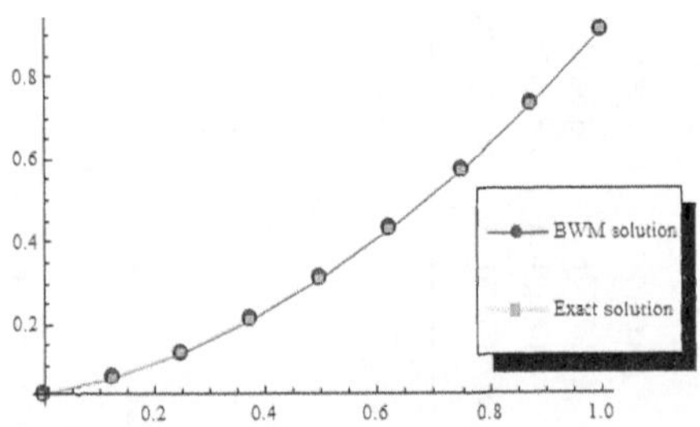

FIGURE 5.13 Comparison of BWM solution with exact solution for function $u(x,t)$.

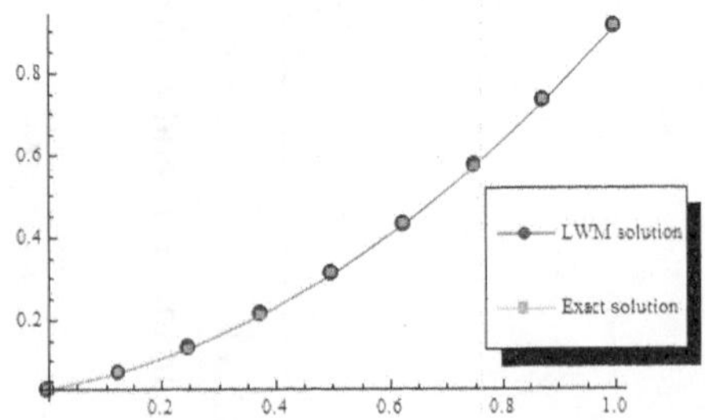

FIGURE 5.14 Comparison of LWM solution with exact solution for function $u(x,t)$.

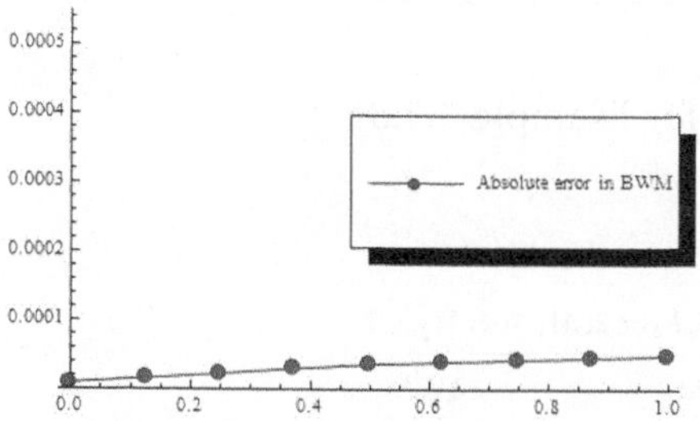

FIGURE 5.15 Error distribution for function $\tilde{e}(x,t)$ in BWM.

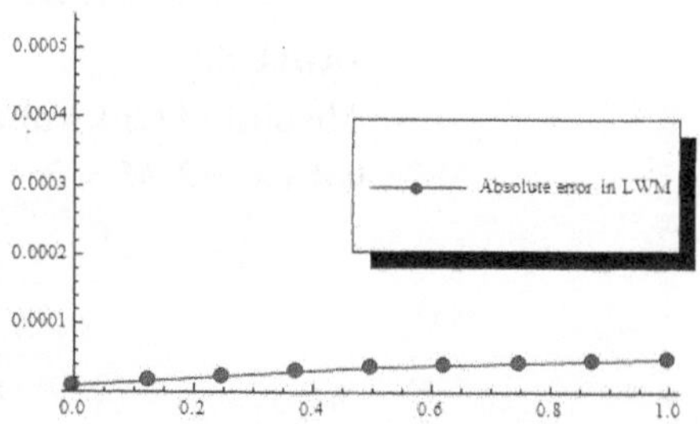

FIGURE 5.16 Error distribution for function $\tilde{e}(x,t)$ in LWM.

Example 5.9.3

Consider the following nonlinear 2D VIDE [49]:

$$u_{tt}(x,t) + u(x,t) - \int_0^x \int_0^t k(x,t,y,z)u^2(y,z)\,dy\,dz = f(x,t), \tag{5.50}$$

where

$$f(x,t) = \frac{1}{8}x^4 \sin t \cos t - \frac{1}{8}x^4 t - \frac{1}{9}x^3 \sin^3 t$$

and

$$k(x,t,y,z) = y + \cos z$$

with the boundary conditions as follows:

$$u(x,0) = 0,\ u_t(x,0) = x.$$

The exact solution is $u(x,t) = x \sin t$.

For $k_1 = 2, k_2 = 2, M_1 = 2, M_2 = 2$, the comparison of approximate solution with exact solution and error analysis is depicted in Figures 5.17, 5.19 and 5.18, 5.20 for BWM and LWM, respectively. The comparison of numerical results with the Tau method [49] has been presented in Table 5.9.

TABLE 5.7
Absolute Error Values for Example 5.9.3 with $k = 2, M = 2$

	$\tilde{e}(x,t)$	
	$k_1 = 2, k_2 = 2, M_1 = 2, M_2 = 2$	
(x,t)	BWM	LWM
(0.125,0.125)	3.4237E-5	3.4237E-5
(0.125,0.375)	2.04238E-4	2.04238E-4
(0.125,0.625)	1.46687E-4	1.46687E-4
(0.125,0.875)	3.41043E-4	3.41043E-4
(0.375,0.125)	1.01702E-4	1.01702E-4
(0.375,0.375)	6.146E-4	6.146E-4
(0.375,0.625)	4.38698E-4	4.38698E-4
(0.375,0.875)	9.87585E-4	9.87585E-4
(0.625,0.125)	1.68792E-4	1.68792E-4
(0.625,0.375)	1.03327E-3	1.03327E-3
(0.625,0.625)	7.60717E-4	7.60717E-4
(0.625,0.875)	1.7471E-3	1.7471E-3
(0.875,0.125)	2.3196E-4	2.3196E-4
(0.875,0.375)	1.46757E-3	1.46757E-3
(0.875,0.625)	1.11187E-3	1.11187E-3
(0.875,0.875)	2.50774E-3	2.50774E-3

TABLE 5.8
Comparison of L^2 and L^∞ Error Values for Example 5.9.3

	BWM		LWM	
	$k = 2, M = 2$		$k = 2, M = 2$	
t	L^2	L^∞	L^2	L^∞
0.2	1.04699E-3	1.60375E-3	1.04699E-3	1.60375E-3
0.4	4.83993E-4	7.50304E-4	4.83993E-4	7.50304E-4
0.6	6.44657E-4	9.61626E-4	6.44657E-4	9.61626E-4
0.8	3.94816E-3	6.06713E-3	3.94816E-3	6.06713E-3

TABLE 5.9
Comparison of Numerical Results for Example 5.9.3

	$k_1 = 2, k_2 = 2, M_1 = 2, M_2 = 2$		
(x,t)	BWM	LWM	Tau Method [49]
(0.01,0.01)	4.71922E-5	4.71922E-5	6.14E-5
(0.02,0.02)	8.5872E-5	8.5872E-5	1.06E-4
(0.1,0.1)	1.14713E-4	1.14713E-4	3.61E-5
(0.2,0.2)	3.63122E-4	3.63122E-4	5.21E-4

For $k = 2, M = 2$, Figures 5.17, 5.19 and 5.18, 5.20 are obtained by BWM and LWM, respectively.

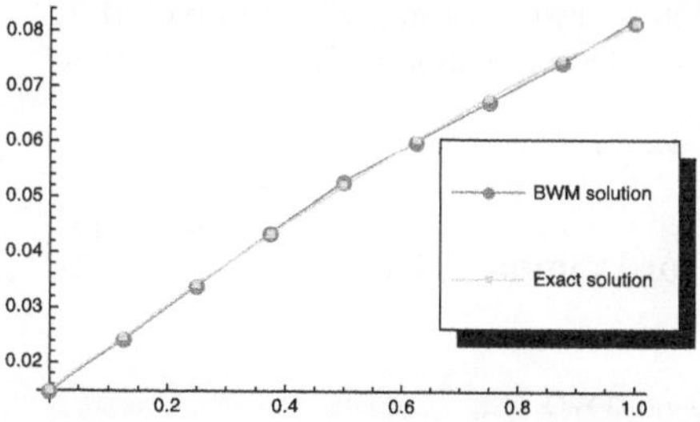

FIGURE 5.17 Comparison of BWM solution with exact solution for function $u(x,t)$.

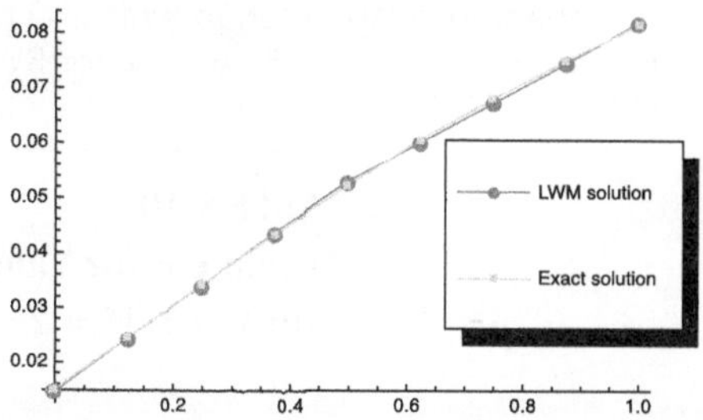

FIGURE 5.18 Comparison of LWM solution with exact solution for function $u(x,t)$.

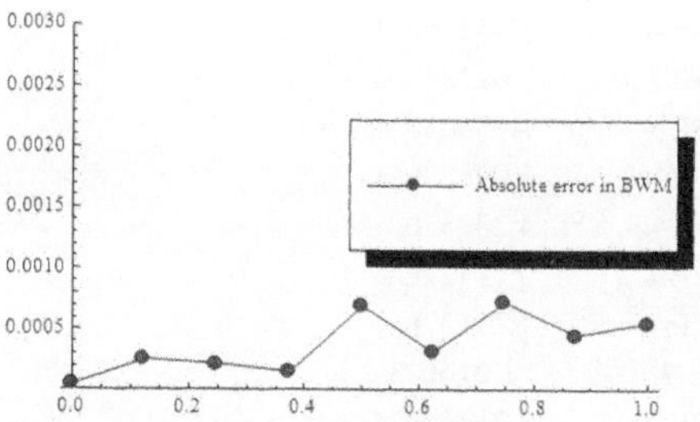

FIGURE 5.19 Error distribution for function $\tilde{e}(x,t)$ in BWM.

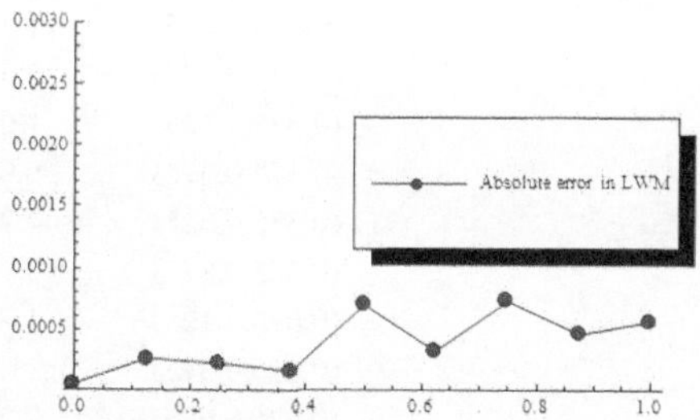

FIGURE 5.20 Error distribution for function $\tilde{e}(x,t)$ in LWM.

Example 5.9.4

Let us consider the following linear 2D VIDE:

$$u_{tt}(x,t) + u(x,t) - \int_0^x \int_0^t k(x,t,y,z)u(y,z)\,dy\,dz = f(x,t), \tag{5.51}$$

where

$$f(x,t) = -\frac{1}{6}x^2(3 + 4x + 3\cos t)\sin^2\frac{t}{2}$$

and

$$k(x,t,y,z) = y + \cos z$$

with the boundary conditions as follows:

$$u(x,0) = 0,\ u_t(x,0) = x.$$

The exact solution is $u(x,t) = x\sin t$.

For $k_1 = 2, k_2 = 2, M_1 = 2, M_2 = 2$, the comparison of approximate solution with exact solution and error analysis is depicted in Figures 5.21, 5.23 and 5.22, 5.24 for BWM and LWM, respectively, and for $k_1 = 2, k_2 = 2, M_1 = 3, M_2 = 3$, the comparison of approximate solution with exact solution and error analysis is depicted in Figures 5.25, 5.27 and 5.26, 5.28 for BWM and LWM, respectively.

TABLE 5.10
Absolute Error Values for Example 5.9.4 with $k = 2, M = 2$

	$\tilde{e}(x,t)$	
	$k_1 = 2, k_2 = 2, M_1 = 2, M_2 = 2$	
(x,t)	BWM	LWM
(0.125,0.125)	3.44586E-5	3.44586E-5
(0.125,0.375)	2.05418E-4	2.05418E-4
(0.125,0.625)	1.53236E-4	1.53236E-4
(0.125,0.875)	3.61284E-4	3.61284E-4
(0.375,0.125)	1.03003E-4	1.03003E-4
(0.375,0.375)	6.09889E-4	6.09889E-4
(0.375,0.625)	4.28957E-4	4.28957E-4
(0.375,0.875)	9.71422E-4	9.71422E-4
(0.625,0.125)	1.73137E-4	1.73137E-4
(0.625,0.375)	1.016E-3	1.016E-3
(0.625,0.625)	7.23468E-4	7.23468E-4
(0.625,0.875)	1.66745E-3	1.66745E-3
(0.875,0.125)	2.4332E-4	2.4332E-4
(0.875,0.375)	1.41412E-3	1.41412E-3
(0.875,0.625)	9.83212E-4	9.83212E-4
(0.875,0.875)	2.24288E-3	2.24288E-3

TABLE 5.11
Absolute Error Values for Example 5.9.4 with $k = 2, M = 3$

	$\tilde{e}(x,t)$	
(x,t)	BWM	LWM
(0.01,0.01)	7.71835E-6	7.71835E-6
(0.02,0.02)	1.12854E-5	1.12854E-5
(0.03,0.03)	1.13787E-5	1.13787E-5
(0.04,0.04)	8.63601E-6	8.63601E-6
(0.05,0.05)	3.65486E-6	3.65486E-6
(0.06,0.06)	3.00702E-6	3.00702E-6
(0.07,0.07)	1.08318E-5	1.08318E-5
(0.08,0.08)	1.93417E-5	1.93417E-5
(0.09,0.09)	2.80985E-5	2.80985E-5

TABLE 5.12
Comparison of L^2 and L^∞ Error Values for Example 5.9.4

	BWM				LWM			
	$k=2, M=2$		$k=2, M=3$		$k=2, M=2$		$k=2, M=3$	
t	L^2	L^∞	L^2	L^∞	L^2	L^∞	L^2	L^∞
0.2	1.03521E-3	1.57976E-3	1.58345E-4	2.52103E-4	1.03521E-3	1.57976E-3	1.58345E-4	2.52103E-4
0.4	4.56083E-4	6.92644E-4	2.08133E-4	3.32198E-4	4.56083E-4	6.92644E-4	2.08133E-4	3.32198E-4
0.6	6.98577E-4	1.07666E-3	1.608E-4	2.54959E-4	6.98577E-4	1.07666E-3	1.608E-4	2.54959E-4
0.8	3.84284E-3	5.84313E-3	1.37818E-4	2.21512E-4	3.84284E-3	5.84313E-3	1.37818E-4	2.21512E-4

For $k=2, M=2$, Figures 5.21, 5.23 and 5.22, 5.24 are obtained by BWM and LWM, respectively.

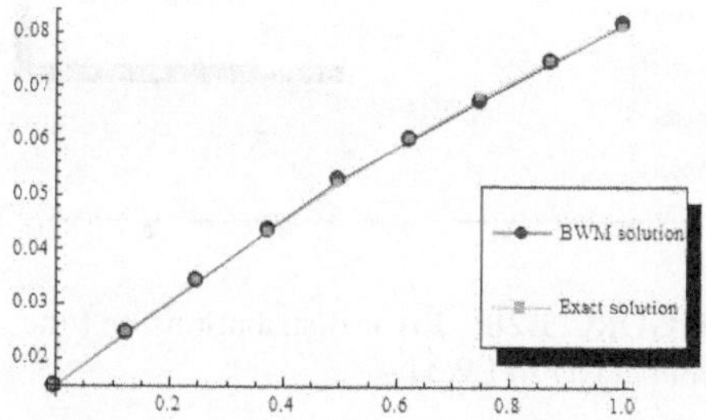

FIGURE 5.21 Comparison of BWM solution with exact solution for function $u(x,t)$.

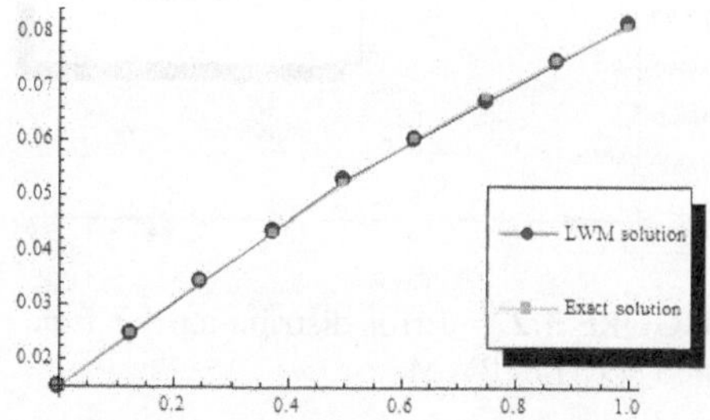

FIGURE 5.22 Comparison of LWM solution with exact solution for function $u(x,t)$.

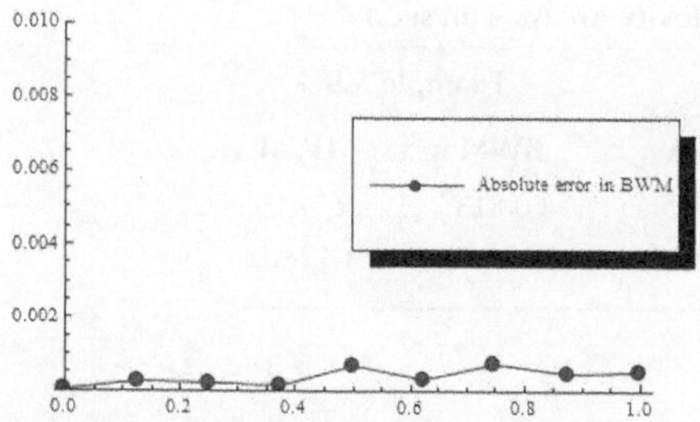

FIGURE 5.23 Error distribution for function $\tilde{e}(x,t)$ in BWM.

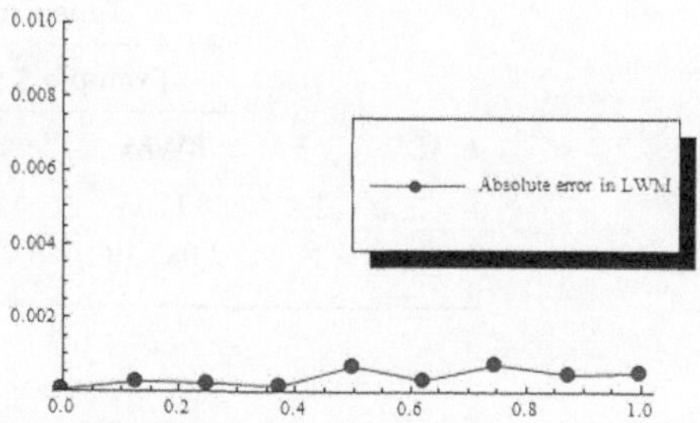

FIGURE 5.24 Error distribution for function $\tilde{e}(x,t)$ in LWM.

For $k=2, M=3$, Figures 5.25, 5.27 and 5.26, 5.28 are obtained by BWM and LWM, respectively.

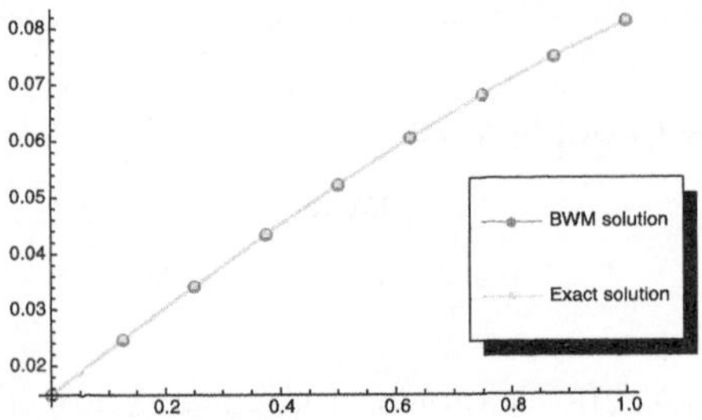

FIGURE 5.25 Comparison of BWM solution with exact solution for function $u(x,t)$.

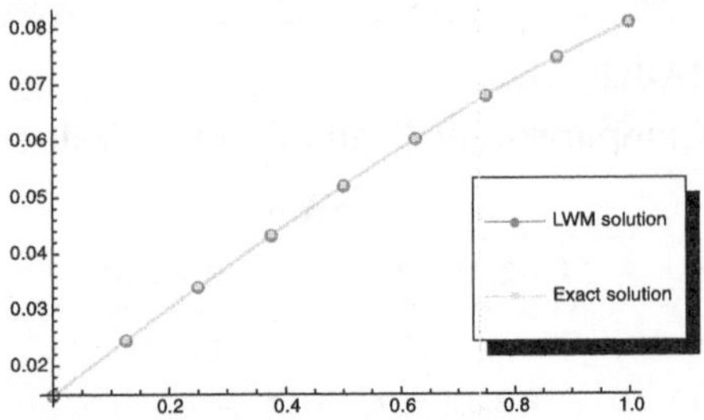

FIGURE 5.26 Comparison of LWM solution with exact solution for function $u(x,t)$.

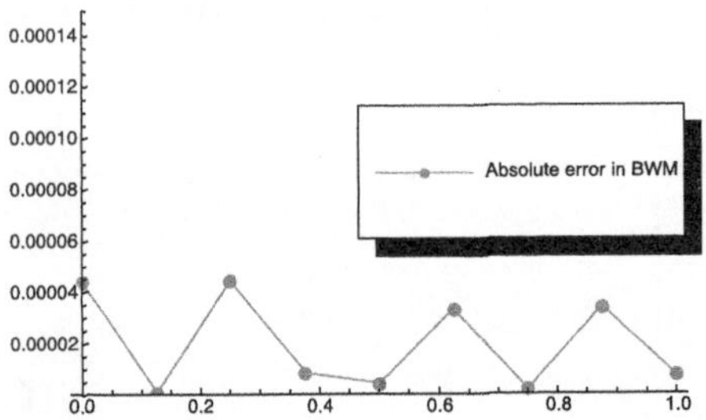

FIGURE 5.27 Error distribution for function $\tilde{e}(x,t)$ in BWM.

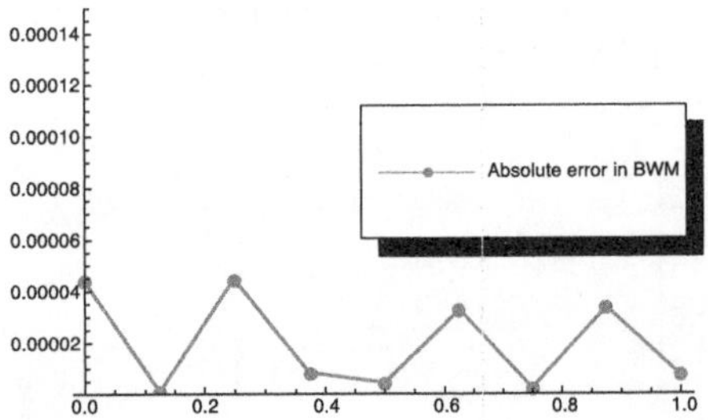

FIGURE 5.28 Error distribution for function $\tilde{e}(x,t)$ in LWM.

TABLE 5.13
Time Complexity Analysis of Numerical Experiments

	Time Complexity Analysis (in sec.)			
	Example 5.9.2		Example 5.9.4	
k, M	BWM	LWM	BWM	LWM
$k=2, M=2$	80.263	87.625	60.813	63.109
$k=2, M=3$	2,083.19	2,375.11	1,171.75	1,178.81

5.10 CONCLUSION

In this chapter, BWM and LWM have been applied to approximate the solution of linear and nonlinear VIDEs. An OM method has been introduced based on 2D BWs and LWs for approximating the solution of VIDEs with the given initial conditions. In the proposed method, by applying the OMs derived from 2D wavelets, VIDEs convert to a system of linear or nonlinear algebraic equations and that algebraic system has been solved by an appropriate numerical method. Furthermore, uniform convergence and error analysis are also studied. Some illustrative examples are investigated in order to show the effectiveness and accuracy of the proposed scheme. Moreover, the L^2 and L^∞ error analyses have been presented to demonstrate the validity and applicability of the present method. The obtained numerical results and error graphs well establish that the proposed scheme is an effective scheme to solve the VIDEs.

REFERENCES

1. Jerri A., 1999, *Introduction to Integral Equations with Applications*, John Wiley & Sons, Hoboken, NJ.
2. Burner H., 2004, *Collocation Methods for Volterra Integral and Related Functional Equations*, Cambridge University Press, Cambridge.
3. Linz P., 1985, *Analytical and Numerical Methods for Volterra Equations: 7*, SIAM, Philadelphia.
4. Wazwaz A.M., 2011, *Linear and Nonlinear Integral Equations: Methods and Applications*, Higher Education Press, Beijing and Springer-Verlag, Berlin.
5. MacCamy R.C., 1977, "An inegro-differential equation with application heat flow", *Quarterly of Applied Mathematics*, **35**(1), pp. 715–723.
6. Shukla A., Sukavanam N., and Pandey D.N., 2015, "Approximate controllability of semilinear system with state delay using sequence method", *Journal of the Franklin Institute*, **352**(11), pp. 5380–5392.
7. Gurtin M.E., and Pipkin A.C., 1968, "A general theory of heat conduction with finite wave speeds", *Archiye for Rational Mechanics and Analysis*, **31**(2), pp. 113–126.
8. Xu L., He J.H., and Liu Y., 2007, "Electrospun nanoporous spheres with Chinese drug", *International Journal of Nonlinear Sciences and Numerical Simulation*, **8**(2), pp. 199–202.
9. Bo T.L., Xie L., and Zheng X.J., 2007, "Numerical approach to wind ripple in desert", *International Journal of Nonlinear Sciences and Numerical Simulation*, **8**(2), pp. 223–228.
10. Wang H., Fu H.M., Zhang H.F., and Hu Z.Q., 2007, "A practical thermodynamic method to calculate the best glass-forming composition for bulk metallic glasses", *International Journal of Nonlinear Sciences and Numerical Simulation*, **8**(2), pp. 171–178.
11. Sun F.Z., Gao M., Lei S.H., Zhao Y.B., Wang K., Shi Y.T., and Wang N.H., 2007, "The fractal dimension of the fractal model of dropwise condensation and its experimental study", *International Journal of Nonlinear Sciences and Numerical Simulation*, **8**(2), pp. 211–222.
12. Shakeri F., and Dehghan M., 2008, "Solution of a model describing biological species living together using the variational iteration method", *Mathematical and Computer Modelling*, **48**(5–6), pp. 685–699.
13. Postnikov E.B., Lebedeva E.A., and Lavrova A.I., 2016, "Computational implementation of the inverse continuous wavelet transform without a requirement of the admissibility condition", *Applied Mathematics and Computation*, **282**, pp. 128–136.
14. Babaaghaie A., and Maleknejad K., 2017, "Numerical solutions of nonlinear two-dimensional partial Volterra integro-differential equations by Haar wavelet", *Journal of Computational and Applied Mathematics*, **317**, pp. 643–651.
15. Araghi M.F., and Behzadi S.S., 2009, "Solving nonlinear Volterral Fredholm integro-differential equations using the modified adomian decomposition method", *Computational Methods in Applied Mathematics*, **9**(4), pp. 321–331.
16. Mahmoudi Y., 2005, "Taylor polynomial solution of non-linear Volterra–Fredholm integral equation", *International Journal of Computer Mathematics*, **82**(7), pp. 881–887.

17. Pour-Mahmoud J., Rahimi-Ardabili M.Y., and Shahmorad S., 2005, "Numerical solution of Volterra integro-differential equations by the Tau method with the Chebyshev and Legendre bases, *Applied Mathematics and Computation*, **170**(1), pp. 314–338.
18. Makroglou A., 1982, "Hybrid methods in the numerical solution of Volterra integro-differential equations", *IMA Journal of Numerical Analysis*, **2**(1), pp. 21–35.
19. Fazeli S., and Hojjati G., 2015, "Numerical solution of Volterra integro-differential equations by superimplicit multistep collocation methods", *Numerical Algorithms*, **68**(4), pp. 741–768.
20. Saberi-Nadjafi J., and Tamamgar M., 2008, "The variational iteration method: a highly promising method for solving the system of integro-differential equations", *Computers & Mathematics with Applications*, **56**(2), pp. 346–351.
21. Saeedi H., and Moghadam M.M., 2011, "Numerical solution of nonlinear Volterra integro-differential equations of arbitrary order by CAS wavelets", *Communications in Nonlinear Science and Numerical Simulation*, **16**(3), pp. 1216–1226.
22. Sahu P.K., and Saha Ray S., 2015, "Legendre wavelets operational method for the numerical solutions of nonlinear Volterra integro-differential equations system", *Applied Mathematics and Computation*, **256**, pp. 715–723.
23. Venkatesh S.G., Ayyaswamy S.K., Balachandar S.R., and Kannan K., 2013, "Wavelet solution for class of nonlinear integro-diferential equations", *Indian Journal of Science and Technology*, **6**(6), pp. 4670–4677.
24. Biazar J., Ghazvini H., and Eslami M., 2009, "He's homotopy perturbation method for systems of integro-differential equations", *Chaos, Solitons & Fractals*, **39**(3), pp. 1253–1258.
25. Nemati S., Lima P.M., and Ordokhani Y., 2013, "Numerical solution of a class of two-dimensional nonlinear Volterra integral equations using Legendre polynomials", *Journal of Computational and Applied Mathematics*, **242**, pp. 53–69.
26. Babolian E., and Dastani N., 2011, "Numerical solutions of two-dimensional linear and nonlinear Volterra integral equations: homotopy perturbation method and differential transform method", *International Journal of Industrial Mathematics*, **3**(3), pp. 157–167.
27. Sahu P.K., and Saha Ray S., 2015, "Legendre spectral collocation method for Fredholm integro-differential-difference equation with variable coefficients and mixed conditions", *Applied Mathematics and Computation*, **268**, pp. 575–580.
28. Maleknejad K., Almasieh H., and Roodaki M., 2010, "Triangular functions (TF) method for the solution of nonlinear Volterra–Fredholm integral equations", *Communications in Nonlinear Science and Numerical Simulation*, **15**(11), pp. 3293–3298.
29. Tari A., Rahimi M.Y., Shahmorad S., and Talati F., 2009, "Development of the tau method for the numerical solution of two-dimensional linear Volterra integro-differential equations", *Computational Methods in Applied Mathematics*, **9**(4), pp. 421–435.
30. Aghazadeh N., and Khajehnasiri A.A., 2013, "Solving nonlinear two-dimensional Volterra integro-differential equations by block-pulse functions", *Mathematical Sciences*, **7**(1), p. 3.
31. Sahu P.K., and Saha Ray S., 2017, "A new Bernoulli wavelet method for numerical solutions of nonlinear weakly singular Volterra integro-differential equations", *International Journal of Computational Methods*, **14**(3), p. 1750022.
32. Das P., and Rana S., 2021, "Theoretical prospects of fractional order weakly singular Volterra integro differential equations and their approximations with convergence analysis", *Mathematical Methods in the Applied Sciences*, **44**(11), pp. 9419–9440.
33. Das P., Rana S., and Ramos H., 2020, "On the approximate solutions of a class of fractional order nonlinear Volterra integro-differential initial value problems and boundary value problems of first kind and their convergence analysis", *Journal of Computational and Applied Mathematics*, **404**, p. 113116.

34. Das P., Rana S., and Ramos H., 2019, "Homotopy perturbation method for solving Caputo-type fractional-order Volterra-Fredholm integro-differential equations", *Computational and Mathematical Methods*, **1**(5), p. e1047.
35. Das P., Rana S., and Ramos H., 2020, "A perturbation-based approach for solving fractional-order Volterra–Fredholm integro differential equations and its convergence analysis", *International Journal of Computer Mathematics*, **97**(10), pp. 1994–2014.
36. Das P., Rana S., and Vigo-Aguiar J., 2020, "Higher order accurate approximations on equidistributed meshes for boundary layer originated mixed type reaction diffusion systems with multiple scale nature", *Applied numerical Mathematics*, **148**, pp. 79–97.
37. Das P., and Natesan S., 2015, "Adaptive mesh generation for singularly perturbed fourth-order ordinary differential equations", *International Journal of Computer Mathematics*, **92**(3), pp. 562–578.
38. Das P., and Vigo-Aguiar J., 2019, "Parameter uniform optimal order numerical approximation of a class of singularly perturbed system of reaction diffusion problems involving a small perturbation parameter", *Journal of Computational and Applied Mathematics*, **354**, pp. 533–544.
39. Das P., 2019, "An a posteriori based convergence analysis for a nonlinear singularly perturbed system of delay differential equations on an adaptive mesh", *Numerical Algorithms*, **81**(2), pp. 465–487.
40. Das P., 2018, "A higher order difference method for singularly perturbed parabolic partial differential equations", *Journal of Difference Equations and Applications*, **24**(3), pp. 452–477.
41. Das P., and Mehrmann V., 2016, "Numerical solution of singularly perturbed convection-diffusion-reaction problems with two small parameters", *BIT Numerical Mathematics*, **56**(1), pp. 51–76.
42. Shakti D., Mohapatra J., Das P., and Vigo-Aguiar J., 2020, "A moving mesh refinement based optimal accurate uniformly convergent computational method for a parabolic system of boundary layer originated reaction–diffusion problems with arbitrary small diffusion terms", *Journal of Computational and Applied Mathematics*, **404**, p. 113167.
43. Kumar K., Podila P.C., Das P., and Ramos H., 2021, "A graded mesh refinement approach for boundary layer originated singularly perturbed time-delayed parabolic convection diffusion problems", *Mathematical Methods in the Applied Sciences*, **44**(16), pp. 12332–12350, https://doi.org/10.1002/mma.7358.
44. EL- Azab M.S., EL- Kalla I.L., and EL- Morsy S.A., 2013, "Solution of KDVB equation via Block-Pulse functions method", *Electronic Journal of Mathematical Analysis and Applications*, **1**, pp. 361–367.
45. Mohammadi F., and Adhami P., 2016, "Numerical study of stochastic Volterra–Fredholm integral equations by using second kind Chebyshev wavelets", *Random Operators and Stochastic Equations*, **24**(2), pp. 129–141.
46. Behera S., and Saha Ray S., 2020, "An operational matrix based scheme for numerical solutions of nonlinear weakly singular partial integro-differential equations", *Applied Mathematics and Computation*, **367**, p. 124771. https://doi.org/10.1016/j.amc.2019.124771.
47. Narumi S., 1920, "Some formulas in the theory of interpolation of many independent variables", *Tohoku Mathematical Journal, First Series*, **18**, pp. 309–321.
48. Rahimkhani P., and Ordokhani Y., 2018, "Numerical solution a class of 2D fractional optimal control problems by using 2D Müntz-Legendre wavelets", *Optimal Control Applications and Methods*, **39**(6), pp. 1916–1934.
49. Ordokhani Y., and Samari F., 2016, "Numerical solution of two-dimensional nonlinear Volterra integro-differential equations by Tau method", *Journal of Information and Computing Science*, **11**(2), pp. 101–110.

6 Numerical Solutions of Volterra Weakly Singular Partial Integro-Differential Equations

6.1 INTRODUCTION

Mathematical modelling of real-world phenomena often leads to functional equations, like integral and integro-differential equations, etc. In wide areas of physics and engineering, the functions of time and space are generally illustrated by PDEs. Sometimes, such a formation may not represent a purely modeled physical system, since expressing the system as a function often fails to capture or explain the influence of its past history. An integro-differential equation involves both integral and derivative of a function. First of all, Volterra studied IDEs and published his lectures on integral and IDEs in 1913 and 1937, and he provided a classification of IDEs. Volterra IDEs are very important effective mathematical models to describe many real-world phenomena concerning economy, atomic energy, biology, chemistry, control theory, physics, information theory, medicine, population dynamics, and engineering technique fields [1–11]. A partial integro-differential equation (PIDE) is obtained when the unknown function or its derivatives, or both, appear under the sign of integration. PIDEs are widely used to model several physical systems in science and engineering such as reaction–diffusion problems in the small cell [12], reactor dynamics [13], flow in fractured biomaterials [14], electricity swaptions [15], population dynamics [16], financial option pricing [17], diffusion [18], viscoelasticity [19], control system theory [20], thermoelasticity [21], etc. There are some different forms of PIDEs, and we consider the second-kind Volterra integral PIDEs. Volterra-type integral equations with a weakly singular kernel appears as many problems of mathematical physics, theory of elasticity, hydrodynamics, heat conduction [22], stereology [23], heat flow [24], etc.

To treat the PIDEs, a substantial number of methods have been applied. For example, the pseudo-spectral Legendre–Galerkin method for solving a parabolic PIDE with convolution-type kernel was presented in a previous study [25]. Combination of radial basis functions and finite difference for solving nonlinear-type PIDEs with smooth kernel containing an unknown function was considered in another study [26]. Furthermore, a spectral method was proposed in a study [27] for the PIDEs with a weakly singular kernel.

DOI: 10.1201/9781003453451-6

Wavelets are very effective mathematical devices used in several numerical schemes. The wavelets theory is used mostly in the field of engineering and mathematics. The applicability of wavelets has been found in many applications [28,29], and particularly, wavelets are also implemented in the signal analysis [30].

Many numerical methods have been used to solve the weakly singular integral equations such as modified Euler method [31], Haar wavelet method [32], Galerkin method [33], quadrature method [34], finite-difference methods [35], compact difference schemes [36], spectral collocation methods [37], orthogonal spline collocation methods [38], variational iteration and Adomian decomposition methods [39], and quasi-wavelet methods [40]. Nemati et al. [41] presented an effective scheme based on operational matrices to obtain the numerical solution of the integral equation. Mirzaee et al. [42] employed Legendre wavelet collocation method for solving the nonlinear Stratonovich Volterra integral equation. Sahu et al. [43] have solved nonlinear Volterra integro-differential equation by Legendre wavelets operational method. Singh et al. [44] applied operational matrix scheme for the approximate solution of PIDE. Mirazee et al. [45] have solved the fractional stochastic integro-differential equation by using the orthonormal Bernstein polynomial. Han et al. [46] introduced the extrapolation method to find the approximate solution of the system of nonlinear weakly singular Volterra integro-differential equation. Mirazee et al. [47,48] developed an operational matrix scheme for solving the system of linear Stratonovich Volterra integral equation. Stochastic operational matrix scheme based on delta functions has been employed to obtain the numerical solution of linear and nonlinear stochastic quadratic integral equations [49]. A numerical method for the solution of two-dimensional (2D) linear stochastic integral equations on non-rectangular domains is proposed in an earlier study [50]. In addition, Mirazee et al. [51] used a numerical approach based on Bernstein polynomials for solving fractional-order 2D nonlinear integral equations. Tohidi et al. [52] introduced a new Bernoulli matrix method to find the approximate solution of the second-order linear partial differential equation. Tohidi et al. [53] employed an operational matrix approach for the solution of second-order linear matrix partial differential equation. Bhrawy et al. [54] have solved higher-order linear and nonlinear Fredholm integro-differential equations by a new Bernoulli matrix method. Tang [55] discussed the application of a class of spline collocation methods to weakly singular Volterra IDEs. Tohidi et al. [56] have analysed the convergence of the Bernoulli matrix approach for one-dimensional matrix hyperbolic equation of the first order. Zogheib and Tohidi [57] developed a new operational matrix scheme for solving the 2D time-dependent diffusion equation.

6.2 OUTLINE OF THE PRESENT STUDY

In this chapter, an operational matrix (OM) scheme based on 2D wavelets has been introduced for the solution of weakly singular linear partial integro-differential equations (LPIDEs) and nonlinear partial integro-differential equations (NLPIDEs). By implementing 2D wavelets approximations and its OMs of integration and differentiation along with collocation points, the linear and nonlinear weakly singular PIDEs are reduced into the system of linear and nonlinear algebraic equations, respectively. Moreover, the Bernoulli wavelet approximation (BWA) and the

Legendre wavelet approximation (LWA) have been used for inspecting the errors and convergence analysis of the given problems. Some numerical examples are included to establish the accuracy of the proposed scheme via BWA and LWA, respectively. Additionally, comparisons of error values between the two wavelets have been presented. Furthermore, some figures are plotted to demonstrate the error analysis of the proposed scheme.

6.3 TWO-DIMENSIONAL LINEAR VOLTERRA WEAKLY SINGULAR PIDEs

In this chapter, Volterra weakly singular LPIDE of the following form has been considered:

$$v_{xt}(x,t)+v_t(x,t)=v(x,t)+h(x,t)+\int_0^x\int_0^t \Phi(x,s)\tilde{\mathrm{H}}(v(s,y))dyds, \quad (x,t)\in\Delta, \tag{6.1}$$

having the following boundary conditions:

$$v(x,0)=v_0(x),$$

$$v(0,t)=v_0(t),$$

where $\tilde{\mathrm{H}}(v(x,t)), h(x,t), v_0(x), v_0(t)$ are known functions and $v(x,t)$ is unknown function. In Eq. (6.1), $\Delta=[0,1]\times[0,1]$, $\Phi(x,s)=\dfrac{1}{(x-s)^{\theta}}$, $0<\theta<1$ is the kernel function which has a weak singularity at $s=x$. Moreover, $\tilde{\mathrm{H}}$ is a differential operator that is considered as $\tilde{\mathrm{H}}(v(x,t))=v_{tt}(x,t)$, and $\tilde{\mathrm{H}}(v(x,t))=v_{xt}(x,t)$ for the linear case as the first and second kinds, respectively. For the existence of the solution $v(x,t)\in C(\Delta)$, it is assumed that $h(x,t)$ and $v(x,t)$ are sufficiently smooth functions.

The known and unknown functions are approximated by means of 2D Bernoulli wavelets (BWs) along with the corresponding operational matrices. Using the operational scheme, the proposed technique transforms the PIDEs into the system of algebraic equation. The solution of Eq. (6.1) by the Bernoulli wavelet method (BWM) is calculated after solving the system of algebraic equations.

6.3.1 Function Approximation by BWs

A function $h(x,t)\in L^2(\Delta)$ can be approximated as follows:

$$h(x,t)=\sum_{n=1}^{2^{k1-1}}\sum_{n'=1}^{2^{k2-1}}\sum_{m=0}^{M_1-1}\sum_{m'=0}^{M_2-1} h_{n,m,n',m'}.\psi_{n,m,n',m'}(x,t)=\Psi^T(x)H\Psi(t)=\Psi^T(t)H^T\Psi(x), \tag{6.2}$$

where

$$\begin{cases} \Psi(x) = [\psi_{1,0}(x), \psi_{1,1}(x), \ldots, \psi_{1,M_1-1}(x), \ldots, \psi_{2^{k_1-1},0}(x), \psi_{2^{k_1-1},1}(x), \ldots, \psi_{2^{k_1-1},M_1-1}(x)]^T, \\ \Psi(t) = [\psi_{1,0}(t), \psi_{1,1}(t), \ldots, \psi_{1,M_2-1}(t), \ldots, \psi_{2^{k_2-1},0}(t), \psi_{2^{k_2-1},1}(t), \ldots, \psi_{2^{k_2-1},M_2-1}(t)]^T, \end{cases} \tag{6.3}$$

where $\psi_{n,m}(x)$, $n = 1,2,\ldots,2^{k_1-1}$, $m = 0,1,\ldots,M_1-1$ is the BWs defined as follows:

$$\psi_{n,m}(x) = \begin{cases} 2^{\frac{k-1}{2}} \mathcal{B}_m(2^{k-1}x - n + 1), & \frac{n-1}{2^{k-1}} \le x < \frac{n}{2^{k-1}}, \\ 0, & \text{otherwise}, \end{cases} \tag{6.4}$$

with

$$\mathcal{B}_m(x) = \begin{cases} 1, & m = 0, \\ \dfrac{1}{\sqrt{\dfrac{(-1)^{m-1}(m!)^2}{(2m!)^2}\widehat{\alpha}_{2m}}} \widehat{\beta}_m(x), & m > 0. \end{cases}$$

The 2D BWs can be defined over the domain $[0,1) \times [0,1)$ as follows:

$$\psi_{n,m,n',m'}(x,t) = \begin{cases} 2^{\frac{k_1+k_2}{2}} \mathcal{B}_m(2^{k_1-1}x - n + 1).\mathcal{B}_{m'}(2^{k_2-1}t - n' + 1), & \text{for } \frac{n-1}{2^{k_1-1}} \le x < \frac{n}{2^{k_1-1}}, \\ & \frac{n'-1}{2^{k_2-1}} \le t < \frac{n'}{2^{k_2-1}}, \\ 0, & \text{otherwise}, \end{cases} \tag{6.5}$$

where $\mathcal{B}_m(x) = \begin{cases} 1, & m = 0 \\ \dfrac{1}{\sqrt{\dfrac{(-1)^{m-1}(m!)^2}{(2m)!}\beta_{2m}}} \beta_m(x), & m > 0 \end{cases}$

and $\mathcal{B}_{m'}(t) = \begin{cases} 1, & m' = 0 \\ \dfrac{1}{\sqrt{\dfrac{(-1)^{m'-1}(m'!)^2}{(2m')!}\beta_{2m'}}} \beta_{m'}(t), & m' > 0, \end{cases}$

where $n = 1,2,\ldots,2^{k_1-1}$, $n' = 1,2,\ldots,2^{k_2-1}$, $m = 0,1,2,\ldots,M_1-1$, and $m' = 0,1,2,\ldots,$ M_2-1. Here, k_1 and k_2 are any positive integers. $\mathcal{B}_m$ and $\mathcal{B}_{m'}$ are Bernoulli polynomials of order m and m', respectively.

Moreover, H is the $(2^{k_1-1}M_1 \times 2^{k_2-1}M_2)$ coefficient matrix with components as follows:

$$h_{i,j} = \int_0^1 \Psi(x).\left(\int_0^1 h(x,t).\Psi(t)dt\right)dx.$$

6.3.2 Operational Matrices

Now, we derive the OMs of integration, differentiation, and almost OM of integration with singularity for the proposed wavelet methods to solve the weakly singular LPIDEs.

6.3.2.1 Operational Matrix of Integration for 2D BWs

In this part, we describe the construction of OM of integration for BWs. The integration of the vector $\Psi(x)$ defined in Eq. (6.3) is given by [58]

$$\int_0^x \Psi(\varsigma)d\varsigma \approx \widehat{P}\Psi(x), \tag{6.6}$$

where $\widehat{P}$ is the $(2^{k-1}M \times 2^{k-1}M)$ OM of integration. To explain the calculation procedure, we choose $k=1$ and $M=3$. We have

$$\Psi(x) = [\psi_{1,0}(x), \psi_{1,1}(x), \psi_{1,2}(x)]^T,$$

$$\left.\begin{aligned} &\psi_{1,0}(x) = 1 \\ &\psi_{1,1}(x) = \sqrt{3}(-1+2x) \\ &\psi_{1,2}(x) = \sqrt{5}(1-6x+6x^2) \end{aligned}\right\}; \quad 0 \le x < 1. \tag{6.7}$$

For $k=1$ and $M=3$,

$$\begin{aligned} f(x) &= \sum_{n=1}^{2^{k-1}} \sum_{m=0}^{M-1} c_{n,m}\psi_{n,m}(x) \\ &= \sum_{n=1}^{1} \sum_{m=0}^{2} c_{n,m}\psi_{n,m}(x) \\ &= c_{1,0}\psi_{1,0}(x) + c_{1,1}\psi_{1,1}(x) + c_{1,2}\psi_{1,2}(x). \end{aligned}$$

By integrating Eq. (6.7) from 0 to x, we obtain

$$\int_0^x \psi_{1,0}(\varsigma)\,d\varsigma = x,\ \ 0 \le x < 1$$

$$= c_{1,0}\psi_{1,0}(x) + c_{1,1}\psi_{1,1}(x) + c_{1,2}\psi_{1,2}(x) \tag{6.8}$$

$$= \frac{1}{2}\psi_{1,0}(x) + \frac{1}{2\sqrt{3}}\psi_{1,1}(x).$$

$$c_{1,0} =< f(x), \psi_{1,0}(x) >= \int_0^1 x.1\,dx = \frac{1}{2},$$

$$c_{1,1} =< f(x), \psi_{1,1}(x) >= \int_0^1 x.\sqrt{3}(-1+2x)\,dx = \frac{1}{2\sqrt{3}},$$

$$c_{1,2} =< f(x), \psi_{1,2}(x) >= \int_0^1 x.\sqrt{5}(1-6x+6x^2)\,dx = 0.$$

$$\int_0^x \psi_{1,1}(\varsigma)\,d\varsigma = \sqrt{3}(x^2 - x),\ \ 0 \le x < 1$$

$$= c_{1,0}\psi_{1,0}(x) + c_{1,1}\psi_{1,1}(x) + c_{1,2}\psi_{1,2}(x) \tag{6.9}$$

$$= -\frac{1}{2\sqrt{3}}\psi_{1,0}(x) + \frac{1}{2\sqrt{15}}\psi_{1,2}(x).$$

$$c_{1,0} =< f(x), \psi_{1,0}(x) >= \int_0^1 \sqrt{3}(x^2 - x).1\,dx = -\frac{1}{2\sqrt{3}},$$

$$c_{1,1} =< f(x), \psi_{1,1}(x) >= \int_0^1 \sqrt{3}(x^2 - x).\sqrt{3}(-1+2x)\,dx = 0,$$

$$c_{1,2} =< f(x), \psi_{1,2}(x) >= \int_0^1 \sqrt{3}(x^2 - x).\sqrt{5}(1-6x+6x^2)\,dx = \frac{1}{2\sqrt{15}}. \tag{6.10}$$

$$\int_0^x \psi_{1,2}(\varsigma)\,d\varsigma = \sqrt{5}(2x^3 - 3x^2 + x),\ \ \ 0 \le x <$$

$$= c_{1,0}\psi_{1,0}(x) + c_{1,1}\psi_{1,1}(x) + c_{1,2}\psi_{1,2}(x)$$

$$= -\frac{1}{2\sqrt{15}}\psi_{1,1}(x).$$

$$c_{1,0} = < f(x), \psi_{1,0}(x) > = \int_0^1 \sqrt{5}(2x^3 - 3x^2 + x).1\,dx = 0,$$

$$c_{1,1} = < f(x), \psi_{1,1}(x) > = \int_0^1 \sqrt{5}(2x^3 - 3x^2 + x).\sqrt{3}(-1+2x)\,dx = -\frac{1}{2\sqrt{15}},$$

$$c_{1,2} = < f(x), \psi_{1,2}(x) > = \int_0^1 \sqrt{5}(2x^3 - 3x^2 + x).\sqrt{5}(1-6x+6x^2)\,dx = 0.$$

By using Eqs. (6.8)–(6.10), we get

$$\int_0^x \Psi(\varsigma)\,d\varsigma = \begin{bmatrix} \int_0^x \psi_{1,0}(\varsigma)\,d\varsigma \\ \int_0^x \psi_{1,1}(\varsigma)\,d\varsigma \\ \int_0^x \psi_{1,2}(\varsigma)\,d\varsigma \end{bmatrix} = \begin{bmatrix} \frac{1}{2} & \frac{1}{2\sqrt{3}} & 0 \\ \frac{-1}{2\sqrt{3}} & 0 & \frac{1}{2\sqrt{15}} \\ 0 & -\frac{1}{2\sqrt{15}} & 0 \end{bmatrix}.\Psi(x) = \widehat{P}.\Psi(x),$$

where

$$\widehat{P} = \begin{bmatrix} \frac{1}{2} & \frac{1}{2\sqrt{3}} & 0 \\ \frac{-1}{2\sqrt{3}} & 0 & \frac{1}{2\sqrt{15}} \\ 0 & -\frac{1}{2\sqrt{15}} & 0 \end{bmatrix}.$$

Similarly, for $k = 2$ and $M = 3$, we get

$$\widehat{P} = \begin{bmatrix} \frac{1}{4} & \frac{1}{4\sqrt{3}} & 0 & \frac{1}{2} & 0 & 0 \\ -\frac{1}{4\sqrt{3}} & 0 & \frac{1}{4\sqrt{15}} & 0 & 0 & 0 \\ 0 & -\frac{1}{4\sqrt{15}} & 0 & 0 & 0 & 0 \\ 0 & 0 & 0 & \frac{1}{4} & \frac{1}{4\sqrt{3}} & 0 \\ 0 & 0 & 0 & -\frac{1}{4\sqrt{3}} & 0 & \frac{1}{4\sqrt{15}} \\ 0 & 0 & 0 & 0 & -\frac{1}{4\sqrt{15}} & 0 \end{bmatrix}.$$

6.3.2.2 Operational Matrix of Differentiation for 2D BWs

Theorem 6.3.1

Let $\Psi(x)$ be the BWs vector defined in Eq. (6.3). The derivative of the vector $\Psi(x)$ can be expressed by

$$\frac{d\Psi(x)}{dx} = \widehat{D}\Psi(x), \tag{6.11}$$

where $\widehat{D}$ is the $(2^{k-1}M \times 2^{k-1}M)$ OM of derivative defined as follows:

$$\widehat{D} = 2^{k-1}.\begin{bmatrix} Q & 0 & . & . & . & 0 \\ 0 & Q & . & . & . & 0 \\ . & . & . & . & . & . \\ . & . & . & . & . & . \\ . & . & . & . & . & . \\ 0 & 0 & . & . & . & Q \end{bmatrix},$$

where Q is $(M \times M)$ matrix and its (i, j)th element is defined as follows:

$$Q_{i,j} = \begin{cases} j\dfrac{\lambda_{i-1}}{\lambda_{j-1}}, & i-j=1,\, i=2,\dots,M,\, j=1,\dots,M-1, \\ 0, & \text{otherwise}, \end{cases} \tag{6.12}$$

where

$$\lambda_m = \begin{cases} 1, & m=0, \\ \dfrac{1}{\sqrt{\dfrac{(-1)^{m-1}(m!)^2}{(2m!)^2}\widehat{\alpha}_{2m}}}, & m>0. \end{cases}$$

Proof:

It may be referred to Ref. [59].

6.3.2.3 Almost Operational Matrix of Integration with Singularity for 2D BWs and LWs

In this section, we discuss the construction of almost OMs of integration for BWs and LWs.

Let $\Psi(x)$ be wavelets vector defined in Eq. (6.3) for $0<\theta<1$, then we get

$$\int_0^x \frac{\Psi(\varsigma)}{(x-\varsigma)^\theta}\, d\varsigma \approx \widehat{E}\Psi(x), \tag{6.13}$$

where $\widehat{E}$ is the $(2^{k-1}M \times 2^{k-1}M)$ OM of integration. To explain the calculation procedure, we choose $k=1$ and $M=3$. We have

$$\Psi(x) = [\psi_{1,0}(x), \psi_{1,1}(x), \psi_{1,2}(x)]^T,$$

$$\left.\begin{aligned} &\psi_{1,0}(x) = 1 \\ &\psi_{1,1}(x) = \sqrt{3}(-1+2x) \\ &\psi_{1,2}(x) = \sqrt{5}(1-6x+6x^2) \end{aligned}\right\}; \quad 0 \le x < 1. \tag{6.14}$$

For $k=1$ and $M=3$,

$$\begin{aligned} f(x) &= \sum_{n=1}^{2^{k-1}} \sum_{m=0}^{M-1} c_{n,m}\psi_{n,m}(x) \\ &= \sum_{n=1}^{1} \sum_{m=0}^{2} c_{n,m}\psi_{n,m}(x) \\ &= c_{1,0}\psi_{1,0}(x) + c_{1,1}\psi_{1,1}(x) + c_{1,2}\psi_{1,2}(x). \end{aligned}$$

By integrating Eq. (6.14) from 0 to x and putting $\theta = 0.5$, we obtain

$$\begin{aligned} \int_0^x \frac{\psi_{1,0}(\varsigma)}{(x-\varsigma)^{0.5}}\, d\varsigma &= 2\sqrt{x}, \quad 0 \le x < 1 \\ &= c_{1,0}\psi_{1,0}(x) + c_{1,1}\psi_{1,1}(x) + c_{1,2}\psi_{1,2}(x) \\ &= \frac{4}{3}\psi_{1,0}(x) + \frac{4}{5\sqrt{3}}\psi_{1,1}(x) - \frac{4}{21\sqrt{5}}\psi_{1,2}(x). \end{aligned} \tag{6.15}$$

$$c_{1,0} =< f(x),\psi_{1,0}(x) >= \int_0^1 2\sqrt{x}.1\,dx = \frac{4}{3},$$

$$c_{1,1} =< f(x),\psi_{1,1}(x) >= \int_0^1 2\sqrt{x}.\sqrt{3}(-1+2x)\,dx = \frac{4}{5\sqrt{3}},$$

$$c_{1,2} =< f(x),\psi_{1,2}(x) >= \int_0^1 2\sqrt{x}.\sqrt{5}(1-6x+6x^2)\,dx = -\frac{4}{21\sqrt{5}}.$$

$$\begin{aligned}\int_0^x \frac{\psi_{1,1}(\varsigma)}{(x-\varsigma)^{0.5}}d\varsigma &= \frac{2\sqrt{x}}{\sqrt{3}}(-3+4x),\ \ 0 \le x < 1\\ &= c_{1,0}\psi_{1,0}(x)+c_{1,1}\psi_{1,1}(x)+c_{1,2}\psi_{1,2}(x)\\ &= -\frac{4}{5\sqrt{3}}\psi_{1,0}(x)+\frac{4}{7}\psi_{1,1}(x)+\frac{4}{3\sqrt{15}}\psi_{1,2}(x).\end{aligned} \tag{6.16}$$

$$c_{1,0} =< f(x),\psi_{1,0}(x) >= \int_0^1 \frac{2\sqrt{x}}{\sqrt{3}}(-3+4x).1\,dx = -\frac{4}{5\sqrt{3}},$$

$$c_{1,1} =< f(x),\psi_{1,1}(x) >= \int_0^1 \frac{2\sqrt{x}}{\sqrt{3}}(-3+4x).\sqrt{3}(-1+2x)\,dx = \frac{4}{7},$$

$$c_{1,2} =< f(x),\psi_{1,2}(x) >= \int_0^1 \frac{2\sqrt{x}}{\sqrt{3}}(-3+4x).\sqrt{5}(1-6x+6x^2)\,dx = \frac{4}{3\sqrt{15}}.$$

$$\begin{aligned}\int_0^x \frac{\psi_{1,2}(\varsigma)}{(x-\varsigma)^{0.5}}d\varsigma &= \frac{2\sqrt{x}}{\sqrt{5}}((5+4x(-5+4x)),\quad 0 \le x < 1\\ &= c_{1,0}\psi_{1,0}(x)+c_{1,1}\psi_{1,1}(x)+c_{1,2}\psi_{1,2}(x)\\ &= -\frac{4}{21\sqrt{5}}\psi_{1,0}(x)-\frac{4}{3\sqrt{15}}\psi_{1,1}(x)+\frac{100}{231}\psi_{1,2}(x).\end{aligned} \tag{6.17}$$

$$c_{1,0} =< f(x),\psi_{1,0}(x) >= \int_0^1 \frac{2\sqrt{x}}{\sqrt{5}}(5+4x(-5+4x)).1\,dx = -\frac{4}{21\sqrt{5}},$$

$$c_{1,1} =< f(x),\psi_{1,1}(x) >= \int_0^1 \frac{2\sqrt{x}}{\sqrt{5}}(5+4x(-5+4x)).\sqrt{3}(-1+2x)\,dx = -\frac{4}{3\sqrt{15}},$$

$$c_{1,2} =< f(x),\psi_{1,2}(x) >= \int_0^1 \frac{2\sqrt{x}}{\sqrt{5}}(5+4x(-5+4x)).\sqrt{5}(1-6x+6x^2)\,dx = \frac{100}{231}.$$

By using Eqs. (6.15)–(6.17), we get

$$\int_0^x \frac{\Psi(\varsigma)}{(x-\varsigma)^{0.5}}d\varsigma = \begin{bmatrix} \int_0^x \frac{\psi_{1,0}(\varsigma)}{(x-\varsigma)^{0.5}}d\varsigma \\ \int_0^x \frac{\psi_{1,1}(\varsigma)}{(x-\varsigma)^{0.5}}d\varsigma \\ \int_0^x \frac{\psi_{1,2}(\varsigma)}{(x-\varsigma)^{0.5}}d\varsigma \end{bmatrix} = \begin{bmatrix} \frac{4}{3} & \frac{4}{5\sqrt{3}} & -\frac{4}{21\sqrt{5}} \\ -\frac{4}{5\sqrt{3}} & \frac{4}{7} & \frac{4}{3\sqrt{15}} \\ -\frac{4}{21\sqrt{5}} & -\frac{4}{3\sqrt{15}} & \frac{100}{231} \end{bmatrix}.\Psi(x) = \widehat{E}.\Psi(x),$$

where

$$\widehat{E} = \begin{bmatrix} \frac{4}{3} & \frac{4}{5\sqrt{3}} & -\frac{4}{21\sqrt{5}} \\ -\frac{4}{5\sqrt{3}} & \frac{4}{7} & \frac{4}{3\sqrt{15}} \\ -\frac{4}{21\sqrt{5}} & -\frac{4}{3\sqrt{15}} & \frac{100}{231} \end{bmatrix}.$$

Similarly, for $k = 2$ and $M = 3$, we get

$$\widehat{E} = \begin{bmatrix} \frac{2\sqrt{2}}{3} & \frac{2\sqrt{2}}{5\sqrt{3}} & -\frac{2\sqrt{2}}{21\sqrt{5}} & -\frac{4}{3}(-2+\sqrt{2}) & \frac{8}{5\sqrt{3}}(-3+2\sqrt{2}) & \frac{4\sqrt{2}}{21\sqrt{5}}(-83+59\sqrt{2}) \\ -\frac{2\sqrt{2}}{5\sqrt{3}} & \frac{2\sqrt{2}}{7} & \frac{2\sqrt{2}}{3\sqrt{15}} & -\frac{8}{5\sqrt{3}}(-3+2\sqrt{2}) & -\frac{4}{35}(-26+19\sqrt{2}) & -\frac{4\sqrt{2}}{21\sqrt{5}}(-104+573\sqrt{2}) \\ -\frac{2\sqrt{2}}{21\sqrt{5}} & -\frac{2\sqrt{2}}{3\sqrt{15}} & \frac{50\sqrt{2}}{231} & -\frac{4}{21\sqrt{5}}(-118+83\sqrt{2}) & -\frac{8}{21\sqrt{15}}(-73+52\sqrt{2}) & -\frac{4}{231}(-162+113\sqrt{2}) \\ 0 & 0 & 0 & \frac{2\sqrt{2}}{3} & \frac{2\sqrt{2}}{5\sqrt{3}} & -\frac{2\sqrt{2}}{21\sqrt{5}} \\ 0 & 0 & 0 & -\frac{2\sqrt{2}}{5\sqrt{3}} & \frac{2\sqrt{2}}{7} & \frac{2\sqrt{2}}{3\sqrt{15}} \\ 0 & 0 & 0 & -\frac{2\sqrt{2}}{21\sqrt{5}} & -\frac{2\sqrt{2}}{3\sqrt{15}} & \frac{50\sqrt{2}}{231} \end{bmatrix}.$$

6.3.3 BWM FOR VOLTERRA WEAKLY SINGULAR LPIDEs

In this part, we apply the OM scheme based on proposed 2D wavelets for solving Volterra weakly singular LPIDEs of first and second kinds.

6.3.3.1 1st kind

For the first kind of singular LPIDEs, we consider $\tilde{\mathrm{H}}(v(x,t)) = v_{tt}(x,t)$ in Eq. (6.1). Therefore, we have

$$v_{xt}(x,t)+v_t(x,t) = v(x,t)+h(x,t)+\int_0^x\int_0^t \frac{v_{tt}(s,y)}{(x-s)^{\theta}}\,dyds, \quad (x,t)\in\Delta. \tag{6.18}$$

To demonstrate the proposed method for solving Eq. (6.18), we start with approximations of $v_{tt}(x,t), v_t(x,t), h(x,t), v(x,0)$ and $v(0,t)$ as follows:

$$v_{xt}(x,t) = \Psi^T(x)C\Psi(t) = \Psi^T(t)C^T\Psi(x), \tag{6.19}$$

$$h(x,t) = \Psi^T(x)H\Psi(t) = \Psi^T(t)H^T\Psi(x), \tag{6.20}$$

$$v_t(0,t) = \Psi^T(x)Z\Psi(t) = \Psi^T(t)Z^T\Psi(x), \tag{6.21}$$

$$v(x,0) = \Psi^T(x)U\Psi(t) = \Psi^T(t)U^T\Psi(x), \tag{6.22}$$

$$v(0,t) = \Psi^T(x)V\Psi(t) = \Psi^T(t)V^T\Psi(x), \tag{6.23}$$

$$v(0,0) = \Psi^T(x)W\Psi(t) = \Psi^T(t)W^T\Psi(x). \tag{6.24}$$

By integrating Eq. (6.19) with respect to x and by using Eqs. (6.21) and (6.6), we have

$$\begin{aligned} v_t(x,t) &= v_t(0,t)+\Psi^T(t)C^T\widehat{P}\Psi(x) \\ &= \Psi^T(x)Z\Psi(t)+\Psi^T(x)\widehat{P}^TC\Psi(t). \end{aligned} \tag{6.25}$$

Again integrating Eq. (6.25) with respect to t and using Eqs. (6.22) and (6.6), we have

$$\begin{aligned} v(x,t) &= v(x,0)+\Psi^T(x)Z\widehat{P}\Psi(t)+\Psi^T(x)\widehat{P}^TC\widehat{P}\Psi(t) \\ &= \Psi^T(x)U\Psi(t)+\Psi^T(x)Z\widehat{P}\Psi(t)+\Psi^T(x)\widehat{P}^TC\widehat{P}\Psi(t). \end{aligned} \tag{6.26}$$

By differentiating Eq. (6.21) with respect to t and using Eq. (6.11), we get

$$\begin{aligned} v_{tt}(x,t) &= \Psi^T(x)Z\hat{D}\Psi(t) + \Psi^T(x)\hat{P}^T C\hat{D}\Psi(t) \\ &= \Psi^T(t)\hat{D}^T Z^T\Psi(x) + \Psi^T(t)\hat{D}^T C^T\hat{P}\Psi(x). \end{aligned} \tag{6.27}$$

Now by inserting the Eqs. (6.19)–(6.27) into Eq. (6.18) and using Eq. (6.13), it follows

$$\begin{aligned} &\Psi^T(x)C\Psi(t) + \Psi^T(x)Z\Psi(t) + \Psi^T(x)\hat{P}^T C\Psi(t) = \Psi^T(x)U\Psi(t) + \Psi^T(x)Z\hat{P}\Psi(t) \\ &\quad + \Psi^T(x)\hat{P}^T C\hat{P}\Psi(t) + \Psi^T(x)H\Psi(t) + \Psi^T(x)\hat{E}^T Z\hat{D}\hat{P}\Psi(t) + \Psi^T(x)\hat{E}^T\hat{P}^T C\hat{D}\hat{P}\Psi(t). \end{aligned}$$

whence

$$C + Z - H - U + \hat{P}^T C - Z\hat{P} - \hat{P}^T C\hat{P} - \hat{E}^T Z\hat{D}\hat{P} - \hat{E}^T\hat{P}^T C\hat{D}\hat{P} = 0. \tag{6.28}$$

Equation (6.28) gives a linear system of $2^{k_1-1}2^{k_2-1}M_1M_2$-algebraic equations. After solving the above system (6.28), we can obtain the value of C immediately.

Hence, by substituting the value of C in Eq. (6.26), we can obtain the approximate solution of Eq. (6.18).

6.3.3.2 Second Kind

For the second kind of singular LPIDEs, we consider $\tilde{\mathrm{H}}(v(x,t)) = v_{xt}(x,t)$ in Eq. (6.1). Therefore, we have

$$v_{xt}(x,t) + v_t(x,t) = v(x,t) + h(x,t) + \int_0^x\int_0^t \frac{v_{xt}(s,y)}{(x-s)^\theta}\,dyds, \quad (x,t)\in\Delta. \tag{6.29}$$

Now by inserting the Eqs. (6.19)–(6.26) into Eq. (6.29), it follows

$$\begin{aligned} &\Psi^T(x)C\Psi(t) + \Psi^T(x)Z\Psi(t) + \Psi^T(x)\hat{P}^T C\Psi(t) = \Psi^T(x)U\Psi(t) + \Psi^T(x)Z\hat{P}\Psi(t) \\ &\quad + \Psi^T(x)\hat{P}^T C\hat{P}\Psi(t) + \Psi^T(x)H\Psi(t) + \Psi^T(x)\hat{E}^T C\hat{P}\Psi(t). \end{aligned}$$

whence

$$C + Z - H - U + \hat{P}^T C - Z\hat{P} - \hat{P}^T C\hat{P} - \hat{E}^T C\hat{P} = 0. \tag{6.30}$$

Equation (6.30) gives a linear system of $2^{k_1-1}2^{k_2-1}M_1M_2$-algebraic equations. After solving the above system Eq. (6.30), we can obtain the value of C immediately.

Hence, by substituting the value of C in Eq. (6.26), we can obtain the approximate solution of Eq. (6.29).

6.4 TWO-DIMENSIONAL NONLINEAR VOLTERRA WEAKLY SINGULAR PIDEs

This section is based on Volterra weakly singular NLPIDEs which is of the following form:

$$v_{xt}(x,t)+v_t(x,t)=v(x,t)+h(x,t)+\int_0^x\int_0^t \Phi(x,s)\tilde{H}(v(s,y))dyds, \quad (x,t)\in\Delta, \tag{6.31}$$

having following boundary conditions:

$$v(x,0)=v_0(x),$$

$$v(0,t)=v_0(t).$$

In Eq. (6.31), $\tilde{H}(v(x,t)), h(x,t), v_0(x), v_0(t)$ are known functions and $v(x,t)$ is an unknown function which should be determined. The domain $\Delta=[0,1]\times[0,1]$, $\Phi(x,s)=\dfrac{1}{(x-s)^{\theta}}$, $0<\theta<1$, and the differential operator $\tilde{H}$ is considered as $\tilde{H}(v(x,t))=[v(x,t)]^{\rho}$, and $\tilde{H}(v(x,t))=[\eta_{i_1,i_2}v(x,t)]^{\rho}$ for the nonlinear case as the first and second kinds, respectively, where ρ is a positive number and η_{i_1,i_2} is a differential operator. For the existence of the solution $v(x,t)\in C(\Delta)$, it is assumed that $h(x,t)$ and $v(x,t)$ are sufficiently smooth functions.

The known and unknown functions are approximated by means of 2D BWs along with the corresponding OMs. In the proposed technique, the OM scheme transforms the PIDEs into the algebraic system of equation. The solution of Eq. (6.31) by BWM is calculated after solving the system of algebraic equations. In Eq. (6.31), it is seen that the kernel function $\Phi(x,s)$ has a weak singularity at $s=x$.

6.4.1 Function Approximation by BWs

A function $h(x,t)$ defined over $[0,1)\times[0,1)$ is approximated as follows:

$$h(x,t)=\sum_{n=1}^{2^{k_1-1}}\sum_{n'=1}^{2^{k_2-1}}\sum_{m=0}^{M_1-1}\sum_{m'=0}^{M_2-1}\overline{w}_{n,m,n',m'}.\psi_{n,m,n',m'}(x,t)=W^T.\Psi(x,t), \tag{6.32}$$

where W and $\Psi(x,t)$ are $(2^{k_1-1}2^{k_2-1}M_1M_2\times 1)$ vectors given by

$$W=[\overline{w}_{1,0,1,0},\ldots,\overline{w}_{1,0,1,M_2-1},\ldots,\overline{w}_{1,0,2^{k_2-1},M_2-1},\ldots,\overline{w}_{2^{k_1-1},M_1-1,2^{k_2-1},M_2-1}]^T, \tag{6.33}$$

$$\Psi(x,t)=[\psi_{1,0,1,0}(x,t),\ldots,\psi_{1,0,1,M_2-1}(x,t),\ldots,\psi_{1,0,2^{k_2-1},M_2-1}(x,t),\ldots,\psi_{2^{k_1-1},M_1-1,2^{k_2-1},M_2-1}(x,t)]^T. \tag{6.34}$$

The $2^{k_1-1}2^{k_2-1}M_1M_2$ vector W contains the coefficient $\overline{w}_{n,m,n',m'}$ that is defined as follows:

$$\overline{w}_{n,m,n',m'} = \frac{<\psi_{n,m}(x),<h(x,t),\psi'_{n',m'}(t)>>}{<\psi_{n,m}(x),\psi_{n,m}(x)>.<\psi'_{n',m'}(t),\psi'_{n',m'}(t)>},$$

where <.,.> denotes the inner product and

$$<\psi_{n,m}(x),\psi_{n,m}(x)>= \int_{\frac{n-1}{2^{k_1-1}}}^{\frac{n}{2^{k_1-1}}} \psi_{n,m}(x).\psi_{n,m}(x)\,dx = 1,$$

$$<\psi'_{n',m'}(t),\psi'_{n',m'}(t)>= \int_{\frac{n'-1}{2^{k_2-1}}}^{\frac{n'}{2^{k_2-1}}} \psi'_{n',m'}(t),\psi'_{n',m'}(t)\,dt = 1..$$

We have

$$\overline{w}_{n,m,n',m'} = \int_{\frac{n-1}{2^{k_1-1}}}^{\frac{n}{2^{k_1-1}}} \psi_{n,m}(x) \int_{\frac{n'-1}{2^{k_2-1}}}^{\frac{n'}{2^{k_2-1}}} h(x,t).\psi'_{n',m'}(t)\,dt\,dx. \tag{6.35}$$

6.4.2 Operational Matrices

Now we derive the OMs of integration, differentiation, and almost integrational OM for the proposed wavelet methods to solve the weakly singular NLPIDEs.

6.4.2.1 Operational Matrix of Integration with Respect to *x*

The integration of 2D BWs $\Psi(x,t)$ defined in Eq. (6.34) is given by [60]

$$\int_0^x \Psi(\varsigma,t)\,d\varsigma = \int_0^x \Psi(\varsigma)\otimes\Psi(t)\,d\varsigma$$

$$= \left(\int_0^x \Psi(\varsigma)d\varsigma\right)\otimes\Psi(t)$$

$$= \begin{pmatrix} \int_0^x \psi_{1,0}(\varsigma)\,d\varsigma \\ \int_0^x \psi_{1,1}(\varsigma)\,d\varsigma \\ . \\ . \\ . \\ \int_0^x \psi_{2^{k_1-1},M_1-1}(\varsigma)\,d\varsigma \end{pmatrix} \otimes \Psi(t)$$

$$= \hat{P}\Psi(x) \otimes I\Psi(t)$$

$$= (\hat{P} \otimes I)(\Psi(x) \otimes \Psi(t))$$

$$= \bar{P}_x \Psi(x,t).$$

Thus, we have

$$\int_0^x \Psi(\varsigma,t)\,d\varsigma = \bar{P}_x \Psi(x,t), \tag{6.36}$$

where I denotes the identity matrix of order $2^{k_1-1}M_1$ and $\bar{P}_x$ is a $(2^{k_1-1}2^{k_2-1}M_1M_2 \times 2^{k_1-1}2^{k_2-1}M_1M_2)$ matrix in which $\hat{P}$ is a matrix of dimension $2^{k_1-1}M_1$ given as follows:

$$\hat{P} = \begin{bmatrix} L & F & F & . & . & . & F \\ 0 & L & F & . & . & . & F \\ 0 & 0 & L & . & . & . & F \\ . & . & . & . & . & . & . \\ . & . & . & . & . & . & . \\ . & . & . & . & . & . & F \\ 0 & 0 & 0 & . & . & . & L \end{bmatrix},$$

where L and F are matrices of order $2^{k_1-2}M_1$.

For $k = 2$ and $M = 3$, we have

$$\widehat{P} = \begin{bmatrix} \frac{1}{4} & \frac{1}{4\sqrt{3}} & 0 & \frac{1}{2} & 0 & 0 \\ -\frac{1}{4\sqrt{3}} & 0 & \frac{1}{4\sqrt{15}} & 0 & 0 & 0 \\ 0 & -\frac{1}{4\sqrt{15}} & 0 & 0 & 0 & 0 \\ 0 & 0 & 0 & \frac{1}{4} & \frac{1}{4\sqrt{3}} & 0 \\ 0 & 0 & 0 & -\frac{1}{4\sqrt{3}} & 0 & \frac{1}{4\sqrt{15}} \\ 0 & 0 & 0 & 0 & -\frac{1}{4\sqrt{15}} & 0 \end{bmatrix},$$

where $L = \begin{bmatrix} \frac{1}{4} & \frac{1}{4\sqrt{3}} & 0 \\ -\frac{1}{4\sqrt{3}} & 0 & \frac{1}{4\sqrt{15}} \\ 0 & -\frac{1}{4\sqrt{15}} & 0 \end{bmatrix}$ and $F = \begin{bmatrix} \frac{1}{2} & 0 & 0 \\ 0 & 0 & 0 \\ 0 & 0 & 0 \end{bmatrix}$.

6.4.2.2 Operational Matrix of Integration with Respect to *t*

OM of integration for the variable t is defined as follows:

$$\int_0^t \Psi(x,\chi)\,d\chi = \int_0^t \Psi(x)\otimes\Psi(\chi)\,d\chi$$

$$= \Psi(x)\otimes\left(\int_0^t \Psi(\chi)d\chi\right)$$

$$= \Psi(x)\otimes\begin{pmatrix} \int_0^t \psi_{1,0}(\chi)\,d\chi \\ \int_0^t \psi_{1,1}(\chi)\,d\chi \\ \cdot \\ \cdot \\ \cdot \\ \int_0^t \psi_{2^{k_2-1},M'-1}(\chi)\,d\chi \end{pmatrix}$$

$$= I\Psi(x) \otimes \widehat{P}'\Psi(t)$$
$$= (I \otimes \widehat{P}')(\Psi(x) \otimes \Psi(t))$$
$$= \bar{P}_t\Psi(x,t).$$

Thus, we have

$$\int_0^t \Psi(x,\chi)\,d\chi = \bar{P}_t\Psi(x,t), \tag{6.37}$$

where I denotes the identity matrix of order $2^{k_2-1}M_2$ and $\bar{P}_t$ is a $(2^{k_1-1}2^{k_2-1}M_1M_2 \times 2^{k_1-1}2^{k_2-1}M_1M_2)$ matrix in which $\widehat{P}'$ is a matrix of dimension $2^{k_2-1}M_2$.

6.4.2.3 Differentiation Operational Matrix with Respect to *x*

The derivative matrix of 2D BWs vector $\Psi(x,t)$ defined in Eq. (6.34) is derived as follows:

$$\frac{\partial}{\partial x}\Psi(x,t) = \frac{\partial}{\partial x}(\Psi(x) \otimes \Psi(t))$$
$$= (\bar{D}\Psi(x)) \otimes (I\Psi(t))$$
$$= (\bar{D} \otimes I)(\Psi(x) \otimes \Psi(t))$$
$$= \bar{D}_x\Psi(x,t).$$

So,

$$\frac{\partial}{\partial x}\Psi(x,t) = \bar{D}_x\Psi(x,t), \tag{6.38}$$

where I is the identity matrix of order $2^{k_1-1}M_1$ and $\bar{D}_x$ is a $(2^{k_1-1}2^{k_2-1}M_1M_2 \times 2^{k_1-1}2^{k_2-1}M_1M_2)$ matrix in which $\bar{D}$ is a matrix of dimension $2^{k_1-1}M_1$ given as follows:

$$\bar{D} = \begin{bmatrix} L_1 & F_1 & F_1 & . & . & . & F_1 \\ 0 & L_1 & F_1 & . & . & . & F_1 \\ 0 & 0 & L_1 & . & . & . & F_1 \\ . & . & . & . & . & . & . \\ . & . & . & . & . & . & . \\ . & . & . & . & . & . & F_1 \\ 0 & 0 & 0 & . & . & . & L_1 \end{bmatrix},$$

where L_1 and F_1 are matrices of order $2^{k_1-2}M_1$.

For $k = 2$ and $M = 3$, we have

$$\bar{D} = \begin{bmatrix} 0 & 0 & 0 & 0 & 0 & 0 \\ 4\sqrt{3} & 0 & 0 & 0 & 0 & 0 \\ 0 & 4\sqrt{15} & 0 & 0 & 0 & 0 \\ 0 & 0 & 0 & 0 & 0 & 0 \\ 0 & 0 & 0 & 4\sqrt{3} & 0 & 0 \\ 0 & 0 & 0 & 0 & 4\sqrt{15} & 0 \end{bmatrix},$$

where $L_1 = \begin{bmatrix} 0 & 0 & 0 \\ 4\sqrt{3} & 0 & 0 \\ 0 & 4\sqrt{15} & 0 \end{bmatrix}$ and $F_1 = \begin{bmatrix} 0 & 0 & 0 \\ 0 & 0 & 0 \\ 0 & 0 & 0 \end{bmatrix}$.

Similarly, OM for differentiation with respect to t can be computed as follows:

$$\begin{aligned} \frac{\partial}{\partial t}\Psi(x,t) &= \frac{\partial}{\partial t}(\Psi(x) \otimes \Psi(t)) \\ &= (I\Psi(x)) \otimes (\bar{D}'\Psi(t)) \\ &= (I \otimes \bar{D}')(\Psi(x) \otimes \Psi(t)) \\ &= \bar{D}_t \Psi(x,t). \end{aligned}$$

Thus,

$$\frac{\partial}{\partial t}\Psi(x,t) = \bar{D}_t \Psi(x,t), \tag{6.39}$$

where I is the identity matrix of order $2^{k_2-1}M_2$ and $\bar{D}_t$ is a $(2^{k_1-1}2^{k_2-1}M_1M_2 \times 2^{k_1-1}2^{k_2-1}M_1M_2)$ matrix in which $\bar{D}'$ is a matrix of dimension $2^{k_2-1}M_2$.

6.4.2.4 Product Operational Matrix

The following property of the product of two vectors $\Psi(x,t)$ and $\Psi^T(x,t)$ defined in Eq. (6.34) is used [60].

Let

$$\Psi(x,t)\Psi^T(x,t)W \cong \tilde{W}\Psi(x,t), \tag{6.40}$$

where W is defined in Eq. (6.33) and $\tilde{W}$ is a $(2^{k_1-1}2^{k_2-1}M_1M_2 \times 2^{k_1-1}2^{k_2-1}M_1M_2)$-dimensional product OM.

6.4.2.5 Almost Operational Matrix of Integration with Singularity

Let $\Psi(x,t)$ be 2D BWs defined as in Eq. (6.34) for $0<\theta<1$, and then we get

$$\begin{aligned}\int_0^x \frac{\Psi(\varsigma,t)}{(x-\varsigma)^\theta}d\varsigma &= \int_0^x \frac{\Psi(\varsigma)\otimes\Psi(t)}{(x-\varsigma)^\theta}d\varsigma \\ &= \int_0^x \frac{\Psi(\varsigma)}{(x-\varsigma)^\theta}d\varsigma\otimes\Psi(t) \\ &= (E\Psi(x))\otimes(I\Psi(t)) \\ &= (E\otimes I)(\Psi(x)\otimes\Psi(t)) \\ &= \bar{E}\Psi(x,t).\end{aligned}$$

Thus,

$$\int_0^x \frac{\Psi(\varsigma,t)}{(x-\varsigma)^\theta}d\varsigma = \bar{E}\Psi(x,t), \tag{6.41}$$

where $\bar{E}$ is the almost OM of dimension $(2^{k_1-1}2^{k_2-1}M_1M_2\times 2^{k_1-1}2^{k_2-1}M_1M_2)$.

6.4.3 BWM for Volterra Weakly Singular NLPIDEs

In this section, we use the OM method derived by proposed 2D wavelets for solving Volterra weakly singular NLPIDEs of first and second kinds.

6.4.3.1 1st kind

For the first kind of weakly singular NLPIDEs, let $\tilde{H}(v(x,t))=[v(x,t)]^p$, in Eq. (6.31). Therefore, we have

$$v_{xt}(x,t)+v_t(x,t)=v(x,t)+h(x,t)+\int_0^x\int_0^t \frac{[v(s,y)]^p}{(x-s)^\theta}dyds, \quad (x,t)\in\Delta. \tag{6.42}$$

To demonstrate the proposed method for solving Eq. (6.42), we start with approximations of $v_{xt}(x,t), v_t(x,t), h(x,t), v(x,0)$, and $v(0,t)$ as follows:

$$v_{xt}(x,t)=\Psi^T(x,t)\bar{C}=\bar{C}^T\Psi(x,t), \tag{6.43}$$

$$h(x,t)=\Psi^T(x,t)\bar{H}=\bar{H}^T\Psi(x,t), \tag{6.44}$$

$$v_t(0,t)=\Psi^T(x,t)\bar{F}=\bar{F}^T\Psi(x,t), \tag{6.45}$$

$$v(x,0)=\Psi^T(x,t)\bar{U}=\bar{U}^T\Psi(x,t), \tag{6.46}$$

$$v(0,t)=\Psi^T(x,t)\bar{V}=\bar{V}^T\Psi(x,t). \tag{6.47}$$

By integrating Eq. (6.43) from both sides with respect to x and by using Eqs. (6.45) and (6.36), we have

$$\begin{aligned} v_t(x,t) &= v_t(0,t)+\bar{C}^T\bar{P}_x\Psi(x,t) \\ &= \Psi^T(x,t)\bar{F}+\Psi^T(x,t)\bar{P}_x{}^T\bar{C} \\ &= \Psi^T(x,t)(\bar{F}+\bar{P}_x{}^T\bar{C}). \end{aligned} \tag{6.48}$$

Again integrating Eq. (6.43) from both sides with respect to t and using Eqs. (6.46) and (6.37), we have

$$\begin{aligned} v(x,t) &= v(x,0)+(\bar{F}^T+\bar{C}^T\bar{P}_x)\bar{P}_t\Psi(x,t) \\ &= \bar{U}^T\Psi(x,t)+\Psi^T(x,t)\bar{P}_t^T\bar{F}+\Psi^T(x,t)\bar{P}_t^T\bar{P}_x{}^T\bar{C} \\ &= \Psi^T(x,t)(\bar{U}+\bar{P}_t^T\bar{F}+\bar{P}_t^T\bar{P}_x{}^T\bar{C}). \end{aligned} \tag{6.49}$$

Thus, $v(x,t)=(\bar{U}^T+\bar{F}^T\bar{P}_t+\bar{C}^T\bar{P}_x\bar{P}_t)\Psi(x,t)$

$=\bar{R}^T\Psi(x,t),$

where $\bar{R}^T$ is a $(1\times 2^{k_1-1}2^{k_2-1}M_1M_2)$-dimensional vector.

Now, in the case of $\rho=2$, we have

$$\begin{aligned} [v(x,t)]^2 &= (\bar{R}^T\Psi(x,t)).(\bar{R}^T\Psi(x,t)) \\ &= \bar{R}^T\Psi(x,t)\Psi^T(x,t)\bar{R} \\ &= \bar{R}^T\hat{R}\Psi(x,t) \\ &= \Psi^T(x,t)\hat{R}^T\bar{R}, \end{aligned} \tag{6.50}$$

where $\Psi(x,t)\Psi^T(x,t)\bar{R}=\hat{R}\Psi(x,t)$ and $\hat{R}$ is a matrix of dimension $(2^{k_1-1}2^{k_2-1}M_1M_2\times 2^{k_1-1}2^{k_2-1}M_1M_2)$.

Now by inserting the Eqs. (6.43)–(6.50) into Eq. (6.42) and using Eq. (6.41), we have

$$\begin{aligned} \bar{C}^T\Psi(x,t)+(\bar{F}^T+\bar{C}^T\bar{P}_x)\Psi(x,t) &= (\bar{U}^T+\bar{F}^T\bar{P}_t+\bar{C}^T\bar{P}_x\bar{P}_t)\Psi(x,t) \\ &\quad +\bar{H}^T\Psi(x,t)+\bar{R}^T\hat{R}\bar{E}\Psi(x,t). \end{aligned}$$

Therefore,

$$\Psi^T(x,t)\bar{C}+\Psi^T(x,t)(\bar{F}+\bar{P}_x{}^T\bar{C})=\Psi^T(x,t)(\bar{U}+\bar{P}_t^T\bar{F}+\bar{P}_t^T\bar{P}_x{}^T\bar{C})+\Psi^T(x,t)\bar{H}$$
$$+\Psi^T(x,t)\bar{E}^T\hat{R}^T\bar{R}. \tag{6.51}$$

We now collocate Eq. (6.51) at $2^{k_1-1}M_1\times 2^{k_2-1}M_2$ points (x_p,t_q) as follows:

$$\Psi^T(x_p,t_q)\bar{C}+\Psi^T(x_p,t_q)(\bar{F}+\bar{P}_x{}^T\bar{C})=\Psi^T(x_p,t_q)(\bar{U}+\bar{P}_t^T\bar{F}+\bar{P}_t^T\bar{P}_x{}^T\bar{C})+\Psi^T(x_p,t_q)\bar{H}$$
$$+\Psi^T(x_p,t_q)\bar{E}^T\hat{R}^T\bar{R}, \tag{6.52}$$

where the collocation points x_p and t_q are

$$x_p=\frac{2p-1}{2.2^{k_1-1}.M_1},\quad p=1,2,\ldots,2^{k_1-1}.M_1,$$

$$t_q=\frac{2q-1}{2.2^{k_2-1}.M_2},\quad q=1,2,\ldots,2^{k_2-1}.M_2.$$

Plugging the collocation points in Eq. (6.52), it gives a nonlinear system of $2^{k_1-1}2^{k_2-1}M_1M_2$-algebraic equations. After solving the algebraic system of equations, we can obtain the value of $\bar{C}$ immediately.

Hence, by substituting the value of $\bar{C}$ in Eq. (6.49), we can obtain the numerical solution of Eq. (6.42).

6.4.3.2 Second Kind

For the second kind of weakly singular NLPIDEs, let $\tilde{H}(v(x,t))=[\eta_{i_1,i_2}v(x,t)]^{\rho}$, in Eq. (6.31). Therefore, we have

$$v_{xt}(x,t)+v_t(x,t)=v(x,t)+h(x,t)+\int_0^x\int_0^t\frac{[\eta_{i_1,i_2}v(s,y)]^{\rho}}{(x-s)^{\theta}}dyds,\quad (x,t)\in\Delta. \tag{6.53}$$

Let

$$v(x,t)=\bar{R}^T\Psi(x,t), \tag{6.54}$$

where $\bar{R}^T=(\bar{U}^T+\bar{F}^T\bar{P}_t+\bar{C}^T\bar{P}_x\bar{P}_t)$.

Differentiating Eq. (6.54) 'i_1' times w.r.t x and 'i_2' times w.r.t t, we get

$$\eta_{i_1,i_2}v(x,t)=\bar{R}^T\bar{D}_x^{i_1}\bar{D}_t^{i_2}\Psi(x,t)$$
$$=\bar{R}_1{}^T\bar{\Psi}(x,t),$$

where $\overline{R}_1^T = \overline{R}^T \overline{D}_x^{i_1} \overline{D}_t^{i_2}$ is a vector of dimension $(1\times 2^{k_1-1}2^{k_2-1}M_1M_2)$.

Now in the case of $\rho = 2$, we have

$$\begin{aligned}[\eta_{i,j}v(x,t)]^2 &= (\overline{R}_1^T\Psi(x,t)).(\overline{R}_1^T\Psi(x,t))\\ &= \overline{R}_1^T\Psi(x,t)\Psi^T(x,t)\overline{R}_1\\ &= \overline{R}_1^T\hat{R}_1\Psi(x,t)\\ &= \Psi^T(x,t)\hat{R}_1^T\overline{R}_1,\end{aligned} \tag{6.55}$$

where $\Psi(x,t)\Psi^T(x,t)\overline{R}_1 = \hat{R}_1\Psi(x,t)$ and $\hat{R}_1$ is a matrix of dimension $(2^{k_1-1}2^{k_2-1}M_1M_2\times 2^{k_1-1}2^{k_2-1}M_1M_2)$.

Now by inserting the Eqs. (6.43)–(6.49) and Eq. (6.55) into Eq. (6.53) and using Eq. (6.38), we have

$$\begin{aligned}\Psi^T(x,t)\overline{C}+\Psi^T(x,t)(\overline{F}+\overline{P}_x{}^T\overline{C}) &= \Psi^T(x,t)(\overline{U}+\overline{P}_t^T\overline{F}+\overline{P}_t^T\overline{P}_x{}^T\overline{C})+\Psi^T(x,t)\overline{H}\\ &\quad +\Psi^T(x,t)\overline{E}^T\hat{R}_1^T\overline{R}_1.\end{aligned} \tag{6.56}$$

We now collocate Eq. (6.56) at $2^{k_1-1}M_1\times 2^{k_2-1}M_2$ points (x_p,t_q) as follows:

$$\begin{aligned}\Psi^T(x_p,t_q)\overline{C}+\Psi^T(x_p,t_q)(\overline{F}+\overline{P}_x{}^T\overline{C}) &= \Psi^T(x_p,t_q)(\overline{U}+\overline{P}_t^T\overline{F}+\overline{P}_t^T\overline{P}_x{}^T\overline{C})+\Psi^T(x_p,t_q)\overline{H}\\ &\quad +\Psi^T(x_p,t_q)\overline{E}^T\hat{R}_1^T\overline{R}_1,\end{aligned} \tag{6.57}$$

where the collocation points x_p and t_q are

$$x_p = \frac{2p-1}{2.2^{k_1-1}.M_1},\quad p = 1,2,\ldots,2^{k_1-1}.M_1,$$

$$t_q = \frac{2q-1}{2.2^{k_2-1}.M_2},\quad q = 1,2,\ldots,2^{k_2-1}.M_2.$$

Plugging the collocation points in Eq. (6.57), it gives a nonlinear system of $2^{k_1-1}2^{k_2-1}M_1M_2$ -algebraic equations. After solving the algebraic system of equations, we can obtain the value of $\overline{C}$ immediately.

Hence, by substituting the value of $\overline{C}$ in Eq. (6.49), we get the numerical solution of Eq. (6.53).

6.5 CONVERGENCE AND ERROR ANALYSIS

Theorem 6.5.1

Let $\left(\frac{\partial^\tau v(x,t)}{\partial t^\tau}\right)_N$ be the BWA of $\left(\frac{\partial^\tau v(x,t)}{\partial t^\tau}\right)$ and assumed that the mixed second derivative of $\left(\frac{\partial^\tau v(x,t)}{\partial t^\tau}\right)$ is bounded by a constant $\widehat{\mathcal{K}}_1$, i.e., $\left|\frac{\partial^{\tau+4} v(x,t)}{\partial x^2 \partial t^{\tau+2}}\right| < \widehat{\mathcal{K}}_1$, and then we have the following bound of error:

$$\left\|\left(\frac{\partial^\tau v}{\partial t^\tau}\right)-\left(\frac{\partial^\tau v}{\partial t^\tau}\right)_N\right\|_2^2 \leq \sum_{n=2^{k_1-1}+1}^{\infty} \sum_{n'=2^{k_2-1}+1}^{\infty} \sum_{m=M_1}^{\infty} \sum_{m'=M_2}^{\infty} \left(\frac{A_1}{2^{\frac{k_1-1}{2}}} \frac{A_2}{2^{\frac{k_2-1}{2}}} \widehat{\mathcal{K}}_1 \frac{16m!}{(2\pi)^{m+1}} \frac{16m'!}{(2\pi)^{m'+1}}\right)^2,$$

where $A_1 = \dfrac{1}{\sqrt{\dfrac{(-1)^{m-1}(m!)^2}{(2m)!}\tilde{\alpha}_{2m}}}$ and $A_2 = \dfrac{1}{\sqrt{\dfrac{(-1)^{m'-1}(m'!)^2}{(2m')!}\tilde{\alpha}_{2m'}}}$.

Moreover, the BW expansion of $h(x,t)$ defined in Eq. (6.2) converges uniformly [61].

Proof:

Let $\frac{\partial^\tau v}{\partial t^\tau} = \sum_{n=1}^{\infty}\sum_{n'=1}^{\infty}\sum_{m=0}^{\infty}\sum_{m'=0}^{\infty} h_{n,m,n',m'}\psi_{n,m,n',m'}(x,t)$.

By truncating the above series, we get

$$\left(\frac{\partial^\tau v}{\partial t^\tau}\right)_N \approx \sum_{n=1}^{2^{k_1-1}}\sum_{n'=1}^{2^{k_2-1}}\sum_{m=0}^{M_1-1}\sum_{m'=0}^{M_2-1} h_{n,m,n',m'}\psi_{n,m,n',m'}(x,t).$$

Now,

$$\left(\frac{\partial^\tau v}{\partial t^\tau}\right)-\left(\frac{\partial^\tau v}{\partial t^\tau}\right)_N = \sum_{n=2^{k_1-1}+1}^{\infty} \sum_{n'=2^{k_2-1}+1}^{\infty} \sum_{m=M}^{\infty} \sum_{m'=M'}^{\infty} h_{n,m,n',m'}\psi_{n,m,n',m'}(x,t).$$

Taking L^2 norm both sides, we get

$$\left\|\left(\frac{\partial^\tau v}{\partial t^\tau}\right)-\left(\frac{\partial^\tau v}{\partial t^\tau}\right)_N\right\|_2^2=\int_0^1\int_0^1\left(\left(\frac{\partial^\tau v}{\partial t^\tau}\right)-\left(\frac{\partial^\tau v}{\partial t^\tau}\right)_N\right)^2 dxdt$$

$$=\int_0^1\int_0^1\left(\sum_{n=2^{k_1-1}+1}^{\infty}\sum_{n'=2^{k_2-1}+1}^{\infty}\sum_{m=M_1}^{\infty}\sum_{m'=M_2}^{\infty} h_{n,m,n',m'}\psi_{n,m,n',m'}(x,t)\right)^2 dxdt$$

$$=\sum_{n=2^{k_1-1}+1}^{\infty}\sum_{n'=2^{k_2-1}+1}^{\infty}\sum_{m=M_1}^{\infty}\sum_{m'=M_2}^{\infty} h^2_{n,m,n',m'}, \tag{6.58}$$

where

$$h_{n,m,n',m'}=\int_0^1\int_0^1\left(\frac{\partial^\tau v}{\partial t^\tau}\right).\psi_{n,m,n',m'}(x,t)\,dx\,dt$$

$$=\int_0^1\left[\int_0^1\left(\frac{\partial^\tau v}{\partial t^\tau}\right)\psi_{n,m}(x)\,dx\right]\psi'_{n',m'}(t)\,dt$$

$$=\int_{\frac{n'-1}{2^{k_2-1}}}^{\frac{n'}{2^{k_2-1}}}\left[\int_{\frac{n-1}{2^{k_1-1}}}^{\frac{n}{2^{k_1-1}}}\left(\frac{\partial^\tau v}{\partial t^\tau}\right)\psi_{n,m}(x)\,dx\right]\psi'_{n',m'}(t)\,dt$$

$$=2^{\frac{k_1-1}{2}}2^{\frac{k_2-1}{2}}A_1A_2\int_{\frac{n'-1}{2^{k_2-1}}}^{\frac{n'}{2^{k_2-1}}}\left[\int_{\frac{n-1}{2^{k_1-1}}}^{\frac{n}{2^{k_1-1}}}\left(\frac{\partial^\tau v}{\partial t^\tau}\right)\mathcal{B}_m(2^{k_1-1}x-n+1)\,dx\right]\mathcal{B}_{m'}(2^{k_2-1}t-n'+1)\,dt.$$

Now substituting $2^{k_1-1}x-n+1=u$, we have

$$h_{n,m,n',m'}=\frac{1}{2^{\frac{k_1-1}{2}}}2^{\frac{k_2-1}{2}}A_1A_2\int_{\frac{n'-1}{2^{k_2-1}}}^{\frac{n'}{2^{k_2-1}}}\left[\int_0^1\left(\frac{\partial^\tau v\left(\frac{u+n-1}{2^{k_1-1}},t\right)}{\partial t^\tau}\right)\mathcal{B}_m(u)\,du\right]\mathcal{B}_{m'}(2^{k_2-1}t-n'+1)\,dt.$$

Similarly, substituting $2^{k_2-1}t-n'+1=w$, we have

$$h_{n,m,n',m'} = \frac{A_1}{2^{\frac{k_1-1}{2}}}\frac{A_2}{2^{\frac{k_2-1}{2}}}\int_0^1\left[\int_0^1\left(\frac{\partial^\tau v\left(\frac{u+n-1}{2^{k_1-1}},\frac{w+n'-1}{2^{k_1-1}}\right)}{\partial t^\tau}\right)\mathcal{B}_m(u)\,du\right]\mathcal{B}_{m'}(w)\,dw$$

$$\leq \frac{A_1}{2^{\frac{k_1-1}{2}}}\frac{A_2}{2^{\frac{k_2-1}{2}}}\int_0^1\left[\int_0^1\left(\frac{\partial^\tau v\left(\frac{u+n-1}{2^{k_1-1}},\frac{w+n'-1}{2^{k_1-1}}\right)}{\partial t^\tau}\right)\left|\mathcal{B}_m(u)\right|du\right]\left|\mathcal{B}_{m'}(w)\right|dw$$

$$\leq \frac{A_1}{2^{\frac{k_1-1}{2}}}\frac{A_2}{2^{\frac{k_2-1}{2}}}\widehat{\mathcal{K}}_1\left(\int_0^1 |\,\mathcal{B}_m(u)\,|\,du\right)\left(\int_0^1 |\,\mathcal{B}_{m'}(w)\,|\,dw\right)$$

$$\leq \frac{A_1}{2^{\frac{k_1-1}{2}}}\frac{A_2}{2^{\frac{k_2-1}{2}}}\widehat{\mathcal{K}}_1\frac{16m!}{(2\pi)^{m+1}}\frac{16m'!}{(2\pi)^{m'+1}},\quad \text{using the property of Bernoulli polynomials.} \tag{6.59}$$

Equations (6.58) and (6.59) together imply

$$\left\|\left(\frac{\partial^\tau v}{\partial t^\tau}\right)-\left(\frac{\partial^\tau v}{\partial t^\tau}\right)_N\right\|_2^2 \leq \sum_{n=2^{k_1-1}+1}^{\infty}\sum_{n'=2^{k_2-1}+1}^{\infty}\sum_{m=M_1}^{\infty}\sum_{m'=M_2}^{\infty}\left(\frac{A_1}{2^{\frac{k_1-1}{2}}}\frac{A_2}{2^{\frac{k_2-1}{2}}}\widehat{\mathcal{K}}_1\frac{16m!}{(2\pi)^{m+1}}\frac{16m'!}{(2\pi)^{m'+1}}\right)^2.$$

Again, it is clear that at $m=0, m'=0$, $\{\psi_{n,m,n',m'}\}$ forms an orthonormal system composed by the Haar scaling function with respect to the weight function and thus the series

$$\sum_{n=1}^{2^{k_1-1}}\sum_{n'=1}^{2^{k_2-1}} h_{n,0,n',0}\psi_{n,0,n',0}(x,t)$$

is convergent. So,

$$\left|\sum_{n=1}^{\infty}\sum_{n'=1}^{\infty}\sum_{m=0}^{\infty}\sum_{m'=0}^{\infty} h_{n,m,n',m'}\psi_{n,m,n',m'}(x,t)\right| \leq \left|\sum_{n=1}^{2^{k_1-1}}\sum_{n'=1}^{2^{k_2-1}} h_{n,0,n',0}\psi_{n,0,n',0}(x,t)\right|$$

$$+\left|\sum_{n=1}^{\infty}\sum_{n'=1}^{\infty}\sum_{m=0}^{\infty}\sum_{m'=0}^{\infty} h_{n,m,n',m'}\psi_{n,m,n',m'}(x,t)\right|.$$

$$<\infty.$$

This implies that $\left|h_{n,m,n',m'}\right|$ is absolutely convergent, and hence the series

$$h(x,t)=\sum_{n=1}^{2^{k_1-1}}\sum_{n'=1}^{2^{k_2-1}}\sum_{m=0}^{M_1-1}\sum_{m'=0}^{M_2-1}h_{n,m,n',m'}\cdot\psi_{n,m,n',m'}(x,t)$$

is uniformly convergent.

Theorem 6.5.2

Let $\left(\dfrac{\partial^\tau v(x,t)}{\partial x^\tau}\right)_N$ be the BWA of $\left(\dfrac{\partial^\tau v(x,t)}{\partial x^\tau}\right)$ and assumed that the mixed second derivative of $\left(\dfrac{\partial^\tau v(x,t)}{\partial x^\tau}\right)$ is bounded by a constant $\widehat{\mathcal{K}}_2$, i.e., $\left|\dfrac{\partial^{\tau+4} v(x,t)}{\partial t^2\,\partial x^{\tau+2}}\right| < \widehat{\mathcal{K}}_2$, and then we have the following bound of error:

$$\left\|\left(\frac{\partial^\tau v}{\partial x^\tau}\right)-\left(\frac{\partial^\tau v}{\partial x^\tau}\right)_N\right\|_2^2 \le \sum_{n=2^{k_1-1}+1}^{\infty}\ \sum_{n'=2^{k_2-1}+1}^{\infty}\ \sum_{m=M_1}^{\infty}\sum_{m'=M_2}^{\infty}\left(\frac{A_1}{2^{\frac{k_1-1}{2}}}\frac{A_2}{2^{\frac{k_2-1}{2}}}\widehat{\mathcal{K}}_2\frac{16m!}{(2\pi)^{m+1}}\frac{16m'!}{(2\pi)^{m'+1}}\right)^2,$$

where $A_1=\dfrac{1}{\sqrt{\dfrac{(-1)^{m-1}(m!)^2}{(2m)!}\tilde{\alpha}_{2m}}}$ and $A_2=\dfrac{1}{\sqrt{\dfrac{(-1)^{m'-1}(m'!)^2}{(2m')!}\tilde{\alpha}_{2m'}}}$.

Proof: Proof is similar to Theorem 6.5.1.

Theorem 6.5.3

Let $\left(\dfrac{\partial^2 v(x,t)}{\partial t^2}\right)_N$ be the BWA of $\left(\dfrac{\partial^2 v(x,t)}{\partial t^2}\right)$ and assumed that the mixed second derivative of $\left(\dfrac{\partial^2 v(x,t)}{\partial t^2}\right)$ is bounded by a constant $\widehat{\mathcal{K}}_3$, i.e., $\left|\dfrac{\partial^6 v(x,t)}{\partial x^2\,\partial t^4}\right| < \widehat{\mathcal{K}}_3$, and then we have the following bound of error:

$$\left\|\left(\frac{\partial^2 v}{\partial t^2}\right)-\left(\frac{\partial^2 v}{\partial t^2}\right)_N\right\|_2^2 \le \sum_{n=2^{k_1-1}+1}^{\infty}\ \sum_{n'=2^{k_2-1}+1}^{\infty}\ \sum_{m=M_1}^{\infty}\sum_{m'=M_2}^{\infty}\left(\frac{A_1}{2^{\frac{k_1-1}{2}}}\frac{A_2}{2^{\frac{k_2-1}{2}}}\widehat{\mathcal{K}}_3\frac{16m!}{(2\pi)^{m+1}}\frac{16m'!}{(2\pi)^{m'+1}}\right)^2,$$

where $A_1=\dfrac{1}{\sqrt{\dfrac{(-1)^{m-1}(m!)^2}{(2m)!}\tilde{\alpha}_{2m}}}$ and $A_2=\dfrac{1}{\sqrt{\dfrac{(-1)^{m'-1}(m'!)^2}{(2m')!}\tilde{\alpha}_{2m'}}}$.

Proof: Proof is similar as Theorem 6.5.2, if $\tau=2$.

Theorem 6.5.4

Let $\left(\frac{\partial^2 v(x,t)}{\partial x^2}\right)_N$ be the with respect to of $\left(\frac{\partial^2 v(x,t)}{\partial x^2}\right)$ and assumed that the mixed second derivative of second derivative of $\left(\frac{\partial^2 v(x,t)}{\partial x^2}\right)$ is bounded by a constant $\widehat{\mathcal{K}}_4$, i.e., $\left|\frac{\partial^6 v(x,t)}{\partial t^2 \partial x^4}\right| < \widehat{\mathcal{K}}_4$, and then we have the following bound of error:

$$\left\|\left(\frac{\partial^2 v}{\partial x^2}\right)-\left(\frac{\partial^2 v}{\partial x^2}\right)_N\right\|_2^2 \leq \sum_{n=2^{k_1-1}+1}^{\infty} \sum_{n'=2^{k_2-1}+1}^{\infty} \sum_{m=M_1}^{\infty} \sum_{m'=M_2}^{\infty} \left(\frac{A_1}{2^{\frac{k_1-1}{2}}} \frac{A_2}{2^{\frac{k_2-1}{2}}} \widehat{\mathcal{K}}_4 \frac{16m!}{(2\pi)^{m+1}} \frac{16m'!}{(2\pi)^{m'+1}}\right)^2,$$

where $A_1 = \frac{1}{\sqrt{\frac{(-1)^{m-1}(m!)^2}{(2m)!}\tilde{\alpha}_{2m}}}$ and $A_2 = \frac{1}{\sqrt{\frac{(-1)^{m'-1}(m'!)^2}{(2m')!}\tilde{\alpha}_{2m'}}}$.

Proof:
Proof is similar as Theorem 6.5.2, if $\tau = 2$.

6.6 NUMERICAL EXPERIMENTS

This section deals with some numerical problems of weakly singular LPIDEs and NLPIDEs of both the first and second kinds to examine the efficiency and accuracy of the presented method. We use the absolute error definition to measure the efficiency of the proposed technique, i.e.,

$$\widehat{e}(x,t) = |v(x,t) - \widehat{v}(x,t)|,$$

where $v(x,t)$ and $\widehat{v}(x,t)$ are the exact and approximate solutions, respectively.

The values of $\widehat{e}(x,t)$ are evaluated for different values of k_1, k_2, M_1, and M_2 as demonstrated in the following tables. Furthermore, we compare the errors obtained by BWs and LWs presented in Tables 6.2, 6.5, 6.8, 6.11, 6.13, and 6.15, respectively.

Furthermore, L_2 and L_∞ error norms for the of 2D Volterra weakly singular LPIDEs and NLPIDEs are presented in Tables 6.3, 6.6, 6.9, 6.12, 6.14, and 6.16 to verify the accuracy of the proposed method. For a fixed value of x, the L_2 and L_∞ error norms at the collocation points for BWs and LWs are presented as follows:

$$L_2 \text{ error} = \left[\frac{1}{2^{k-1}M} \sum_{i=1}^{2^{k-1}M} \left(v(x,t_i) - v_N(x,t_i)\right)^2\right]^{\frac{1}{2}}$$

and

$$L_\infty \text{ error} = \operatorname*{Max}_{1\le i\le 2^{k-1}M} |v(x,t_i) - v_N(x,t_i)|,$$

respectively, where $v(x,t_i)$ is the exact solution and $v_N(x,t_i)$ is the approximate solution.

Example 6.6.1

Let us consider the first-kind weakly singular LPIDEs with $\theta = \frac{1}{2}$ and boundary conditions as follows [62]:

$$\begin{cases} v_{xt}(x,t) + v_t(x,t) = v(x,t) + h(x,t) + \int_0^x \int_0^t \frac{u_{tt}(s,y)}{(x-s)^\theta} dy\, ds, \\ v(x,0) = 0, \\ v(0,t) = 0, \\ h(x,t) = 2t + 2xt - xt^2 - \frac{8}{3} x^{\frac{3}{2}} t. \end{cases} \tag{6.60}$$

The exact solution of Eq. (6.60) is $v(x,t) = xt^2$. The values of absolute error for $k_1 = 2, k_2 = 2, M_1 = 3, M_2 = 3$ and $k_1 = 3, k_2 = 3, M_1 = 3, M_2 = 3$ are shown in Table 6.1. Moreover, for $k_1 = 2, k_2 = 2, M_1 = 3, M_2 = 3$, the numerical results and error analysis obtained from BWM and Legendre wavelet method (LWM) are plotted in Figures 6.1, 6.3 and 6.2, 6.4, respectively, and for $k_1 = 3, k_2 = 3, M_1 = 3, M_2 = 3$, the numerical results and error analysis obtained from BWs method and LWs method are plotted in Figures 6.5, 6.7 and 6.6, 6.8, respectively.

TABLE 6.1
Numerical Results of the Absolute Error Function $\widehat{e}(x,t)$ for Example 6.6.1

Error Function $\widehat{e}(x,t)$

(x,t)	$k_1 = 2, k_2 = 2, M_1 = 3, M_2 = 3$	$k_1 = 3, k_2 = 3, M_1 = 3, M_2 = 3$	(x,t)	$k_1 = 2, k_2 = 2, M_1 = 3, M_2 = 3$	$k_1 = 3, k_2 = 3, M_1 = 3, M_2 = 3$
(0.1,0.1)	2.1684E-19	6.50521E-19	(0.5,0.6)	0	5.55112E-17
(0.1,0.2)	8.67362E-19	8.67362E-19	(0.5,0.7)	2.77556E-17	5.55112E-17
(0.1,0.3)	1.73742E-18	0	(0.5,0.8)	5.55112E-17	1.66533E-16
(0.1,0.4)	3.46945E-18	6.93889E-18	(0.5,0.9)	5.55112E-17	1.66533E-16
(0.1,0.5)	3.46945E-18	1.04083E-17	(0.6,0.1)	4.33681E-18	4.33681E-18
(0.1,0.6)	0	6.93889E-18	(0.6,0.2)	0	3.46945E-18
(0.1,0.7)	6.93889E-18	1.38778E-17	(0.6,0.3)	6.93889E-18	0
(0.1,0.8)	1.38778E-17	1.38778E-17	(0.6,0.4)	0	2.77556E-17
(0.1,0.9)	1.38778E-17	4.16334E-17	(0.6,0.5)	2.77556E-17	2.77556E-17
(0.2,0.1)	4.77049E-18	1.30104E-18	(0.6,0.6)	2.77556E-17	8.32667E-17
(0.2,0.2)	5.20417E-18	1.73472E-18	(0.6,0.7)	5.55112E-17	5.55112E-17
(0.2,0.3)	3.46945E-18	0	(0.6,0.8)	0	1.66533E-16
(0.2,0.4)	6.93889E-18	1.38778E-17	(0.6,0.9)	1.11022E-16	2.22045E-16
(0.2,0.5)	2.77556E-17	2.08167E-17	(0.7,0.1)	1.73472E-18	5.20417E-18
(0.2,0.6)	2.77556E-17	1.38778E-17	(0.7,0.2)	3.46945E-18	3.46945E-18
(0.2,0.7)	2.77556E-17	2.77556E-17	(0.7,0.3)	1.38778E-17	1.38778E-17
(0.2,0.8)	2.77556E-17	2.77556E-17	(0.7,0.4)	1.38778E-17	1.38778E-17
(0.2,0.9)	2.77556E-17	8.32667E-17	(0.7,0.5)	5.55112E-17	2.77556E-17

(Continued)

TABLE 6.1 (*Continued*)
Numerical Results of the Absolute Error Function $\widehat{e}(x,t)$ for Example 6.6.1

Error Function $\widehat{e}(x,t)$

(x,t)	$k_1=2,k_2=2,$ $M_1=3,M_2=3$	$k_1=3,k_2=3,$ $M_1=3,M_2=3$	(x,t)	$k_1=2,k_2=2,$ $M_1=3,M_2=3$	$k_1=3,k_2=3,$ $M_1=3,M_2=3$
(0.3,0.1)	1.73472E-18	1.73472E-18	(0.7,0.6)	5.55112E-17	5.55112E-17
(0.3,0.2)	3.46945E-18	0	(0.7,0.7)	0	0
(0.3,0.3)	6.93889E-18	0	(0.7,0.8)	0	5.55112E-17
(0.3,0.4)	1.38778E-17	1.38778E-17	(0.7,0.9)	0	0
(0.3,0.5)	2.77556E-17	1.38778E-17	(0.8,0.1)	0	0
(0.3,0.6)	4.16334E-17	2.77556E-17	(0.8,0.2)	6.93889E-18	1.38778E-17
(0.3,0.7)	5.55112E-17	2.77556E-17	(0.8,0.3)	1.38778E-17	1.38778E-17
(0.3,0.8)	5.55112E-17	2.77556E-17	(0.8,0.4)	0	2.77556E-17
(0.3,0.9)	5.55112E-17	8.32667E-17	(0.8,0.5)	5.55112E-17	0
(0.4,0.1)	9.54098E-18	2.60209E-18	(0.8,0.6)	5.55112E-17	0
(0.4,0.2)	1.04083E-17	3.46945E-18	(0.8,0.7)	0	0
(0.4,0.3)	6.93889E-18	6.93889E-18	(0.8,0.8)	1.11022E-16	0
(0.4,0.4)	1.38778E-17	2.77556E-17	(0.8,0.9)	0	1.11022E-16
(0.4,0.5)	5.55112E-17	4.16334E-17	(0.9,0.1)	2.08167E-17	3.46945E-18
(0.4,0.6)	5.55112E-17	2.77556E-17	(0.9,0.2)	2.77556E-17	0
(0.4,0.7)	5.55112E-17	5.55112E-17	(0.9,0.3)	4.16334E-17	0
(0.4,0.8)	5.55112E-17	1.11022E-16	(0.9,0.4)	5.55112E-17	2.77556E-17
(0.4,0.9)	5.55112E-17	1.11022E-16	(0.9,0.5)	5.55112E-17	5.55112E-17
(0.5,0.1)	3.46945E-18	3.46945E-18	(0.9,0.6)	5.55112E-17	0
(0.5,0.2)	0	0	(0.9,0.7)	1.11022E-16	0
(0.5,0.3)	0	0	(0.9,0.8)	1.11022E-16	0
(0.5,0.4)	0	1.38778E-17	(0.9,0.9)	1.11022E-16	1.11022E-16
(0.5,0.5)	0	4.16334E-17			

TABLE 6.2
Comparison of the Absolute Error Function $\widehat{e}(x,t)$ for Example 6.6.1

	Error Function $\widehat{e}(x,t)$			
	$k_1=2,k_2=2,M_1=3,M_2=3$		$k_1=3,k_2=3,M_1=3,M_2=3$	
(x,t)	BWM	LWM	BWM	LWM
(0.1,0.1)	2.1684E-19	2.1684E-19	6.50521E-19	6.50521E-19
(0.2,0.2)	5.20417E-18	5.20417E-18	1.73472E-18	1.73472E-18
(0.3,0.3)	6.93889E-18	1.04083E-17	0	3.46945E-18
(0.4,0.4)	1.38778E-17	1.38778E-17	2.77556E-17	2.77556E-17
(0.5,0.5)	0	5.55112E-17	4.16334E-17	4.16334E-17
(0.6,0.6)	2.77556E-17	0	8.32667E-17	8.32667E-17
(0.7,0.7)	0	0	0	0
(0.8,0.8)	1.11022E-16	0	0	0
(0.9,0.9)	1.11022E-16	1.11022E-16	1.11022E-16	1.11022E-16

TABLE 6.3
Comparison of L_2 and L_∞ Error Norms for Example 6.6.1

	BWs				LWs			
	$k_1 = 2, k_2 = 2, M_1 = 3, M_2 = 3$		$k_1 = 3, k_2 = 3, M_1 = 3, M_2 = 3$		$k_1 = 2, k_2 = 2, M_1 = 3, M_2 = 3$		$k_1 = 3, k_2 = 3, M_1 = 3, M_2 = 3$	
x	L_2	L_∞	L_2	L_∞	L_2	L_∞	L_2	L_∞
0.2	9.06936E-18	2.08167E-17	3.51935E-18	6.93889E-18	9.06936E-18	2.08167E-17	5.23757E-18	1.38778E-17
0.4	1.26884E-17	2.77556E-17	2.4037E-17	5.55112E-17	2.40787E-17	5.55112E-17	2.44508E-17	5.55112E-17
0.6	5.3524E-17	1.11022E-16	4.86986E-17	1.11022E-16	3.98861E-17	5.55112E-17	4.66793E-17	1.11022E-16
0.8	5.61579E-17	1.11022E-16	1.10737E-16	2.22045E-16	3.58323E-17	5.55112E-17	1.10737E-16	2.22045E-16

For $k_1 = 2, k_2 = 2, M_1 = 3, M_2 = 3$, we have Figures 6.1 and 6.3 for BWM and Figures 6.2 and 6.4 for LWM.

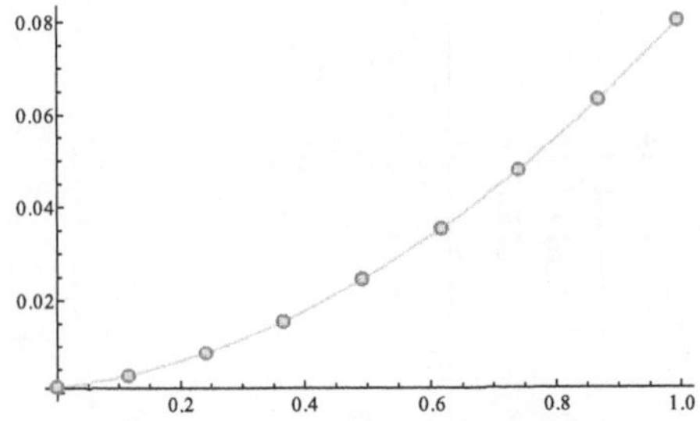

FIGURE 6.1 Exact and approximate solutions of BWM.

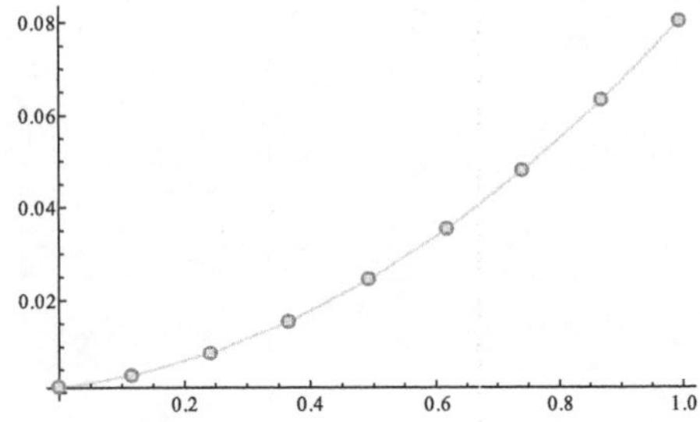

FIGURE 6.2 Exact and approximate solutions of LWM.

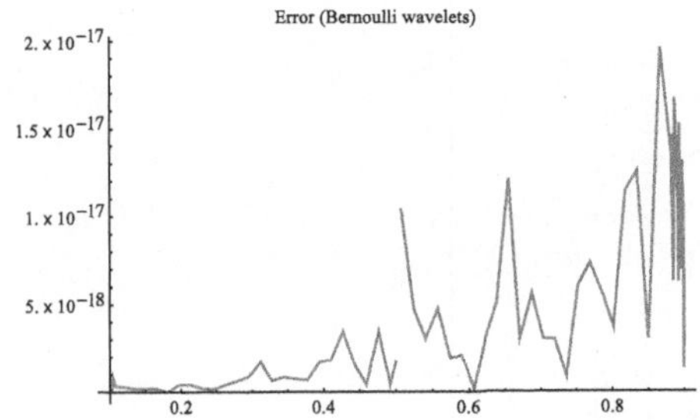

FIGURE 6.3 Error analysis of BWM.

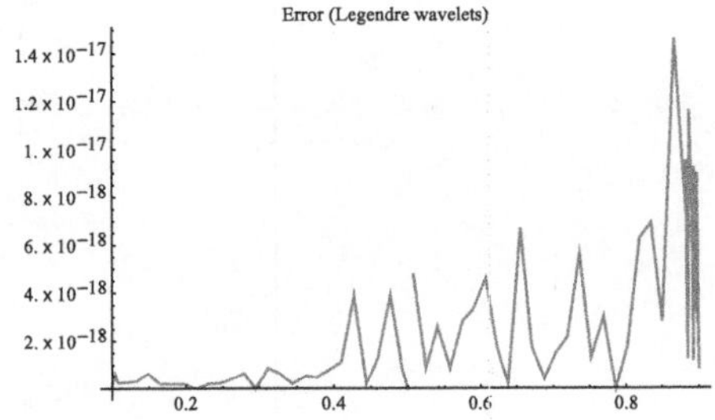

FIGURE 6.4 Error analysis of LWM.

For $k_1 = 3, k_2 = 3, M_1 = 3, M_2 = 3$, we have Figures 6.5 and 6.7 for BWM and Figures 6.6 and 6.8 for LWM.

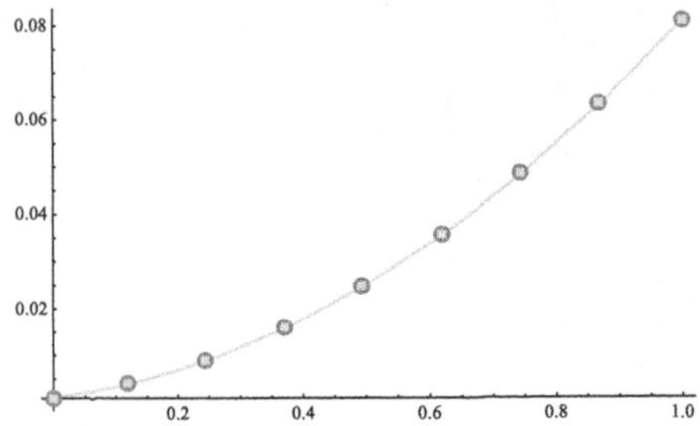

FIGURE 6.5 Exact and approximate solutions of BWM.

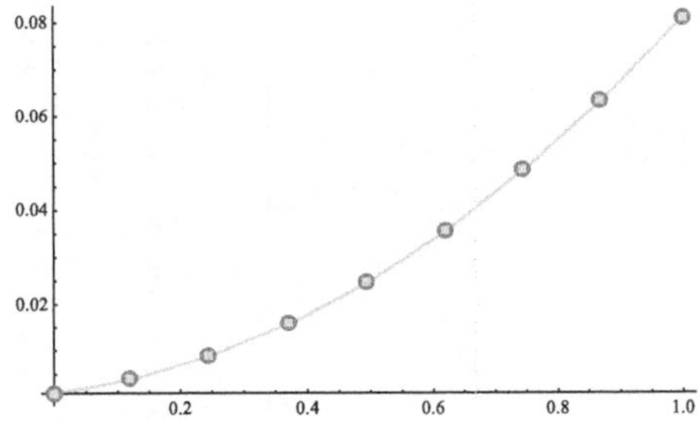

FIGURE 6.6 Exact and approximate solutions of LWM.

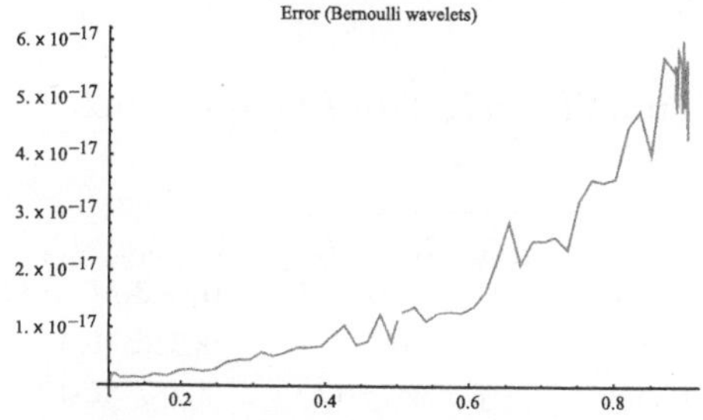

FIGURE 6.7 Error analysis of BWM.

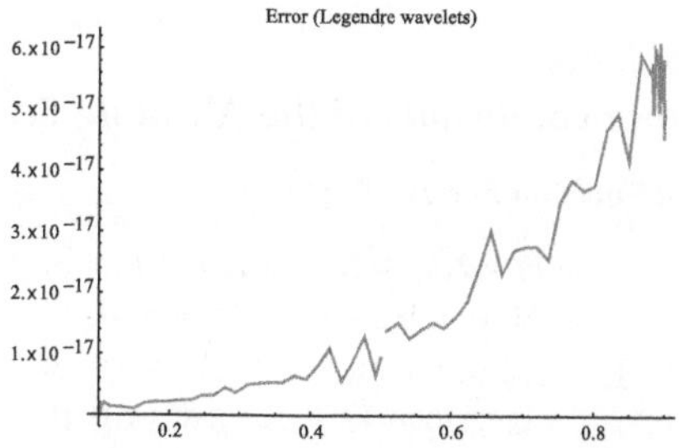

FIGURE 6.8 Error analysis of LWM.

Example 6.6.2

Considering the second-kind weakly singular LPIDEs with $\theta = \frac{1}{2}$ and boundary conditions as follows [62]:

$$\begin{cases} v_{xt}(x,t) + v_t(x,t) = v(x,t) + h(x,t) + \int_0^x \int_0^t \frac{v_{xt}(s,y)}{(x-s)^{\theta}} dy\, ds, \\ v(x,0) = 0, \\ v(0,t) = 0, \\ h(x,t) = 2t + 2xt - xt^2 - 2x^{\frac{1}{2}}t^2 \end{cases} \tag{6.61}$$

with the exact solution. $v(x,t) = xt^2$. The values of absolute error for $k_1 = 2, k_2 = 2, M_1 = 3, M_2 = 3$ and $k_1 = 3, k_2 = 3, M_1 = 3, M_2 = 3$ are shown in Table 6.4. Also, for $k_1 = 2, k_2 = 2, M_1 = 3, M_2 = 3$, the numerical results and error analysis obtained from BWM and LWM are plotted in Figures 6.9, 6.11 and 6.10, 6.12, respectively, and for $k_1 = 3, k_2 = 3, M_1 = 3, M_2 = 3$, the numerical results and error analysis obtained from BWM and LWM are plotted in Figures 6.13, 6.15 and 6.14, 6.16, respectively.

TABLE 6.4
Numerical Results of the Absolute Error Function $\widehat{e}(x,t)$ for Example 6.6.2

Error Function $\widehat{e}(x,t)$

(x,t)	$k_1 = 2, k_2 = 2,$ $M_1 = 3, M_2 = 3$	$k_1 = 3, k_2 = 3,$ $M_1 = 3, M_2 = 3$	(x,t)	$k_1 = 2, k_2 = 2,$ $M_1 = 3, M_2 = 3$	$k_1 = 3, k_2 = 3,$ $M_1 = 3, M_2 = 3$
(0.1,0.1)	1.0842E-18	6.50521E-19	(0.5,0.6)	0	8.32667E-17
(0.1,0.2)	2.60209E-18	8.67362E-19	(0.5,0.7)	8.32667E-17	1.38778E-17
(0.1,0.3)	5.20417E-18	1.73472E-18	(0.5,0.8)	1.11022E-17	2.22045E-16
(0.1,0.4)	1.04083E-18	0	(0.5,0.9)	1.66533E-17	2.77556E-17
(0.1,0.5)	6.93889E-18	3.46945E-18	(0.6,0.1)	9.54098E-18	4.33681E-18
(0.1,0.6)	1.38778E-17	0	(0.6,0.2)	1.73472E-17	3.46945E-18
(0.1,0.7)	2.77556E-17	6.93889E-18	(0.6,0.3)	3.46945E-18	0
(0.1,0.8)	2.77556E-17	0	(0.6,0.4)	4.16334E-17	1.38778E-17
(0.1,0.9)	5.55112E-17	0	(0.6,0.5)	2.77556E-17	0
(0.2,0.1)	4.33681E-19	1.30104E-18	(0.6,0.6)	8.32667E-17	5.55112E-17
(0.2,0.2)	3.46945E-18	1.73472E-18	(0.6,0.7)	1.11022E-17	5.55112E-17
(0.2,0.3)	6.93889E-18	3.46945E-18	(0.6,0.8)	1.66533E-17	1.11022E-17
(0.2,0.4)	6.93889E-18	0	(0.6,0.9)	1.66533E-17	1.66533E-17
(0.2,0.5)	1.38778E-17	6.93889E-18	(0.7,0.1)	8.67362E-18	5.20417E-18
(0.2,0.6)	2.77556E-17	0	(0.7,0.2)	2.08167E-18	3.46945E-18
(0.2,0.7)	4.16334E-17	1.38778E-17	(0.7,0.3)	2.77556E-17	0
(0.2,0.8)	2.77556E-17	0	(0.7,0.4)	5.55112E-17	0
(0.2,0.9)	8.32667E-17	0	(0.7,0.5)	5.55112E-17	2.77556E-17
(0.3,0.1)	1.73472E-18	1.73472E-18	(0.7,0.6)	1.11022E-16	5.55112E-17
(0.3,0.2)	1.73472E-18	0	(0.7,0.7)	1.11022E-16	0
(0.3,0.3)	6.93889E-18	0	(0.7,0.8)	2.22045E-16	5.55112E-17
(0.3,0.4)	6.93889E-18	0	(0.7,0.9)	2.22045E-16	0
(0.3,0.5)	1.38778E-17	1.38778E-17	(0.8,0.1)	6.93889E-18	3.46945E-18
(0.3,0.6)	1.38778E-17	1.38778E-17	(0.8,0.2)	1.38778E-17	0
(0.3,0.7)	2.77556E-17	2.77556E-17	(0.8,0.3)	2.77556E-17	1.38778E-17
(0.3,0.8)	5.55112E-17	2.77556E-17	(0.8,0.4)	5.55112E-17	2.77556E-17
(0.3,0.9)	5.55112E-17	5.55112E-17	(0.8,0.5)	2.77556E-17	2.77556E-17
(0.4,0.1)	8.67362E-19	2.60209E-18	(0.8,0.6)	1.11022E-16	0
(0.4,0.2)	6.93889E-18	3.46945E-18	(0.8,0.7)	1.11022E-16	0
(0.4,0.3)	1.38778E-17	0	(0.8,0.8)	1.11022E-16	0

(Continued)

TABLE 6.4 (*Continued*)
Numerical Results of the Absolute Error Function $\widehat{e}(x,t)$ for Example 6.6.2

Error Function $\widehat{e}(x,t)$

(x,t)	$k_1=2,k_2=2,$ $M_1=3,M_2=3$	$k_1=3,k_2=3,$ $M_1=3,M_2=3$	(x,t)	$k_1=2,k_2=2,$ $M_1=3,M_2=3$	$k_1=3,k_2=3,$ $M_1=3,M_2=3$
(0.4,0.4)	1.38778E-17	0	(0.8,0.9)	2.22045E-16	1.11022E-16
(0.4,0.5)	2.77556E-17	1.38778E-17	(0.9,0.1)	2.08167E-17	3.46945E-18
(0.4,0.6)	5.55112E-17	0	(0.9,0.2)	4.16334E-17	0
(0.4,0.7)	8.32667E-17	2.77556E-17	(0.9,0.3)	6.93889E-17	0
(0.4,0.8)	5.55112E-17	5.55112E-17	(0.9,0.4)	8.32667E-17	2.77556E-17
(0.4,0.9)	1.66533E-17	1.11022E-16	(0.9,0.5)	1.66533E-17	5.55112E-17
(0.5,0.1)	3.46945E-18	2.60209E-18	(0.9,0.6)	1.66533E-17	0
(0.5,0.2)	6.93889E-18	6.93889E-18	(0.9,0.7)	1.66533E-17	0
(0.5,0.3)	1.38778E-17	6.93889E-18	(0.9,0.8)	2.22045E-16	0
(0.5,0.4)	2.77556E-17	2.77556E-17	(0.9,0.9)	3.33067E-16	1.11022E-16
(0.5,0.5)	5.55112E-17	6.93889E-18			

TABLE 6.5
Comparison of the Absolute Error Function $\widehat{e}(x,t)$ for Example 6.6.2

	Error Function $\widehat{e}(x,t)$			
	$k_1=2,k_2=2,M_1=3,M_2=3$		$k_1=3,k_2=3,M_1=3,M_2=3$	
(x,t)	BWM	LWM	BWM	LWM
(0.1,0.1)	1.0842E-18	1.0842E-18	6.50521E-19	6.50521E-19
(0.2,0.2)	3.46945E-18	3.46945E-18	1.73472E-18	1.73472E-18
(0.3,0.3)	6.93889E-18	0	0	3.46945E-18
(0.4,0.4)	1.38778E-17	1.38778E-17	0	1.38778E-17
(0.5,0.5)	5.55112E-17	2.77556E-17	6.93889E-18	6.93889E-18
(0.6,0.6)	8.32667E-17	8.32667E-17	5.55112E-17	5.55112E-17
(0.7,0.7)	1.11022E-16	1.11022E-16	0	0
(0.8,0.8)	1.11022E-16	1.11022E-16	0	0
(0.9,0.9)	3.33067E-16	3.33067E-16	1.11022E-16	1.11022E-16

TABLE 6.6
Comparison of L_2 and L_∞ Error Norms for Example 6.6.2

	BWs				LWs			
	$k_1 = 2, k_2 = 2, M_1 = 3, M_2 = 3$		$k_1 = 3, k_2 = 3, M_1 = 3, M_2 = 3$		$k_1 = 2, k_2 = 2, M_1 = 3, M_2 = 3$		$k_1 = 3, k_2 = 3, M_1 = 3, M_2 = 3$	
x	L_2	L_∞	L_2	L_∞	L_2	L_∞	L_2	L_∞
0.2	1.61736E-17	2.77556E-17	3.51935E-18	1.04083E-17	1.61736E-17	2.77556E-17	4.40538E-18	1.04083E-17
0.4	4.24919E-17	8.32667E-17	2.24176E-17	5.55112E-17	4.51084E-17	8.32667E-17	2.25069E-17	5.55112E-17
0.6	6.53387E-17	1.11022E-16	2.02302E-17	5.55112E-17	6.92877E-17	1.11022E-16	2.83986E-17	5.55112E-17
0.8	2.06172E-16	3.33067E-16	7.55086E-17	1.66533E-16	2.06172E-16	3.33067E-16	8.20288E-17	1.66533E-16

For $k_1 = 2, k_2 = 2, M_1 = 3, M_2 = 3$, we have Figures 6.9 and 6.11 for BWM and Figures 6.10 and 6.12 for LWM.

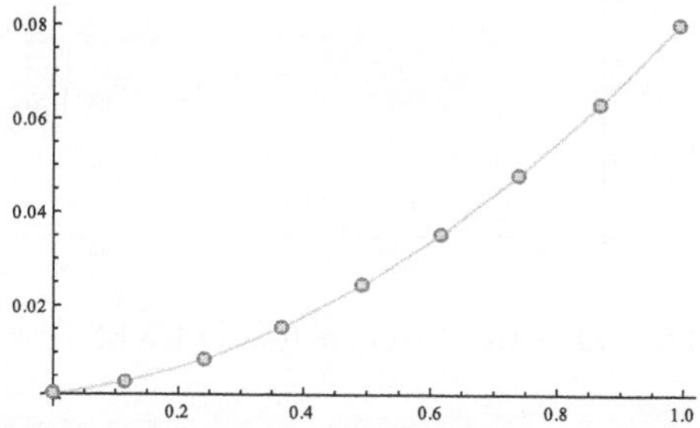

FIGURE 6.9 Exact and approximate solutions of BWM.

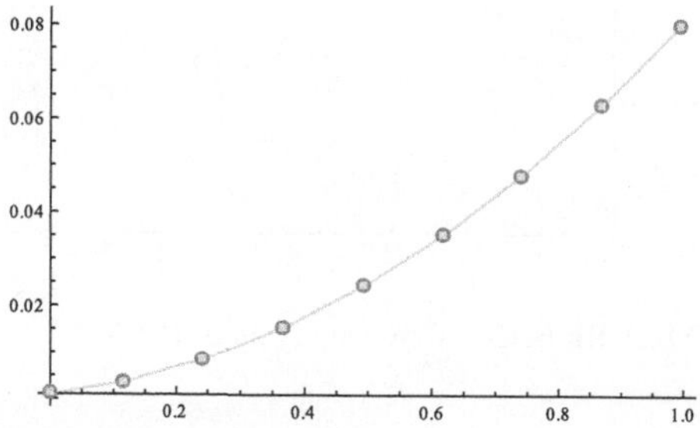

FIGURE 6.10 Exact and approximate solutions of LWM.

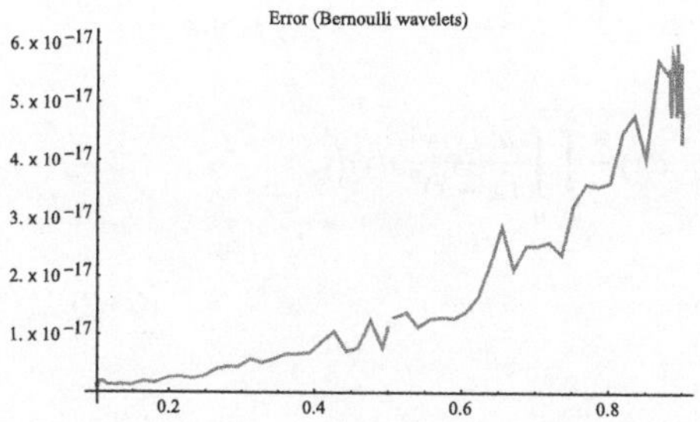

FIGURE 6.11 Error analysis of BWM.

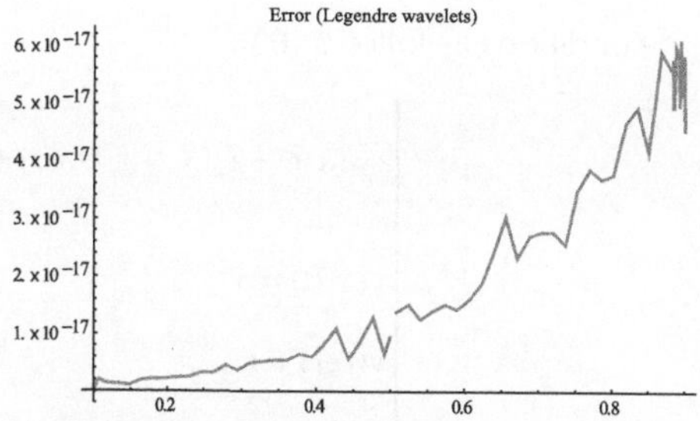

FIGURE 6.12 Error analysis of LWM.

For $k_1 = 3, k_2 = 3, M_1 = 3, M_2 = 3$, we have Figures 6.13 and 6.15 for BWM and Figures 6.14 and 6.16 for LWM.

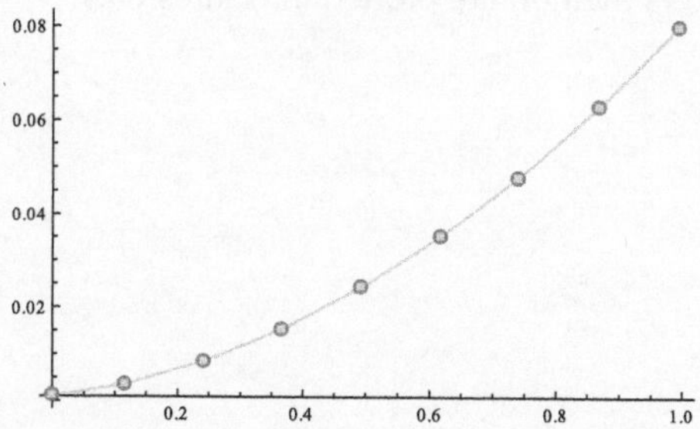

FIGURE 6.13 Exact and approximate solutions of BWM.

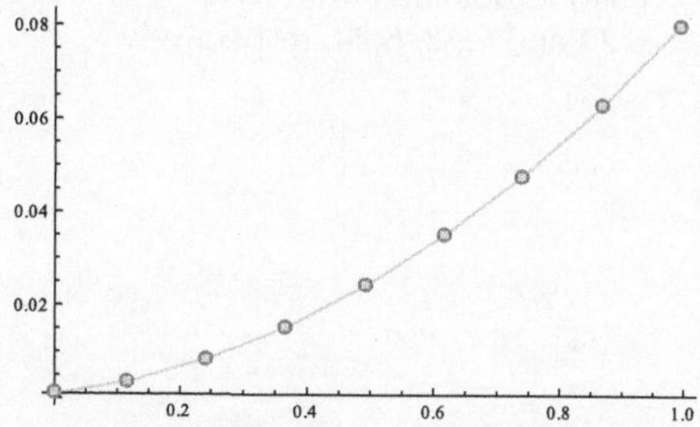

FIGURE 6.14 Exact and approximate solutions of LWM.

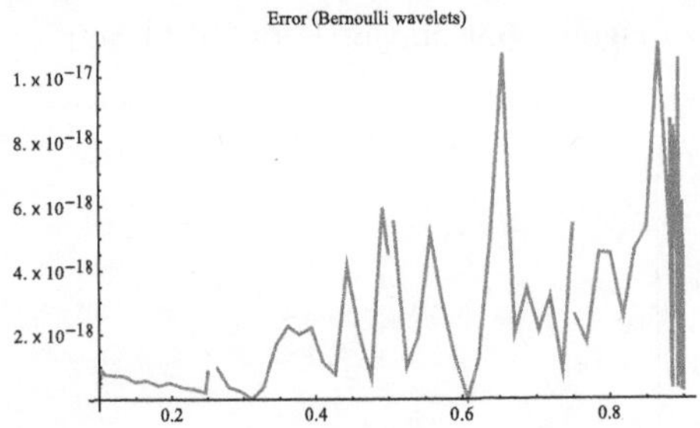

FIGURE 6.15 Error analysis of BWM.

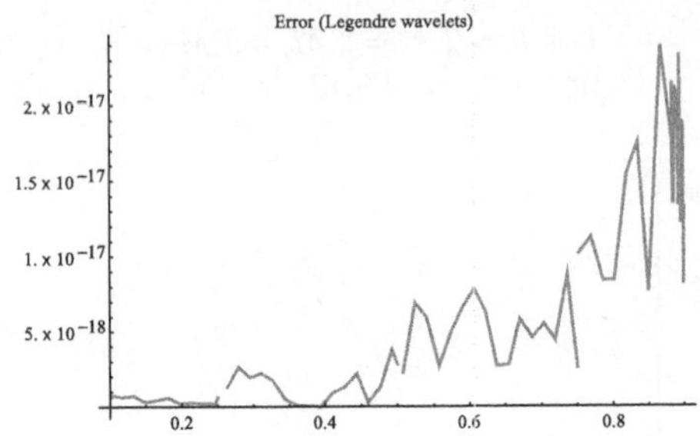

FIGURE 6.16 Error analysis of LWM.

Example 6.6.3

Let us consider the first-kind weakly singular LPIDEs with $\theta = \frac{1}{4}$ and boundary conditions as follows [63]:

$$\begin{cases} v_{xt}(x,t) + v_t(x,t) = v(x,t) + h(x,t) + \int_0^x \int_0^t \frac{u_{tt}(s,y)}{(x-s)^{\theta}} dy\, ds, \\ v(x,0) = x^2, \\ v(0,t) = t^2, \\ h(x,t) = 2t - x^2 - t^2 - \frac{8}{3} x^{\frac{3}{4}} t. \end{cases} \tag{6.62}$$

The exact solution of Eq. (6.62) is $v(x,t) = x^2 + t^2$. The values of absolute error for $k_1 = 2, k_2 = 2, M_1 = 3, M_2 = 3$ and $k_1 = 3, k_2 = 3, M_1 = 3, M_2 = 3$ are shown in Table 6.7. Moreover, for $k_1 = 2, k_2 = 2, M_1 = 3, M_2 = 3$, the numerical results and error analysis obtained from BWM and LWM are plotted in Figures 6.17, 6.19 and 6.18, 6.20, respectively, and for $k_1 = 3, k_2 = 3, M_1 = 3, M_2 = 3$, the numerical results and error analysis obtained from BWs method and LWs method are plotted in Figures 6.21, 6.23 and 6.22, 6.24, respectively.

TABLE 6.7
Numerical Results of the Absolute Error Function $\widehat{e}(x,t)$ for Example 6.6.3

Error Function $\widehat{e}(x,t)$

(x,t)	$k_1=2,k_2=2,$ $M_1=3,M_2=3$	$k_1=3,k_2=3,$ $M_1=3,M_2=3$	(x,t)	$k_1=2,k_2=2,$ $M_1=3,M_2=3$	$k_1=3,k_2=3,$ $M_1=3,M_2=3$
(0.1,0.1)	1.38778E-17	6.93889E-18	(0.5,0.6)	1.11022E-16	1.11022E-16
(0.1,0.2)	2.08167E-17	6.93889E-18	(0.5,0.7)	0	1.11022E-16
(0.1,0.3)	2.77556E-17	2.77556E-17	(0.5,0.8)	0	1.11022E-16
(0.1,0.4)	2.77556E-17	0	(0.5,0.9)	0	0
(0.1,0.5)	0	0	(0.6,0.1)	0	5.55112E-17
(0.1,0.6)	0	5.55112E-17	(0.6,0.2)	0	1.11022E-16
(0.1,0.7)	5.55112E-17	5.55112E-17	(0.6,0.3)	5.55112E-17	5.55112E-17
(0.1,0.8)	1.11022E-16	0	(0.6,0.4)	0	0
(0.1,0.9)	0	0	(0.6,0.5)	0	1.11022E-16
(0.2,0.1)	6.93889E-18	6.93889E-18	(0.6,0.6)	0	1.11022E-16
(0.2,0.2)	1.38778E-17	0	(0.6,0.7)	0	1.11022E-16
(0.2,0.3)	2.77556E-17	2.77556E-17	(0.6,0.8)	1.11022E-16	0
(0.2,0.4)	2.77556E-17	0	(0.6,0.9)	2.22045E-17	2.22045E-17
(0.2,0.5)	1.11022E-16	5.55112E-17	(0.7,0.1)	5.55112E-17	0
(0.2,0.6)	1.11022E-16	5.55112E-17	(0.7,0.2)	0	0
(0.2,0.7)	1.11022E-16	0	(0.7,0.3)	0	0
(0.2,0.8)	0	0	(0.7,0.4)	0	1.11022E-16
(0.2,0.9)	0	0	(0.7,0.5)	1.11022E-16	1.11022E-16
(0.3,0.1)	0	1.38778E-17	(0.7,0.6)	1.11022E-16	0
(0.3,0.2)	0	2.77556E-17	(0.7,0.7)	0	0
(0.3,0.3)	0	5.55112E-17	(0.7,0.8)	0	0
(0.3,0.4)	0	0	(0.7,0.9)	0	2.22045E-17
(0.3,0.5)	5.55112E-17	0	(0.8,0.1)	0	0
(0.3,0.6)	5.55112E-17	0	(0.8,0.2)	0	0
(0.3,0.7)	0	0	(0.8,0.3)	0	0
(0.3,0.8)	1.11022E-16	0	(0.8,0.4)	0	1.11022E-16
(0.3,0.9)	0	0	(0.8,0.5)	1.11022E-16	0
(0.4,0.1)	2.77556E-17	0	(0.8,0.6)	0	1.11022E-16
(0.4,0.2)	2.77556E-17	0	(0.8,0.7)	0	2.22045E-17
(0.4,0.3)	5.55112E-17	5.55112E-17	(0.8,0.8)	0	4.44089E-16
(0.4,0.4)	0	0	(0.8,0.9)	0	2.22045E-17
(0.4,0.5)	0	5.55112E-17	(0.9,0.1)	1.11022E-16	0
(0.4,0.6)	0	0	(0.9,0.2)	1.11022E-16	0
(0.4,0.7)	0	1.11022E-16	(0.9,0.3)	1.11022E-16	1.11022E-16

(Continued)

TABLE 6.7 (*Continued*)
Numerical Results of the Absolute Error Function $\hat{e}(x,t)$ for Example 6.6.3

Error Function $\hat{e}(x,t)$					
(x,t)	$k_1 = 2, k_2 = 2, M_1 = 3, M_2 = 3$	$k_1 = 3, k_2 = 3, M_1 = 3, M_2 = 3$	(x,t)	$k_1 = 2, k_2 = 2, M_1 = 3, M_2 = 3$	$k_1 = 3, k_2 = 3, M_1 = 3, M_2 = 3$
(0.4,0.8)	1.11022E-16	0	(0.9,0.4)	1.11022E-16	0
(0.4,0.9)	1.11022E-16	0	(0.9,0.5)	0	0
(0.5,0.1)	0	0	(0.9,0.6)	2.22045E-16	2.22045E-16
(0.5,0.2)	0	5.55112E-17	(0.9,0.7)	0	0
(0.5,0.3)	5.55112E-17	5.55112E-17	(0.9,0.8)	0	0
(0.5,0.4)	0	1.11022E-16	(0.9,0.9)	0	2.22045E-16
(0.5,0.5)	1.11022E-16	5.55112E-17			

TABLE 6.8
Comparison of the Absolute Error Function $\hat{e}(x,t)$ for Example 6.6.3

	Error Function $\hat{e}(x,t)$			
	$k_1 = 2, k_2 = 2, M_1 = 3, M_2 = 3$		$k_1 = 3, k_2 = 3, M_1 = 3, M_2 = 3$	
(x,t)	BWM	LWM	BWM	LWM
(0.1,0.1)	1.38778E-17	1.38778E-17	6.93889E-18	6.93889E-18
(0.2,0.2)	1.38778E-17	1.38778E-17	0	0
(0.3,0.3)	0	0	5.55112E-17	2.77556E-17
(0.4,0.4)	0	0	0	0
(0.5,0.5)	1.11022E-16	1.11022E-16	5.55112E-17	5.55112E-17
(0.6,0.6)	0	0	1.11022E-16	1.11022E-16
(0.7,0.7)	0	0	0	0
(0.8,0.8)	0	0	4.44089E-16	4.44089E-16
(0.9,0.9)	0	0	2.22045E-16	2.22045E-16

TABLE 6.9
Comparison of L_2 and L_∞ Error Norms for Example 6.6.3

	BWs				LWs			
	$k_1 = 2, k_2 = 2, M_1 = 3, M_2 = 3$		$k_1 = 3, k_2 = 3, M_1 = 3, M_2 =$		$k_1 = 2, k_2 = 2, M_1 = 3, M_2 =$		$k_1 = 3, k_2 = 3, M_1 = 3, M_2 =$	
x	L_2	L_∞	L_2	L_∞	L_2	L_∞	L_2	L_∞
0.2	5.46369E-17	1.11022E-16	6.0525E-17	1.11022E-16	5.46369E-17	1.11022E-16	4.01118E-17	1.11022E-16
0.4	1.02686E-16	2.22045E-16	8.01234E-17	2.22045E-16	1.02608E-16	2.22045E-16	6.60715E-17	2.22045E-16
0.6	1.35974E-16	2.22045E-16	1.60247E-16	4.44089E-16	1.01349E-16	1.11022E-16	1.15556E-16	2.22045E-16
0.8	1.11022E-16	2.22045E-16	2.6037E-16	6.66134E-16	1.01349E-16	2.22045E-16	2.68144E-16	6.66134E-16

For $k_1 = 2, k_2 = 2, M_1 = 3, M_2 = 3$, we have Figures 6.17 and 6.19 for BWM and Figures 6.18 and 6.20 for LWM.

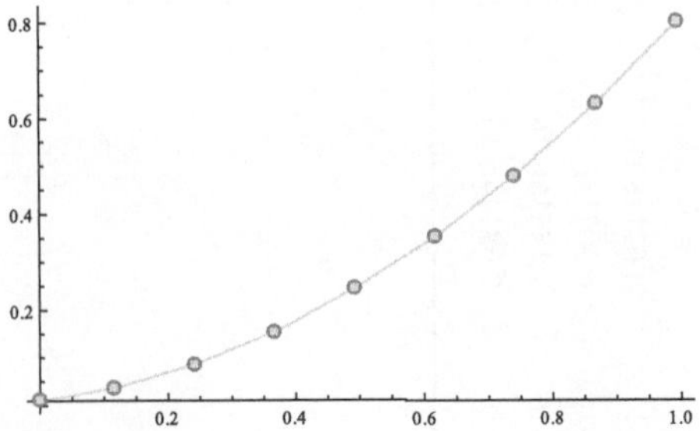

FIGURE 6.17 Exact and approximate solutions of BWM.

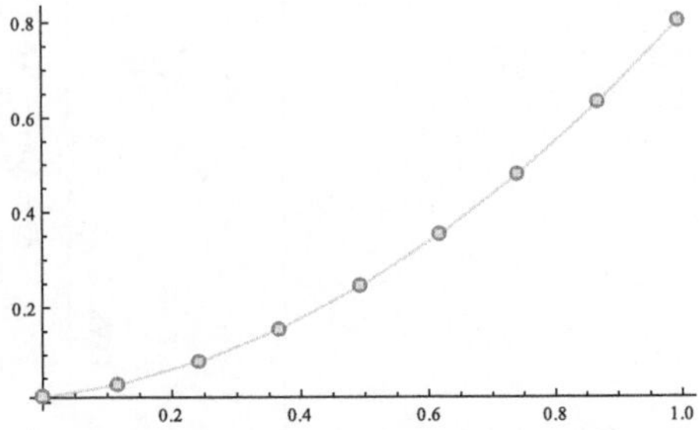

FIGURE 6.18 Exact and approximate solutions of LWM.

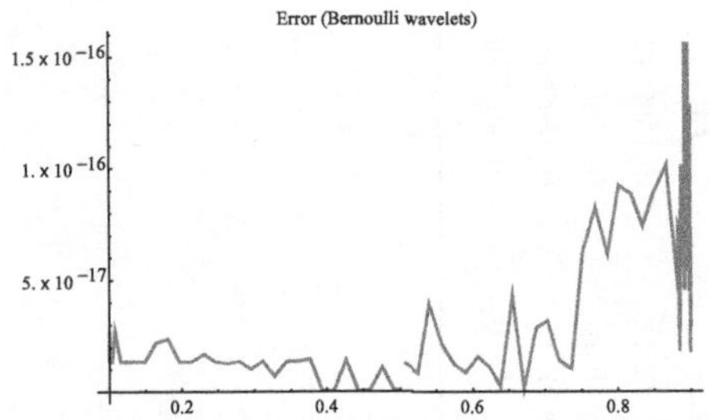

FIGURE 6.19 Error analysis of BWM.

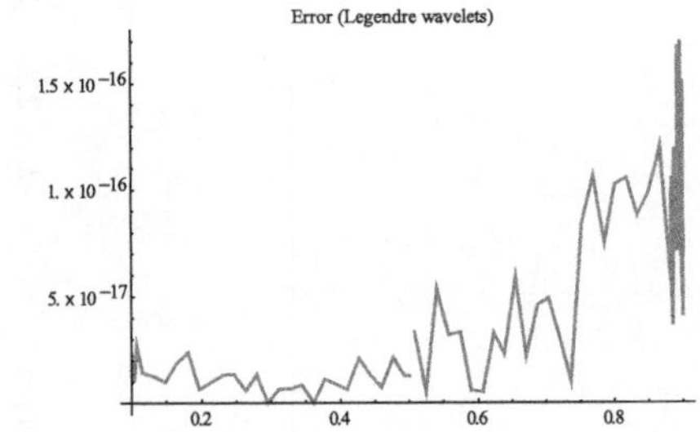

FIGURE 6.20 Error analysis of LWM.

For $k_1 = 3, k_2 = 3, M_1 = 3, M_2 = 3$, we have Figures 6.21 and 6.23 for BWM and Figures 6.22 and 6.24 for LWM.

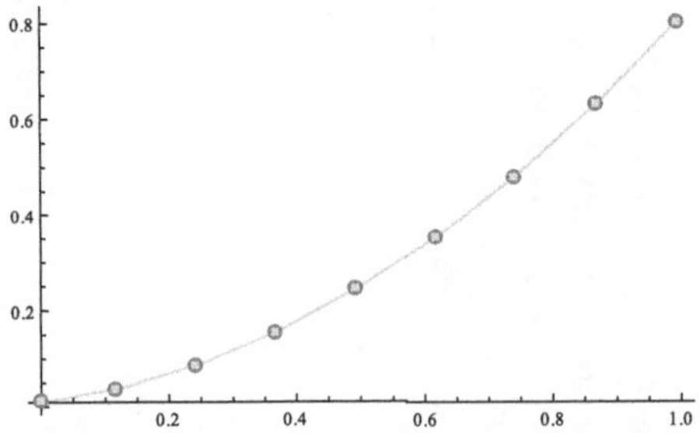

FIGURE 6.21 Exact and approximate solutions of BWM.

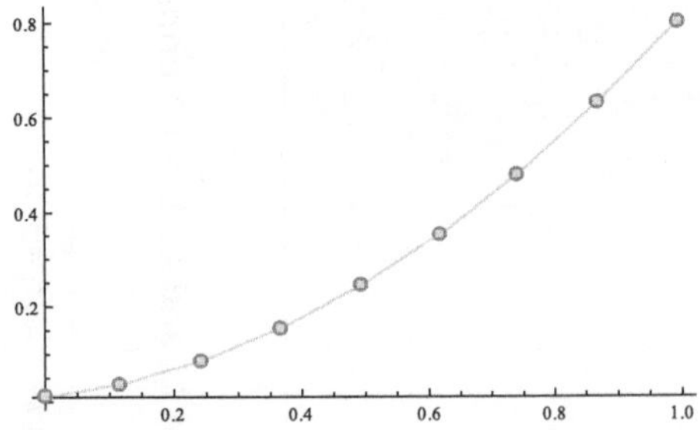

FIGURE 6.22 Exact and approximate solutions of LWM.

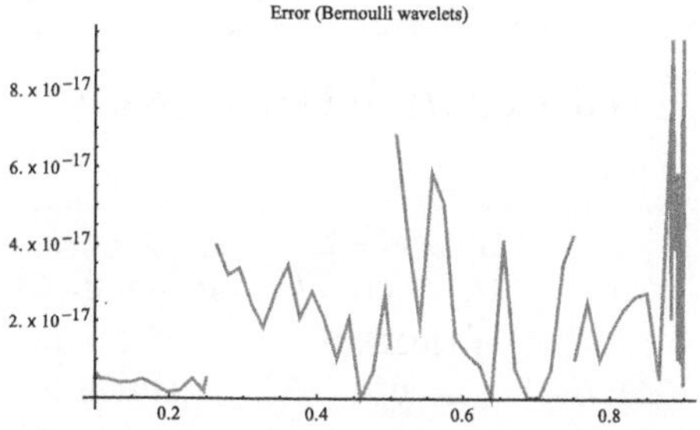

FIGURE 6.23 Error analysis of BWM.

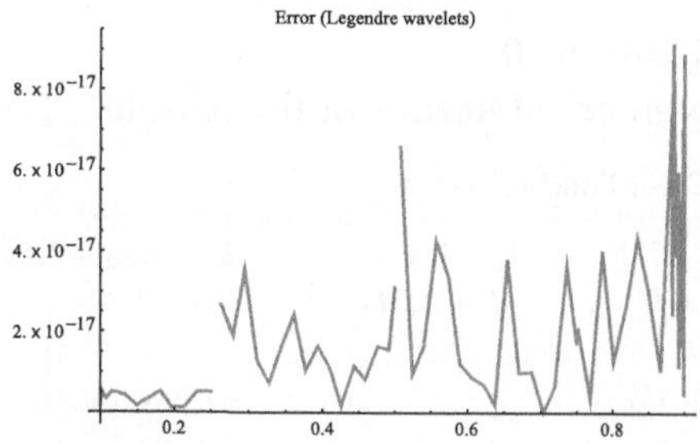

FIGURE 6.24 Error analysis of LWM.

Example 6.6.4

Let us consider the second-kind weakly singular LPIDEs with $\theta = \frac{1}{4}$ and boundary conditions as follows [63]:

$$\begin{cases} v_{xt}(x,t) + v_t(x,t) = v(x,t) + h(x,t) + \int_0^x \int_0^t \frac{u_{xt}(s,y)}{(x-s)^\theta} dy\, ds, \\ v(x,0) = x^2, \\ v(0,t) = t^2, \\ h(x,t) = 2x - x^2 - t^2. \end{cases} \tag{6.63}$$

The exact solution of Eq. (6.63) is $v(x,t) = x^2 + t^2$. The values of absolute error for $k_1 = 2, k_2 = 2, M_1 = 3, M_2 = 3$ and $k_1 = 3, k_2 = 3, M_1 = 3, M_2 = 3$ are shown in Table 6.10. Moreover, for $k_1 = 2, k_2 = 2, M_1 = 3, M_2 = 3$, the numerical results and error analysis obtained from BWM and LWM are plotted in Figures 6.25, 6.27 and 6.26, 6.28, respectively, and for $k_1 = 3, k_2 = 3, M_1 = 3, M_2 = 3$, the numerical results and error analysis obtained from BWs method and LWs method are plotted in Figures 6.29, 6.31 and 6.30, 6.32, respectively.

TABLE 6.10
Numerical Results of the Absolute Error Function $\widehat{e}(x,t)$ for Example 6.6.4

Error Function $\widehat{e}(x,t)$					
(x,t)	$k_1=2,k_2=2,$ $M_1=3,M_2=3$	$k_1=3,k_2=3,$ $M_1=3,M_2=3$	(x,t)	$k_1=2,k_2=2,$ $M_1=3,M_2=3$	$k_1=3,k_2=3,$ $M_1=3,M_2=3$
(0.1,0.1)	1.38778E-17	6.93889E-18	(0.5,0.6)	1.11022E-16	1.11022E-16
(0.1,0.2)	2.08167E-17	6.93889E-18	(0.5,0.7)	0	1.11022E-16
(0.1,0.3)	2.77556E-17	2.77556E-17	(0.5,0.8)	0	1.11022E-16
(0.1,0.4)	2.77556E-17	0	(0.5,0.9)	0	0
(0.1,0.5)	0	0	(0.6,0.1)	0	5.55112E-17
(0.1,0.6)	0	5.55112E-17	(0.6,0.2)	0	1.11022E-16
(0.1,0.7)	5.55112E-17	5.55112E-17	(0.6,0.3)	5.55112E-17	5.55112E-17
(0.1,0.8)	1.11022E-16	0	(0.6,0.4)	0	0
(0.1,0.9)	0	0	(0.6,0.5)	1.11022E-16	1.11022E-16
(0.2,0.1)	6.93889E-18	6.93889E-18	(0.6,0.6)	0	1.11022E-16
(0.2,0.2)	1.38778E-17	0	(0.6,0.7)	0	1.11022E-16
(0.2,0.3)	2.77556E-17	2.77556E-17	(0.6,0.8)	1.11022E-16	0
(0.2,0.4)	2.77556E-17	0	(0.6,0.9)	2.22045E-17	2.22045E-17
(0.2,0.5)	1.11022E-16	0	(0.7,0.1)	5.55112E-17	0
(0.2,0.6)	1.11022E-16	5.55112E-17	(0.7,0.2)	0	0
(0.2,0.7)	1.11022E-16	0	(0.7,0.3)	0	0
(0.2,0.8)	0	0	(0.7,0.4)	0	0
(0.2,0.9)	0	0	(0.7,0.5)	0	1.11022E-16
(0.3,0.1)	0	1.38778E-17	(0.7,0.6)	1.11022E-16	0
(0.3,0.2)	0	2.77556E-17	(0.7,0.7)	0	0
(0.3,0.3)	0	5.55112E-17	(0.7,0.8)	0	0
(0.3,0.4)	0	0	(0.7,0.9)	0	2.22045E-17
(0.3,0.5)	5.55112E-17	5.55112E-17	(0.8,0.1)	0	0
(0.3,0.6)	5.55112E-17	0	(0.8,0.2)	0	0
(0.3,0.7)	0	0	(0.8,0.3)	0	0
(0.3,0.8)	1.11022E-16	0	(0.8,0.4)	0	1.11022E-16
(0.3,0.9)	0	0	(0.8,0.5)	0	1.11022E-16
(0.4,0.1)	2.77556E-17	0	(0.8,0.6)	0	1.11022E-16
(0.4,0.2)	2.77556E-17	0	(0.8,0.7)	0	0
(0.4,0.3)	5.55112E-17	5.55112E-17	(0.8,0.8)	0	0

(Continued)

TABLE 6.10 (*Continued*)
Numerical Results of the Absolute Error Function $\hat{e}(x,t)$ for Example 6.6.4

Error Function $\hat{e}(x,t)$					
(x,t)	$k_1=2,k_2=2,$ $M_1=3,M_2=3$	$k_1=3,k_2=3,$ $M_1=3,M_2=3$	(x,t)	$k_1=2,k_2=2,$ $M_1=3,M_2=3$	$k_1=3,k_2=3,$ $M_1=3,M_2=3$
(0.4,0.4)	0	0	(0.8,0.9)	0	2.22045E-17
(0.4,0.5)	0	5.55112E-17	(0.9,0.1)	1.11022E-16	0
(0.4,0.6)	0	0	(0.9,0.2)	1.11022E-16	0
(0.4,0.7)	0	1.11022E-16	(0.9,0.3)	1.11022E-16	0
(0.4,0.8)	1.11022E-16	0	(0.9,0.4)	1.11022E-16	1.11022E-16
(0.4,0.9)	1.11022E-16	0	(0.9,0.5)	0	0
(0.5,0.1)	0	0	(0.9,0.6)	2.22045E-16	2.22045E-16
(0.5,0.2)	0	5.55112E-17	(0.9,0.7)	0	0
(0.5,0.3)	5.55112E-17	0	(0.9,0.8)	0	0
(0.5,0.4)	0	5.55112E-17	(0.9,0.9)	0	2.22045E-16
(0.5,0.5)	1.11022E-16	5.55112E-17			

TABLE 6.11
Comparison of the Absolute Error Function $\hat{e}(x,t)$ for Example 6.6.4

	Error Function $\hat{e}(x,t)$			
	$k_1=2,k_2=2,M_1=3,M_2=3$		$k_1=3,k_2=3,M_1=3,M_2=3$	
(x,t)	BWM	LWM	BWM	LWM
(0.1,0.1)	1.38778E-17	1.38778E-17	6.93889E-18	6.93889E-18
(0.2,0.2)	1.38778E-17	1.38778E-17	0	0
(0.3,0.3)	0	0	5.55112E-17	2.77556E-17
(0.4,0.4)	0	0	0	0
(0.5,0.5)	1.11022E-16	1.11022E-16	5.55112E-17	5.55112E-17
(0.6,0.6)	0	0	1.11022E-16	1.11022E-16
(0.7,0.7)	0	0	0	0
(0.8,0.8)	0	0	0	0
(0.9,0.9)	0	0	2.22045E-16	2.22045E-16

TABLE 6.12
Comparison of L_2 and L_∞ Error Norms for Example 6.6.4

	BWs				LWs			
	$k_1 = 2, k_2 = 2, M_1 = 3, M_2 = 3$		$k_1 = 3, k_2 = 3, M_1 = 3, M_2 = 3$		$k_1 = 2, k_2 = 2, M_1 = 3, M_2 = 3$		$k_1 = 3, k_2 = 3, M_1 = 3, M_2 = 3$	
x	L_2	L_∞	L_2	L_∞	L_2	L_∞	L_2	L_∞
0.2	5.46369E-17	1.11022E-16	6.0525E-17	1.11022E-16	5.46369E-17	1.11022E-16	4.01118E-17	1.11022E-16
0.4	1.02608E-16	2.22045E-16	8.62956E-17	2.22045E-16	1.02608E-16	2.22045E-16	3.58323E-17	1.11022E-16
0.6	1.35974E-16	2.22045E-16	8.47947E-17	2.22045E-16	9.06493E-17	1.11022E-16	8.47947E-17	2.22045E-16
0.8	1.11022E-16	2.22045E-16	1.35974E-16	2.22045E-16	1.11022E-16	2.22045E-16	1.50325E-16	2.22045E-16

For $k_1 = 2, k_2 = 2, M_1 = 3, M_2 = 3$, we have Figures 6.25 and 6.27 for BWM and Figures 6.26 and 6.28 for LWM.

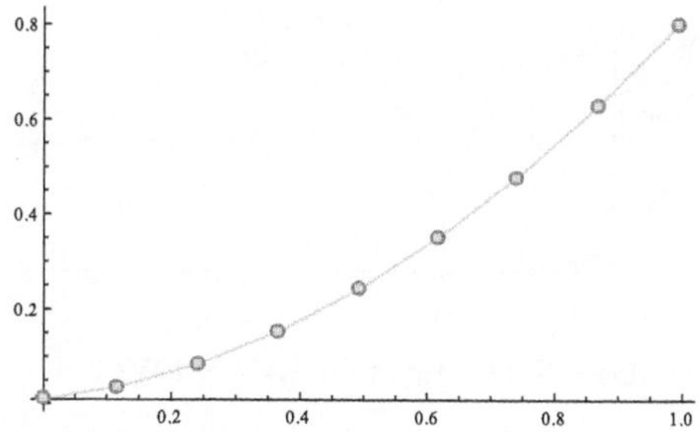

FIGURE 6.25 Exact and approximate solutions of BWM.

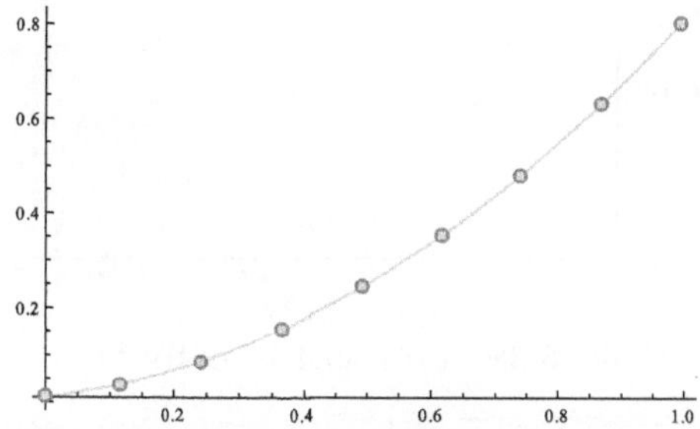

FIGURE 6.26 Exact and approximate solutions of LWM.

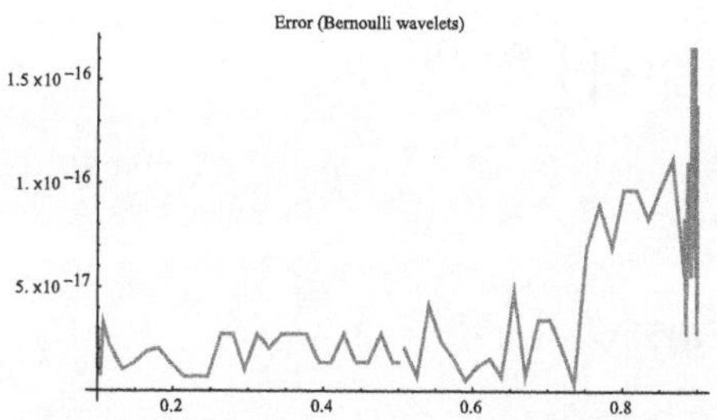

FIGURE 6.27 Error analysis of BWM.

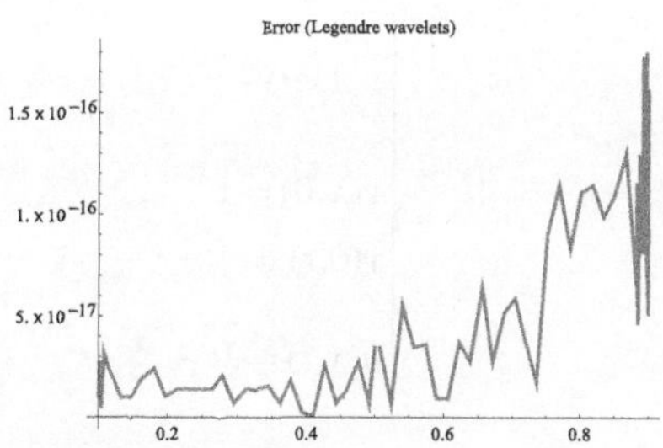

FIGURE 6.28 Error analysis of LWM.

For $k_1 = 3, k_2 = 3, M_1 = 3, M_2 = 3$, we have Figures 6.29 and 6.31 for BWM and Figures 6.30 and 6.32 for LWM.

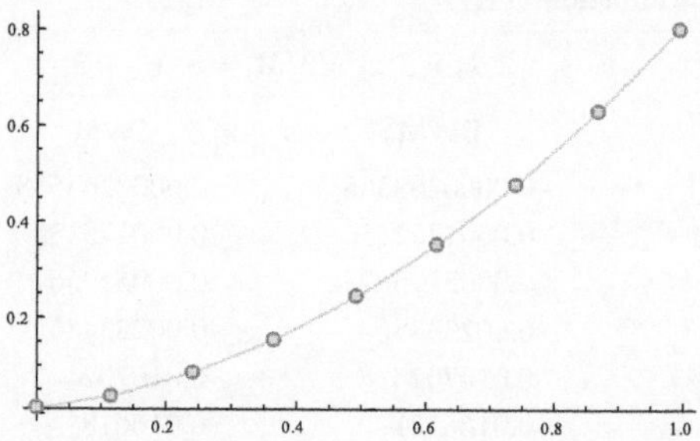

FIGURE 6.29 Exact and approximate solutions of BWM.

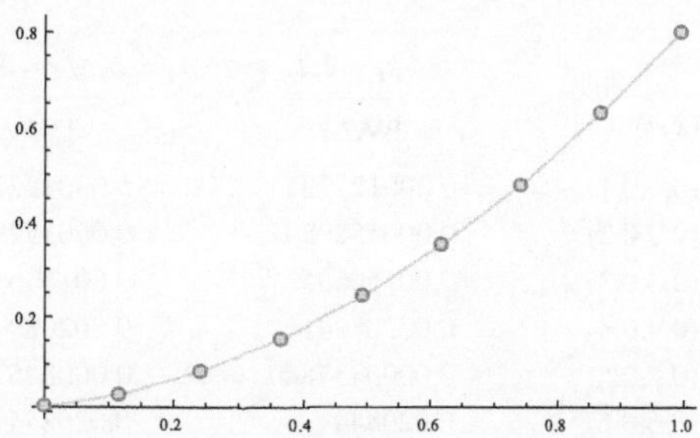

FIGURE 6.30 Exact and approximate solutions of LWM.

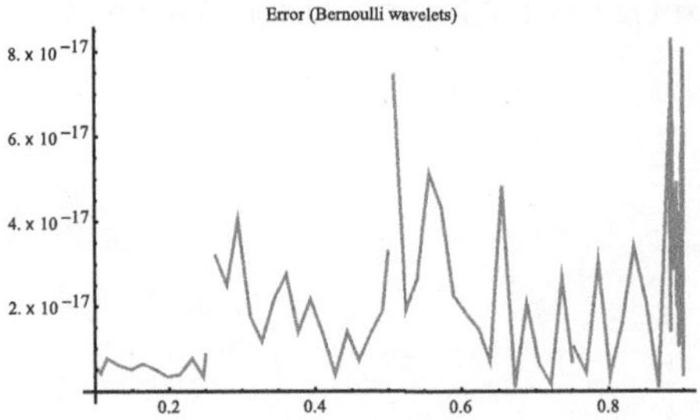

FIGURE 6.31 Error analysis of BWM.

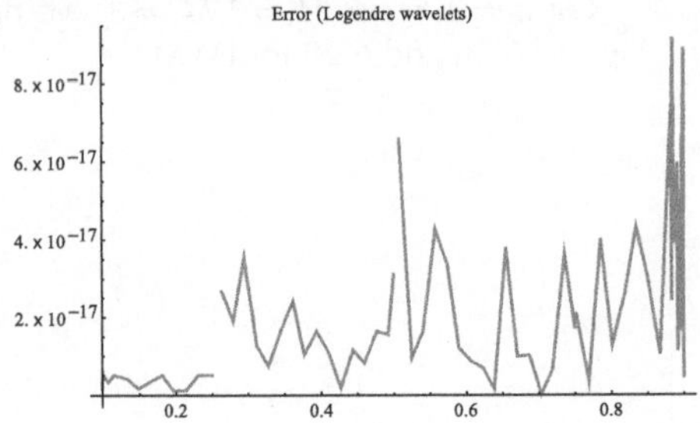

FIGURE 6.32 Error analysis of LWM.

Example 6.6.5

Let us consider the following weakly singular NLPIDEs of the first kind with $\theta = \frac{1}{2}$ and $\rho = 2$ and following boundary conditions [63]:

$$\begin{cases} v_{xt}(x,t) + v_t(x,t) = v(x,t) + h(x,t) + \int\limits_0^x \int\limits_0^t \frac{[\varphi(s,y)]^\rho}{(x-s)^\theta} dyds, \\ v(x,0) = x^2, \\ v(0,t) = t^2, \\ h(x,t) = 2t - x^2 - t^2 - \frac{2}{315} x^{\frac{1}{2}} t(128x^4 + 112x^2t^2 + 63t^4) \end{cases} \tag{6.64}$$

with the exact solution $v(x,t) = x^2 + t^2$.

The values of absolute error for $k_1 = 2, k_2 = 2, M_1 = 3, M_2 = 3$ and $k_1 = 3, k_2 = 3, M_1 = 3, M_2 = 3$ are shown in Table 6.13. Moreover, for

TABLE 6.13
Comparison of the Absolute Error Function $\widehat{e}(x,t)$ for Example 6.6.5

	Error Function $\widehat{e}(x,t)$			
	$k_1 = 2, k_2 = 2, M_1 = 3, M_2 = 3$		$k_1 = 3, k_2 = 3, M_1 = 3, M_2 = 3$	
(x,t)	BWM	LWM	BWM	LWM
(0.1,0.1)	0.000127731	0.000127731	0.0000263558	0.0000263558
(0.2,0.2)	0.000552979	0.000552979	0.000175235	0.000175235
(0.3,0.3)	0.00150695	0.00150695	0.000519567	0.000519567
(0.4,0.4)	0.00208941	0.00208941	0.00233486	0.00233486
(0.5,0.5)	0.0000457847	0.0000457847	0.0107024	0.0107024
(0.6,0.6)	0.020844	0.020844	0.0166181	0.0166181
(0.7,0.7)	0.0259186	0.0259186	0.0253296	0.0253296
(0.8,0.8)	0.0136683	0.0136683	0.0348035	0.0348035
(0.9,0.9)	0.0174007	0.0174007	0.0534762	0.0534762

TABLE 6.14
Comparison of L_2 and L_∞ Error Norms for Example 6.6.5

	BWM				LWM			
	$k_1 = 2, k_2 = 2, M_1 = 3, M_2 = 3$		$k_1 = 3, k_2 = 3, M_1 = 3, M_2 = 3$		$k_1 = 2, k_2 = 2, M_1 = 3, M_2 = 3$		$k_1 = 3, k_2 = 3, M_1 = 3, M_2 = 3$	
x	L^2	L^∞	L^2	L^∞	L^2	L^∞	L^2	L^∞
0.2	0.0044742	0.00848429	0.00317381	0.00825382	0.00447415	0.00848429	0.00317381	0.00825382
0.4	0.0119603	0.0232343	0.0143585	0.0318414	0.0119603	0.0232343	0.0143585	0.0318414
0.6	0.0151017	0.0302681	0.0254689	0.0595949	0.0151017	0.0302681	0.0254689	0.0595949
0.8	0.0098789	0.018344	0.0270321	0.0641273	0.00987886	0.018344	0.0270321	0.0641273

$k_1 = 2, k_2 = 2, M_1 = 3, M_2 = 3$, the numerical results and error analysis obtained from BWM and LWM are plotted in Figures 6.33, 6.35 and 6.34, 6.36, respectively, and for $k_1 = 3, k_2 = 3, M_1 = 3, M_2 = 3$, the numerical results and error analysis obtained from BWM and LWM are plotted in Figures 6.37, 6.39 and 6.38, 6.40, respectively.

For $k_1 = 2, k_2 = 2, M_1 = 3, M_2 = 3$, we have Figures 6.33 and 6.35 for BWM and Figures 6.34 and 6.36 for LWM.

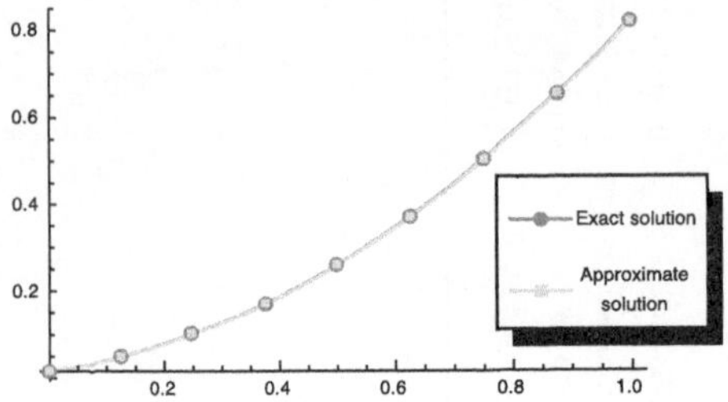

FIGURE 6.33 Exact and approximate solutions of BWM.

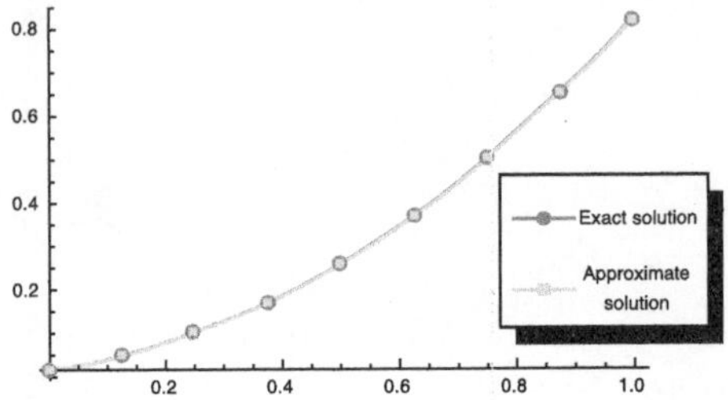

FIGURE 6.34 Exact and approximate solutions of LWM.

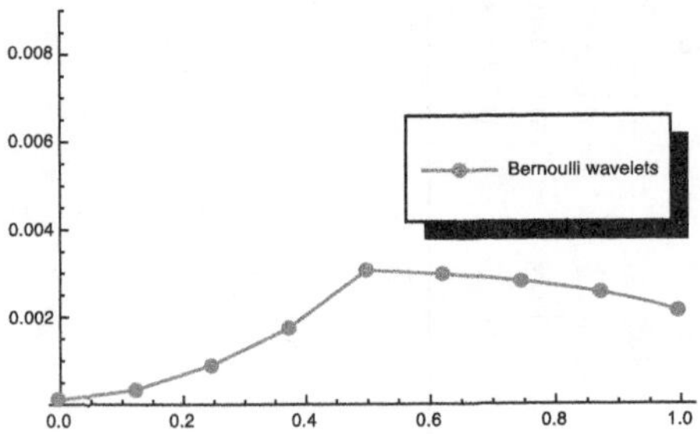

FIGURE 6.35 Error analysis of BWM.

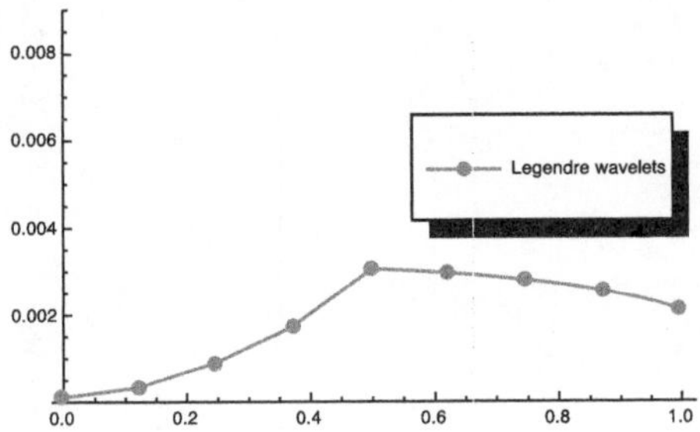

FIGURE 6.36 Error analysis of LWM.

For $k_1 = 3, k_2 = 3, M_1 = 3, M_2 = 3$, we have Figures 6.37 and 6.39 for BWM and Figures 6.38 and 6.40 for LWM.

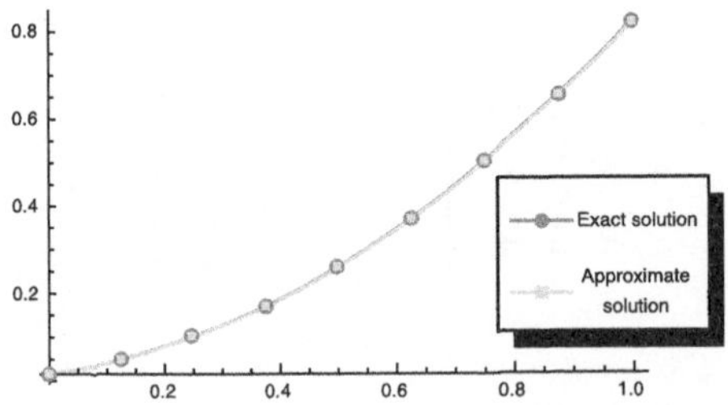

FIGURE 6.37 Exact and approximate solutions of BWM.

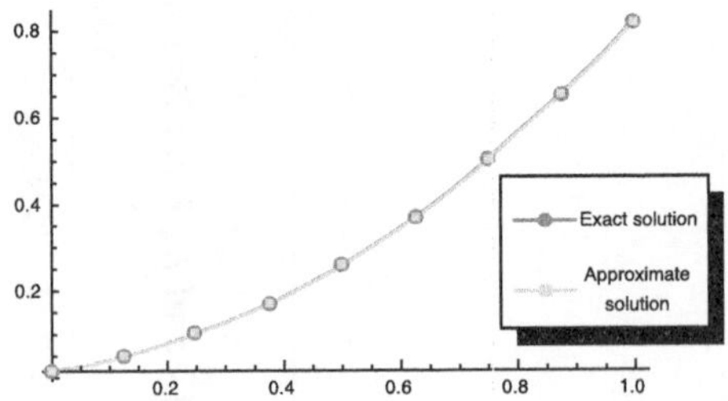

FIGURE 6.38 Exact and approximate solutions of LWM.

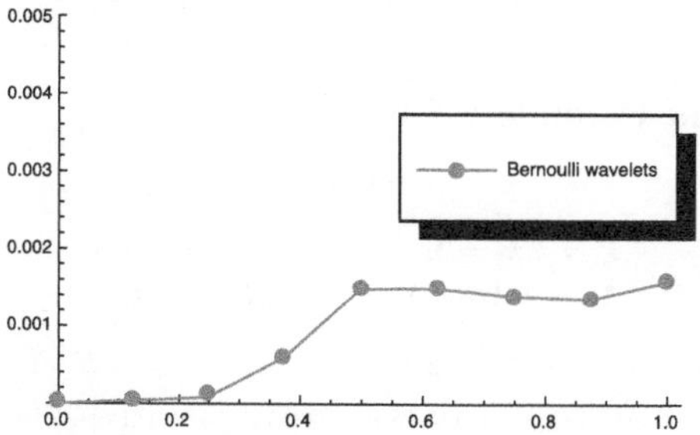

FIGURE 6.39 Error analysis of BWM.

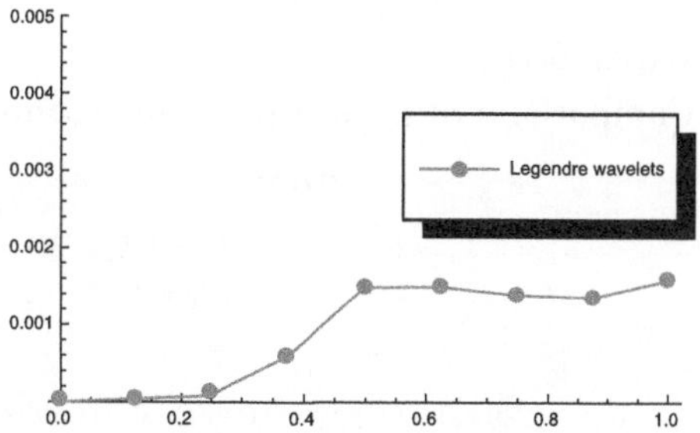

FIGURE 6.40 Error analysis of LWM.

Example 6.6.6

Let us consider the following weakly singular NLPIDEs of the second kind with $\theta=\frac{1}{2}$, $i_1=1$, $i_2=0$, and $\rho=2$ and following boundary conditions [62]:

$$\begin{cases} v_{xt}(x,t)+v_t(x,t)=v(x,t)+h(x,t)+\int_0^x\int_0^t \frac{[\eta_{i_1,i_2}v(s,y)]^\rho}{(x-s)^\theta}dyds, \\ v(x,0)=0, \\ v(0,t)=0, \\ h(x,t)=2x\cos t+x^2\cos t-x^2\sin t-\frac{32}{15}x^{\frac{5}{2}}(t-\cos t\sin t) \end{cases} \quad (6.65)$$

with the exact solution $v(x,t)=x^2\sin t$.

For $k_1=2, k_2=2, M_1=3, M_2=3$, the error analysis obtained from BWM and LWM are plotted in Figures 6.41 and 6.42, respectively, and for $k_1=3, k_2=3, M_1=3, M_2=3$, the error analysis obtained from BWM and LWM are plotted in Figures 6.43 and 6.44, respectively.

TABLE 6.15
Comparison of the Absolute Error Function $\widehat{e}(x,t)$ for Example 6.6.6

	Error Function $\widehat{e}(x,t)$			
	$k_1=2,k_2=2,M_1=3,M_2=3$		$k_1=3,k_2=3,M_1=3,M_2=3$	
(x,t)	BWM	LWM	BWM	LWM
(0.1,0.1)	0.000136481	0.000136481	0.000163778	0.000163778
(0.2,0.2)	0.00325562	0.00325562	0.00198743	0.00198743
(0.3,0.3)	0.00755051	0.00755051	0.0100557	0.0100557
(0.4,0.4)	0.0116827	0.0116827	0.0126752	0.0126752
(0.5,0.5)	0.00964363	0.00964363	0.025302	0.025302
(0.6,0.6)	0.0291125	0.0291125	0.0482585	0.0482585
(0.7,0.7)	0.0744001	0.0744001	0.117988	0.117988

TABLE 6.16
Comparison of L_2 and L_∞ Error Norms for Example 6.6.6

	BWM				LWM			
	$k_1 = 2, k_2 = 2, M_1 = 3, M_2 = 3$		$k_1 = 3, k_2 = 3, M_1 = 3, M_2 = 3$		$k_1 = 2, k_2 = 2, M_1 = 3, M_2 = 3$		$k_1 = 3, k_2 = 3, M_1 = 3, M_2 = 3$	
x	L^2	L^∞	L^2	L^∞	L^2	L^∞	L^2	L^∞
0.2	0.0144789	0.0263542	0.0390245	0.0698858	0.0144789	0.0263542	0.0390245	0.0698858
0.4	0.0232057	0.0396978	0.0375404	0.0831098	0.0232057	0.0396978	0.0375404	0.0831098
0.6	0.0481244	0.0909529	0.073104	0.170468	0.0481244	0.0909529	0.073104	0.170468

For $k_1 = 2, k_2 = 2, M_1 = 3, M_2 = 3$, we have Figure 6.41 for BWM and Figure 6.42 for LWM.

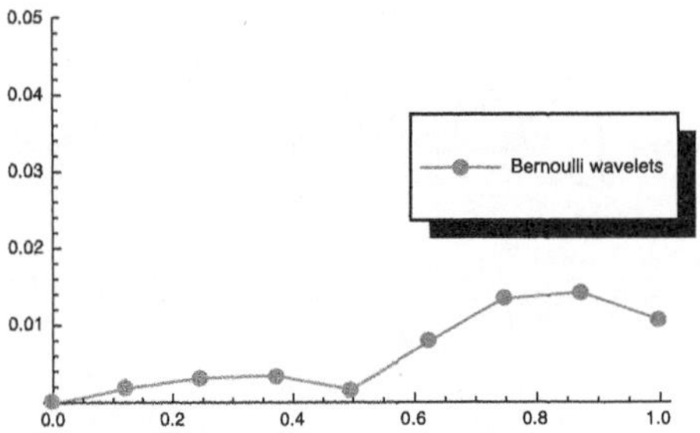

FIGURE 6.41 Error analysis of BWM.

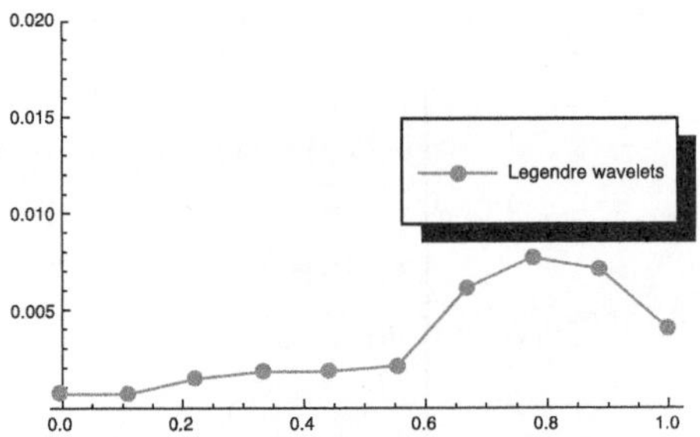

FIGURE 6.42 Error analysis of LWM.

For $k_1 = 3, k_2 = 3, M_1 = 3, M_2 = 3$, we have Figure 6.43 for BWM and Figure 6.44 for LWM.

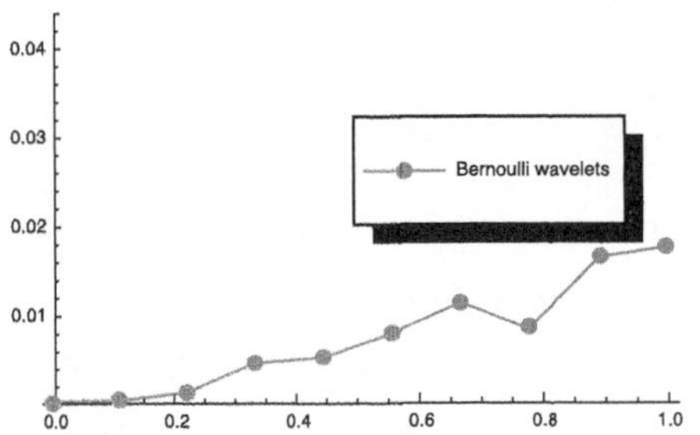

FIGURE 6.43 Error analysis of BWM.

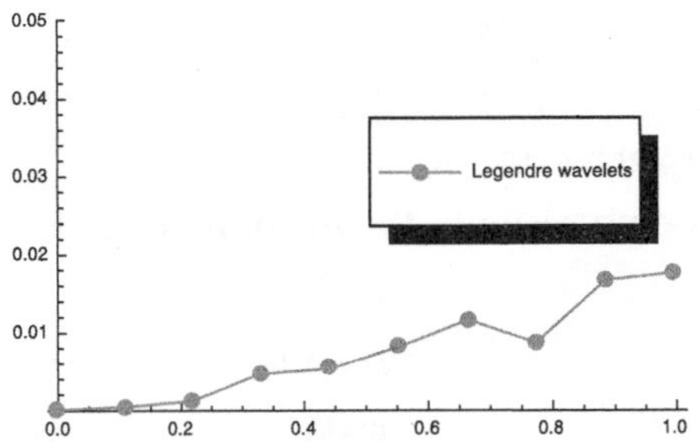

FIGURE 6.44 Error analysis of LWM.

6.7 CONCLUSION

In this study, an OM scheme based on 2D BWs and LWs has been established to solve the weakly singular Volterra LPIDEs and NLPIDEs with the given boundary conditions. In our scheme, by employing operational matrices of integration and differentiation and OM of product, weakly singular LPIDEs and NLPIDEs convert to a system of linear and nonlinear algebraic equations, respectively, which can be solved by using an appropriate numerical method. Furthermore, the convergence and error analyses of the proposed technique are presented for the given function of approximation. Moreover, some numerical examples are presented which manifest the applicability and plausibility of the proposed scheme. The absolute error analyses for BWs and LWs are also presented in this chapter. Furthermore, the L^2 and L^∞ error norms for both the 2D wavelets method have been presented to establish the accuracy of the proposed technique.

REFERENCES

1. Wazwaz A.M., 2011, *Linear and Nonlinear Integral Equations Methods and Applications*, Higher Education Press, Beijing and Springer-Verlag, Berlin.
2. Saha Ray S., and Sahu P.K., 2013, "Numerical methods for solving Fredholm integral equations of second kind", *Abstract and Applied Analysis*, **2013**(426916), p. 17.
3. Wazwaz A.M., 1997, *A First Course in Integral Equations*, World Scientific, Singapore and River Edge, NJ.
4. Norkin B.V., 2004, "Method of Successive approximation for solving integral equations of the Theory of Risk Processes", *Cybernetics and Systems Analysis*, **40**(4), pp. 517–526.
5. Ioakimidis N.I., 1982, "On the Quadrature methods for the numerical solution of singular integral equations", *Journal of Computational and Applied Mathematics*, **8**(2), pp. 81–86.
6. Rashed M.T., 2004, "Lagrange interpolation to compute the numerical solutions to differential, integral and integro-differential equations", *Applied Mathematics and Computations*, **151**(3), pp. 869–878.
7. Avazzadeh Z., and Heydari M., 2012, "Chebyshev polynomials for solving two dimensional linear and nonlinear integral equations of the second kind", *Computational and Applied Mathematics*, **31**(1), pp. 127–142.
8. Wang X.T., and Li Y.M., 2012, "Numerical solution of Fredholm integral equations of the second kind by General Legendre wavelets", *International Journal of Innovative computing, Information and Control*, **8**(1B), pp. 799–805.
9. Babolian E., and Shahsavaran A., 2009, "Numerical solution of nonlinear Fredholm integral equations of the second kind using Haar wavelets", *Journal of Computational and Applied Mathematics*, **225**(1), pp. 87–95.
10. Thomas K.S., 1981, "Galerkin methods for singular integral equations", *Mathematics of Computation*, **36**(153), pp. 193–205.
11. Saeed R.K., and Hassan J.S., 2014, "Solving singular integral equations by using Collocation method", *Mathematical Sciences Letters*, **3**(3), pp. 185–187.
12. Dixon J.A., 1987, "A nonlinear weakly singular Volterra integro-differential equation arising from a reaction-diffusion study in a small cell," *Journal of Computational and Applied Mathematics*, **18**(3), pp. 289–305.
13. Pao C.V., Payne L., and Amann H., 1979, "Bifurcation analysis on a nonlinear diffusion system in reactor dynamics", *Applicable Analysis*, **9**(2), pp. 107–119.

14. Zadeh K.S., 2011, "An integro-partial differential equation for modeling biofluids flow in fractured biomaterials", *Journal of Theoretical Biology*, **273**(1), pp. 72–79.
15. Hepperger P., 2012, "Hedging electricity swaptions using partial integro-differential equations", *Stochastic Processes and their Applications*, **122**(2), pp. 600–622.
16. Fakhar-Izadi F., and Dehghan M., 2013, "An efficient pseudo-spectral Legendre–Galerkin method for solving a nonlinear partial integro-differential equation arising in population dynamics", *Mathematical Methods in the Applied Sciences*, **36**(12), pp. 1485–1511.
17. Lee Y., 2014, "Financial options pricing with regime-switching jump-diffusions", *Computers & Mathematics with Applications*, **68**(3), pp. 392–404.
18. Ali A., Rahman E., Jan Z., Hussain I., and Ahmad S., 2016, "A Meshless collocation method for the approximate solution of a partial integro-differential equation", *Sindh University Research Journal-SURJ (Science Series)*, **48**(3), pp. 589–594.
19. Gurtin M.E., and Pipkin A.C., 1968, "A general theory of heat conduction with finite wave speeds", *Archive for Rational Mechanics and Analysis*, **31**(2), pp. 113–126.
20. Shukla A., Sukavanam N., and Pandey D.N., 2015, "Approximate controllability of semilinear system with state delay using sequence method", *Journal of the Franklin Institute*, **352**(11), pp. 5380–5392.
21. Allegretto W., Cannon J.R., and Lin Y., 1997, "A parabolic integro-differential equation arising from thermoelastic contact", *Discrete and Continuous Dynamical Systems-A*, **3**(2), pp. 217–234.
22. Wazwaz A.M., 2011. *Linear and Nonlinear Integral Equations: 639*, Springer, Heidelberg.
23. Linz P., 1985, *Analytical and Numerical Methods for Volterra Equations: 7*, SIAM, Philadelphia, PA.
24. MacCamy R.C., 1977, "An integro-differential equation with application in heat flow", *Quarterly of Applied Mathematics*, **35**(1), pp. 1–19.
25. Fakhar-Izadi F., and Dehghan M., 2013, "An efficient pseudo-spectral Legendre–Galerkin method for solving a nonlinear partial integro-differential equation arising in population dynamics", *Mathematical Methods in the Applied Sciences*, **36**(12), pp. 1485–1511.
26. Avazzadeh Z., Rizi Z.B., Ghaini F.M., and Loghmani G.B., 2012, "A numerical solution of nonlinear parabolic-type Volterra partial integro-differential equations using radial basis functions", *Engineering Analysis with Boundary Elements*, **36**(5), pp. 881–893.
27. Fakhar-Izadi F., and Dehghan M., 2014, "Space–time spectral method for a weakly singular parabolic partial integro-differential equation on irregular domains", *Computers & Mathematics with Applications*, **67**(10), pp. 1884–1904.
28. Sahu P.K., and Ray S.S., 2016, "Legendre spectral collocation method for the solution of the model describing biological species living together", *Journal of Computational and Applied Mathematics*, **296**, pp. 47–55.
29. Sahu P.K., and Ray S.S., 2016, "Comparative experiment on the numerical solutions of Hammerstein integral equation arising from chemical phenomenon", *Journal of Computational and Applied Mathematics*, **291**, pp. 402–409.
30. Postnikov E.B., Lebedeva E.A., and Lavrova A.I., 2016, "Computational implementation of the inverse continuous wavelet transform without a requirement of the admissibility condition", *Applied Mathematics and Computation*, **282**, pp. 128–136.
31. Majidian H., 2014, "Modified Euler's method with a graded mesh for a class of Volterra integral equations with weakly singular kernel", *Numerical Algorithms*, **67**(2), pp. 405–422.
32. Shahsavaran A., 2011, "Haar wavelet method to solve Volterra integral equations with weakly singular kernel by collocation method", *Applied Mathematical Sciences*, **5**(65), pp. 3201–3210.

33. Maleknejad K., Nosrati M., and Najafi E., 2012, "Wavelet Galerkin method for solving singular integral equations", *Computational & Applied Mathematics*, **31**, pp. 373–390.
34. Pan L., He X., and Lü T., 2010, "High accuracy combination method for solving the systems of nonlinear Volterra integral and integro-differential equations with weakly singular kernels of the second kind", *Mathematical Problems in Engineering*, **2010**, pp. 1–21.
35. Dehghan M., 2006, "Solution of a partial integro-differential equation arising from viscoelasticity", *International Journal of Computer Mathematics*, **83**(1), pp. 123–129.
36. Luo M., Xu D., and Li L., 2015, "A compact difference scheme for a partial integro-differential equation with a weakly singular kernel", *Applied Mathematical Modelling*, **39**(2), pp. 947–954.
37. Kim C.H., and Choi U.J., 1998, "Spectral collocation methods for a partial integro-differential equation with a weakly singular kernel", *The ANZIAM Journal*, **39**(3), pp. 408–430.
38. Bialecki B., and Fairweather G., 2001, "Orthogonal spline collocation methods for partial differential equations", *Journal of Computational and Applied Mathematics*, **128**(1–2), pp. 55–82.
39. Yoon J.M., Xie S., and Hrynkiv V., 2012, "Two numerical algorithms for solving a partial integro-differential equation with a weakly singular kernel", *Applications and Applied Mathematics*, **7**(1), pp. 133–141.
40. Long W., Xu D., and Zeng X., 2012, "Quasi wavelet based numerical method for a class of partial integro-differential equation", *Applied Mathematics and Computation*, **218**(24), pp. 11842–11850.
41. Nemati S., Lima P.M., and Ordokhani Y., 2013, "Numerical solution of a class of two-dimensional nonlinear Volterra integral equations using Legendre polynomials", *Journal of Computational and Applied Mathematics*, **242**, pp. 53–69.
42. Mirzaee F., and Samadyar N., 2018, "Convergence of Legendre wavelet collocation method for solving nonlinear Stratonovich Volterra integral equations", *Computational Methods for Differential Equations*, **6**(1), pp. 80–97.
43. Sahu P.K., and Saha Ray S., 2015, "Legendre wavelets operational method for the numerical solutions of nonlinear Volterra integro-differential equations system", *Applied Mathematics and Computation*, **256**, pp. 715–723.
44. Singh S., Patel V.K., Singh V.K., and Tohidi E., 2017, "Numerical solution of nonlinear weakly singular partial integro-differential equation via operational matrices", *Applied Mathematics and Computation*, **298**, pp. 310–321.
45. Mirzaee F., and Samadyar N., 2017, "Application of orthonormal Bernstein polynomials to construct an efficient scheme for solving fractional stochastic integro-differential equation", *Optik*, **132**, pp. 262–273.
46. Han H., Liu Y., Lu T., and He X., 2010, "New algorithm for the system of nonlinear weakly singular Volterra integral equations of the second kind and integro-differential equations", *Journal of Information and Computational Science*, **7**(6), pp. 1229–1235.
47. Mirzaee F., and Samadyar N., 2017, "Application of operational matrices for solving system of linear Stratonovich Volterra integral equation", *Journal of Computational and Applied Mathematics*, **320**, pp. 164–175.
48. Mirzaee F., Samadyar N., and Hoseini S.F., 2017, "A new scheme for solving nonlinear Stratonovich Volterra integral equations via Bernoulli's approximation", *Applicable Analysis*, **96**(13), pp. 2163–2179.
49. Mirzaee F., and Samadyar N., 2018, "On the numerical solution of stochastic quadratic integral equations via operational matrix method", *Mathematical Methods in the Applied Sciences*, **41**(12), pp. 4465–4479.

50. Mirzaee F., and Samadyar N., 2018, "Using radial basis functions to solve two-dimensional linear stochastic integral equations on non-rectangular domains", *Engineering Analysis with Boundary Elements*, **92**, pp. 180–195.
51. Mirzaee F., and Samadyar N., 2019, "Numerical solution based on two-dimensional orthonormal Bernstein polynomials for solving some classes of two-dimensional non-linear integral equations of fractional order", *Applied Mathematics and Computations*, **344**, pp. 192–203.
52. Toutounian F., and Tohidi E., 2013, "A new Bernoulli matrix method for solving second order linear partial differential equations with the convergence analysis", *Applied Mathematics and Computation*, **223**, pp. 298–310.
53. Tohidi E., and Zak M.K., 2016, "A new matrix approach for solving second-order linear matrix partial differential equations", *Mediterranean Journal of Mathematics*, **13**(3), pp. 1353–1376.
54. Bhrawy A.H., Tohidi E., and Soleymani F., 2012, "A new Bernoulli matrix method for solving high-order linear and nonlinear Fredholm integro-differential equations with piecewise intervals", *Applied Mathematics and Computation*, **219**(2), pp. 482–497.
55. Tang T., 1992, "Superconvergence of numerical solutions to weakly singular Volterra integro-differential equations", *Numerische Mathematik*, **61**, pp. 373–382.
56. Tohidi E., and Toutounian F., 2014, "Convergence analysis of Bernoulli matrix approach for one-dimensional matrix hyperbolic equations of the first order", *Computers & Mathematics with Applications*, **68**(1–2), pp. 1–12.
57. Zogheib B., and Tohidi E., 2016, "A new matrix method for solving two-dimensional time-dependent diffusion equations with Dirichlet boundary conditions", *Applied Mathematics and Computation*, **291**, pp. 1–13.
58. Keshavarz E., Ordokhani Y., and Razzaghi M., 2019, "The Bernoulli wavelets operational matrix of integration and its applications for the solution of linear and nonlinear problems in calculus of variations", *Applied Mathematics and Computation*, **351**, pp. 83–98.
59. Bokhari A., Amir A., Bahri S.M., and Belgacem F.B.M., 2017, "A generalized Bernoulli wavelet operational matrix of derivative applications to optimal control problems", *Nonlinear Stud*, **24**(4), pp. 775–790.
60. Singh S., Patel V.K., and Singh V.K., 2018, "Convergence rate of collocation method based on wavelet for nonlinear weakly singular partial integro-differential equation arising from viscoelasticity", *Numerical Methods for Partial Differential Equations*, **34**(5), pp. 1781–1798.
61. Patel V.K., Singh S., and Singh V.K., 2017, "Two-dimensional wavelets collocation method for electromagnetic waves in dielectric media", *Journal of Computational and Applied Mathematics*, **317**, pp. 307–330.
62. Mirzaee F., Alipour S., and Samadyar N., 2019, "A numerical approach for solving weakly singular partial integro-differential equations via two-dimensional-orthonormal Bernstein polynomials with the convergence analysis", *Numerical Methods for Partial Differential Equations*, **35**(2), pp. 615–637.
63. Patel V.K., Singh S., Singh V.K., and Tohidi E., 2018, "Two dimensional wavelets collocation scheme for linear and nonlinear Volterra weakly singular partial integro-differential equations", *International Journal of Applied and Computational Mathematics*, **4**(5), pp. 1–27.

7 Numerical Solutions of Pantograph Volterra Delay-Integro-Differential Equations

7.1 INTRODUCTION

In recent years, integro-differential equations (IDEs) play a vital role in the modelling of physical phenomena in engineering and science [1]. Fractional calculus is a rapidly increasing mathematical tool that aids in the differentiation and integration of functions with non-integer orders [2,3]. In 1695, two mathematicians, Leibnitz and L'Hopital, investigated the notion of fractional derivative [4]. Fractional calculus has been widely used in recent years in the fields of physical systems [5], electromagnetic waves [6], signal processing [7], viscoelasticity [8], nonlinear dynamical systems [9], continuum and statistical mechanics [10], fluid-dynamic traffic models [11], population dynamics model [12], diffusion problem [13], and optimal control [14].

The term pantograph originated from the work of Ockendon and Tayler [15]. The pantograph equations are usually referred to as IDEs with proportional delays. Pantograph Volterra delay-integro-differential equations (PVDIDEs) play a major role in the field of economy, control, biology, and astrophysics [16]. Several methods have been implemented for the solution of pantograph equation such as the Hermite polynomial method [17], Adomian decomposition method [18], spectral tau method [19], Taylor polynomial method [20], Laguerre matrix method [21], operational tau method [22], and Haar wavelet method [23]. Volterra delay-integro-differential equations (VDIDEs) have been studied with stability and convergence properties of various numerical methods in previous studies [24,25]. Moreover, the qualitative analysis of the IDEs has been discussed in other studies [26,27].

The numerical solution of integral and differential equations with delays has been studied by several authors such as Yao et al. [28], Song et al. [29], Aimene et al. [30], Dehestani et al. [31], Linz et al. [32], Burner [33], and Sahu et al. [34]. In a study [35], Sahu and Saha Ray have applied the Legendre spectral collocation method for the numerical solution of the Fredholm integro-differential-difference equation. Ezz-Eldien and Doha have applied a new collocation spectral method for solving the pantograph-type Volterra IDEs [36]. In another study [37], the Euler wavelets method is applied to solve fractional-order PVDIDEs. Tang et al. [38] introduced the Laguerre spectral collocation method for solving the Volterra delay IDEs. In their study [39], Tunç et al. have studied the stability of the systems of IDEs with time-delay retardation. Guoting et al. [40] have solved pantograph-type

DOI: 10.1201/9781003453451-7

Volterra IDEs with singular kernels by applying pseudo-spectral Galerkin method. Saha Ray and Behera [41] have developed an operational matrix method for the solution of Volterra weakly singular partial IDEs. The Legendre collocation methods have been introduced for the numerical solution of VDIDEs [42]. The Legendre spectral collocation method has been introduced for the numerical solution PVDIDEs [43]. Mirzaee et al. [44] have introduced a numerical framework for solving higher order of PVDIDE.

Many numerical techniques have been developed to solve fractional integral equations and IDEs. Sahu and Saha Ray [45] have applied the Legendre wavelets method for the numerical solution of nonlinear fractional Volterra–Fredholm IDEs with mixed boundary conditions. Zhu et al. [46,47] have solved nonlinear fractional Fredholm IDEs and nonlinear fractional Volterra integro-differential equations by using second-kind Chebyshev wavelet. Ganji et al. [48] have studied the numerical solutions of time-fractional Klein–Gordon equations. Tohidi et al. [49] have introduced a numerical method via Bernoulli polynomials for the solution of nonlinear fractional Volterra IDEs. In a study [50], the Crank–Nicolson method was applied to analyse the time-space fractional Bloch–Torrey equation. Tuan et al. [51] have employed the spectral collocation method for the numerical solution of fractional rheological models and fractional Newell–Whitehead–Segel equation. Ganji et al. [52] have introduced a new approach for solving multi variable orders differential equations by using fifth-kind Chebyshev polynomials. Sayevand et al. [53] have analysed the fractional Navier–Stokes equations by applying non-standard finite difference method. Yang and Huang [54] proposed a spectral-collocation method for fractional pantograph delay IDEs. Saha Ray and Behera [55] established an operational matrix approach for solving Volterra weakly singular partial IDEs. Ali et al. [56] have solved pantograph-type differential and integral equations with multiple delays by using spectral methods. Jafari et al. [57] have applied the Transferred Legendre pseudospectral method for the solution of pantograph delay differential equations.

7.2 OUTLINE OF THE PRESENT STUDY

In this chapter, an effective numerical framework has been developed to obtain the solution of the PVDIDE. By utilizing the Mu¨ntz–Legendre wavelets (MLWs)–based operational matrices along with the collocation points, the one-dimensional (1D) PVDIDEs are reduced into an explicit system of algebraic equations. Furthermore, the convergence analysis and error estimation of the proposed scheme have been analysed. Finally, some numerical examples are examined to illustrate the applicability and accuracy of the proposed method.

In addition, a numerical operational matrix approach based on Euler wavelets (EWs) is proposed to solve the nonlinear PVDIDE of fractional order. The operational matrix (OM) of fractional integration (FI) based on EWs has been constructed. Then, the nonlinear PVDIDEs are reduced to algebraic equations by using the fractional integral OM. Several theorems are presented to establish the convergence and error analyses of the proposed method. Additionally, some numerical problems are solved to justify the applicability and validity of the presented technique.

7.3 INTEGER-ORDER LINEAR PANTOGRAPH VOLTERRA DELAY-INTEGRO-DIFFERENTIAL EQUATION

The main objective of this study is to find the numerical solution of 1D PVDIDEs in the following form:

$$u'(x) = a(x)u(x) + b(x)u(qx) + f(x) + \int_0^x k_1(x,z)u(z)\,dz + \int_0^{qx} k_2(x,z)u(z)\,dz, \tag{7.1}$$

where $u(0) = 0,\ x \in \hat{I} := [0,T]$, and qx is the proportional delays $(0 < q < 1)$.

Here, $a(x)$, $b(x)$, and $f(x)$ are known functions defined on $\hat{I}$ and $u(x)$ is the unknown function to be determined. The kernel functions $k_1(x,z)$ and $k_2(x,z)$ are defined on $\hat{D}_1 := \{(x,z): 0 < x < T, 0 < z < x\}$ and $\hat{D}_2 := \{(x,z): 0 < x < T, 0 < z < qx\}$, respectively.

7.3.1 Function Approximation by MLWs

Consider a function $g(x) \in L^2[0,1)$ which can be expressed as follows:

$$g(x) \approx \sum_{n=1}^{2^{k-1}} \sum_{m=0}^{M-1} \tilde{c}_{n,m}\psi_{n,m}(x) = \tilde{C}^T\Psi(x), \tag{7.2}$$

where $\tilde{C}$ and $\Psi(x)$ are vectors of dimension $(2^{k-1}M \times 1)$ given by

$$\tilde{C} = [\tilde{c}_{1,0}, \tilde{c}_{1,1}, \ldots, \tilde{c}_{1,M-1}, \ldots, \tilde{c}_{2^{k-1},0}, \tilde{c}_{2^{k-1},1}, \ldots \tilde{c}_{2^{k-1},M-1}]^T \tag{7.3}$$

and

$$\Psi(x) = [\psi_{1,0}(x), \psi_{1,1}(x), \ldots, \psi_{1,M-1}(x), \ldots, \psi_{2^{k-1},0}(x), \psi_{2^{k-1},1}(x), \ldots \psi_{2^{k-1},M-1}(x)]^T, \tag{7.4}$$

where $\psi_{n,m}(x)$, $n = 1,2,\ldots,2^{k-1}$, $m = 0,1,\ldots,M-1$ is the MLWs defined on $[0,1)$ as follows [58]:

$$\psi_{n,m}(x) = \begin{cases} \sqrt{(2\tau_m + 1)}\,2^{\frac{k-1}{2}}\,\tilde{L}_m(2^{k-1}x - n + 1), & \dfrac{n-1}{2^{k-1}} \le x < \dfrac{n}{2^{k-1}}, \\ 0, & \text{otherwise.} \end{cases} \tag{7.5}$$

The coefficients $\tilde{c}_{n,m}$ of $2^{k-1}M$-dimensional vector $\tilde{C}$ are given as follows:

$$\tilde{c}_{n,m} = \int_0^1 g(x)\psi_{n,m}(x)\,dx.$$

The kernel functions $k_1(x,z)$ and $k_2(x,z)$ can be approximated as follow:

$$k_1(x,z) \approx \Psi^T(x)K_1\Psi(z)$$

and

$$k_2(x,z) \approx \Psi^T(x) K_2 \Psi(z),$$

where K_1 and K_2 are the coefficient matrix of dimension $(2^{k-1}M \times 2^{k-1}M)$ and its components are defined as follows:

$$k_{1_{i,j}} =< \psi_{n,m}(x), < k_1(x,z), \psi_{n_1,m_1}(z) >>$$

and

$$k_{2_{i,j}} =< \psi_{n,m}(x), < k_2(x,z), \psi_{n_1,m_1}(z) >>,$$

in which

$$i = n.M - (M - (m+1)),$$

$$j = n_1.M - (M - (m_1 + 1)),$$

where $n, n_1 = 1, 2, ..., 2^{k-1}$ and $m, m_1 = 0, 1, ..., M-1$.

7.3.2 Operational Matrices of Integration for MLWs

This section deals with the derivation of OMs of each integral term of Eq. (7.1).

The integration from 0 to x of 1D MLWs $\Psi(z)$ defined in Eq. (7.4) is given by

$$\int_0^x \Psi(z)\,dz = \begin{pmatrix} \int_0^x \psi_{1,0}(z)\,dz \\ \int_0^x \psi_{1,1}(z)\,dz \\ . \\ . \\ . \\ \int_0^x \psi_{2^{k-1},M-1}(z)\,dz \end{pmatrix}$$

$$= P\Psi(x).$$

Thus,

$$\int_0^x \Psi(z)\,dz = P\Psi(x), \tag{7.6}$$

where P is the OM of dimension $2^{k-1}M$.

For $k = 2, M = 3$ and $\gamma = 1$, we have

$$P = \begin{bmatrix} \frac{1}{4} & \frac{1}{4\sqrt{3}} & 0 & \frac{1}{2} & 0 & 0 \\ -\frac{1}{4\sqrt{3}} & 0 & \frac{1}{4\sqrt{15}} & 0 & 0 & 0 \\ 0 & -\frac{1}{4\sqrt{15}} & 0 & 0 & 0 & 0 \\ 0 & 0 & 0 & \frac{1}{4} & \frac{1}{4\sqrt{3}} & 0 \\ 0 & 0 & 0 & -\frac{1}{4\sqrt{3}} & 0 & \frac{1}{4\sqrt{15}} \\ 0 & 0 & 0 & 0 & -\frac{1}{4\sqrt{15}} & 0 \end{bmatrix}.$$

Similarly, the integration from 0 to qx $(0 < q < 1)$ of 1D MLWs $\Psi(z)$ is given by

$$\int_0^{qx} \Psi(z)\,dz = \begin{pmatrix} \int_0^{qx} \psi_{1,0}(z)\,dz \\ \int_0^{qx} \psi_{1,1}(z)\,dz \\ \cdot \\ \cdot \\ \cdot \\ \int_0^{qx} \psi_{2^{k-1},M-1}(z)\,dz \end{pmatrix}$$

$$= \hat{P}\Psi(x).$$

Thus,

$$\int_0^{qx} \Psi(z)\,dz = \hat{P}\Psi(x), \tag{7.7}$$

where $\hat{P}$ is the OM of dimension $2^{k-1}M$.

For $k = 2, M = 3$ and $\gamma = 1$, we have

$$\hat{P} = \begin{bmatrix} \frac{q}{4} & \frac{q}{4\sqrt{3}} & 0 & \frac{3q}{4} & \frac{q}{4\sqrt{3}} & 0 \\ -\frac{q}{2\sqrt{3}} & -\frac{q}{8} & \frac{q}{8\sqrt{15}} & -\frac{q}{2\sqrt{3}} & \frac{q}{8} & \frac{q}{8\sqrt{15}} \\ \frac{q\sqrt{5}}{16} & -\frac{q}{16\sqrt{15}} & -\frac{q}{16} & -\frac{q\sqrt{5}}{16} & -\frac{q}{16\sqrt{15}} & \frac{q}{16} \\ 0 & 0 & 0 & 0 & 0 & 0 \\ 0 & 0 & 0 & 0 & 0 & 0 \\ 0 & 0 & 0 & 0 & 0 & 0 \end{bmatrix}.$$

7.3.3 Müntz–Legendre Wavelet Method for PVDIDE

In this part, a new proposed technique is used to turn the PVDIDE Eq. (7.1) into a system of algebraic equations. Firstly, consider the approximation of following functions:

$$u'(x) \approx \Psi^T(x)\tilde{C}_1 = \tilde{C}_1^T\Psi(x), \tag{7.8}$$

$$a(x) \approx \Psi^T(x)A = A^T\Psi(x), \tag{7.9}$$

$$b(x) \approx \Psi^T(x)B = B^T\Psi(x), \tag{7.10}$$

$$f(x) \approx \Psi^T(x)F = F^T\Psi(x), \tag{7.11}$$

$$u(0) \approx \Psi^T(x)U = U^T\Psi(x), \tag{7.12}$$

$$k_1(x,z) \approx \Psi^T(x)K_1\Psi(z) = \Psi^T(z)K_1^T\Psi(x), \tag{7.13}$$

$$k_2(x,z) \approx \Psi^T(x)K_2\Psi(z) = \Psi^T(z)K_2^T\Psi(x). \tag{7.14}$$

Using Eq. (7.6) and integrating Eq. (7.8) with respect to x, we now obtain

$$\begin{aligned} u(x) &= u(0) + \tilde{C}_1^T P\Psi(x) \\ &= U^T\Psi(x) + \tilde{C}_1^T P\Psi(x) \\ &= (U^T + \tilde{C}_1^T P)\Psi(x) \\ &= \Psi^T(x)\tilde{C}_2, \end{aligned} \tag{7.15}$$

where $\tilde{C}_2 = U + P^T\tilde{C}_1$.

Let us consider the approximation of the function $u(qx)$ $(0 < q < 1)$ as follows:

$$u(qx) \approx \Psi^T(x)\tilde{C}_3 = \tilde{C}_3^T \Psi(x), \tag{7.16}$$

where $\tilde{C}_3^T = \int_0^1 \Psi^T(qx)\tilde{C}_2\Psi^T(x)\,dx$.

Now the following approximations for the term $a(x)u(x)$ and $b(x)u(qx)$ of Eq. (7.1) are determined:

$$\begin{aligned} a(x)u(x) &= \Psi^T(x)A.\tilde{C}_2^T\Psi(x) \\ &= \Psi^T(x)A_1\Psi(x), \end{aligned}$$

where $A_1 = A.\tilde{C}_2^T$,

$$a(x)u(x) = \tilde{A}_1^T \Psi(x), \tag{7.17}$$

where

$$\tilde{A}_1^T = \int_0^1 \Psi^T(x)A_1\Psi(x)\Psi^T(x)\,dx.$$

Again,

$$\begin{aligned} b(x)u(qx) &= \Psi^T(x)B.\tilde{C}_3^T\Psi(x) \\ &= \Psi^T(x)B_1\Psi(x), \end{aligned}$$

where $B_1 = B.\tilde{C}_3^T$,

$$b(x)u(qx) = \tilde{B}_1^T \Psi(x), \tag{7.18}$$

where

$$\tilde{B}_1^T = \int_0^1 \Psi^T(x)B_1\Psi(x)\Psi^T(x)\,dx.$$

In Eq. (7.1), the integral portions can be interpreted as follows:

$$\begin{aligned} \int_0^x k_1(x,z)u(z)\,dz &= \int_0^x \Psi^T(x)K_1\Psi(z)\tilde{C}_2^T\Psi(z)\,dz \\ &= \Psi^T(x)K_1\int_0^x \Psi(z)\Psi^T(z)\tilde{C}_2\,dz \\ &= \Psi^T(x)K_1\int_0^x \tilde{C}_{22}^T\Psi(z)\,dz, \end{aligned}$$

where $\tilde{C}_{22}^{T} = \int_{0}^{1} \Psi(z)\Psi^{T}(z)\tilde{C}_{2}\Psi^{T}(z)\,dz$

$$
\begin{aligned}
&= \Psi^{T}(x)K_{1}\tilde{C}_{22}^{T}P\Psi(x) \\
&= \tilde{C}_{33}^{T}\Psi(x),
\end{aligned}
\tag{7.19}
$$

where $\tilde{C}_{33}^{T} = \int_{0}^{1} \Psi^{T}(x)K_{1}\tilde{C}_{22}^{T}P\Psi(x)\Psi^{T}(x)\,dx$.

Similarly,

$$
\begin{aligned}
\int_{0}^{qx} k_{2}(x,z)u(z)\,dz &= \int_{0}^{qx} \Psi^{T}(x)K_{2}\Psi(z)\tilde{C}_{2}^{T}\Psi(z)\,dz \\
&= \tilde{\Psi}^{T}(x)K_{2}\int_{0}^{qx} \Psi(z)\Psi^{T}(z)\tilde{C}_{2}\,dz \\
&= \Psi^{T}(x)K_{2}\int_{0}^{qx} \tilde{C}_{22}^{T}\Psi(z)\,dz,
\end{aligned}
$$

where $\tilde{C}_{22}^{T} = \int_{0}^{1} \Psi(z)\Psi^{T}(z)\tilde{C}_{2}\Psi^{T}(z)\,dz$

$$
\begin{aligned}
&= \Psi^{T}(x)K_{2}\tilde{C}_{22}^{T}\hat{P}\Psi(x) \\
&= \tilde{C}_{44}^{T}\Psi(x),
\end{aligned}
\tag{7.20}
$$

where $\tilde{C}_{44}^{T} = \int_{0}^{1} \Psi^{T}(x)K_{2}\tilde{C}_{22}^{T}\hat{P}\Psi(x)\Psi^{T}(x)\,dx$.

Equations (7.8)–(7.20) are now substituted into Eq. (7.1), yielding

$$
\Psi^{T}(x)\tilde{C}_{1} = \Psi^{T}(x)\tilde{A}_{1} + \Psi^{T}(x)\tilde{B}_{1} + \Psi^{T}(x)F + \Psi^{T}(x)\tilde{C}_{33} + \Psi^{T}(x)\tilde{C}_{44}. \tag{7.21}
$$

Collocating Eq. (7.21) at $2^{k-1}M$-points x_i yields

$$
\Psi^{T}(x_i)\tilde{C}_{1} = \Psi^{T}(x_i)\tilde{A}_{1} + \Psi^{T}(x_i)\tilde{B}_{1} + \Psi^{T}(x_i)F + \Psi^{T}(x_i)\tilde{C}_{33} + \Psi^{T}(x_i)\tilde{C}_{44}, \tag{7.22}
$$

where the collocation points x_i are given as follows:

$$
x_i = \frac{2i-1}{2.2^{k-1}M}, \quad i = 1,2,\ldots,2^{k-1}M.
$$

Equation (7.22) generates $2^{k-1}M$ - system of algebraic equations after substituting the collocation points. The value of $\tilde{C}_1$ can be determined by solving the above set of equations. Furthermore, the value of $\tilde{C}_2$ is calculated by using the value of $\tilde{C}_1$.

The approximate solution of PVDIDE Eq. (7.1) can thus be determined numerically by plugging the value of $\tilde{C}_2$ into Eq. (7.15).

7.4 FRACTIONAL-ORDER PANTOGRAPH VOLTERRA DELAY-INTEGRO-DIFFERENTIAL EQUATION

The primary goal of this study is to solve the 1D fractional PVDIDEs in the following form:

$$\begin{cases} {}_0^C D_x^\alpha u(x) = a(x)u(x) + b(x)u(qx) + f(x) + \int\limits_0^x k_1(x,z)u(z)\,dz + \int\limits_0^{qx} k_2(x,z)u(z)\,dz \\ u(0) = 0 \end{cases}, \tag{7.23}$$

where $x \in \tilde{I} := [0,T]$ and qx is the proportional delays $(0 < q < 1)$.

Here, $a(x)$, $b(x)$, and $f(x)$ are known functions defined on $\tilde{I}$ and $u(x)$ is the unknown function to be determined. The kernel functions $k_1(x,z)$ and $k_2(x,z)$ are defined on $\Omega_1 := \{(x,z) : 0 < x < T, 0 < z < x\}$ and $\Omega_2 := \{(x,z) : 0 < x < T, 0 < z < qx\}$, respectively, and ${}_0^C D_x^\alpha$ denotes the Caputo fractional derivative of fractional order α, $0 < \alpha \leq 1$.

7.4.1 Function Approximation by EWs

If $g(x)$ is any function defined on $L^2[0,1)$, it can be approximated as follows:

$$g(x) \approx \sum_{n=1}^{2^{k-1}} \sum_{m=0}^{M-1} \tilde{c}_{n,m} \psi_{n,m}(x) = \tilde{C}^T \Psi(x), \tag{7.24}$$

where $\tilde{C}$ and $\Psi(x)$ are vectors of dimension $(2^{k-1}M \times 1)$ given by

$$\tilde{C} = [\tilde{c}_{1,0}, \tilde{c}_{1,1}, \ldots, \tilde{c}_{1,M-1}, \ldots, \tilde{c}_{2^{k-1},0}, \tilde{c}_{2^{k-1},1}, \ldots \tilde{c}_{2^{k-1},M-1}]^T \tag{7.25}$$

and

$$\Psi(x) = [\psi_{1,0}(x), \psi_{1,1}(x), \ldots, \psi_{1,M-1}(x), \ldots, \psi_{2^{k-1},0}(x), \psi_{2^{k-1},1}(x), \ldots \psi_{2^{k-1},M-1}(x)]^T, \tag{7.26}$$

where $\psi_{n,m}(x)$, $n = 1, 2, \ldots, 2^{k-1}$, $m = 0, 1, \ldots, M-1$ is the EWs defined on $[0,1)$ as follows [59]:

$$\psi_{n,m}(x) = \begin{cases} 2^{\frac{k-1}{2}} \bar{E}_m(2^{k-1}x - n + 1), & \frac{n-1}{2^{k-1}} \leq x < \frac{n}{2^{k-1}}, \\ 0, & \text{otherwise}, \end{cases} \tag{7.27}$$

with

$$\bar{E}_m(x) = \begin{cases} 1, & m = 0, \\ \dfrac{1}{\sqrt{\dfrac{2(-1)^{m-1}(m!)^2}{(2m)!} E_{2m+1}(0)}} E_m(x), & m > 0. \end{cases}$$

The coefficient vector $\tilde{C}$ can be determined from Eq. (7.24) as follows:

$$\tilde{C}^T = \left(\int_0^1 g(x)\Psi^T(x)dx \right).\tilde{D}^{-1}, \tag{7.28}$$

where $\tilde{D}$ is a $(2^{k-1}M \times 2^{k-1}M)$-dimensional matrix and is given by

$$\tilde{D} = \int_0^1 \Psi(x)\Psi^T(x)\,dx. \tag{7.29}$$

For $k = 2$, $M = 3$, we have

$$\tilde{D} = \begin{bmatrix} 1 & 0 & -\frac{1}{\sqrt{6}} & 0 & 0 & 0 \\ 0 & \frac{1}{3} & 0 & 0 & 0 & 0 \\ -\frac{1}{\sqrt{6}} & 0 & \frac{1}{5} & 0 & 0 & 0 \\ 0 & 0 & 0 & 1 & 0 & -\frac{1}{\sqrt{6}} \\ 0 & 0 & 0 & 0 & \frac{1}{3} & 0 \\ 0 & 0 & 0 & -\frac{1}{\sqrt{6}} & 0 & \frac{1}{5} \end{bmatrix}.$$

Similarly, $k(x,z)$ can be approximated as follows:

$$k(x,z) = \Psi^T(x).\tilde{\Pi}.\Psi(z), \tag{7.30}$$

where $\tilde{\Pi}$ is a square matrix of dimension-$2^{k-1}M$ and is given by

$$\tilde{\Pi} = \tilde{D}^{-1}.\left(\int_0^1 \Psi(x).\left(\int_0^1 k(x,z)\Psi^T(z)dz \right) dx \right).\tilde{D}^{-1}.$$

7.4.2 Operational Matrices for EWs

This section deals with the construction of the FI's OM for the proposed method to solve the fractional PVDIDEs.

7.4.2.1 OM of the FI

Let $\Psi(x)$ be the EWs vector introduced in Eq. (7.26). Then, the fractional integral OM is given by

$$\int_0^x \Psi(\xi)\,d\xi = P_1\Psi(x), \tag{7.31}$$

where P_1 is the $(2^{k-1}M \times 2^{k-1}M)$-dimensional OM of integration [60].

Similarly, the integration from 0 to qx $(0 < q < 1)$ of EWs vector $\Psi(x)$ is given by

$$\int_0^{qx} \Psi(\xi)\,d\xi = P_2\Psi(x), \tag{7.32}$$

where P_2 is the $(2^{k-1}M \times 2^{k-1}M)$-dimensional OM of integration.

For $k = 2$, $M = 3$, we have

$$P_2 = \begin{bmatrix} \frac{q}{4} & \frac{q}{4} & 0 & \frac{3q}{4} & \frac{q}{4} & 0 \\ -\frac{q}{8} & -\frac{q}{8} & \frac{7\sqrt{3}q}{4\sqrt{2}} - \frac{5q}{\sqrt{6}} & -\frac{q}{8} & \frac{q}{8} & \frac{7\sqrt{3}q}{4\sqrt{2}} - \frac{5q}{\sqrt{6}} \\ -\frac{3\sqrt{3}q}{8\sqrt{2}} + \frac{7q}{8\sqrt{6}} & -\frac{7\sqrt{3}q}{80} & -\frac{q}{16} & -\frac{q\sqrt{3}}{4\sqrt{2}} & -\frac{7\sqrt{3}q}{80} & \frac{q}{16} \\ 0 & 0 & 0 & 0 & 0 & 0 \\ 0 & 0 & 0 & 0 & 0 & 0 \\ 0 & 0 & 0 & 0 & 0 & 0 \end{bmatrix}.$$

Now, the FI of the EWs vector $\Psi(x)$ is defined as follows:

$$\tilde{J}^{\alpha}\Psi(x) = P^{\alpha}\Psi(x), \tag{7.33}$$

where P^{α} is the $(2^{k-1}M \times 2^{k-1}M)$-dimensional OM of the FI.

Now using the Riemann–Liouville (RL) fractional integral definition, i.e.,

$$\tilde{J}^{\alpha}v(x) = \frac{1}{\Gamma(\alpha)}\int_0^x (x-z)^{\alpha-1}v(z)\,dz, \quad x > 0,\ \alpha \in \mathbb{R}^+,$$

and from Eq. (7.33), the matrix P^{α} can be obtained as follows:

$$P^{\alpha} = \left(\int_0^1 \tilde{J}^{\alpha}\Psi(x).\Psi^T(x)\,dx\right).\tilde{D}^{-1}. \tag{7.34}$$

For $k = 2$, $M = 2$ and $\alpha = 0.5$, we have

$$P^{0.5} = \begin{bmatrix} 0.531923 & 0.319154 & 0.440659 & -0.154879 \\ -0.106385 & 0.227967 & 0.0516265 & -0.0561003 \\ 0 & 0 & 0.531923 & 0.319154 \\ 0 & 0 & -0.106385 & 0.227967 \end{bmatrix}.$$

7.4.3 Euler Wavelet Method for Fractional PVDIDE

In this section, the fractional PVDIDE Eq. (7.23) converts into an algebraic system of equations by implementing a newly proposed technique. Firstly, consider the approximation of the following functions:

$$ {}_0^C D_x^{\alpha} u(x) \approx \Psi^T(x)\tilde{C}_1 = \tilde{C}_1{}^T \Psi(x), \tag{7.35}$$

$$a(x) \approx \Psi^T(x)\tilde{A} = \tilde{A}^T\Psi(x), \tag{7.36}$$

$$b(x) \approx \Psi^T(x)\tilde{B} = \tilde{B}^T\Psi(x), \tag{7.37}$$

$$f(x) \approx \Psi^T(x)\tilde{F} = \tilde{F}^T\Psi(x), \tag{7.38}$$

$$u(0) \approx \Psi^T(x)\tilde{U} = \tilde{U}^T\Psi(x), \tag{7.39}$$

$$k_1(x,z) \approx \Psi^T(x)\tilde{K}_1\Psi(z) = \Psi^T(z)\tilde{K}_1^T\Psi(x), \tag{7.40}$$

$$k_2(x,z) \approx \Psi^T(x)\tilde{K}_2\Psi(z) = \Psi^T(z)\tilde{K}_2^T\Psi(x). \tag{7.41}$$

Now integrating Eq. (7.35) with respect to x and by using Eq. (7.33), we have

$$\begin{aligned} u(x) &= u(0) + \tilde{C}_1^T P^{\alpha}\Psi(x) \\ &= \tilde{U}^T\Psi(x) + \tilde{C}_1^T P^{\alpha}\Psi(x) \\ &= \tilde{U}^T\Psi(x) + \tilde{C}_1^T P^{\alpha}\Psi(x) \\ &= \Psi^T(x)\tilde{C}_2, \end{aligned} \tag{7.42}$$

where $\tilde{C}_2 = \tilde{U} + P^{\alpha T}\tilde{C}_1$.

Let us consider the approximation of the function $u(qx)$ $(0 < q < 1)$ as follows:

$$u(qx) \approx \Psi^T(x)\tilde{C}_3 = \tilde{C}_3^T\Psi(x), \tag{7.43}$$

where $\tilde{C}_3^T = \left(\int_0^1 \Psi^T(qx)\tilde{C}_2\Psi^T(x)\,dx \right).\tilde{D}^{-1}$.

Now the following approximations for the term $a(x)u(x)$ and $b(x)u(qx)$ of Eq. (7.23) are determined:

$$\begin{aligned} a(x)u(x) &= \Psi^T(x)\tilde{A}.\tilde{C}_2^T\Psi(x) \\ &= \Psi^T(x)A_1\Psi(x), \end{aligned}$$

where $A_1 = \tilde{A}.\tilde{C}_2^T$,

$$a(x)u(x) = \tilde{A}_1^T\Psi(x), \tag{7.44}$$

where

$$\tilde{A}_1^T = \left(\int_0^1 \Psi^T(x)A_1\Psi(x)\Psi^T(x)\,dx\right).\tilde{D}^{-1}.$$

Again,

$$\begin{aligned} b(x)u(qx) &= \Psi^T(x)\tilde{B}.\tilde{C}_3^T\Psi(x) \\ &= \tilde{\psi}^T(x)B_1\Psi(x), \end{aligned}$$

where $B_1 = \tilde{B}.\tilde{C}_3^T$,

$$b(x)u(qx) = \tilde{B}_1^T\Psi(x), \tag{7.45}$$

where

$$\tilde{B}_1^T = \left(\int_0^1 \Psi^T(x)B_1\Psi(x)\Psi^T(x)\,dx\right).\tilde{D}^{-1}.$$

The integral parts of Eq. (7.23) can be expressed as follows:

$$\begin{aligned} \int_0^x k_1(x,z)u(z)\,dz &= \int_0^x \Psi^T(x)\tilde{K}_1\Psi(z)\tilde{C}_2^T\Psi(z)\,dz \\ &= \Psi^T(x)\tilde{K}_1\int_0^x \Psi(z)\Psi^T(z)\tilde{C}_2\,dz \\ &= \Psi^T(x)\tilde{K}_1\int_0^x \tilde{C}_{22}^T\Psi(z)\,dz, \end{aligned}$$

where $\tilde{C}_{22}^T = \left(\int_0^1 \Psi(z)\Psi^T(z)\tilde{C}_2\Psi^T(z)\,dz \right).\tilde{D}^{-1}$

$$= \Psi^T(x)\tilde{K}_1\tilde{C}_{22}^T P_1\Psi(x)$$

$$= \tilde{C}_{33}^T\Psi(x), \tag{7.46}$$

where $\tilde{C}_{33}^T = \left(\int_0^1 \Psi^T(x)\tilde{K}_1\tilde{C}_{22}^T P_1\Psi(x)\Psi^T(x)\,dx \right).\tilde{D}^{-1}.$

Similarly,

$$\int_0^{qx} k_2(x,z)u(z)\,dz = \int_0^{qx} \Psi^T(x)\tilde{K}_2\Psi(z)\tilde{C}_2^T\Psi(z)\,dz$$

$$= \Psi^T(x)\tilde{K}_2 \int_0^{qx} \Psi(z)\Psi^T(z)\tilde{C}_2\,dz$$

$$= \Psi^T(x)\tilde{K}_2 \int_0^{qx} \tilde{C}_{22}^T\Psi(z)\,dz \tag{7.47}$$

$$= \Psi^T(x)\tilde{K}_2\tilde{C}_{22}^T P_2\Psi(x)$$

$$= \tilde{C}_{44}^T\Psi(x),$$

where $\tilde{C}_{44}^T = \left(\int_0^1 \Psi^T(x)\tilde{K}_2\tilde{C}_{22}^T P_2\Psi(x)\Psi^T(x)\,dx \right).\tilde{D}^{-1}.$

Now by substituting Eqs. (7.35)–(7.47) into Eq. (7.23), it follows

$$\Psi^T(x)\tilde{C}_1 = \Psi^T(x)\tilde{A}_1 + \Psi^T(x)\tilde{B}_1 + \Psi^T(x)\tilde{F} + \Psi^T(x)\tilde{C}_{33} + \Psi^T(x)\tilde{C}_{44}. \tag{7.48}$$

Collocating Eq. (7.48) at $2^{k-1}M$-points x_i yields

$$\Psi^T(x_i)\tilde{C}_1 = \Psi^T(x_i)\tilde{A}_1 + \Psi^T(x_i)\tilde{B}_1 + \Psi^T(x_i)\tilde{F} + \Psi^T(x_i)\tilde{C}_{33} + \Psi^T(x_i)\tilde{C}_{44}, \tag{7.49}$$

where the collocation points x_i are given as follows:

$$x_i = \frac{2i-1}{2.2^{k-1}.M}, \quad i = 1,2,\dots,2^{k-1}.M.$$

After substituting the collocation points, Eq. (7.49) gives $2^{k-1}M$ - system of algebraic equations. By solving the above system of equations, the value of $\tilde{C}_1$ can be obtained. Furthermore, the value of $\tilde{C}_2$ is calculated by using the value of $\tilde{C}_1$.

Therefore, the approximate solution of PVDIDE Eq. (7.23) can be computed numerically by plugging the value of $\tilde{C}_2$ in Eq. (7.42).

7.5 CONVERGENCE AND ERROR ANALYSIS

This section is intended to investigate the error bound and the convergence of the present method in the Sobolev space. In this context, the error estimates of the EWs expansion truncated error $u - P_{M-1}^{2^{k-1}} u$ are obtained in the sense of Sobolev norms. To achieve the desired objective, we introduce certain lemmas and notations used in the convergence analysis. The Sobolev norm of integer order $\mu \geq 0$ in the interval (a,b) is defined by [61]

$$\begin{aligned} \|u\|_{H^{\mu}(a,b)} &= \left[\sum_{j=0}^{\mu} \int_a^b \left| u^{(j)}(x) \right|^2 dx \right]^{\frac{1}{2}} \\ &= \left[\sum_{j=0}^{\mu} \left\| u^{(j)}(x) \right\|_{L^2[a,b]}^2 \right]^{\frac{1}{2}}, \end{aligned} \tag{7.50}$$

where $u^{(j)}(x)$ denotes the j-th distributional derivative of u. Moreover, let us define the following symbol [61]:

$$|u|_{H^{\mu;M}(0,1)} = \left[\sum_{j=\min(\mu,M+1)}^{\mu} \left\| u^{(j)}(x) \right\|_{L^2[0,1]}^2 \right]^{\frac{1}{2}}. \tag{7.51}$$

The seminorm is defined [62] as follows:

$$|u|_{H^{r,\mu;M;N}(0,1)} = \left[\sum_{j=\min(\mu,M+1)}^{\mu} N^{2r-2j} \left\| u^{(j)}(x) \right\|_{L^2(0,1)}^2 \right]^{\frac{1}{2}}, \tag{7.52}$$

$$|u|_{H^{r;\mu;M;N}(0,1)} = N^{r-\mu} \left\| u^{(\mu)} \right\|_{L^2(0,1)}, \tag{7.53}$$

where $M \geq \mu - 1$.

Theorem 7.5.1

Suppose that $v \in H^{\mu}(0,1), (\mu \geq 0)$ and $P_M v = \sum_{i=0}^{M} c_i \beta_i$ is the best approximation of u, and then

$$\|v - P_M v\|_{L^2(0,1)} \leq c M^{-\mu} |v|_{H^{\mu;M}(0,1)}, \tag{7.54}$$

and for $1 \le \tau \le \mu$,

$$\|v - P_M v\|_{H^\tau(0,1)} \le cM^{2\tau - \frac{1}{2} - \mu} |v|_{H^{\mu;M}(0,1)}, \tag{7.55}$$

where c depends on μ.

Proof:

It may be referred to Ref. [63]. ■

Lemma 7.5.1

For $n = 1, 2, ..., 2^{k-1}$, suppose $z_n : \left(\frac{n-1}{2^{k-1}}, \frac{n}{2^{k-1}}\right) \to \mathbb{R}$ is a function in $H^\mu\left(\frac{n-1}{2^{k-1}}, \frac{n}{2^{k-1}}\right)$. Consider the function $F_n z_n(\zeta) : (0,1) \to \mathbb{R}$ such that $(F_n z_n)(\zeta) = z_n\left(\frac{\zeta + n - 1}{2^{k-1}}\right)$ for all $\zeta \in (0,1)$, and then for $0 \le r \le \mu$, it follows

$$\left\|(F_n z_n)^{(r)}\right\|_{L^2(0,1)} = 2^{(k-1)(\frac{1}{2} - r)} \left\|z_n^{(r)}\right\|_{L^2\left(\frac{n-1}{2^{k-1}}, \frac{n}{2^{k-1}}\right)}. \tag{7.56}$$

Proof:

It is quite easy to prove. ■

The upper bound of the error for the approximation function is presented in the following theorem.

Theorem 7.5.2

Suppose that $u \in H^\mu(0,1), (\mu \ge 0)$ and $P_{M-1}^{2^{k-1}} u = \sum_{n=1}^{2^{k-1}} \sum_{m=0}^{M-1} a_{n,m} \psi_{n,m}$ is the best approximation of u, and then

$$\left\|u - P_{M-1}^{2^{k-1}} u\right\|_{L^2(0,1)} \le c(M-1)^{-\mu} |u|_{H^{0;\mu;M-1;2^{k-1}}(0,1)}, \tag{7.57}$$

and for $1 \le r \le \mu$,

$$\left\|u - P_{M-1}^{2^{k-1}} u\right\|_{H^r(0,1)} \le c(M-1)^{2r - \frac{1}{2} - \mu} |u|_{H^{r;\mu;M-1;2^{k-1}}(0,1)}. \tag{7.58}$$

Proof:

For $n = 1, 2, ..., 2^{k-1}$, let us define a function $z_n : \left(\frac{n-1}{2^{k-1}}, \frac{n}{2^{k-1}}\right) \to \mathbb{R}$ such that $z_n(x) = u(x)$ for all $x \in \left(\frac{n-1}{2^{k-1}}, \frac{n}{2^{k-1}}\right)$.

Using Lemma 7.5.1, Eqs. (7.51), (7.52), and (7.54), it follows

$$\left\| u - P_{M-1}^{2^{k-1}} u \right\|_{L^2(0,1)}^2 = \sum_{n=1}^{2^{k-1}} \left\| z_n - \sum_{m=0}^{M-1} a_{n,m} \psi_{n,m} \right\|_{L^2\left(\frac{n-1}{2^{k-1}}, \frac{n}{2^{k-1}}\right)}^2$$

$$= \sum_{n=1}^{2^{k-1}} \int_{\frac{n-1}{2^{k-1}}}^{\frac{n}{2^{k-1}}} \left| z_n(x) - \sum_{m=0}^{M-1} a_{n,m} \psi_{n,m} \right|^2 dx.$$

Using Theorem 7.5.1 and by changing the variable, $x = \frac{\zeta + n - 1}{2^{k-1}}$, it follows

$$\left\| u - P_{M-1}^{2^{k-1}} u \right\|_{L^2(0,1)}^2 = 2^{-(k-1)} \sum_{n=1}^{2^{k-1}} \left\| F_n z_n - P_{M-1}(F_n z_n) \right\|_{L^2(0,1)}^2$$

$$\leq c^2 (M-1)^{-2\mu} 2^{-(k-1)} \sum_{n=1}^{2^{k-1}} \left(\sum_{j=\min(\mu,M)}^{\mu} \left\| (F_n z_n)^{(j)} \right\|_{L^2(0,1)}^2 \right)$$

$$= c^2 (M-1)^{-2\mu} \sum_{j=\min(\mu,M)}^{\mu} 2^{(k-1)(-2j)} \sum_{n=1}^{2^{k-1}} \left\| z_n^{(j)} \right\|_{L^2\left(\frac{n-1}{2^{k-1}}, \frac{n}{2^{k-1}}\right)}^2$$

$$= c^2 (M-1)^{-2\mu} \sum_{j=\min(\mu,M)}^{\mu} 2^{(k-1)(-2j)} \left\| u^{(j)} \right\|_{L^2(0,1)}^2$$

$$= c^2 (M-1)^{-2\mu} |u|_{H^{0;\mu;M-1;2^{k-1}}(0,1)}^2 .$$

Also, for $1 \leq r \leq \mu$, again using Lemma 7.5.1, Eqs. (7.51), (7.52) and (7.55), it follows

$$
\begin{aligned}
\left\|u-P_{M-1}^{2^{k-1}}u\right\|_{H^r(0,1)}^2 &= \sum_{n=1}^{2^{k-1}}\left\|z_n-\sum_{m=0}^{M-1}a_{n,m}\psi_{n,m}\right\|_{H^r\left(\frac{n-1}{2^{k-1}},\frac{n}{2^{k-1}}\right)}^2 \\
&= \sum_{n=1}^{2^{k-1}}\sum_{j=0}^{\mu}\left\|z_n^{(j)}-\left(\sum_{m=0}^{M-1}a_{n,m}\psi_{n,m}\right)^{(j)}\right\|_{L^2\left(\frac{n-1}{2^{k-1}},\frac{n}{2^{k-1}}\right)}^2 \\
&= \sum_{n=1}^{2^{k-1}}\sum_{j=0}^{\mu}2^{(k-1)(2j-1)}\left\|(F_nz_n)^{(j)}-P_{M-1}((F_nz_n)^{(j)})\right\|_{L^2(0,1)}^2 \\
&\le \sum_{n=1}^{2^{k-1}}2^{(k-1)(2r-1)}\sum_{j=0}^{\mu}\left\|(F_nz_n)^{(j)}-P_{M-1}((F_nz_n)^{(j)})\right\|_{L^2(0,1)}^2 \\
&= \sum_{n=1}^{2^{k-1}}2^{(k-1)(2r-1)}\left\|(F_nz_n)-P_{M-1}(F_nz_n)\right\|_{H^{\mu}(0,1)}^2 \\
&\le \sum_{n=1}^{2^{k-1}}2^{(k-1)(2r-1)}c^2(M-1)^{4r-1-2\mu}\left|F_nz_n\right|_{H^{\mu;M-1}(0,1)}^2 \\
&= \sum_{n=1}^{2^{k-1}}2^{(k-1)(2r-1)}c^2(M-1)^{4r-1-2\mu}\sum_{i=\min(\mu,M)}^{\mu}\left\|(F_nz_n)^{(i)}\right\|_{L^2(0,1)}^2 \\
&= \sum_{n=1}^{2^{k-1}}2^{(k-1)(2r-1)}c^2(M-1)^{4r-1-2\mu}\sum_{i=\min(\mu,M)}^{\mu}\int_0^1\left|(F_nz_n)^{(i)}\right|^2 d\zeta \\
&= \sum_{n=1}^{2^{k-1}}2^{(k-1)(2r-1)}c^2(M-1)^{4r-1-2\mu}\sum_{i=\min(\mu,M)}^{\mu}\int_0^1\left|z_n^{(i)}\left(\frac{x+n-1}{2^{k-1}}\right)\right|^2 d\zeta \\
&= \sum_{n=1}^{2^{k-1}}2^{(k-1)(2r-1)}c^2(M-1)^{4r-1-2\mu}\sum_{i=\min(\mu,M)}^{\mu}2^{(k-1)(1-2i)}\int_{\frac{n-1}{2^{k-1}}}^{\frac{n}{2^{k-1}}}\left|z_n^{(i)}(x)\right|^2 dx \\
&= 2^{(k-1)(2r)}c^2(M-1)^{4r-1-2\mu}\sum_{i=\min(\mu,M)}^{\mu}2^{(k-1)(1-2i)}\left\|u^{(i)}\right\|_{L^2(0,1)}^2 \\
&= c^2(M-1)^{4r-1-2\mu}\left|u\right|_{H^{r;\mu;M-1;2^{k-1}}(0,1)}^2 .
\end{aligned}
$$

Hence, for $1 \le r \le \mu$,

$$\left\| u - P_{M-1}^{2^{k-1}} u \right\|_{H^r(0,1)} \le c(M-1)^{2r - \frac{1}{2} - \mu} |u|_{H^{r;\mu;M-1;2^{k-1}}(0,1)}.$$

Also, we have proved

$$\left\| u - P_{M-1}^{2^{k-1}} u \right\|_{L^2(0,1)} \le c(M-1)^{-\mu} |u|_{H^{0;\mu;M-1;2^{k-1}}(0,1)}. \blacksquare$$

Corollary 7.5.1

Suppose $u \in H^{\mu}(0,1)$ with $\mu \ge 0$, and $M \ge \mu$, and then using Theorem 7.5.2 and Eq. (7.53), we obtain

$$\left\| u - P_{M-1}^{2^{k-1}} u \right\|_{L^2(0,1)} \le c(M-1)^{-\mu} 2^{(k-1)(-\mu)} \left\| u^{(\mu)} \right\|_{L^2(0,1)},$$

and for $r \ge 1$, we have

$$\left\| u - P_{M-1}^{2^{k-1}} u \right\|_{H^r(0,1)} \le c(M-1)^{2r - \frac{1}{2} - \mu} (2^{k-1})^{r-\mu} \left\| u^{(\mu)} \right\|_{L^2(0,1)}.$$

This result shows that the rate of convergence of $P_{M-1}^{2^{k-1}} u$ to u is faster than $\frac{1}{2^{k-1}}$ to the power of $M - r$ and any power of $\frac{1}{M-1}$, which establishes the supremacy of the proposed method.

Now, the error bound for the fractional derivative of u is obtained by the following theorem.

Theorem 7.5.3

Suppose $u \in H^{\mu}(0,1)$ with $\mu \ge 0$, $M \ge r$ and $0 < \alpha \le 1$, and then

$$\left\| {}_0^C D_x^{\alpha} u - {}_0^C D_x^{\alpha} (P_{M-1}^{2^{k-1}} u) \right\|_{L^2(0,1)} \le \frac{c(M-1)^{2r - \frac{1}{2} - \mu} (2^{k-1})^{r-\mu} \left\| u^{(\mu)} \right\|_{L^2(0,1)}}{\Gamma(2-\alpha)},$$

where $1 \le r \le \mu$.

Proof:

Using the definition of fractional RL integral and the norm of convolution product [64],

$$\|g * h\|_p \le \|g\|_1 \|h\|_p.$$

Therefore, we obtain

$$\begin{aligned}
\left\| {}_0^C D_x^\alpha u - {}_0^C D_x^\alpha (P_{M-1}^{2^{k-1}} u) \right\|_{L^2(0,1)}^2 &= \left\| {}_0 J_x^{1-\alpha} \left({}_0^C D_x^{(1)} u - {}_0^C D_x^{(1)} (P_{M-1}^{2^{k-1}} u) \right) \right\|_{L^2(0,1)}^2 \\
&= \left\| \frac{x^{-\alpha}}{\Gamma(1-\alpha)} * \left({}_0^C D_x^{(1)} u - {}_0^C D_x^{(1)} (P_{M-1}^{2^{k-1}} u) \right) \right\|_{L^2(0,1)}^2 \\
&\le \left(\int_0^1 \frac{x^{-\alpha}}{\Gamma(1-\alpha)} dx \right)^2 \left\| {}_0^C D_x^{(1)} u - {}_0^C D_x^{(1)} (P_{M-1}^{2^{k-1}} u) \right\|_{L^2(0,1)}^2 \\
&= \left(\frac{1}{(1-\alpha)\Gamma(1-\alpha)} \right)^2 \left\| {}_0^C D_x^{(1)} u - {}_0^C D_x^{(1)} (P_{M-1}^{2^{k-1}} u) \right\|_{L^2(0,1)}^2 \\
&\le \left(\frac{1}{\Gamma(2-\alpha)} \right)^2 \left\| u - P_{M-1}^{2^{k-1}} u \right\|_{H^\mu(0,1)}^2 \\
&\le \left(\frac{1}{\Gamma(2-\alpha)} \right)^2 c^2 (M-1)^{4r-1-2\mu} (2^{k-1})^{2r-2\mu} \left\| u^{(\mu)} \right\|_{L^2(0,1)}^2 .
\end{aligned}$$

Thus, using Theorem 7.5.2, it implies that

$$\left\| {}_0^C D_x^\alpha u - {}_0^C D_x^\alpha (P_{M-1}^{2^{k-1}} u) \right\|_{L^2(0,1)}^2 \le \left(\frac{1}{\Gamma(2-\alpha)} \right)^2 c^2 (M-1)^{4r-1-2\mu} (2^{k-1})^{2r-2\mu} \left\| u^{(\mu)} \right\|_{L^2(0,1)}^2 .$$

Hence,

$$\left\| {}_0^C D_x^\alpha u - {}_0^C D_x^\alpha (P_{M-1}^{2^{k-1}} u) \right\|_{L^2(0,1)} \le \frac{c(M-1)^{2r-\frac{1}{2}-\mu} (2^{k-1})^{r-\mu} \left\| u^{(\mu)} \right\|_{L^2(0,1)}}{\Gamma(2-\alpha)} . \blacksquare$$

Finally, the error bound for the Euler wavelet expansion for the function $u(qx)$ is obtained by the following last theorem.

Theorem 7.5.4

Suppose that $u(qx) \in H^\mu(0,1), (\mu \ge 0)$ and $P_{M-1}^{2^{k-1}} u(qx) = \sum_{n=1}^{2^{k-1}} \sum_{m=0}^{M-1} \tilde{a}_{n,m} \psi_{n,m}$ is the best approximation of $u(qx)$, and then,

$$\left\| u(qx) - P_{M-1}^{2^{k-1}} u(qx) \right\|_{L^2(0,1)} \le c(M-1)^{-\mu} 2^{(k-1)(-\mu)} q^\mu \left\| u^{(\mu)} \right\|_{L^2(0,1)} .$$

Proof:

$$\left\|u(qx)-P_{M-1}^{2^{k-1}}u(qx)\right\|_{L^2(0,1)}=\sum_{n=1}^{2^{k-1}}\left\|z_n(qx)-\sum_{m=0}^{M-1}\tilde{a}_{n,m}\psi_{n,m}\right\|^2_{L^2\left(\frac{n-1}{2^{k-1}},\frac{n}{2^{k-1}}\right)}$$

$$=\sum_{n=1}^{2^{k-1}}\frac{1}{2^{k-1}}\int_0^1\left|F_n z_n-P_{M-1}(F_n z_n)\right|^2dx$$

$$=2^{-(k-1)}\sum_{n=1}^{2^{k-1}}\left\|F_n z_n-P_{M-1}(F_n z_n)\right\|^2_{L^2(0,1)}.$$

Using Theorem 7.5.1, Eqs. (7.52) and (7.53), we get

$$\left\|u(qx)-P_{M-1}^{2^{k-1}}u(qx)\right\|_{L^2(0,1)}\leq c^2(M-1)^{-2\mu}2^{-(k-1)}\sum_{n=1}^{2^{k-1}}\left(\sum_{j=\min(\mu,M)}^{\mu}\left\|(F_n z_n)^{(j)}\right\|^2_{L^2(0,1)}\right)$$

$$\leq c^2(M-1)^{-2\mu}2^{-(k-1)}\sum_{n=1}^{2^{k-1}}\sum_{j=\min(\mu,M)}^{\mu}2^{(k-1)(-2j)}q^{2j}\left\|z_n^{(j)}\right\|^2_{L^2\left(\frac{n-1}{2^{k-1}},\frac{n}{2^{k-1}}\right)}$$

$$=c^2(M-1)^{-2\mu}\sum_{j=\min(\mu,M)}^{\mu}2^{(k-1)(-2j)}q^{2j}\left\|u^{(j)}\right\|^2_{L^2(0,1)}$$

$$=c^2(M-1)^{-2\mu}\left|u\right|^2_{H^{0;\mu;M-1;(2^{k-1})q^{-1}}(0,1)}$$

$$=c^2(M-1)^{-2\mu}(2^{(k-1)}q^{-1})^{-2\mu}\left\|u^{(\mu)}\right\|^2_{L^2(0,1)}$$

$$=c^2(M-1)^{-2\mu}(2^{(k-1)(-2\mu)})q^{2\mu}\left\|u^{(\mu)}\right\|^2_{L^2(0,1)}.$$

Hence,

$$\left\|u(qx)-P_{M-1}^{2^{k-1}}u(qx)\right\|_{L^2(0,1)}\leq c(M-1)^{-\mu}2^{(k-1)(-\mu)}q^{\mu}\left\|u^{(\mu)}\right\|_{L^2(0,1)}.\quad\blacksquare$$

7.6 APPLICATION OF THE PANTOGRAPH EQUATION

The pantograph equation is one of the most significant types of delay-differential equations (DDEs) and is utilized as an essential tool in the study of electrodynamics [65]. A pantograph is a device that gathers electric current from overhead power lines in order to energize electric trains or trams. The term pantograph originates from the resemblance to pantograph systems for writing, copying, and drawings (Figure 7.1, taken from [66], represents a pantograph and a tram line).

(a) (b)

FIGURE 7.1 A 'Z'-shaped pantograph for electrical pickup.

The numerical investigation establishes the long-time dynamical behaviour of the exact solution of the pantograph equation:

$$\begin{cases} u'(x) = y_1 u(x) + y_2 u(qx), \\ u(0) = 0, \end{cases} \tag{7.59}$$

where y_1, y_2, and q are real constants and $q \in (0,1)$.

In particular, this is the case where $|y_1| + y_2 < 0$, which indicates that $u(x) \to 0$ as $x \to \infty$.

Let us consider a nonhomogeneous pantograph equation in the following form:

$$\begin{cases} u'(x) = \sum_{i=1}^{r} y_i(x) u(q_i x) + f(x), \ \ x \in [0,T], \\ u(0) = u_0, \end{cases} \tag{7.60}$$

where $0 < q_i \leq 1$ and $y_i(x)$ are continuous functions on $[0,T]$ for $i = 1,2,...,r$, and f is the Riemann integrable.

The impact of the proposed method on the above model has been describe in Example 7.7.4.

7.7 NUMERICAL EXPERIMENTS

The following examples have been presented to demonstrate the performance of the proposed approach for solving PVDIDEs and to assess the accuracy and efficiency of the scheme. The absolute error definition to measure the validity of the proposed method is given by

$$\tilde{E}(x) = |u(x) - \tilde{u}(x)|,$$

where $\tilde{u}(x)$ is the approximate solution and $u(x)$ is the exact solution.

Furthermore, the $L^2(0,1)$ and $H^1(0,1)$ error norms for the proposed wavelets are defined as follows:

$$L^2(0,1)_{\text{error}} = \left[\int_0^1 |u(x)-\tilde{u}(x)|^2\,dx\right]^{\frac{1}{2}}$$

and

$$H^1(0,1)_{\text{error}} = \left[\sum_{j=0}^{1}\int_0^1 |u^{(j)}(x)-\tilde{u}^{(j)}(x)|^2\,dx\right]^{\frac{1}{2}},$$

respectively, where $u(x)$ and $\tilde{u}(x)$ are the exact and approximate solutions, respectively.

The $L^2(0,1)$ and $H^1(0,1)$ (Sobolev norm) error values have been compared with the Legendre wavelets method (LWM) and the Bernoulli wavelets method (BWM) presented in Tables 7.3, 7.9, 7.19, and 7.23, respectively. Moreover, the $L^2(0,1)$ error values obtained by the Euler wavelet method (EWM) have been presented in Tables 7.28 and 7.36, respectively.

Again, the maximum absolute error on the set of collocation points $\Delta=\{x_{-N},\ldots,x_N\}$, where $x_l=\frac{e^{lh}}{1+e^{lh}}$, $l=-N,\ldots,N$, $h=\pi\left(\frac{1}{N}\right)^{\frac{1}{2}}$, is defined as $\|E(h)\|_{\infty}=\underset{-N\le j\le N}{Max}|u(x_j)-\tilde{u}(x_j)|$.

Here, $u(x_j)$ and $\tilde{u}(x_j)$ are the exact and approximate solutions, respectively. In Table 7.11, the maximum absolute error has been compared to the Sinc collocation method (SCM) [67].

To justify the accuracy of the approached technique, the numerical convergence rate (NCR) analysis has been performed using the following formula:

$$\text{NCR} = \log_2\left(\frac{\text{MAE}_{k=\alpha,M=\beta}}{\text{MAE}_{k=\alpha+1,M=\beta}}\right),\quad \alpha,\beta\in\mathbb{Z}^+,$$

where

$$\text{Maximun Absolute Error(MAE)} = \underset{1\le p\le 2^{k-1}.M}{\text{Max}}|u(x_p)-\tilde{u}(x_p)|,$$

$$x_p=\frac{2.p-1}{2.2^{k-1}.M},\ p=1,2,\ldots,2^{k-1}.M.$$

Example 7.7.1

Consider the following PVDIDE:

$$u'(x)=3x^2-\frac{x^2e^x}{5}u(x)-\frac{x^2}{20}u\left(\frac{x}{2}\right)+\int_0^x e^x zu(z)\,dz+\int_0^{\frac{x}{2}} zu(z)\,dz,\quad x\in[0,1], \tag{7.61}$$

$$u(0)=0,$$

with the exact solution $u(x) = x^3$.

Tables 7.1 and 7.2 illustrate the absolute error values of $\tilde{E}(x)$ computed by Müntz–Legendre wavelet method (MLWM) for different values of γ. The numerical solutions and absolute error estimation by MLWM are shown in Figures 7.2–7.5. For various values of k, M, Table 7.4 displays the maximum absolute error norm and NCR. The increasing amount of k leads to a roughly third-order convergence rate.

For $k = 2, M = 3$, Figures 7.2–7.5 are drawn by MLWM for different values of γ.

TABLE 7.1
$\tilde{E}(x)$ Errors by MLWM for Example 7.7.1 with $k = 2, M = 3$

	$\tilde{E}(x)$				
x	$\gamma = \frac{1}{4}$	$\gamma = \frac{1}{2}$	$\gamma = 1$	$\gamma = \frac{3}{2}$	$\gamma = 2$
0.0833333	1.49118E-2	9.82046E-3	1.608E-3	2.1966E-4	3.22508E-4
0.25	1.19946E-2	4.73438E-3	6.76645E-6	1.54637E-4	7.7009E-4
0.416667	1.74051E-3	2.58399E-3	1.61489E-3	4.74032E-5	8.73004E-4
0.583333	4.6792E-2	2.49373E-2	1.61829E-3	2.21507E-3	2.18247E-3
0.75	4.39471E-2	1.38325E-2	7.04942E-5	1.1255E-3	8.64326E-3
0.916667	5.34072E-3	2.84577E-3	1.54451E-3	1.8456E-3	8.23093E-3

TABLE 7.2
$\tilde{E}(x)$ Errors by MLWM for Example 7.7.1 with $k = 3, M = 3$

	$\tilde{E}(x)$				
x	$\gamma = \frac{1}{4}$	$\gamma = \frac{1}{2}$	$\gamma = 1$	$\gamma = \frac{3}{2}$	$\gamma = 2$
0.0416667	1.855E-3	1.22559E-3	2.02334E-4	2.77778E-5	3.95675E-5
0.125	1.50726E-3	5.98273E-4	1.57125E-7	1.82763E-5	9.7155E-5
0.20833	2.20758E-4	3.31079E-4	2.02483E-4	4.3446E-6	1.08549E-4
0.291667	5.73621E-3	3.08345E-3	2.02515E-4	2.79856E-4	2.58431E-4
0.375	5.55933E-3	1.79072E-3	8.74156E-7	1.4207E-4	1.0948E-3
0.458333	7.00322E-4	3.95735E-4	2.01542E-4	2.29087E-4	1.0188E-3
0.541667	1.09675E-2	4.93541E-3	2.01504E-4	1.50109E-3	1.49145E-4
0.625	1.1276E-2	2.98284E-3	6.29307E-7	6.5146E-4	4.50911E-3
0.708333	2.25019E-3	4.54571E-4	2.02502E-4	1.37665E-3	4.74738E-3
0.791667	1.75969E-2	6.80571E-3	2.01674E-4	3.67695E-3	3.14693E-4
0.875	1.86405E-2	4.14792E-3	2.28873E-6	1.4996E-3	1.00903E-2
0.958333	4.39483E-3	4.98668E-4	1.99532E-4	3.44829E-3	1.11142E-2

TABLE 7.3
L^2-Error Norm and H^1-Error Norm Obtained by MLWM for Example 7.7.1

k, M	$L^2(0,1)$	$H^1(0,1)$
$k=2, M=3$	2.36268E-3	6.13421E-2
$k=3, M=3$	2.95288E-4	1.5316E-2
$k=4, M=3$	3.69106E-5	3.82775E-3
$k=5, M=3$	4.61382E-6	9.56859E-4
$k=6, M=3$	5.76728E-7	2.3921E-4

For $\gamma = \frac{1}{2}$, Figure 7.2 depicts for the comparison between exact solution and MLWM solution, and also the corresponding error distribution has been presented.

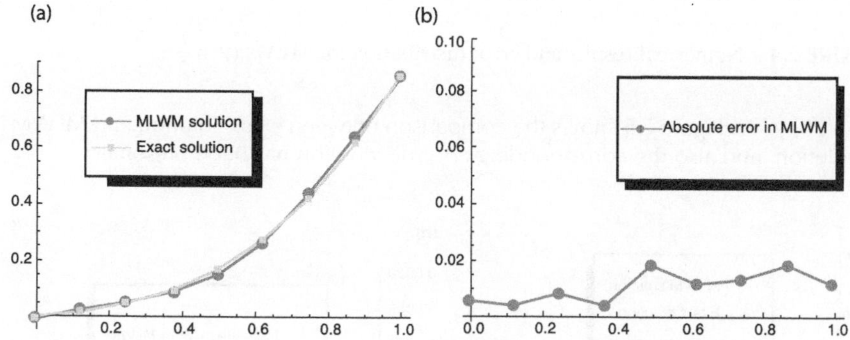

FIGURE 7.2 Numerical results and error distribution for MLWs ($\gamma = \frac{1}{2}$).

For $\gamma = 1$, Figure 7.3 displays the comparison between exact solution and MLWM solution, and also the corresponding error distribution has been presented.

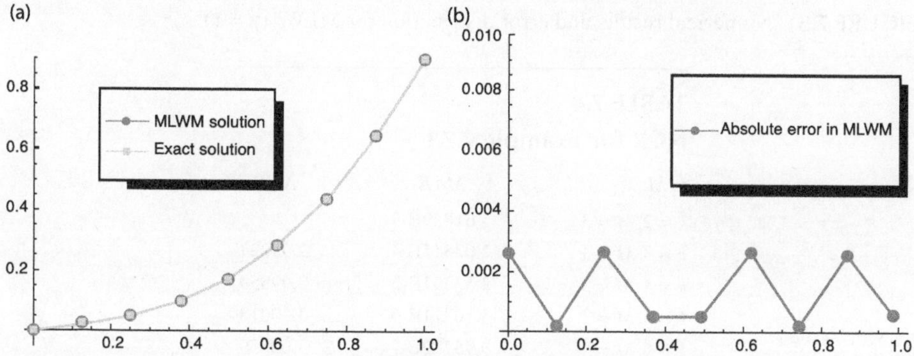

FIGURE 7.3 Numerical results and error distribution for MLWs ($\gamma = 1$).

For $k = 3, M = 3$, Figures 7.4–7.5 are drawn by MLWM for different values of γ.

For $\gamma = \frac{1}{2}$, Figure 7.4 displays the comparison between exact solution and MLWM solution, and also the corresponding error distribution has been presented.

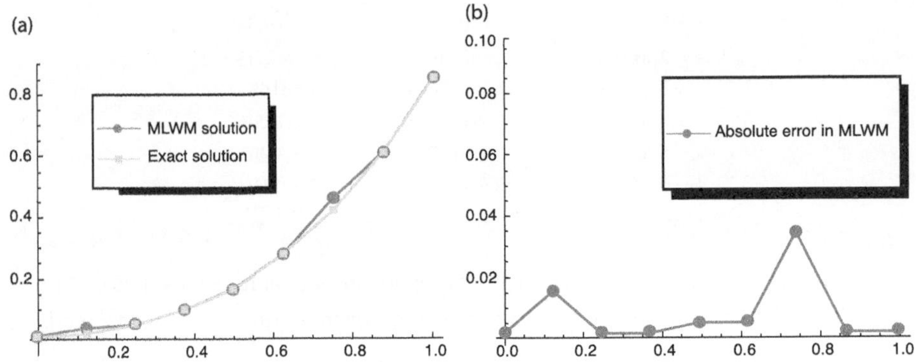

FIGURE 7.4 Numerical results and error distribution for MLWs ($\gamma = \frac{1}{2}$).

For $\gamma = 1$, Figure 7.5 shows the comparison between exact solution and MLWM solution, and also the corresponding error distribution has been presented.

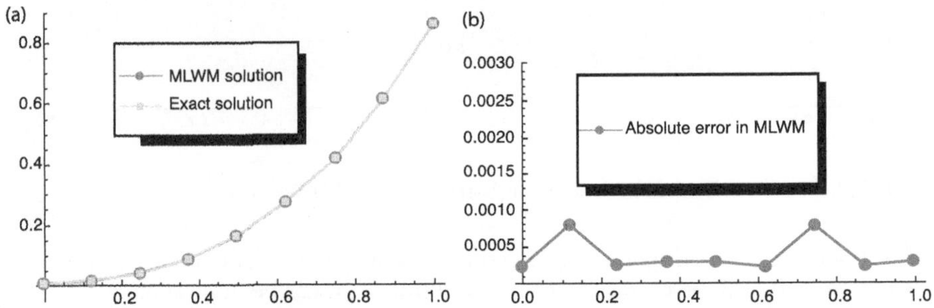

FIGURE 7.5 Numerical results and error distribution for MLWs ($\gamma = 1$).

TABLE 7.4
NCR for Example 7.7.1

k, M	MAE	NCR
$k = 2, M = 3$	1.61829E-3	–
$k = 3, M = 3$	2.02515E-4	2.99837
$k = 4, M = 3$	2.53391E-5	2.99859
$k = 5, M = 3$	3.16711E-6	3.00013
$k = 6, M = 3$	3.95782E-7	3.00039

Example 7.7.2

Consider the following PVDIDE [67]:

$$u'(x) = u(x) - \frac{1}{2}\ln\left(1+\frac{x}{2}\right)u\left(\frac{x}{2}\right) + \int_0^x \frac{x}{1+z}u(z)\,dz + \int_0^{\frac{x}{2}} \frac{1}{1+z}u(z)\,dz$$

$$+\frac{1}{1+x} - \ln(1+x)\left[\frac{x}{2}\ln(1+x)+1\right],\ x \in [0,1], \tag{7.62}$$

$$u(0) = 0,$$

with the exact solution $u(x) = \ln(1+x)$.

Tables 7.5 and 7.7 illustrate the absolute error values of $\tilde{E}(x)$ computed by MLWM for different values of γ. The absolute errors obtained by MLWM and LWM for $k = 2, M = 3$ and $k = 3, M = 3$ have been compared and presented in

TABLE 7.5
$\tilde{E}(x)$ Errors by MLWM for Example 7.7.2 with $k = 2, M = 3$

	$\tilde{E}(x)$			
x	$\gamma = \frac{1}{2}$	$\gamma = 1$	$\gamma = \frac{3}{2}$	$\gamma = 2$
0.0833333	2.76028E-3	3.31972E-4	5.88175E-3	3.72703E-3
0.25	2.02308E-3	1.06862E-4	1.63441E-3	1.13212E-2
0.416667	2.54517E-4	2.53932E-4	5.85376E-3	1.47408E-2
0.583333	1.59816E-3	1.08735E-4	3.38986E-3	1.20109E-3
0.75	8.79113E-4	1.24353E-5	1.02674E-3	7.07966E-3
0.916667	2.11072E-4	8.50155E-5	3.77179E-3	1.00241E-2

TABLE 7.6
Comparison of $\tilde{E}(x)$ Errors with $k = 2, M = 3$ for Example 7.7.2

	$\tilde{E}(x)$	
x	MLWM ($\gamma = 1$)	LWM
0.0833333	3.31972E-4	3.31972E-4
0.25	1.06862E-4	1.06862E-4
0.416667	2.53932E-4	2.53932E-4
0.583333	1.08735E-4	1.08735E-4
0.75	1.24353E-5	1.24353E-5
0.916667	8.50155E-5	8.50155E-5

TABLE 7.7
$\tilde{E}(x)$ Errors by MLWM for Example 7.7.2 with $k = 3, M = 3$

x	$\tilde{E}(x)$			
	$\gamma = \frac{1}{2}$	$\gamma = 1$	$\gamma = \frac{3}{2}$	$\gamma = 2$
0.0416667	9.07881E-4	5.26971E-5	2.76455E-3	1.48942E-3
0.125	6.28833E-4	8.73144E-6	9.20877E-4	5.87517E-3
0.20833	3.06674E-5	4.36122E-5	2.65065E-3	7.05875E-3
0.291667	6.29157E-4	2.84973E-5	2.15014E-3	1.03078E-3
0.375	4.07844E-4	2.55908E-6	7.4682E-4	4.68058E-3
0.458333	2.09783E-5	2.31449E-5	2.06717E-3	5.60103E-3
0.541667	2.13131E-3	1.68964E-5	7.20099E-4	9.12591E-4
0.625	2.02255E-3	2.24135E-6	1.91339E-3	3.94707E-3
0.708333	1.71214E-3	1.5449E-5	3.73961E-3	4.74541E-3
0.791667	3.66849E-3	1.04306E-5	9.266E-4	2.70769E-4
0.875	1.94172E-2	8.46928E-7	4.19016E-3	3.07807E-3
0.958333	3.72263E-2	9.47339E-6	5.22067E-3	3.97414E-3

TABLE 7.8
Comparison of $\tilde{E}(x)$ Errors with $k = 3, M = 3$ for Example 7.7.2

x	$\tilde{E}(x)$	
	MLWM ($\gamma = 1$)	LWM
0.0416667	5.26971E-5	5.26971E-5
0.125	8.73144E-6	8.73144E-6
0.20833	4.36122E-5	4.36122E-5
0.291667	2.84973E-5	2.84973E-5
0.375	2.55908E-6	2.55908E-6
0.458333	2.31449E-5	2.31449E-5
0.541667	1.68964E-5	1.68964E-5
0.625	2.24135E-6	2.24135E-6
0.708333	1.5449E-5	1.5449E-5
0.791667	1.04306E-5	1.04306E-5
0.875	8.46928E-7	8.46928E-7
0.958333	9.47339E-6	9.47339E-6

Tables 7.6 and 7.8, respectively. The numerical results and absolute error obtained by different wavelets methods are shown in Figures 7.6–7.13. For various values of k, M, Table 7.10 displays the MAE norm and NCR. As the size of k increases, a fourth-order convergence rate is obtained.

TABLE 7.9
L^2-Error Norm and H^1-Error Norm for Example 7.7.2

	MLWM		LWM	
k,M	$L^2(0,1)$	$H^1(0,1)$	$L^2(0,1)$	$H^1(0,1)$
$k=2, M=3$	3.14217E-4	8.16027E-3	3.14217E-4	8.16027E-3
$k=3, M=3$	4.21165E-5	2.18589E-3	4.21165E-5	2.18589E-3
$k=3, M=4$	1.6571E-6	1.25775E-4	1.6571E-6	1.25775E-4

TABLE 7.10
NCR for Example 7.7.2

k,M	MAE	NCR
$k=3, M=4$	3.22398E-6	–
$k=4, M=4$	2.46292E-7	3.7104
$k=5, M=4$	1.71046E-8	3.84791
$k=6, M=4$	1.12848E-9	3.92193

TABLE 7.11
Comparison of Maximum Absolute Errors for Example 7.7.2

		$\|E(h)\|_\infty$	
N	h	MLWM	SCM [67]
5	1.404963	4.28727E-12	0.0017
10	0.993459	8.65695E-12	1.1119E-4
20	0.702481	8.94401E-12	1.9619E-6
30	0.573574	8.94857E-12	8.7010E-8
40	0.496729	8.94882E-12	6.2620E-9
50	0.444288	8.94886E-12	6.1468E-10
60	0.405578	8.94889E-12	7.5222E-11

For $k=2, M=3$, Figures 7.6 and 7.7 are drawn by MLWM for different values of γ.

For $\gamma=1$, Figure 7.6 displays the comparison between exact solution and MLWM solution, and also the corresponding error distribution has been presented.

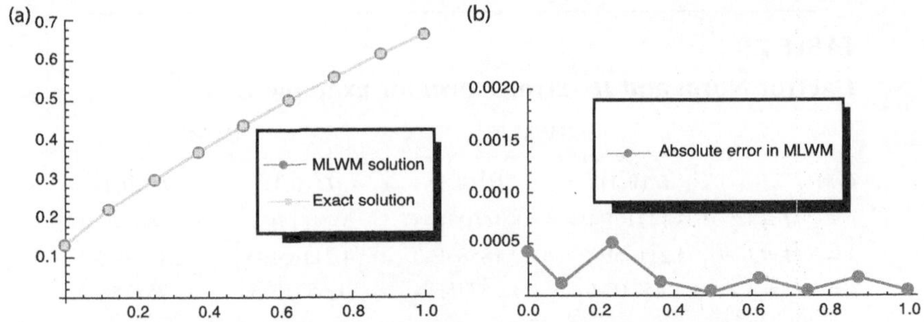

FIGURE 7.6 Numerical results and error distribution for MLWs ($\gamma = 1$).

For $\gamma = \frac{3}{2}$, Figure 7.7 shows the comparison between exact solution and MLWM solution, and also the corresponding error distribution has been presented.

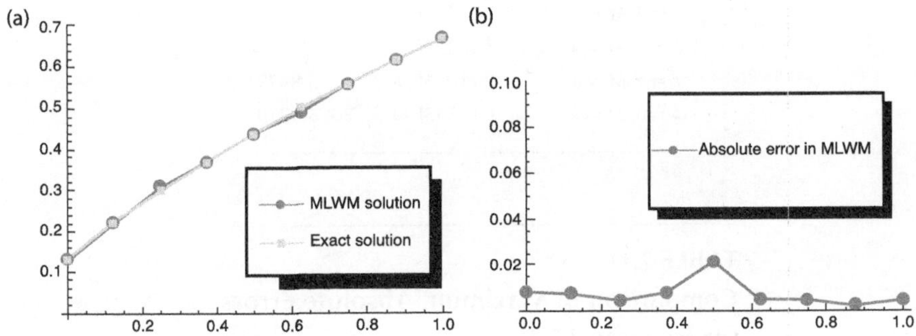

FIGURE 7.7 Numerical results and error distribution for MLWs ($\gamma = \frac{3}{2}$).

Now for $k = 2, M = 3$, Figures 7.8 and 7.9 are also drawn by LWM.

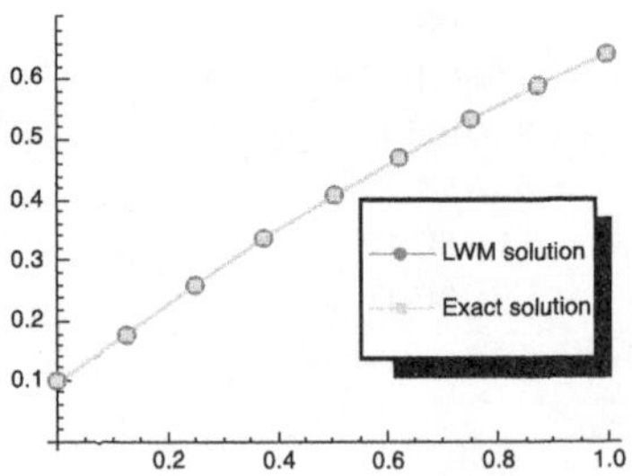

FIGURE 7.8 Numerical results by LWM for function $u(x)$.

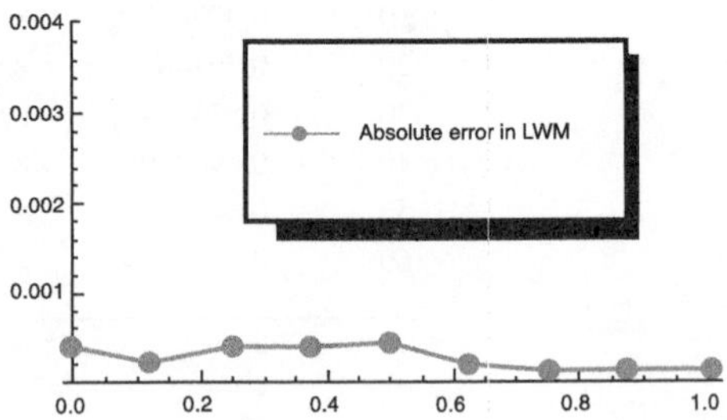

FIGURE 7.9 Error distribution for function $\tilde{E}(x)$ in LWM.

For $k = 3, M = 3$, Figures 7.10 and 7.11 are drawn by MLWM for different values of γ.

For $\gamma = 1$, Figure 7.10 displays the comparison between exact solution and MLWM solution, and also the corresponding error distribution has been presented.

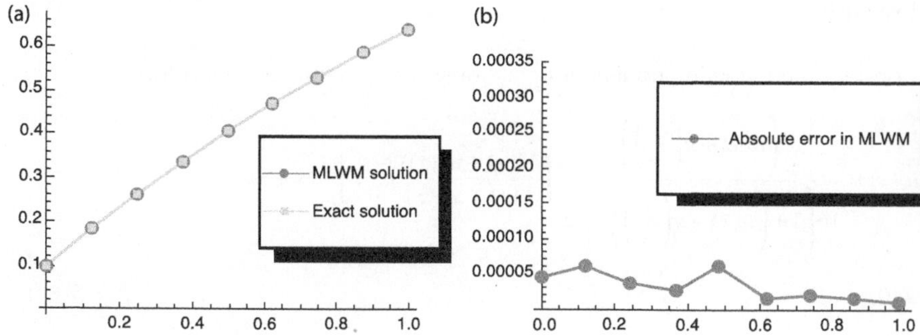

FIGURE 7.10 Numerical results and error distribution for MLWs ($\gamma = 1$).

For $\gamma = \frac{3}{2}$, Figure 7.11 shows the comparison between exact solution and MLWM solution, and also the corresponding error distribution has been presented.

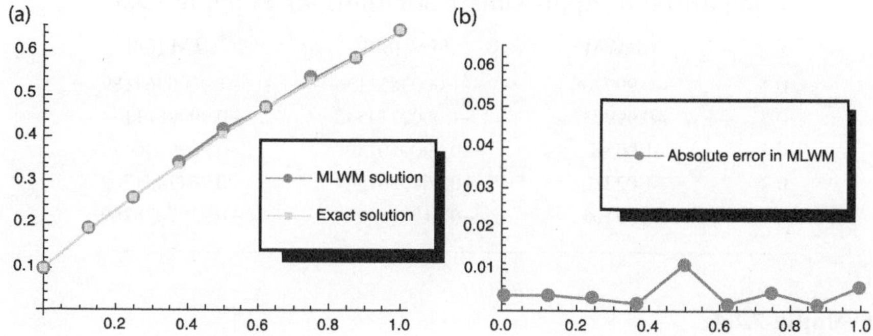

FIGURE 7.11 Numerical results and error distribution for MLWs ($\gamma = \frac{3}{2}$).

Now for $k = 3, M = 3$, Figures 7.12 and 7.13 are also drawn by LWM.

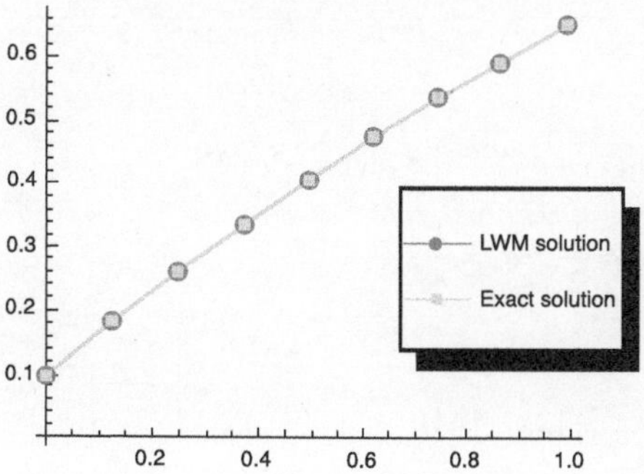

FIGURE 7.12 Numerical results by LWM for function $u(x)$.

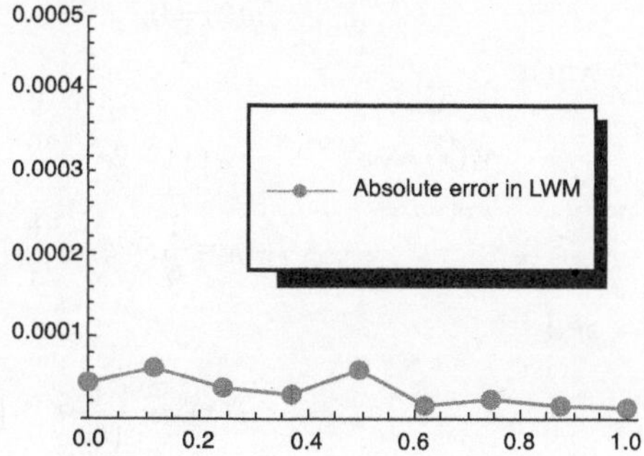

FIGURE 7.13 Error distribution for function $\tilde{E}(x)$ in LWM.

Example 7.7.3

Consider the following nonlinear delay integro-differential equation [68]:

$$u'(x) = \frac{x^2\left(u(x)+u\left(\frac{x}{3}\right)\right)}{16\left(1+\left(u(x)+u\left(\frac{x}{3}\right)\right)^2\right)} + \int_0^x \frac{\ln(1+u^2(z))}{8(1+x+z)}dz + \int_0^{\frac{x}{4}}(x+z)u(z)\,dz, \; x \in [0,1],$$

$$u(0) = -1. \tag{7.63}$$

The comparison of approximate solutions with single exponential method (SEM) and double exponential method (DEM) has been presented in Table 7.12.

TABLE 7.12
Comparison of Approximate Solutions for Example 7.7.3

x	MLWM	SEM [68]	DEM [68]
0.1	−0.999708	−1.0003851136	−1.0005364189
0.3	−0.999731	−1.0002214545	−0.9999801171
0.5	−1.0054	−0.9900501507	−0.9899646019
0.7	−1.02212	−0.9649261544	−0.9647451709
0.9	−1.05506	−0.9198885568	−0.9197278269

Example 7.7.4

Consider the following nonhomogeneous pantograph equation [65]:

$$u'(\xi) = \sum_{i=1}^{4} y_i(x)u(q_i x) + f(x), \; x \in [0,1],$$
$$u(0) = 0, \tag{7.64}$$

where

$$y_1(x) = 3e^{-(5/6)x}, \quad y_2(x) = 2e^{-(2/3)x}, \quad y_3(x) = e^{-(1/2)x}, \quad y_4(x) = -3e^{-(3/4)x},$$

$$q_1 = \frac{1}{6}, \quad q_2 = \frac{1}{3}, \quad q_3 = \frac{1}{2}, \quad q_4 = \frac{1}{4},$$

and

$$f(x) = -\frac{1}{144}e^{-x}(197x^2 - 12x + 144).$$

The exact solution is $u(x) = e^{-x}(x^2 + x)$.

Tables 7.13 and 7.14 illustrate the absolute error values of $\tilde{E}(x)$ computed by MLWM for different values of γ. The numerical results and absolute error

TABLE 7.13
$\tilde{E}(x)$ Errors by MLWM with $k = 2, M = 3$ for Example 7.7.4

	$\tilde{E}(x)$			
x	$\gamma = \frac{1}{2}$	$\gamma = 1$	$\gamma = \frac{3}{2}$	$\gamma = 2$
0.0833333	1.94259E-3	2.53316E-4	2.69697E-4	9.66703E-3
0.25	3.14113E-3	6.0937E-4	9.37295E-3	1.34033E-2
0.416667	1.59626E-3	7.41922E-4	1.8738E-2	4.949E-2
0.583333	2.45817E-3	5.16805E-4	1.06653E-2	3.48702E-2
0.75	4.30445E-3	7.2943E-4	1.66698E-2	3.57785E-2
0.916667	1.97691E-3	7.15768E-4	2.52548E-2	6.41387E-2

TABLE 7.14
$\tilde{E}(x)$ Errors by MLWM with $k = 3, M = 3$ for Example 7.7.4

	$\tilde{E}(x)$			
x	$\gamma = \frac{1}{2}$	$\gamma = 1$	$\gamma = \frac{3}{2}$	$\gamma = 2$
0.0416667	3.9905E-4	5.74996E-5	1.34754E-3	1.57736E-3
0.125	3.46163E-4	5.24735E-5	1.55496E-3	3.79341E-4
0.20833	1.50552E-4	9.96478E-5	5.22385E-3	1.42888E-2
0.291667	8.29462E-4	2.58595E-6	2.51802E-4	5.9178E-3
0.375	5.9377E-4	5.37512E-5	2.23012E-3	2.60138E-3
0.458333	1.09304E-4	7.3852E-5	5.76464E-3	1.57381E-2
0.541667	9.55851E-4	3.87218E-5	1.51225E-3	9.04041E-3
0.625	6.02067E-4	4.33495E-5	2.80372E-3	5.97864E-3
0.708333	7.569E-5	1.93478E-5	4.84636E-3	1.47488E-2
0.791667	1.02064E-3	8.59825E-6	1.07594E-3	7.3878E-3
0.875	5.7816E-4	1.23432E-5	2.63717E-3	5.633078E-3
0.958333	8.10228E-7	2.29608E-5	5.24874E-3	1.3999E-2

estimation by MLWM are shown in Figures 7.14–7.17. Table 7.15 displays the comparison of absolute errors between MLWM and Adomian decomposition method (ADM) [65]. For various values of k, M, Table 7.16 displays the maximum absolute error norm and NCR. As the size of k increases, a third-order convergence rate is obtained.

For $k=2, M=3$, Figures 7.14 and 7.15 are drawn by MLWM for different values of γ.

For $\gamma=\frac{1}{2}$, Figure 7.14 displays the comparison between exact solution and MLWM solution, and also the corresponding error distribution has been presented.

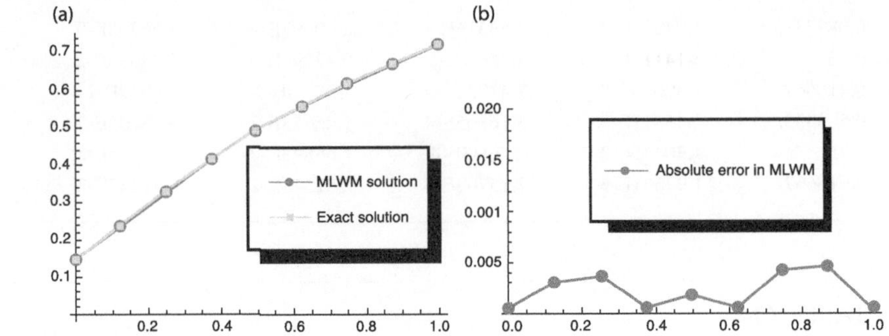

FIGURE 7.14 Numerical results and error distribution for MLWs ($\gamma=\frac{1}{2}$).

For $\gamma=1$, Figure 7.15 displays the comparison between exact solution and MLWM solution, and also the corresponding error distribution has been presented.

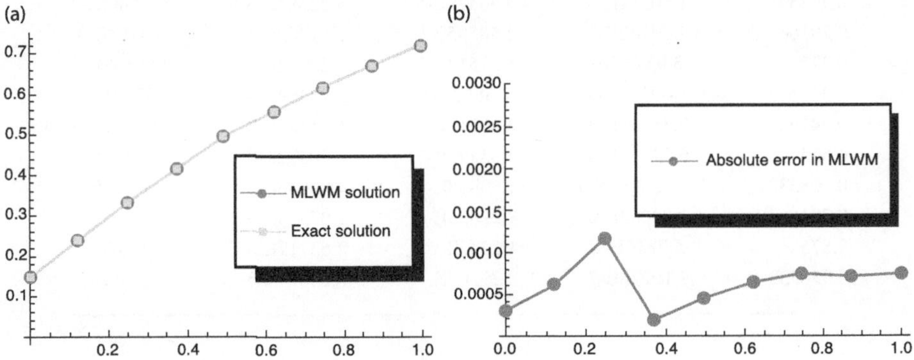

FIGURE 7.15 Numerical results and error distribution for MLWs ($\gamma=1$).

For $k=3, M=3$, Figures 7.16 and 7.17 are drawn by MLWM for different values of γ.

For $\gamma=\frac{1}{2}$, Figure 7.16 displays the comparison between exact solution and MLWM solution, and also the corresponding error distribution has been presented.

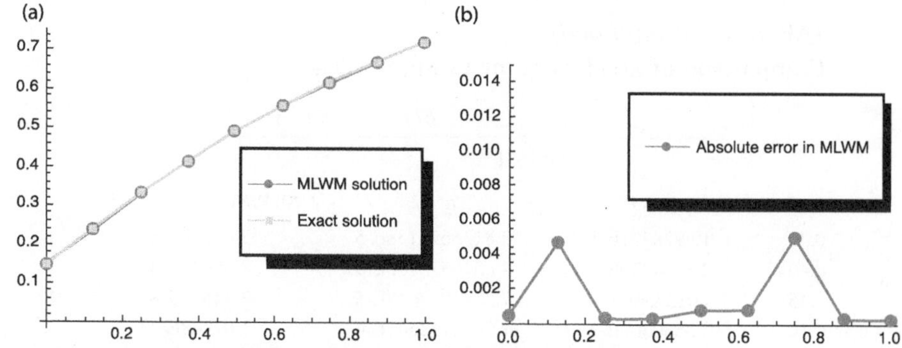

FIGURE 7.16 Numerical results and error distribution for MLWs ($\gamma = \frac{1}{2}$).

For $\gamma = 1$, Figure 7.17 displays the comparison between exact solution and MLWM solution, and also the corresponding error distribution has been presented.

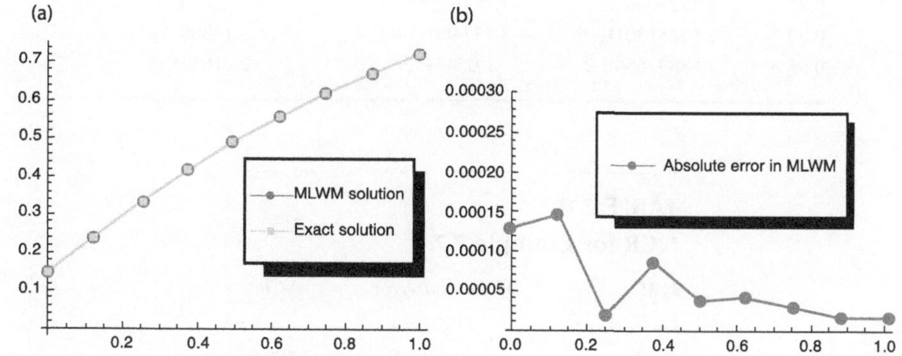

FIGURE 7.17 Numerical results and error distribution for MLWs ($\gamma = 1$).

TABLE 7.15
Comparison of $\tilde{E}(x)$ Errors for Example 7.7.4

	$\tilde{E}(x)$		
	ADM [65]		
x	$n = 1$	$n = 2$	MLWM ($k = 6, M = 3$)
0.05	6.83946425003E-6	1.4140320914E-8	1.99814E-7
0.10	5.3243428947E-5	2.1924392967E-7	2.30231E-7
0.15	1.7486767556E-4	1.0756273063E-6	1.83478E-7
0.20	4.0337747045E-4	3.2946326626E-6	8.24211E-8
0.25	7.6673283637E-4	7.7957596200E-6	3.31804E-7
0.30	1.2894574E-3	1.5668091057E-5	8.63901E-8

(Continued)

TABLE 7.15 (*Continued*)
Comparison of $\tilde{E}(x)$ Errors for Example 7.7.4

	$\tilde{E}(x)$		
	ADM [65]		
x	$n=1$	$n=2$	MLWM ($k=6, M=3$)
0.35	1.9928921E-3	2.8135673138E-5	7.63748E-8
0.40	2.8954330E-3	4.6526532730E-5	8.58923E-8
0.45	4.0127568 E-3	7.2245037612E-5	3.34529E-8
0.50	5.3580309E-3	1.0674732541E-4	1.07948E-7
0.55	6.9421126E-3	1.5151954676E-4	5.56358E-8
0.60	8.7737351E-3	2.0805868574E-4	5.17757E-9
0.65	1.08596826E-2	2.7785573819E-4	7.35624E-9
0.70	1.32049542E-2	3.6238104387E-4	1.13272E-8
0.75	1.58129177E-2	4.6307158331E-4	2.55234E-8
0.80	1.86854538E-2	5.8132006435E-4	9.12178E-9
0.85	2.18230907E-2	7.1846563575E-4	5.5841E-9
0.90	2.52251301E-2	8.7578607794E-4	3.14908E-8
0.95	2.88897654E-2	1.0544913E-3	2.81612E-8

TABLE 7.16
NCR for Example 7.7.4

k, M	*Maximum error*	*NCR*
$k=2, M=3$	7.41922E-4	–
$k=3, M=3$	9.96478E-5	2.89636
$k=4, M=3$	1.25595E-5	2.98806
$k=5, M=3$	1.57438E-6	2.99592
$k=6, M=3$	1.97206E-7	2.99701

Example 7.7.5

Let us consider the following PVDIDE [69]:

$$u'(x) = (4x-1)u(x) - \left(1-\frac{x}{2}\right)u\left(\frac{x}{2}\right) - 2\int_0^x (x+z)u(z)\,dz - \int_0^{\frac{x}{2}} zu(z)\,dz + f(x), \tag{7.65}$$

$$u(0)=1,$$

where $f(x) = 3-2x, x \in [0,1]$ and the exact solution is $u(x) = e^x$.

Tables 7.17 and 7.18 illustrate the absolute error values of $\tilde{E}(x)$ computed by MLWM for different values of γ. The numerical solutions and absolute error estimation by MLWM are shown in Figures 7.18–7.21. For various values of k,M, Table 7.20 displays the maximum absolute error norm and NCR. The increasing amount of k leads to a roughly third-order convergence rate.

TABLE 7.17
$\tilde{E}(x)$ Errors by MLWM with $k = 2, M = 3$ for Example 7.7.5

	$\tilde{E}(x)$					
x	$\gamma = \frac{1}{2}$	$\gamma = \frac{1}{4}$	$\gamma = \frac{4}{3}$	$\gamma = 1$	$\gamma = \frac{3}{2}$	$\gamma = 2$
0.0833333	7.00863E-3	1.67781E-2	4.4637E-3	3.8754E-4	5.90515E-3	4.29663E-3
0.25	4.80822E-3	2.31819E-2	7.19933E-4	6.4447E-5	2.73025E-3	1.55901E-2
0.416667	2.22102E-3	8.5576E-4	3.08526E-3	4.69426E-4	5.26172E-3	1.29017E-2
0.583333	1.07241E-2	3.11346E-2	5.68986E-3	4.42911E-4	6.96584E-3	1.19864E-3
0.75	7.02569E-3	2.95515E-2	1.55693E-4	1.57197E-5	2.42081E-3	1.94246E-2
0.916667	1.15717E-3	1.33808E-2	4.60654E-3	7.18248E-4	8.38055E-3	2.3547E-2

TABLE 7.18
$\tilde{E}(x)$ Errors by MLWM with $k = 3, M = 3$ for Example 7.7.5

	$\tilde{E}(x)$					
x	$\gamma = \frac{1}{2}$	$\gamma = \frac{1}{4}$	$\gamma = \frac{4}{3}$	$\gamma = 1$	$\gamma = \frac{3}{2}$	$\gamma = 2$
0.0416667	1.47287E-3	5.8185E-3	2.14387E-3	3.99528E-5	2.74071E-3	1.52643E-3
0.125	9.63517E-4	7.35358E-3	2.53184E-4	3.52572E-6	1.15772E-3	6.82954E-3
0.20833	2.29298E-4	1.49679E-3	1.36477E-3	4.39284E-5	2.46267E-3	6.63267E-3
0.291667	1.8577E-3	8.1239E-3	2.41975E-3	4.35177E-5	2.96002E-3	7.43793E-4
0.375	1.12824E-3	8.35339E-3	1.75953E-4	2.23564E-6	1.15457E-3	7.63839E-3
0.458333	9.13964E-5	3.13138E-3	1.57041E-3	5.06672E-5	2.9551E-3	8.64293E-3
0.541667	2.46474E-3	1.02242E-2	3.45067E-3	6.4473E-5	4.36648E-3	2.15389E-3
0.625	1.51098E-3	1.14556E-2	4.25368E-4	6.54526E-6	1.88693E-3	1.10667E-2
0.708333	2.74768E-4	3.0784E-3	2.07365E-3	6.81433E-5	3.79338E-3	1.04796E-2
0.791667	3.08051E-3	1.33382E-2	4.18074E-3	7.59965E-5	5.2273E-3	2.17611E-3
0.875	1.89167E-3	1.38979E-2	3.97395E-4	4.93673E-6	2.13996E-3	1.34881E-2
0.958333	1.97384E-4	5.07397E-3	2.52576E-3	8.38079E-5	4.70322E-3	1.34206E-2

For $k = 2, M = 3$, Figures 7.18 and 7.19 are drawn by MLWM for different values of γ.

For $\gamma = \frac{1}{2}$, Figure 7.18 depicts for the comparison between exact solution and MLWM solution, and also the corresponding error distribution has been presented.

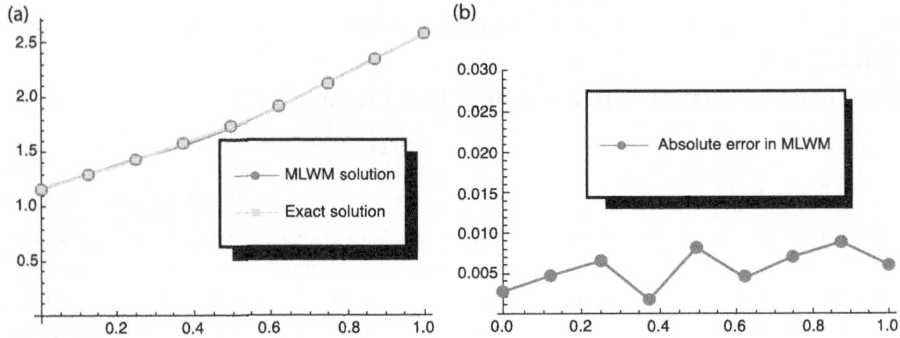

FIGURE 7.18 Numerical results and error distribution for MLWs ($\gamma = \frac{1}{2}$).

For $\gamma = 1$, Figure 7.19 displays the comparison between exact solution and MLWM solution, and also the corresponding error distribution has been presented.

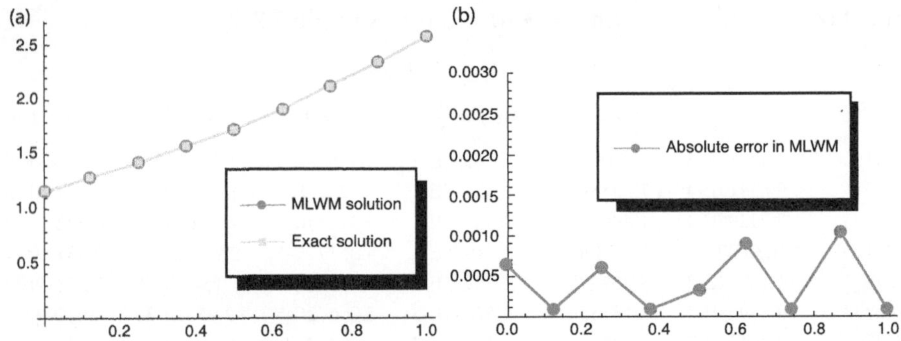

FIGURE 7.19 Numerical results and error distribution for MLWs ($\gamma = 1$).

For $k = 3, M = 3$, Figures 7.20 and 7.21 are drawn by MLWM for different values of γ.

For $\gamma = \frac{1}{2}$, Figure 7.20 displays the comparison between exact solution and MLWM solution, and also the corresponding error distribution has been presented.

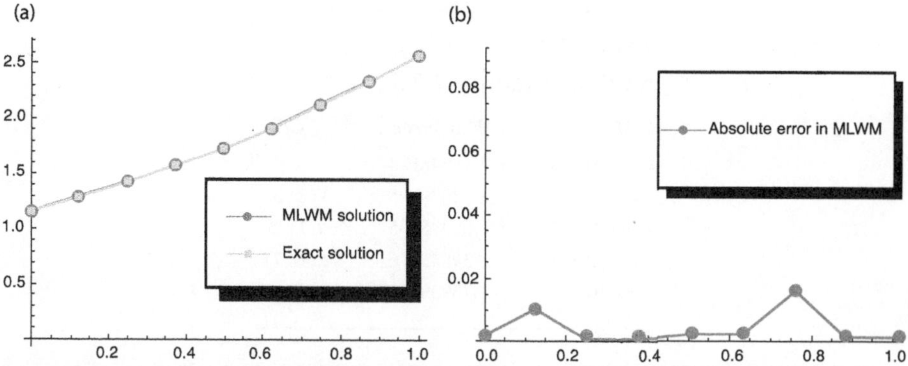

FIGURE 7.20 Numerical results and error distribution for MLWs ($\gamma = \frac{1}{2}$).

For $\gamma = 1$, Figure 7.21 shows the comparison between exact solution and MLWM solution, and also the corresponding error distribution has been presented.

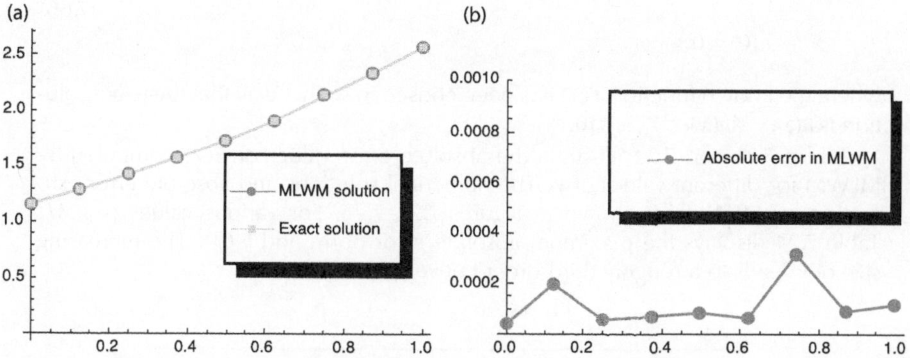

FIGURE 7.21 Numerical results and error distribution for MLWs ($\gamma = 1$).

TABLE 7.19
L^2-Error Norm and H^1-Error Norm Obtained by MLWM for Example 7.7.5

k, M	$L^2(0,1)$	$H^1(0,1)$
$k = 2, M = 3$	6.97935E-4	1.78995E-2
$k = 3, M = 3$	8.76697E-5	4.54088E-3
$k = 4, M = 3$	1.09859E-5	1.13886E-3
$k = 5, M = 3$	1.37411E-6	2.8495E-4

TABLE 7.20
NCR for Example 7.7.5

k, M	*Max error*	*NCR*
$k = 2, M = 3$	7.18248E-4	–
$k = 3, M = 3$	8.38079E-5	3.09932
$k = 4, M = 3$	1.09398E-5	2.9375
$k = 5, M = 3$	1.39843E-6	2.96771
$k = 6, M = 3$	1.76939E-7	2.98248

Example 7.7.6

Let us consider the following PVDIDE [69]:

$$\begin{aligned} u'(x) &= 3u(x) + u\left(\frac{x}{2}\right) + \int_0^x 2^x u(z)\,dz - 8\int_0^{\frac{x}{2}} e^x \cos(z) u(z)\,dz + f(x), \\ u(0) &= 0, \end{aligned} \tag{7.66}$$

where the known function $f(x)$ has been chosen in such a way that the exact solution is $u(x) = \sin(x)e^{2x-1}$, $x \in [0,1]$.

Tables 7.21 and 7.22 illustrate the absolute error values of $\tilde{E}(x)$ computed by MLWM for different values of γ. The numerical solutions and absolute error estimation by MLWM are shown in Figures 7.22–7.25. For various values of k, M, Table 7.24 displays the maximum absolute error norm and NCR. The increasing size of k leads to a roughly third-order convergence rate.

TABLE 7.21
$\tilde{E}(x)$ Errors by MLWM with $k = 2, M = 3$ for Example 7.7.6

	$\tilde{E}(x)$		
x	$\gamma = \frac{1}{2}$	$\gamma = 1$	$\gamma = 2$
0.0833333	1.63475E-2	1.13108E-3	2.63019E-3
0.25	7.08756E-3	3.79739E-6	1.57048E-3
0.416667	6.36764E-5	2.1414E-3	6.05571E-3
0.583333	4.82327E-2	3.67972E-3	6.51909E-3
0.75	2.43364E-2	3.7868E-4	1.30592E-2
0.916667	5.38453E-3	4.76237E-3	3.36044E-2

TABLE 7.22
$\tilde{E}(x)$ Errors by MLWM with $k = 3, M = 3$ for Example 7.7.6

x	$\tilde{E}(x)$ $\gamma = \frac{1}{2}$	$\gamma = 1$	$\gamma = 2$
0.0416667	2.77963E-3	1.47769E-4	3.69524E-4
0.125	1.51988E-3	1.18204E-5	1.76874E-3
0.20833	1.49923E-4	1.81531E-4	2.56834E-3
0.291667	5.09486E-3	2.78153E-4	1.65005E-4
0.375	3.03609E-3	2.31493E-5	4.76852E-3
0.458333	6.40693E-4	3.22382E-4	6.4743E-3
0.541667	8.65747E-3	3.8463E-4	1.10571E-3
0.625	4.79169E-3	2.42155E-5	9.20813E-3
0.708333	4.67103E-4	4.41716E-4	1.32676E-2
0.791667	1.37063E-2	6.11598E-4	4.281E-4
0.875	8.31828E-3	4.11195E-5	1.8054E-2
0.958333	1.50549E-3	6.5637E-4	2.62739E-2

For $k = 2, M = 3$, Figures 7.22 and 7.23 are drawn by MLWM for different values of γ.

For $\gamma = 1$, Figure 7.22 displays the comparison between exact solution and MLWM solution, and also the corresponding error distribution has been presented.

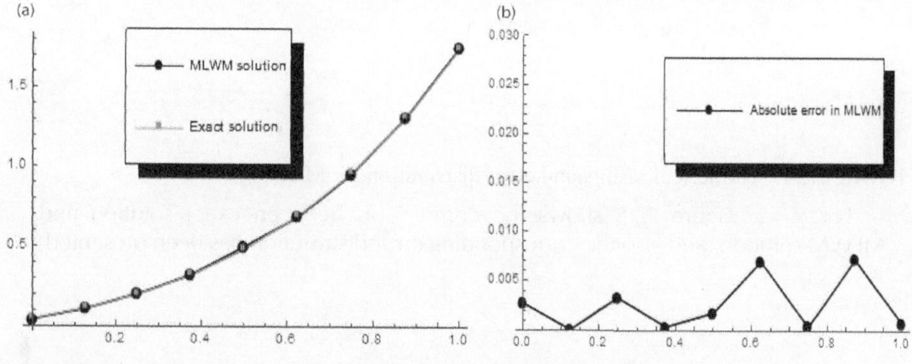

FIGURE 7.22 Numerical results and error distribution for MLWs ($\gamma = 1$).

For $\gamma = 2$, Figure 7.23 represents the comparison between exact solution and MLWM solution, and also the corresponding error distribution has been presented.

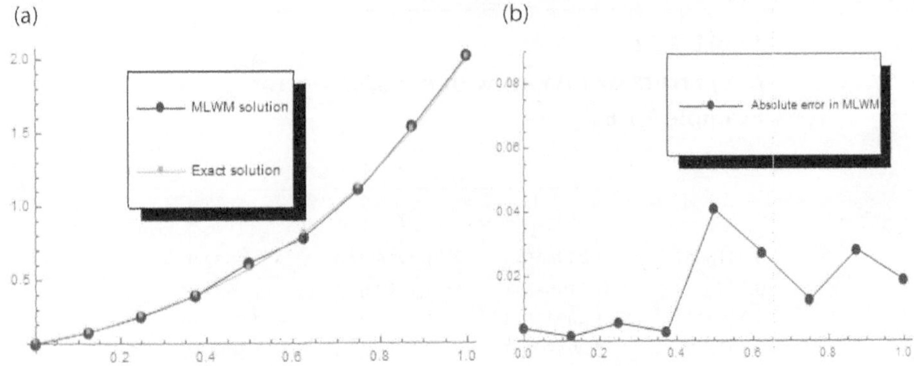

FIGURE 7.23 Numerical results and error distribution for MLWs ($\gamma = 2$).

For $k = 3, M = 3$, Figures 7.24 and 7.25 are drawn by MLWM for different values of γ.

For $\gamma = 1$, Figure 7.24 displays the comparison between exact solution and MLWM solution, and also the corresponding error distribution has been presented.

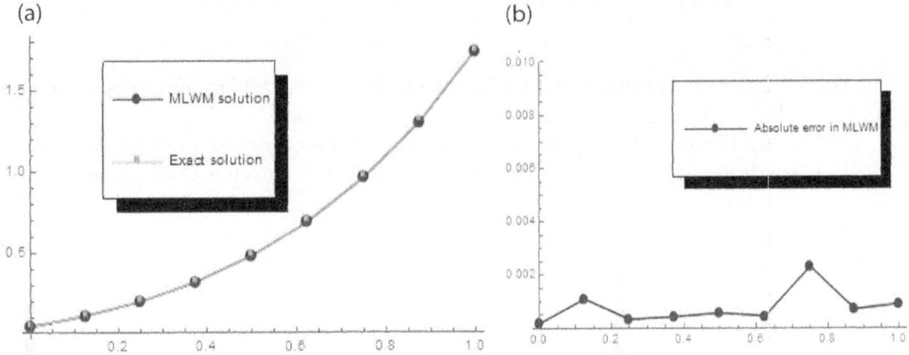

FIGURE 7.24 Numerical results and error distribution for MLWs ($\gamma = 1$).

For $\gamma = 2$, Figure 7.25 shows the comparison between exact solution and MLWM solution, and also the corresponding error distribution has been presented.

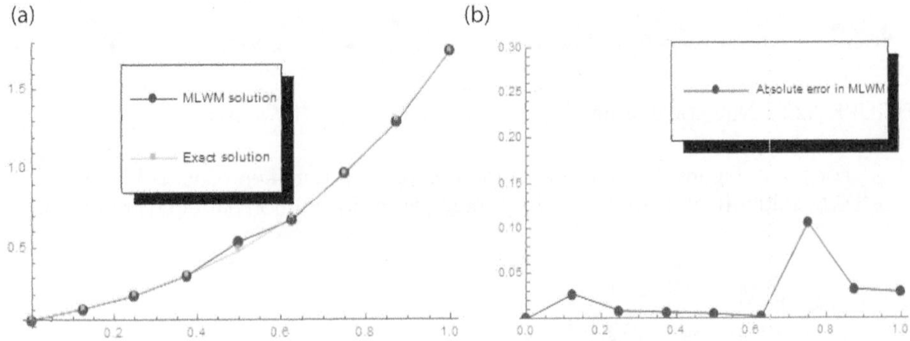

FIGURE 7.25 Numerical results and error distribution for MLWs ($\gamma = 2$).

TABLE 7.23
L^2-Error Norm and H^1-Error Norm Obtained by MLWM for Example 7.7.6

k, M	$L^2(0,1)$	$H^1(0,1)$
$k=2, M=3$	4.73925E-3	1.23004E-1
$k=3, M=3$	6.03029E-4	3.12549E-2
$k=4, M=3$	7.56756E-5	7.84624E-3
$k=5, M=3$	9.46847E-6	1.96356E-3
$k=6, M=3$	1.18384E-6	4.91015E-4

TABLE 7.24
NCR for Example 7.7.6

k, M	*Max error*	*NCR*
$k=2, M=3$	4.76237E-3	–
$k=3, M=3$	6.5637E-4	2.8591
$k=4, M=3$	8.50854E-5	2.94753
$k=5, M=3$	1.08012E-5	2.97772
$k=6, M=3$	1.35905E-6	2.99052

Example 7.7.7

Consider the following fractional PVDIDE [54]:

$$ {}_0^C D_x^{\alpha} u(x) = f(x) - \frac{x^2 e^x}{5} u(x) - \frac{(qx)^2}{5} u(qx) + \int_0^x e^x z u(z)\,dz + \int_0^{qx} z u(z)\,dz, \tag{7.67}$$

$$u(0) = 0.$$

act solution is $u(x) = x^3$ and $q = \frac{1}{2}$. In this problem,

$$f(x) = \begin{cases} \dfrac{6x^{2.5}}{\Gamma(3.5)}, & \text{for } \alpha = 0.5, \\[2mm] \dfrac{6x^{2.25}}{\Gamma(3.25)}, & \text{for } \alpha = 0.75, \\[2mm] \dfrac{6x^{2.2}}{\Gamma(3.2)}, & \text{for } \alpha = 0.8, \\[2mm] \dfrac{6x^{2.1}}{\Gamma(3.1)}, & \text{for } \alpha = 0.9, \\[2mm] \dfrac{6x^{2}}{\Gamma(3)}, & \text{for } \alpha = 1. \end{cases}$$

For different values of k and M, the values of $\tilde{E}(x)$ obtained by the EWM are presented in Tables 7.25–7.27, respectively. Figures 7.26–7.33 display the numerical solutions and absolute error analysis by EWM for different values of α. Table 7.29 presents the NCR for different values of k and M. With the increase of the size of k, the convergence rate approximately reaches to 3 for $\alpha = 1$.

TABLE 7.25
$\tilde{E}(x)$ Errors by EWM for Example 7.7.7 with $k = 2$, $M = 3$

x	$\alpha = 0.5$	$\alpha = 0.75$	$\alpha = 0.8$	$\alpha = 0.9$	$\alpha = 1$
0.1	2.24623E-03	2.23258E-03	2.23182E-03	2.23304E-03	2.23789E-03
0.3	1.4859E-03	1.61453E-03	1.64424E-03	1.70203E-03	1.75379E-03
0.5	6.59677E-03	6.45248E-03	6.42286E-03	6.36554E-03	6.31357E-03
0.7	2.13266E-03	1.92699E-03	1.89525E-03	1.84318E-03	1.80578E-03
0.9	2.25203E-03	2.17684E-03	2.17115E-03	2.16751E-03	2.17165E-03
CPU Time (in sec.)	143.205	149.125	127.156	147.423	85.061

TABLE 7.26
$\tilde{E}(x)$ Errors by EWM for Example 7.7.7 with $k = 3$, $M = 3$

x	$\alpha = 0.5$	$\alpha = 0.75$	$\alpha = 0.8$	$\alpha = 0.9$	$\alpha = 1$
0.1	2.5625E-04	2.37028E-04	2.32845E-04	2.25059E-04	2.18559E-04
0.3	2.94771E-04	2.84255E-04	2.8308E-04	2.81684E-04	2.81344E-04
0.5	8.08764E-04	7.9344E-04	7.90694E-04	7.85964E-04	7.82402E-04
0.7	2.89151E-04	2.8262E-04	2.82046E-04	2.81427E-04	2.81285E-04
0.9	1.97362E-04	2.09293E-04	2.11074E-04	2.13952E-04	2.15988E-04
CPU Time (in sec.)	1,685.47	1,673.28	1,449.64	1,669.31	1,521.86

TABLE 7.27
$\tilde{E}(x)$ Errors by EWM for Example 7.7.7 with $k = 4$, $M = 3$

x	$\alpha = 0.5$	$\alpha = 0.75$	$\alpha = 0.8$	$\alpha = 0.9$	$\alpha = 1$
0.1	3.75463E-05	3.58874E-05	3.56472E-05	3.53077E-05	3.51559E-05
0.3	2.9458E-05	2.79869E-05	2.77899E-05	2.74995E-05	2.73292E-05
0.5	9.97803E-05	9.84679E-05	9.82536E-05	9.79105E-05	9.76814E-05
0.7	2.59193E-05	2.68564E-05	2.69796E-05	2.71635E-05	2.72777E-05
0.9	3.56743E-05	3.51636E-05	3.51336E-05	3.51094E-05	3.51115E-05
CPU Time (in sec.)	2,503.03	3,171.37	3,053.7	3,521.62	1,994.64

For $k = 3$, $M = 3$, Figures 7.26–7.29 represent the numerical results obtained by EWM for $\alpha = 0.75, 0.8, 0.9, 1$.

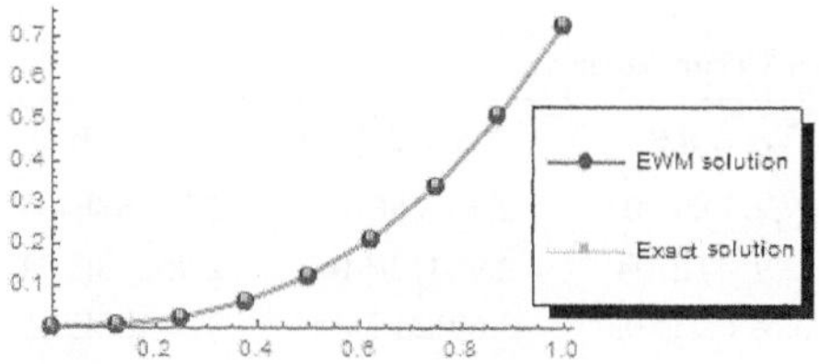

FIGURE 7.26 Numerical results for $\alpha = 0.75$.

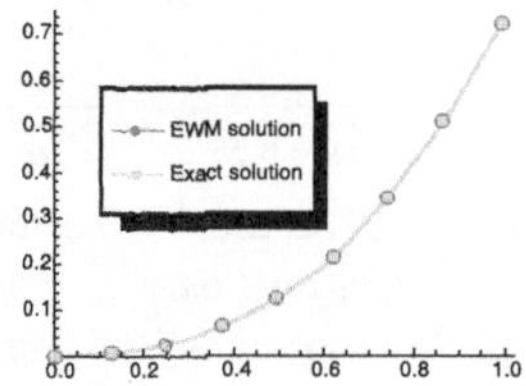

FIGURE 7.27 Numerical results for $\alpha = 0.8$.

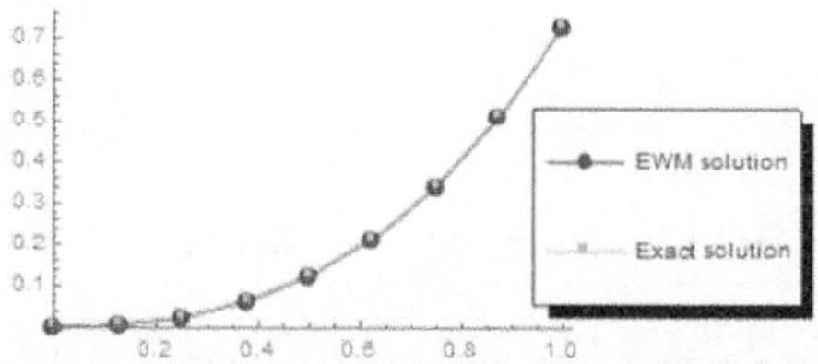

FIGURE 7.28 Numerical results for $\alpha = 0.9$.

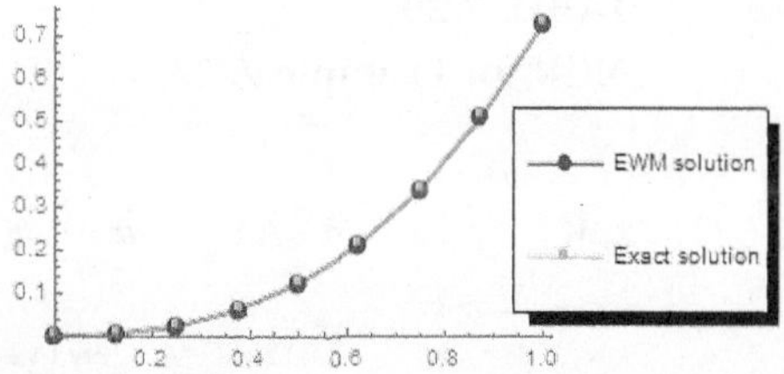

FIGURE 7.29 Numerical results for $\alpha = 1$.

For $k = 3$, $M = 3$, Figures 7.30–7.33 represent the error analysis by EWM for $\alpha = 0.75, 0.8, 0.9, 1$.

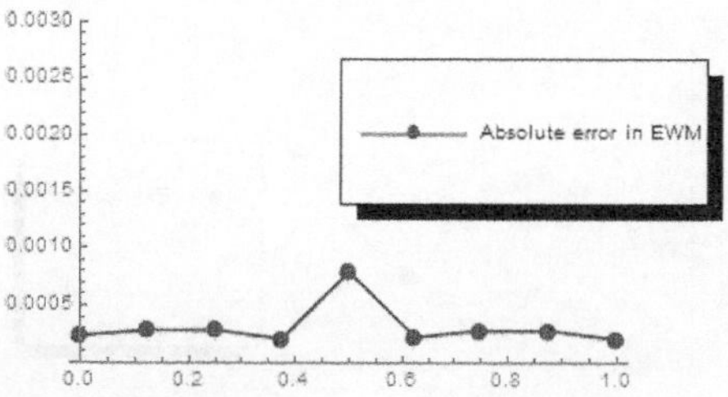

FIGURE 7.30 Error distribution for $\alpha = 0.75$.

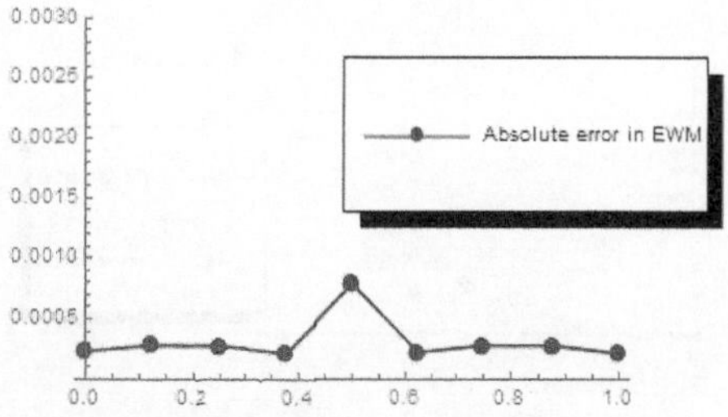

FIGURE 7.31 Error distribution for $\alpha = 0.8$.

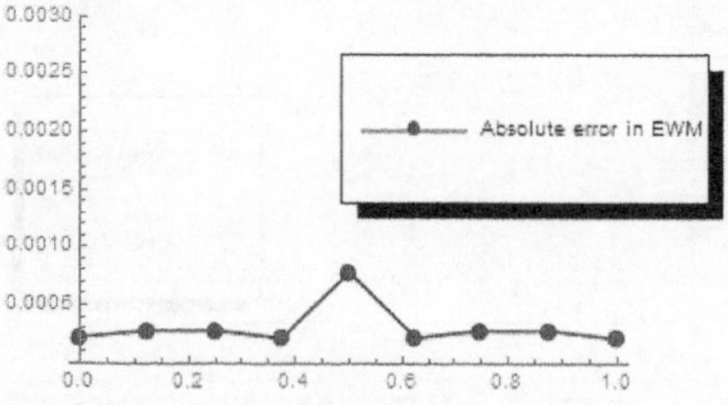

FIGURE 7.32 Error distribution for $\alpha = 0.9$.

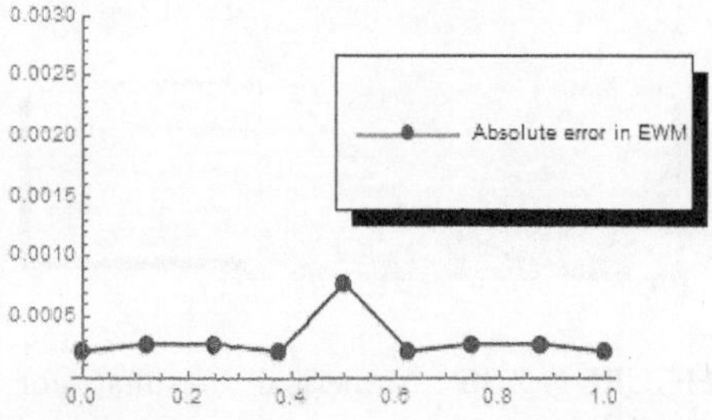

FIGURE 7.33 Error distribution for $\alpha = 1$.

TABLE 7.28
L^2-Error Norm of Example 7.7.7 for $M = 3$ and Different Values of k

k, M	L^2 – Error Norm				
	$\alpha = 0.5$	$\alpha = 0.75$	$\alpha = 0.8$	$\alpha = 0.9$	$\alpha = 1$
$k = 2$, $M = 3$	2.3826E-03	2.36664E-03	2.36507E-03	2.36326E-03	2.36268E-03
$k = 3$, $M = 3$	2.96496E-04	2.955E-04	2.9541E-04	2.95312E-04	2.95288E-04
$k = 4$, $M = 3$	3.69938E-05	3.69239E-05	3.69182E-05	3.69121E-05	3.69106E-05
$k = 5$, $M = 3$	4.61969E-06	4.61468E-06	4.6143E-06	4.61392E-06	4.61382E-06

TABLE 7.29
NCR for Example 7.7.7

k, M	*NCR*				
	$\alpha = 0.5$	$\alpha = 0.75$	$\alpha = 0.8$	$\alpha = 0.9$	$\alpha = 1$
$k = 2$, $M = 3$	–	–	–	–	–
$k = 3$, $M = 3$	3.01265	2.97112	2.96482	2.95603	2.95246
$k = 4$, $M = 3$	3.01571	2.99163	2.98886	2.98575	2.98535
$k = 5$, $M = 3$	3.01224	2.99792	2.99662	2.99538	2.99544

For $k = 4$, $M = 3$, Figures 7.34–7.37 represent the numerical results obtained by EWM for $\alpha = 0.75, 0.8, 0.9, 1$.

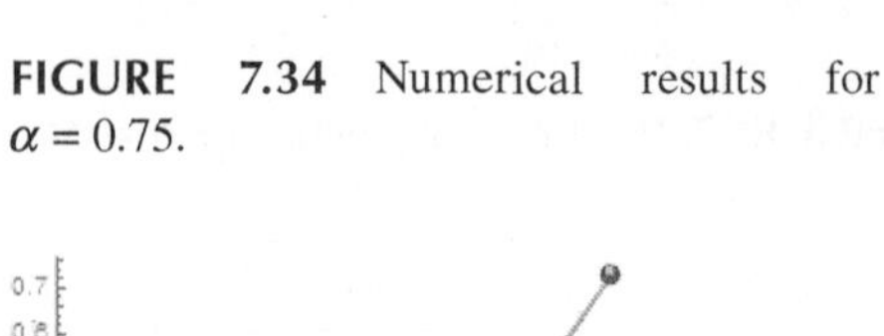

FIGURE 7.34 Numerical results for $\alpha = 0.75$.

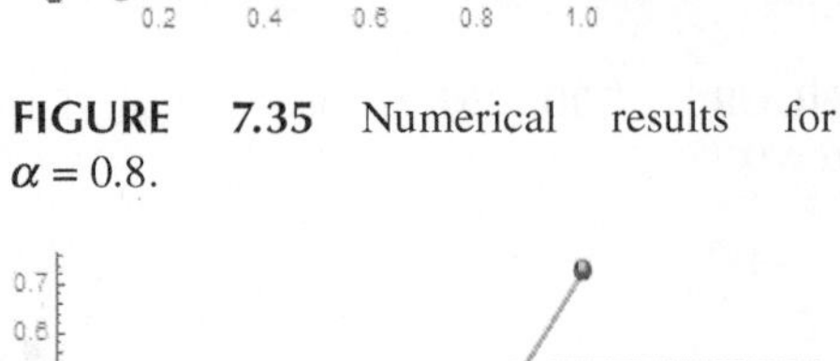

FIGURE 7.35 Numerical results for $\alpha = 0.8$.

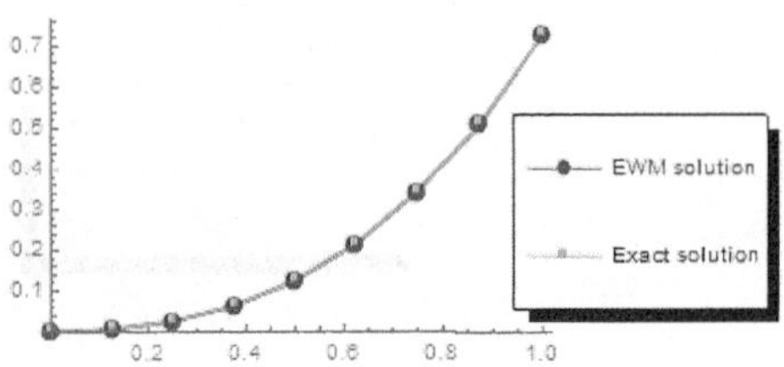

FIGURE 7.36 Numerical results for $\alpha = 0.9$.

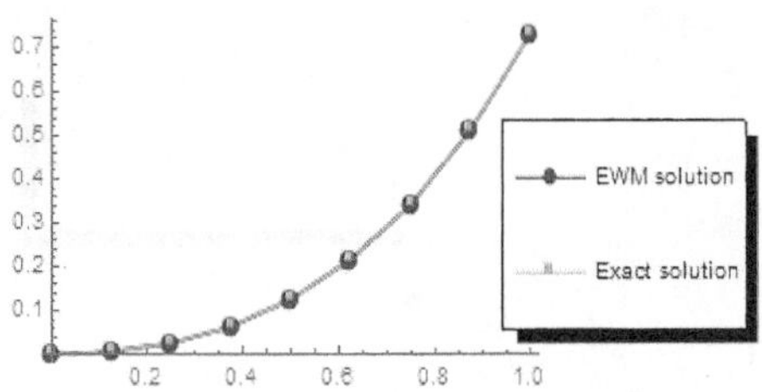

FIGURE 7.37 Numerical results for $\alpha = 1$.

For $k=4, M=3$, Figures 7.38–7.41 represent the error analysis by EWM for $\alpha=0.75,0.8,0.9,1$.

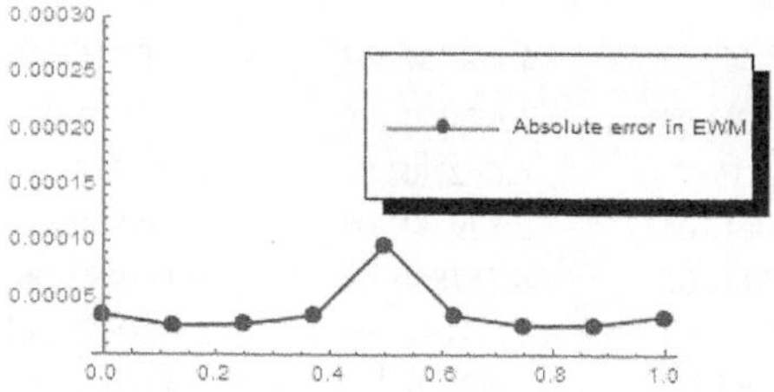

FIGURE 7.38 Error distribution for $\alpha=0.75$.

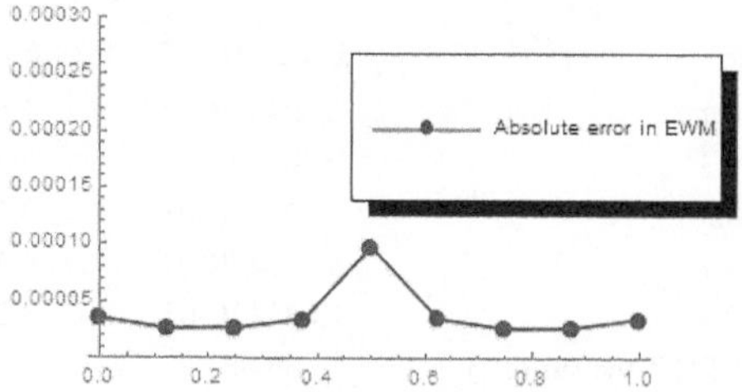

FIGURE 7.39 Error distribution for $\alpha=0.8$.

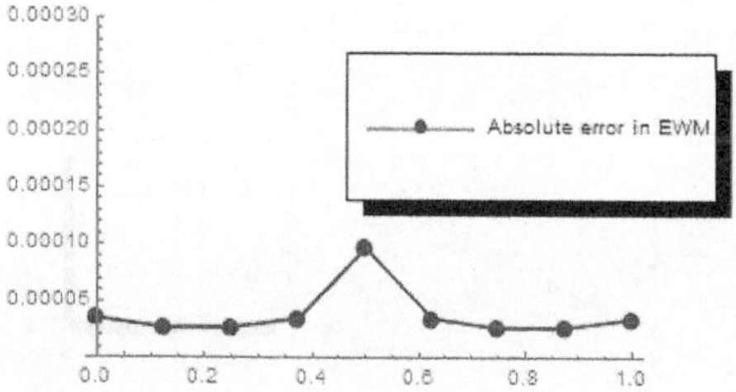

FIGURE 7.40 Error distribution for $\alpha=0.9$.

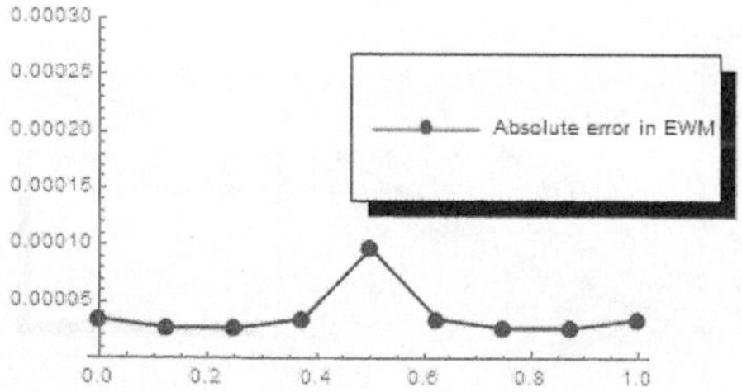

FIGURE 7.41 Error distribution for $\alpha=1$.

Example 7.7.8

Let us consider the following fractional PVDIDE [70]:

$$
{}_0^C D_x^\alpha u(x)=\frac{1}{2}-\frac{x}{4}e^{\frac{x}{4}}+\frac{x^2}{32}-\frac{e^{3x}}{2}+e^{2x}+\frac{1}{2}u(x)+u\left(\frac{x}{4}\right)+\int_0^x e^{x+z}u(z)\,dz+\int_0^{\frac{x}{4}} zu(z)\,dz,
$$

$$
u(0)=0. \tag{7.68}
$$

The exact solution is $u(x)=e^x-1$ when $\alpha=1$.

The values of $\tilde{E}(x)$ obtained by the EWM are presented in Table 7.30 for a fixed M and different values of k. Figures 7.42–7.49 display the numerical solutions and absolute error analysis by EWM. Figures 7.50 and 7.51 depict the comparison of the exact solution with the approximate solution obtained by different values of α. Table 7.31 lists the MAE norm and NCR for different values of k and M. As the size of k increases, it attains an approximately third-order convergence rate. Moreover, the comparison of MAEs is presented in Table 7.32.

TABLE 7.30
$\tilde{E}(x)$ Errors by EWM for Example 7.7.8 with $\alpha = 1$

x	$k = 2, M = 3$	$k = 3, M = 3$	$k = 4, M = 3$	$k = 5, M = 3$
0.1	3.67842E-04	3.51132E-05	6.67464E-06	5.9719E-07
0.3	5.3689E-04	6.5021E-05	6.69278E-06	1.00221E-06
0.5	2.32103E-03	2.36943E-04	2.87939E-05	3.46585E-06
0.7	5.65242E-04	9.37398E-05	8.75788E-06	1.47795E-06
0.9	1.04112E-03	1.00829E-04	1.42666E-05	1.45252E-06
CPU Time (in sec.)	83.485	153.859	826.702	4,088.31

Figures 7.42–7.45 represent the numerical results obtained by EWM for $k = 2,3,4,5$ and $M = 3$.

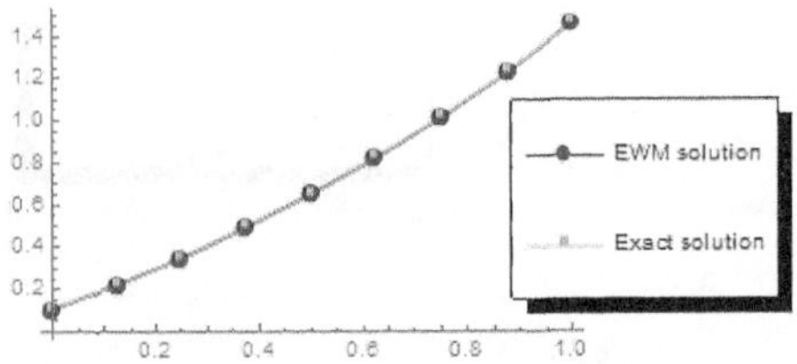

FIGURE 7.42 Numerical results for $k = 2$, $M = 3$.

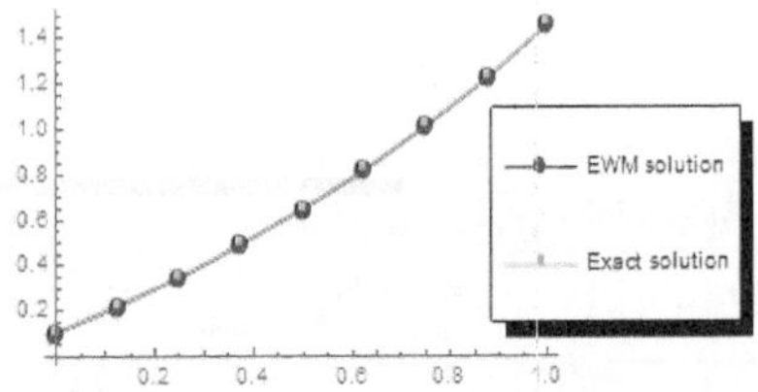

FIGURE 7.43 Numerical results for $k = 3$, $M = 3$.

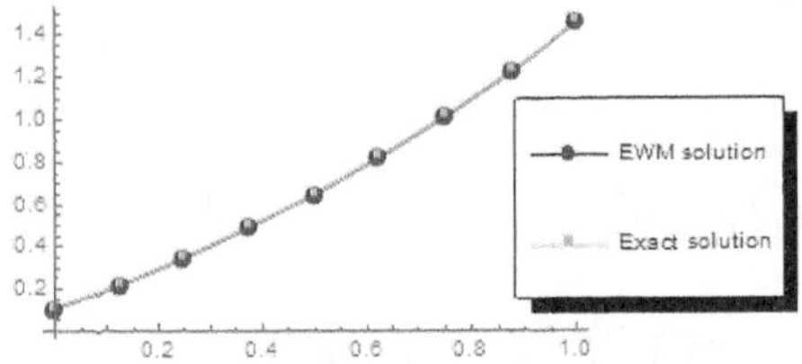

FIGURE 7.44 Numerical results for $k = 4$, $M = 3$.

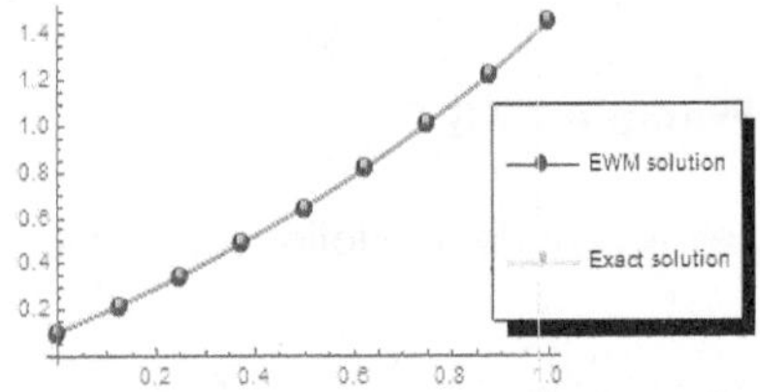

FIGURE 7.45 Numerical results for $k = 5$, $M = 3$.

Figures 7.46–7.49 represent the error analysis by EWM for $k = 2,3,4,5$ and $M = 3$.

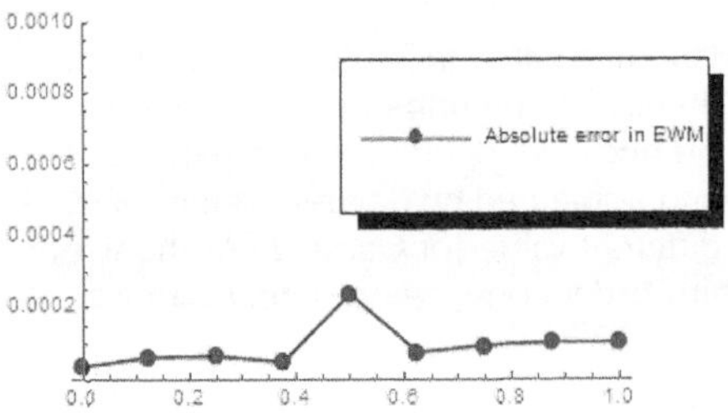

FIGURE 7.46 Error distribution for $k = 2$, $M = 3$.

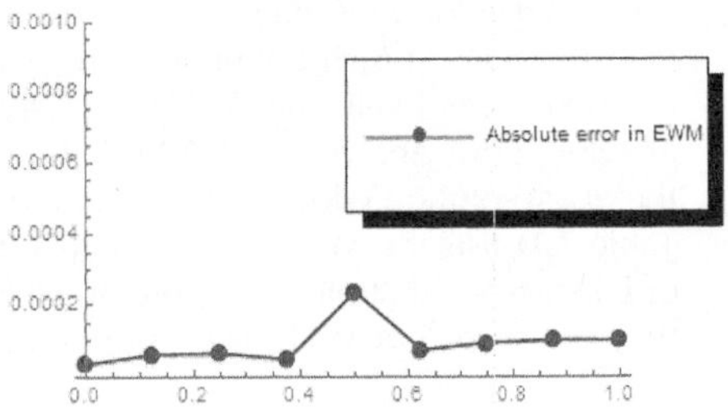

FIGURE 7.47 Error distribution for $k = 3$, $M = 3$.

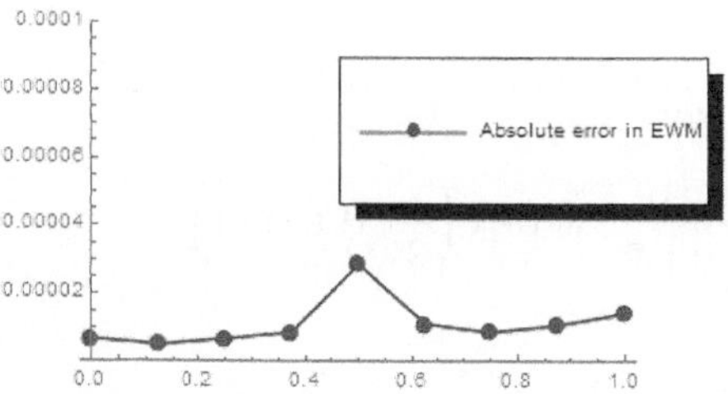

FIGURE 7.48 Error distribution for $k = 4$, $M = 3$.

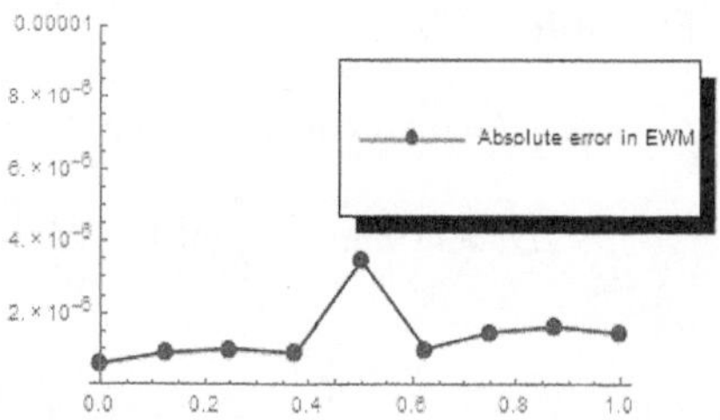

FIGURE 7.49 Error distribution for $k = 5$, $M = 3$.

Figures 7.50 and 7.51 show the comparison of numerical results for different values of α for $k = 2$, $M = 3$ and $k = 3$, $M = 3$.

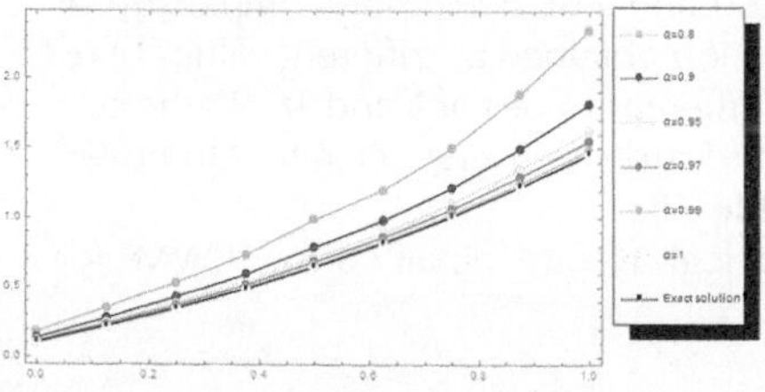

FIGURE 7.50 Comparison of numerical results for different values of α for $k = 2$, $M = 3$.

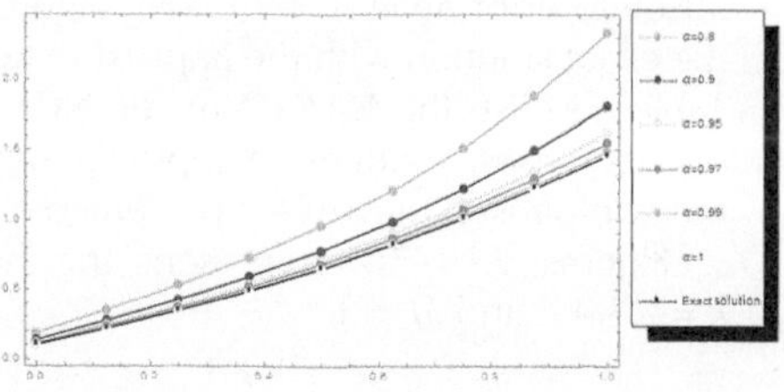

FIGURE 7.51 Comparison of numerical results for different values of α for $k = 3$, $M = 3$.

TABLE 7.31
NCR for Example 7.7.8

k, M	*MAE*	*NCR*
$k = 2$, $M = 3$	8.43157E-04	–
$k = 3$, $M = 3$	1.02972E-04	3.03355
$k = 4$, $M = 3$	1.17739E-05	3.12859
$k = 5$, $M = 3$	1.43379E-06	3.03769

TABLE 7.32
Comparison of MAE for Example 7.7.8

Method of Jafari et al. [70]		EWM	
M	*MAE*	*k, M*	*MAE*
3	2.30E-03	$k = 2$, $M = 4$	2.86599E-05
5	2.83E-05	$k = 3$, $M = 4$	1.9775E-06
7	1.20E-07	$k = 4$, $M = 4$	1.32589E-07

Example 7.7.9

Consider the following fractional PVDIDE [70]:

$$ {}_0^C D_x^{\alpha} u(x) = u(x) - \frac{1}{2}\ln\left(1+\frac{x}{2}\right)u\left(\frac{x}{2}\right) + \frac{1}{1+x} - \ln(1+x)\left(\frac{x}{2}\ln(1+x)+1\right) $$

$$ + \int_0^x \frac{x}{1+z}u(z)\,dz + \int_0^{\frac{x}{2}} \frac{1}{1+z}u(z)\,dz, $$

$$ u(0) = 0. $$

The exact solution is $u(x) = \ln(1+x)$ when $\alpha = 1$.

The values of $\tilde{E}(x)$ obtained by the EWM are presented in Table 7.33 for a fixed M and different values of k. Figures 7.52–7.59 display the numerical solutions and absolute error analysis by EWM. Figures 7.60 and 7.61 depict the comparison of the exact solution with the approximate solution obtained by different values of α. Table 7.34 lists the MAE norm and NCR for different values of k and M. As the size of k increases, it attains an approximately third-order convergence rate. Moreover, the comparison of MAEs is presented in Table 7.35.

Figures 7.44–7.47 represent the numerical results obtained by EWM for $k = 2,3,4,5$ and $M = 3$.

TABLE 7.33
Absolute Error Values by EWM for Example 7.7.9 with $\alpha = 1$

x	$k = 2, M = 3$	$k = 3, M = 3$	$k = 4, M = 3$	$k = 5, M = 3$
0.1	4.25151E-04	4.4648E-05	9.52202E-06	8.9515E-07
0.3	3.93306E-04	3.81815E-05	3.82225E-06	6.89157E-07
0.5	4.21071E-04	6.0926E-05	8.53526E-06	1.13291E-06
0.7	9.95881E-05	2.18374E-05	1.94714E-06	2.90327E-07
0.9	1.26472E-04	1.16326E-05	1.61964E-06	1.62154E-07
CPU Time (in sec.)	109.361	682.737	1,148.7	6,938.34

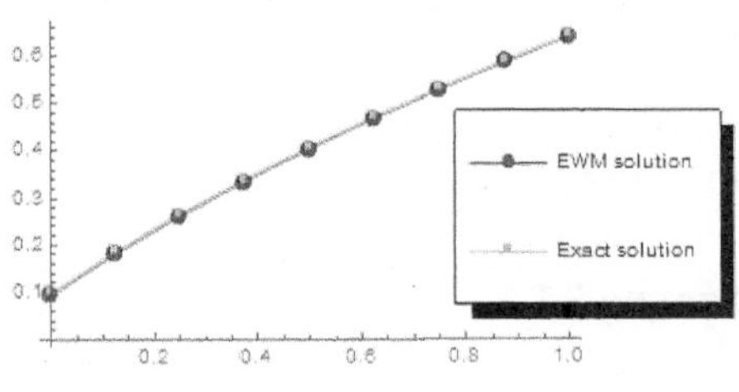

FIGURE 7.52 Numerical results for $k = 2$, $M = 3$.

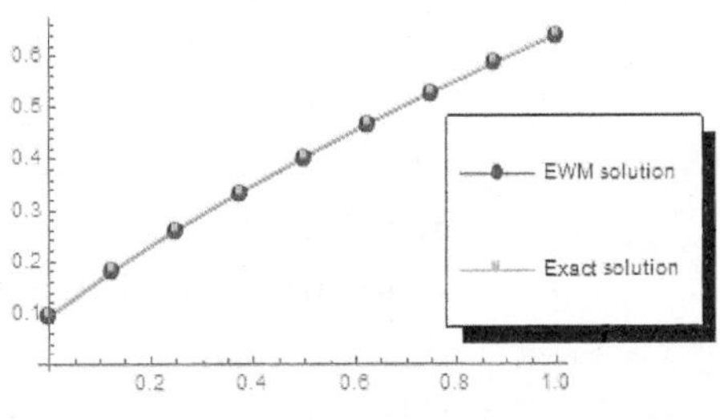

FIGURE 7.53 Numerical results for $k = 3$, $M = 3$.

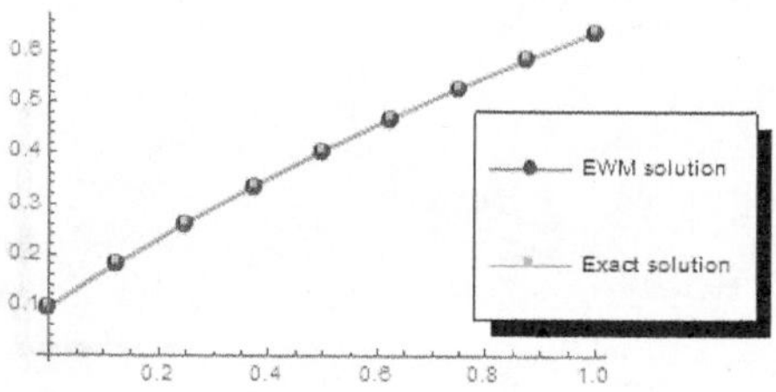

FIGURE 7.54 Numerical results for $k = 4$, $M = 3$.

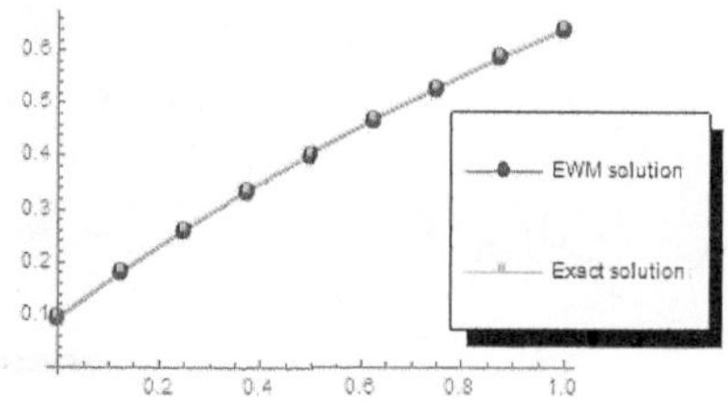

FIGURE 7.55 Numerical results for $k = 5$, $M = 3$.

Figures 7.56 and 7.57 represent the error analysis by EWM for $k = 2,3,4,5$ and $M = 3$.

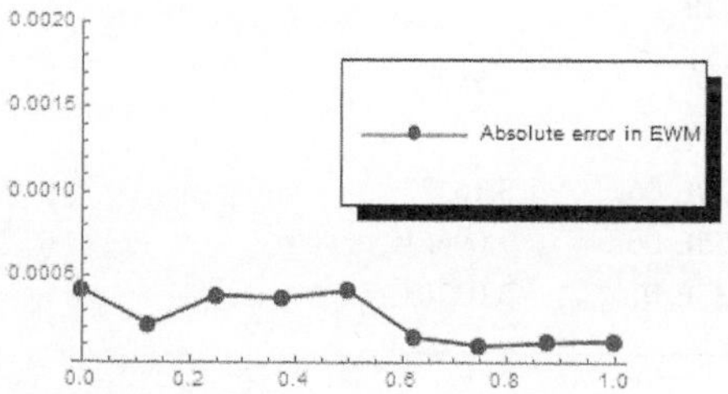

FIGURE 7.56 Error distribution for $k = 2$, $M = 3$.

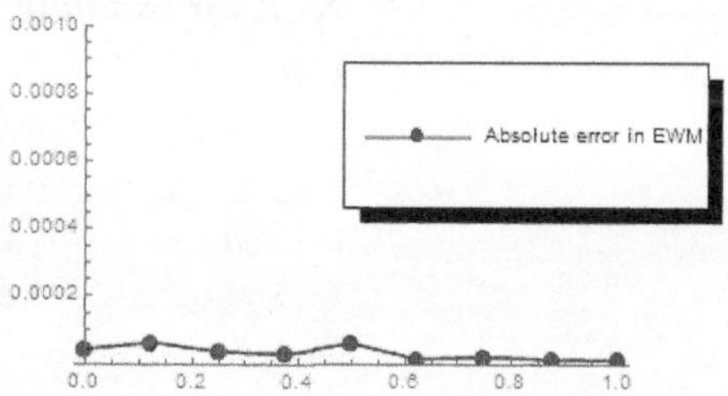

FIGURE 7.57 Error distribution for $k = 3$, $M = 3$.

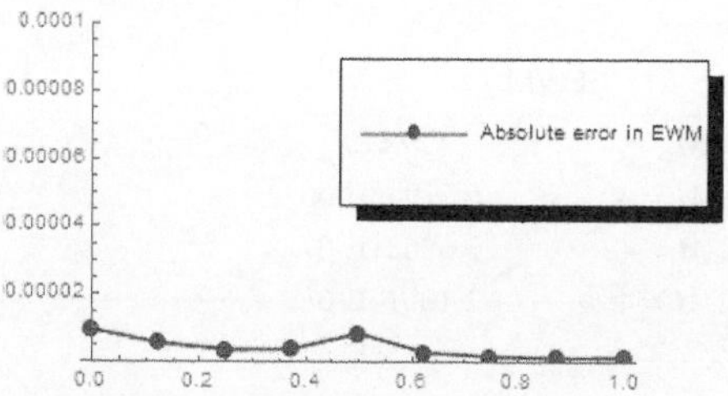

FIGURE 7.58 Error distribution for $k = 4$, $M = 3$.

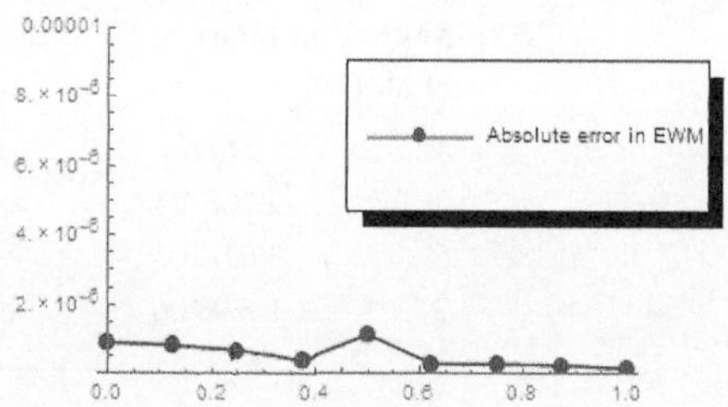

FIGURE 7.59 Error distribution for $k = 5$, $M = 3$.

Figures 7.60 and 7.61 show the comparison of numerical results for different values of α for $k = 2$, $M = 3$ and $k = 3$, $M = 3$.

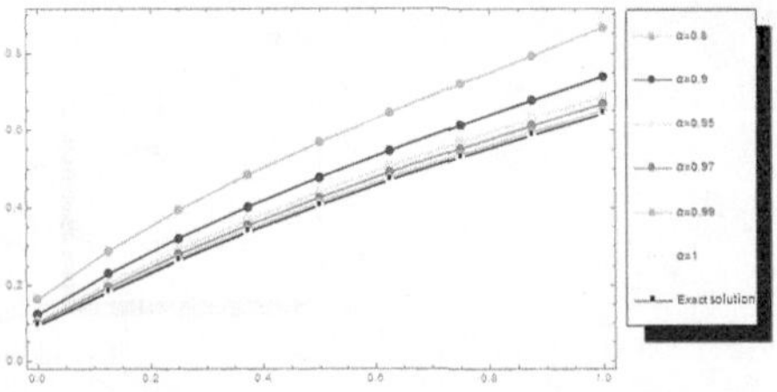

FIGURE 7.60 Comparison of numerical results for different values of α for $k = 2$, $M = 3$.

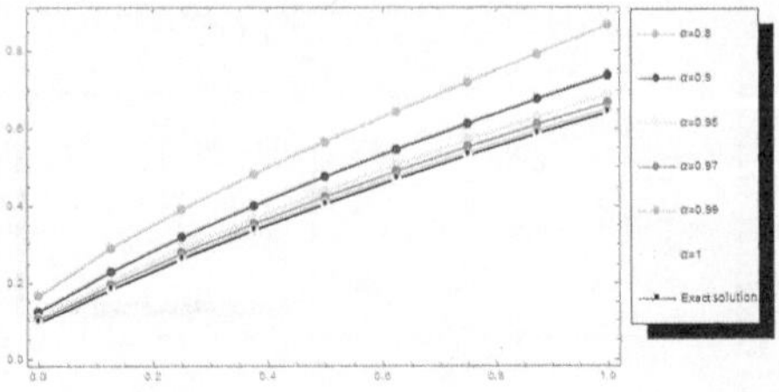

FIGURE 7.61 Comparison of numerical results for different values of α for $k = 3$, $M = 3$.

TABLE 7.34
NCR for Example 7.7.9

k, M	MAE	NCR
$k = 2$, $M = 3$	8.50155E-05	–
$k = 3$, $M = 3$	9.47339E-06	3.16577
$k = 4$, $M = 3$	1.10755E-06	3.09651
$k = 5$, $M = 3$	1.34935E-07	3.03704

TABLE 7.35
Comparison of MAE for Example 7.7.9

Method of Jafari et al. [70]		EWM	
M	MAE	k, M	MAE
3	9.96E-04	$k = 2$, $M = 4$	6.77948E-06
5	3.62E-05	$k = 3$, $M = 4$	3.68621E-07
7	1.57E-06	$k = 4$, $M = 4$	2.10514E-08

TABLE 7.36
L^2-Error Norm of Example 7.7.8 and Example 7.7.9 for $M = 3$ and Different Values of k

	L^2 – Error norm ($\alpha = 1$)	
k, M	Example 7.7.8	Example 7.7.9
$k = 2$, $M = 3$	7.11153E-04	3.14217E-04
$k = 3$, $M = 3$	8.80222E-05	4.21165E-05
$k = 4$, $M = 3$	1.09925E-05	5.37567E-06
$k = 5$, $M = 3$	1.37424E-06	6.75688E-07

7.8 CONCLUSION

In this chapter, a new algorithm has been introduced for the numerical solution of linear PVDIDE. An operational scheme based on MLWs is proposed for obtaining the approximate solution of PVDIDE. The proposed operational scheme is based on a collocation method which transforms the PVDIDEs into a system of explicit algebraic equations and hence results in desirable accurate numerical solution. In addition, the concept of Sobolev norm has been used to analyse the error estimation for the given function approximation. Furthermore, various numerical examples are carried out to demonstrate the effectiveness and performance of the proposed methodology. Furthermore, the absolute error achieved by MLWM has been compared with the other wavelets method. Moreover, the graphical representations of numerical results for different wavelets method are plotted in this chapter to establish the accuracy of the proposed new technique.

Moreover, an efficient technique has been developed for the numerical solution of fractional PVDIDE. An OM method based on EWs is proposed for obtaining the approximate solution of fractional PVDIDE. In the proposed scheme, PVDIDEs are converted to a system of linear algebraic equations by utilizing the OM of FI and properties of EWs, and that algebraic system is solved numerically. In addition, the error analysis for the acquired numerical results has been analysed by the sense of L^2 - norm. The NCR analysis is presented to justify the accuracy of the proposed method. Furthermore, some numerical experiments are conducted to establish the efficiency and validity of the proposed numerical technique. The graphical representations of numerical results indicate that the numerical solutions for the various values of α approach to the exact solutions as α approaches to 1. Moreover, the comparisons of MAE obtained by EWM with the available results of the other method confirm the supremacy of the proposed scheme.

REFERENCES

1. Wazwaz A.M., 2011, *Linear and Nonlinear Integral Equations: Methods and Applications*, Springer-Verlag, Berlin Heidelberg.
2. Milici C., Draganescu G., and Tenreiro Machado J.A., 2019, *Introduction to Fractional Differential Equations*, Springer International Publishing, Cham, Switzerland.
3. Saha Ray S., 2020, *Nonlinear Differential Equations in Physics*, Springer Nature, Singapore.
4. Oldham K., and Spanier J., 1974, *The Fractional Calculus Theory and Applications of Differentiation and Integration to Arbitrary Order*, Elsevier, Amsterdam, Netherlands.
5. Baleanu D., Jajarmi A., Asad J.H., and Blaszczyk T., 2017, "The motion of a bead sliding on a wire in fractional sense", *Acta Physica Polonica A*, **131,** pp. 1561–1564.
6. Ichise M., Nagayanagi Y., and Kojima T., 1971, "An analog simulation of non-integer order transfer functions for analysis of electrode processes", *Journal of Electroanalytical Chemistry and Interfacial Electrochemistry*, **33**(2), pp. 253–265.
7. Povstenko Y., 2010, "Signaling problem for time-fractional diffusion-wave equation in a half-space in the case of angular symmetry", *Nonlinear Dynamics*, **59**(4), pp. 593–605.
8. Saghali S., Javidi M., and Saei F.D., 2019, "Analytical solution of a fractional differential equation in the theory of viscoelastic fluids", *International Journal of Applied and Computational Mathematics*, **5**(3), pp. 1–13.

9. Baleanu D., Jajarmi A., and Hajipour M., 2018, "On the nonlinear dynamical systems within the generalized fractional derivatives with Mittag–Leffler kernel", *Nonlinear Dynamics*, **94**(1), pp. 397–414.
10. Mainardi F., 1997, *Fractional Calculus: Some Basic Problems in Continuum and Statistical Mechanics*, Springer, New York.
11. He J.H., 1999, "Some applications of nonlinear fractional differential equations and their approximations", *Bulletin of Science, Technology & Society*, **15**(2), pp. 86–90.
12. Jafari H., Ganji R.M., Nkomo N.S., and Lv Y.P., 2021, "A numerical study of fractional order population dynamics model", *Results in Physics*, **27**, p. 104456.
13. Karimi M., Zallani F., and Sayevand K., 2020, "Wavelet regularization strategy for the fractional inverse diffusion problem", *Numerical Algorithms*, **87**, pp. 1679–1705.
14. Jafari H., Ganji R.M., Sayevand K., and Baleanu D., 2021, "A numerical approach for solving fractional optimal control problems with mittag-leffler kernel", *Journal of Vibration and Control*, **28**, p. 10775463211016967.
15. Ockendon J.R., and Tayler A.B., 1971, "The dynamics of a current collection system for an electric locomotive", *Proceedings of the Royal Society of London. A. Mathematical and Physical Sciences*, **322**(1551), pp. 447–468.
16. Buhmann M., and Iserles A., 1993. "Stability of the discretized pantograph differential equation", *Mathematics of Computation*, **60**(202), pp. 575–589.
17. Yüzbaşı Ş., Gök E., and Sezer M., 2015, "Residual correction of the Hermite polynomial solutions of the generalized pantograph equations", *New Trends in Mathematical Sciences*, **3**(2), pp. 118–125.
18. Shakeri F., and Dehghan M., 2010, "Application of the decomposition method of Adomian for solving the pantograph equation of order m", *Zeitschrift für Naturforschung A*, **65**(5), pp. 453–460.
19. Ezz-Eldien S.S., 2018, "On solving systems of multi-pantograph equations via spectral tau method", *Applied Mathematics and Computation*, **321**, pp. 63–73.
20. Sezer M., Yalçinbaş S., and Gülsu M., 2008, "A Taylor polynomial approach for solving generalized pantograph equations with nonhomogenous term", *International Journal of Computer Mathematics*, **85**(7), pp. 1055–1063.
21. Yüzbaşı Ş., Gök E., and Sezer M., 2014, "Laguerre matrix method with the residual error estimation for solutions of a class of delay differential equations", *Mathematical Methods in the Applied Sciences*, **37**(4), pp. 453–463.
22. Trif D., 2012, "Direct operatorial tau method for pantograph-type equations", *Applied Mathematics and Computation*, **219**(4), pp. 2194–2203.
23. Ahmad I., Amin R., Abdeljawad T., and Shah K., 2021, "A numerical method for fractional pantograph delay integro-differential equations on haar wavelet", *International Journal of Applied and Computational Mathematics*, **7**(2), pp. 1–13.
24. Refaai D.A., El-Sheikh M.M.A., Ismail G.A., Abdalla B., and Abdeljawad T., 2021, "Hyers–Ulam stability of impulsive Volterra delay integro-differential equations", *Advances in Difference Equations*, **2021**(1), pp. 1–13.
25. Ofem A.E., Udofia U.E., and Igbokwe D.I., 2021, "A robust iterative approach for solving nonlinear volterra delay integro–differential equations", *Ural Mathematical Journal*, **7**(2), pp. 59–85.
26. Bohner M., Tunç O., and Tunç C., 2021, "Qualitative analysis of Caputo fractional integro-differential equations with constant delays", *Computational and Applied Mathematics*, **40**(6), pp. 1–17.
27. Tunç C., and Tunç O., 2018, "New qualitative criteria for solutions of Volterra integro-differential equations", *Arab Journal of Basic and Applied Sciences*, **25**(3), pp. 158–165.
28. Yao G., Tao D., and Zhang C., 2022, "A hybrid spectral method for the nonlinear Volterra integral equations with weakly singular kernel and vanishing delays", *Applied Mathematics and Computation*, **417**, p. 126780.

29. Song H., Xiao Y., and Chen M., 2021, "Collocation methods for third-kind Volterra integral equations with proportional delays", *Applied Mathematics and Computation*, **388**, p. 125509.
30. Aimene D., Baleanu D., and Seba D., 2019, "Controllability of semilinear impulsive Atangana-Baleanu fractional differential equations with delay", *Chaos, Solitons & Fractals*, **128**, pp. 51–57.
31. Dehestani H., Ordokhani Y., and Razzaghi M., 2019, "On the applicability of Genocchi wavelet method for different kinds of fractional-order differential equations with delay", *Numerical Linear Algebra with Applications*, **26**(5), p. e2259.
32. Linz P., and Wang R.L., 1992, "Error bounds for the solution of Volterra and delay equations", *Applied Numerical Mathematics*, **9**(3–5), pp. 201–207.
33. Brunner H., 1994, "Iterated collocation methods for Volterra integral equations with delay arguments", *Mathematics of Computation*, **62**(206), pp. 581–599.
34. Sahu P.K., and Saha Ray S., 2017, "A new Bernoulli wavelet method for accurate solutions of nonlinear fuzzy Hammerstein–Volterra delay integral equations", *Fuzzy Sets and Systems*, **309**, pp. 131–144.
35. Sahu P.K., and Saha Ray S., 2015, "Legendre spectral collocation method for Fredholm integro-differential-difference equation with variable coefficients and mixed conditions", *Applied Mathematics and Computation*, **268**, pp. 575–580.
36. Ezz-Eldien S.S., and Doha E.H., 2019, "Fast and precise spectral method for solving pantograph type Volterra integro-differential equations", *Numerical Algorithms*, **81**(1), pp. 57–77.
37. Behera S., and Saha Ray S., 2022, "An efficient numerical method based on Euler wavelets for solving fractional order pantograph Volterra delay-integro-differential equations", *Journal of Computational and Applied Mathematics*, **406**, p. 113825.
38. Tang Z., Tohidi E., and He F., 2020, "Generalized mapped nodal Laguerre spectral collocation method for Volterra delay integro-differential equations with noncompact kernels", *Computational and Applied Mathematics*, **39**(4), pp. 1–22.
39. Tunç C., and Tunç O., 2021, "On the stability, integrability and boundedness analyses of systems of integro-differential equations with time-delay retardation", *Revista de la Real Academia de Ciencias Exactas, Físicas y Naturales. Serie A. Matemáticas*, **115**(3), pp. 1–17.
40. Deng G., Yang Y., and Tohidi E., 2021, "High accurate pseudo-spectral Galerkin scheme for pantograph type Volterra integro-differential equations with singular kernels", *Applied Mathematics and Computation*, **396**, p. 125866.
41. Saha Ray S., and Behera S., 2020, "Two-dimensional wavelets operational method for solving Volterra weakly singular partial integro-differential equations", *Journal of Computational and Applied Mathematics*, **366**, p. 112411.
42. Zhao J., Cao Y., and Xu Y., 2016, "Legendre spectral collocation methods for Volterra delay-integro-differential equations", *Journal of Scientific Computing*, **67**(3), pp. 1110–1133.
43. Wei Y., and Chen Y., 2012, "Legendre spectral collocation methods for pantograph Volterra delay-integro-differential equations", *Journal of Scientific Computing*, **53**(3), pp. 672–688.
44. Mirzaee F., Bimesl S., and Tohidi E., 2016, "A numerical framework for solving high-order pantograph-delay Volterra integro-differential equations", *Kuwait Journal of Science*, **43**(1), pp. 69–83.
45. Sahu P.K., and Saha Ray S., 2016, "A numerical approach for solving nonlinear fractional Volterra–Fredholm integro-differential equations with mixed boundary conditions", *International Journal of Wavelets Multiresolution and Information Processing*, **14**(05), p. 1650036.
46. Zhu L., and Fan Q., 2013, "Numerical solution of nonlinear fractional-order Volterra integro-differential equations by SCW", *Communications in Nonlinear Science and Numerical Simulation*, **18**(5), pp. 1203–1213.

47. Zhu L., and Fan Q., 2012, "Solving fractional nonlinear Fredholm integro-differential equations by the second kind Chebyshev wavelet", *Communications in Nonlinear Science and Numerical Simulation*, **17**(6), pp. 2333–2341.
48. Ganji R.M., Jafari H., Kgarose M., and Mohammadi A., 2021, "Numerical solutions of time-fractional Klein-Gordon equations by clique polynomials", *Alexandria Engineering Journal*, **60**(5), pp. 4563–4571.
49. Tohidi E., Ezadkhah M.M., and Shateyi S., 2014, "Numerical solution of nonlinear fractional Volterra integro-differential equations via Bernoulli polynomials", *Abstract and Applied Analysis*, **162896**, p. 7.
50. Sayevand K., Ghanbari N., and Masti I., 2021, "A robust computational framework for analyzing the Bloch–Torrey equation of fractional order", *Computational and Applied Mathematics*, **40**(4), pp. 1–21.
51. Tuan N.H., Ganji R.M., and Jafari H., 2020, "A numerical study of fractional rheological models and fractional Newell-Whitehead-Segel equation with non-local and non-singular kernel", *Chinese Journal of Physics*, **68**, pp. 308–320.
52. Ganji R.M., Jafari H., and Baleanu D., 2020, "A new approach for solving multi variable orders differential equations with Mittag–Leffler kernel", *Chaos, Solitons & Fractals*, **130**, p. 109405.
53. Sayevand K., Machado J.T., and Moradi V., 2019, "A new non-standard finite difference method for analyzing the fractional Navier–Stokes equations", *Computers & Mathematics with Applications*, **78**(5), pp. 1681–1694.
54. Yang Y., and Huang Y., 2013, "Spectral-collocation methods for fractional pantograph delay-integrodifferential equations", *Advances in Mathematical Physics*, 2013(821327), p. 14.
55. Saha Ray S., and Behera S., 2020, "Two-dimensional wavelets operational method for solving Volterra weakly singular partial integro-differential equations", *Journal of Computational and Applied Mathematics*, **366**, p. 112411.
56. Ali I., Brunner H., and Tang T., 2009, "Spectral methods for pantograph-type differential and integral equations with multiple delays", *Frontiers of Mathematics in China*, **4**(1), pp. 49–61.
57. Jafari H., Mahmoudi M., and Skandari M.N., 2021, "A new numerical method to solve pantograph delay differential equations with convergence analysis", *Advances in Difference Equations*, 2021(1), pp. 1–12.
58. Rahimkhani P., and Ordokhani Y., 2018, "Numerical solution a class of 2D fractional optimal control problems by using 2D Müntz-Legendre wavelets", *Optimal Control Applications and Methods*, **39**(6), pp. 1916–1934.
59. Behera S., and Saha Ray S., 2022, "An efficient numerical method based on Euler wavelets for solving fractional order pantograph Volterra delay-integro-differential equations", *Journal of Computational and Applied Mathematics*, **406**, p. 113825.
60. Behera S., and Saha Ray S., 2020, "An operational matrix based scheme for numerical solutions of nonlinear weakly singular partial integro-differential equations", *Applied Mathematics and Computation*, **367**, p. 124771.
61. Canuto C., Hussaini M.Y., Quarteroni A., and Zang T.A., 2006, *Spectral Methods: Fundamentals in Single Domains*, Springer, Berlin.
62. Marzban H.R., Tabrizidooz H.R., and Razzaghi M., 2011, "A composite collocation method for the nonlinear mixed Volterra–Fredholm–Hammerstein integral equations", *Communications in Nonlinear Science and Numerical Simulation*, **16**(3), pp. 1186–1194.
63. Mashayekhi S., Razzaghi M., and Wattanataweekul M., 2016, "Analysis of multi-delay and piecewise constant delay systems by hybrid functions approximation", *Differential Equations and Dynamical Systems*, **24**(1), pp. 1–20.

64. Dehestani H., Ordokhani Y., and Razzaghi M., 2020, "Modified wavelet method for solving fractional variational problems", *Journal of Vibration and Control*, **27**(5–6), pp. 582–596.
65. Dehghan M., and Shakeri F., 2008, "The use of the decomposition procedure of Adomian for solving a delay differential equation arising in electrodynamics", *Physica Scripta*, **78**(6), p. 065004.
66. Wikipedia, The Free Encyclopedia. www.wikipedia.org.
67. Zhao J., Cao Y., and Xu Y., 2017, "Sinc numerical solution for pantograph Volterra delay-integro-differential equation", *International Journal of Computer Mathematics*, **94**(5), pp. 853–865.
68. Jalilian Y., and Ghasemi M., 2017, "On the solutions of a nonlinear fractional integro-differential equation of pantograph type", *Mediterranean Journal of Mathematics*, **14**(5), p. 194.
69. Zhao J., Cao Y., and Xu Y., 2020, "Tau approximate solution of linear pantograph Volterra delay-integro-differential equation", *Computational and Applied Mathematics*, **39**(2), pp. 1–15.
70. Jafari H., Tuan N.A., and Ganji R.M., 2021, "A new numerical scheme for solving pantograph type nonlinear fractional integro-differential equations", *Journal of King Saud University-Science*, **33**(1), p. 101185.

8 Numerical Solutions of Fractional-Order Integro-Differential Equations with Weakly Singular Kernels

8.1 INTRODUCTION

Recently, it has been studied that fractional-order derivatives and integrations are very suitable phenomena for modelling various branches of science and engineering [1]. Fractional calculus plays a significant role in the integration and differentiation of functions with non-integer orders [2,3]. The concept of fractional derivative was investigated by two mathematicians, Leibniz and L'Hopital, in 1695 [4]. Fractional integro-differential equations (FIDEs) have emerged in the physical and mathematical modelling of various processes, and therefore, numerical analyses of fractional differential equations and fractional integral equations have been receiving special attention from many researchers [5,6]. FIDEs with weakly singular kernels are applied to solve a wide range of physical problems, including heat conduction problems [7], elasticity and fracture mechanics [8], and radiative equilibrium [9]. Various applications of fractional calculus have been seen in the field of diffusion problems [10,11], physical systems [12], electromagnetic waves [13], optimization [14], viscoelasticity [15], wave equation [16], fluid dynamic traffic [17], time-fractional problems [18,19], nonlinear dynamical systems [20], and viscoelastic fluids [21].

Several numerical methods were developed by many researchers for solving the fractional-order differential and integral equations. Examples include homotopy analysis method [22], reproducing kernel method [23], variational iteration method [24], Adomian decomposition method [25], collocation method [26], discrete Galerkin method [27], and wavelet method [28,29]. Moreover, various numerical techniques have been introduced for the solution of FIDEs such as collocation method [30], Adomian decomposition method [31], operational Jacobi Tau method [32], shifted Lucas polynomials method [33], and wavelet method [34–36].

The numerical solutions of FIDEs have been studied by several authors such as Keshavaraz and Ordokhani [37], Sahu and Saha Ray [38], and Heydari et al. [39]. Saha Ray et al. [40] have solved nonlinear fractional differential equations by the Adomian decomposition method. In an earlier study [41], Behera and Saha Ray have used the Euler wavelets method approach for solving fractional Volterra–Fredholm

DOI: 10.1201/9781003453451-8

integro-differential equation with weakly singular kernels. Zhao et al. [42] have solved the FIDEs with weakly singular kernels by applying collocation method. Yi et al. [43,44] have employed the CAS wavelets and Legendre wavelets method for solving the linear and nonlinear fractional weakly singular integro-differential equations. Nemati et al. [45] have introduced a modification of hat functions for the solution of nonlinear FIDEs with weakly singular kernels. Pedas et al. [46] have solved fractional weakly singular integro-differential equations by using spline collocation method. The Legendre wavelets method has been introduced for the solution of fractional partial integro-differential viscoelastic equations with weakly singular kernel [47].

8.2 OUTLINE OF THE PRESENT STUDY

The main objective of this chapter is to establish a fractional-order operational matrix method based on Euler wavelets (EWs) for solving linear Volterra–Fredholm integro-differential equations (VFIDEs) with a weakly singular kernel. The operational matrix of fractional integration (FI) based on EWs has been constructed, and by using the operational matrix of integration of fractional order, the weakly singular linear fractional Volterra–Fredholm integro-differential equations (FVFIDEs) are reduced to a system of algebraic equations. Moreover, the convergence and error analyses of the proposed technique have been investigated. Furthermore, some numerical examples are solved to illustrate the applicability and efficiency of the proposed approach.

Moreover, a wavelet-based approach has been employed for the solution of linear and nonlinear fractional-order Volterra integro-differential equations (VIDEs) with weakly singular kernels. The fractional-order integral operational matrix has been constructed based on the Taylor wavelets (TWs). The fractional integral operational matrix is utilized to transform linear and nonlinear fractional Volterra integro-differential equations (FVIDEs) to the linear and nonlinear algebraic systems of equations, respectively. Several theorems are analysed for convergence, and also, the error analysis of the proposed scheme has been performed. Additionally, the results of numerical experiments have also been reported in both graphical and tabular forms to demonstrate the efficiency and validity of the proposed technique.

8.3 FRACTIONAL-ORDER LINEAR VOLTERRA–FREDHOLM INTEGRO-DIFFERENTIAL EQUATIONS WITH WEAKLY SINGULAR KERNELS

In this chapter, the one-dimensional (1D) fractional VFIDE involved with a weakly singular kernel has been solved by implementing the EWs algorithm. The 1D-FVFIDE is of the following form:

$$ {}_0^C D_x^{\alpha} u(x) = \lambda_1 \int_0^x \frac{u(z)}{(x-z)^{\beta}} dz + \lambda_2 \int_0^1 k(x,z) u(z) dz + f(x), \tag{8.1} $$

subject to the initial condition

$$u(0) = 0, \tag{8.2}$$

where $u(x), f(x) \in L^2[0,1)$ are unknown and known functions, respectively. $k(x,z)$ is the kernel function defined in $L^2([0,1)\times[0,1))$, and λ_1, λ_2 are real constants. Here, $0 < \alpha, \beta < 1$ and ${}_0^C D_x^\alpha$ denote the Caputo fractional-order derivative of order α.

8.3.1 Function Approximation by EWs

Consider a function $g(x) \in L^2[0,1)$, which can be defined in terms of the EWs as follows:

$$g(x) \approx \sum_{n=1}^{2^{k-1}} \sum_{m=0}^{M-1} \tilde{g}_{n,m} \psi_{n,m}(x) = G^T \Psi(x), \tag{8.3}$$

where G and $\Psi(x)$ are vectors of dimension $(2^{k-1}M \times 1)$ given by

$$G = [\tilde{g}_{1,0}, \tilde{g}_{1,1}, \ldots, \tilde{g}_{1,M-1}, \ldots, \tilde{g}_{2^{k-1},0}, \tilde{g}_{2^{k-1},1}, \ldots \tilde{g}_{2^{k-1},M-1}]^T \tag{8.4}$$

and

$$\Psi(x) = [\psi_{1,0}(x), \psi_{1,1}(x), \ldots, \psi_{1,M-1}(x), \ldots, \psi_{2^{k-1},0}(x), \psi_{2^{k-1},1}(x), \ldots \psi_{2^{k-1},M-1}(x)]^T, \tag{8.5}$$

where $\psi_{n,m}(x)$, $n = 1, 2, \ldots, 2^{k-1}$, $m = 0, 1, \ldots, M-1$ is the EWs defined on $[0,1)$ as follows:

$$\psi_{n,m}(x) = \begin{cases} 2^{\frac{k-1}{2}} \bar{E}_m(2^{k-1}x - n + 1), & \frac{n-1}{2^{k-1}} \le x < \frac{n}{2^{k-1}}, \\ 0, & \text{otherwise}. \end{cases} \tag{8.6}$$

The coefficient vector G can be obtained from Eq. (8.3) as follows:

$$G^T = \left(\int_0^1 g(x) \Psi^T(x) dx \right) . \tilde{D}^{-1}, \tag{8.7}$$

where $\tilde{D}$ is a $(2^{k-1}M \times 2^{k-1}M)$-dimensional matrix and is given by

$$\tilde{D} = \int_0^1 \Psi(x) \Psi^T(x) \, dx. \qquad 8.8)$$

For $k = 2, M = 3$, we have

$$\tilde{D} = \begin{bmatrix} 1 & 0 & -\frac{1}{\sqrt{6}} & 0 & 0 & 0 \\ 0 & \frac{1}{3} & 0 & 0 & 0 & 0 \\ -\frac{1}{\sqrt{6}} & 0 & \frac{1}{5} & 0 & 0 & 0 \\ 0 & 0 & 0 & 1 & 0 & -\frac{1}{\sqrt{6}} \\ 0 & 0 & 0 & 0 & \frac{1}{3} & 0 \\ 0 & 0 & 0 & -\frac{1}{\sqrt{6}} & 0 & \frac{1}{5} \end{bmatrix}.$$

Similarly, the kernel function $k(x,z)$ can be approximated as follows:

$$k(x,z) = \Psi^T(x).\tilde{\mathrm{K}}.\Psi(z), \tag{8.9}$$

where $\tilde{\mathrm{K}}$ is a $(2^{k-1}M \times 2^{k-1}M)$-dimensional matrix and is given by

$$\tilde{\mathrm{K}} = \tilde{D}^{-1}.\left(\int_0^1 \Psi(x).\left(\int_0^1 k(x,z)\Psi^T(z)dz\right)dx\right).\tilde{D}^{-1}.$$

8.3.2 Operational Matrices

Now, we construct the operational matrix (OM) of the FI and operational matrix of integration with singularity for the approached method to solve the FVFIDEs.

8.3.2.1 Operational Matrix of the Fractional Integration

The integration of the EW vector $\Psi(x)$ defined in Eq. (8.5) can be determined as follows:

$$\int_0^x \Psi(z)\,dz = P\Psi(x), \tag{8.10}$$

where P is the $(2^{k-1}M \times 2^{k-1}M)$-dimensional OM of integration [48].

Similarly, the FI of the EW vector $\Psi(x)$ is defined as follows:

$$J^{\alpha}\Psi(x) = P^{\alpha}\Psi(x), \tag{8.11}$$

where P^{α} is the $(2^{k-1}M \times 2^{k-1}M)$-dimensional OM of the FI.

From Riemann–Liouville fractional integral definition [49], it follows that

$$J^{\alpha}\Psi(x)=\frac{1}{\Gamma(\alpha)}\int_0^x (x-z)^{\alpha-1}\Psi(z)\,dz. \tag{8.12}$$

Now from Eqs. (8.11) and (8.12), the matrix P^{α} can be obtained as follows:

$$P^{\alpha}=\left(\int_0^1\left(\frac{1}{\Gamma(\alpha)}\int_0^x (x-z)^{\alpha-1}\Psi(z)\,dz\right).\Psi^T(x)dx\right).\tilde{D}^{-1}. \tag{8.13}$$

For $k=2, M=2$ and $\alpha=0.5$, we have

$$P^{0.5}=\begin{bmatrix} 0.531923 & 0.319154 & 0.440659 & -0.154879 \\ -0.106385 & 0.227967 & 0.0516265 & -0.0561003 \\ 0 & 0 & 0.531923 & 0.319154 \\ 0 & 0 & -0.106385 & 0.227967 \end{bmatrix}.$$

8.3.2.2 Operational Matrix of the Integration with Singularity

Let $\Psi(x)$ be 1D EWs defined as in Eq. (8.5) for $0<\beta<1$, and then, we get

$$\int_0^x \frac{\Psi(z)}{(x-z)^{\beta}}dz=\tilde{P}\Psi(x), \tag{8.14}$$

where $\tilde{P}$ is the $(2^{k-1}M\times 2^{k-1}M)$-dimensional OM of integration with singularity, and it can be obtained as follows:

$$\tilde{P}=\left(\int_0^1\left(\int_0^x \frac{\Psi(z)}{(x-z)^{\beta}}dz\right).\Psi^T(x)dx\right).\tilde{D}^{-1}. \tag{8.15}$$

For $k=2, M=2$, and $\beta=0.5$, we have

$$\tilde{P}=\begin{bmatrix} 0.942809 & 0.565685 & 0.781049 & -0.274517 \\ -0.188562 & 0.404061 & 0.0915055 & -0.0994352 \\ 0 & 0 & 0.942809 & 0.565685 \\ 0 & 0 & -0.188562 & 0.404061 \end{bmatrix}.$$

8.3.3 Euler Wavelet Method for FVFIDE

Here, the FVFIDE Eq. (8.1) is converted into a system of algebraic equations by implementing the proposed 1D EWs.

Now approximating the following functions with respect to 1D EWs yields

$$ {}_{0}^{C}D_{x}^{\alpha}u(x) \approx \Psi^{T}(x)\tilde{C}_{1} = \tilde{C}_{1}^{T}\Psi(x), \tag{8.16}$$

$$ f(x) \approx \Psi^{T}(x)F = F^{T}\Psi(x), \tag{8.17}$$

$$ u(0) \approx \Psi^{T}(x)U = U^{T}\Psi(x), \tag{8.18}$$

$$ k(x,z) \approx \Psi^{T}(x)\tilde{\mathrm{K}}\Psi(z) = \Psi^{T}(z)\tilde{\mathrm{K}}^{T}\Psi(x) \tag{8.19}$$

Now integrating Eq. (8.16) with respect to x and by using Eq. (8.11), we have

$$\begin{aligned} u(x) &= u(0) + \tilde{C}_{1}^{T}P^{\alpha}\Psi(x) \\ &= U^{T}\Psi(x) + \tilde{C}_{1}^{T}P^{\alpha}\Psi(x) \\ &= (U^{T} + \tilde{C}_{1}^{T}P^{\alpha})\Psi(x) \\ &= \Psi^{T}(x)\tilde{C}_{2}, \end{aligned} \tag{8.20}$$

where $\tilde{C}_{2} = U + P^{\alpha T}\tilde{C}_{1}$.

The integral parts of Eq. (8.1) can be defined as follows:

$$\begin{aligned} \int_{0}^{x} \frac{u(z)}{(x-z)^{\beta}}\,dz &= \int_{0}^{x} \frac{\tilde{C}_{2}^{T}\Psi(z)}{(x-z)^{\beta}}\,dz \\ &= \tilde{C}_{2}^{T}.\int_{0}^{\xi} \frac{\Psi(z)}{(x-z)^{\beta}}\,dz \\ &= \tilde{C}_{2}^{T}.\tilde{P}.\Psi(x) \\ &= \tilde{C}_{3}^{T}.\Psi(x), \end{aligned} \tag{8.21}$$

where $\tilde{C}_{3} = \tilde{P}^{T}\tilde{C}_{2}$.

Similarly,

$$\begin{aligned} \int_{0}^{1} k(x,z)u(z)\,dz &= \int_{0}^{1} \Psi^{T}(x)\tilde{\mathrm{K}}\Psi(z)\Psi^{T}(z)\tilde{C}_{2}\,dz \\ &= \Psi^{T}(x)\tilde{\mathrm{K}}.\int_{0}^{1} \Psi(z)\Psi^{T}(z)\,dz.\tilde{C}_{2} \\ &= \Psi^{T}(x).\tilde{\mathrm{K}}.\tilde{D}.\tilde{C}_{2} \\ &= \Psi^{T}(x)\tilde{C}_{4}, \end{aligned} \tag{8.22}$$

where $\tilde{C}_4 = \tilde{\mathrm{K}}.\tilde{D}.\tilde{C}_2$.

Now by substituting Eqs. (8.16)–(8.22) into Eq. (8.1), it follows

$$\Psi^T(x)\tilde{C}_1 = \lambda_1\Psi^T(x)\tilde{C}_3 + \lambda_2\Psi^T(x)\tilde{C}_4 + \Psi^T(x)F. \tag{8.23}$$

This implies that

$$\tilde{C}_1 = \lambda_1\tilde{C}_3 + \lambda_2\tilde{C}_4 + F. \tag{8.24}$$

Equation (8.24) is a linear system of $2^{k-1}M$ - algebraic equations. By solving the above system of equations, the value of $\tilde{C}_1$ can be obtained. Therefore, by plugging the value of $\tilde{C}_1$ in Eq. (8.20), the approximate solution of FVFIDE Eq. (8.1) can be computed numerically.

8.3.4 Error Analysis

This section deals with the error analysis for EWs which have been derived by using the L^2-norms. Let $f(x)$ be any continuous function defined on the interval $[0,1)$. The L^2-norm for the function $f(x)$ is defined as follows:

$$\|f(x)\|_2^2 = \int_0^1 |f(x)|^2\, dx.$$

Theorem 8.3.1

Let us consider the function ${}_0^C D_x^\alpha \tilde{u}_{k,M}(x) \in L^2[0,1)$ obtained by using EWs, as the approximation of ${}_0^C D_x^\alpha u(x)$ and $\left|{}_0^C D_x^\alpha u(x)\right| \le \tilde{\mathfrak{R}}$, where $\tilde{\mathfrak{R}}$ is a positive constant, and then, we have the following estimation:

$$\left\|{}_0^C D_x^\alpha u(x) - {}_0^C D_x^\alpha \tilde{u}_{k,M}(x)\right\|_2^2 \le \sum_{n=2^{k-1}+1}^{\infty}\sum_{m=M}^{\infty}\left(\frac{\tilde{A}}{2^{\frac{k-1}{2}}}\tilde{\mathfrak{R}}\frac{32(m!)(1+2^{m+1})}{(2\pi)^{m+2}}\right)^2.\tilde{D}.$$

Proof:

Any function ${}_0^C D_x^\alpha u(x) \in L^2[0,1)$ can be defined by the EWs as follows:

$${}_0^C D_x^\alpha u(x) = \sum_{n=1}^{\infty}\sum_{m=0}^{\infty}\tilde{c}_{n,m}\psi_{n,m}(x).$$

If ${}_0^C D_x^\alpha \tilde{u}_{k,M}(x)$ be the truncated series, then it follows

$${}_0^C D_x^\alpha u(x) - {}_0^C D_x^\alpha \tilde{u}_{k,M}(x) = \sum_{n=2^{k-1}+1}^{\infty}\sum_{m=M}^{\infty}\tilde{c}_{n,m}\psi_{n,m}(x).$$

This implies that

$$\left\| {}_0^C D_x^{\alpha} u(x) - {}_0^C D_x^{\alpha} \tilde{u}_{k,M}(x) \right\|_2^2 = \left\| \sum_{n=2^{k-1}+1}^{\infty} \sum_{m=M}^{\infty} \tilde{c}_{n,m} \psi_{n,m}(x) \right\|_2^2$$

$$= \int_0^1 \left| \sum_{n=2^{k-1}+1}^{\infty} \sum_{m=M}^{\infty} \tilde{c}_{n,m} \psi_{n,m}(x) \right|^2 dx$$

$$\leq \sum_{n=2^{k-1}+1}^{\infty} \sum_{m=M}^{\infty} \left| \tilde{c}_{n,m} \right|^2 \int_0^1 \left| \psi_{n,m}(x) \right|^2 dx \tag{8.25}$$

$$\leq \sum_{n=2^{k-1}+1}^{\infty} \sum_{m=M}^{\infty} \left| \tilde{c}_{n,m} \right|^2 \int_0^1 \left| \psi_{n,m}(x) \right|^2 dx$$

$$= \sum_{n=2^{k-1}+1}^{\infty} \sum_{m=M}^{\infty} \left| \tilde{c}_{n,m} \right|^2 . \tilde{D}.$$

The term $\tilde{c}_{n,m}$ can be obtained as follows:

$$\tilde{c}_{n,m} = < {}_0^C D_x^{\alpha} u(x), \psi_{n,m}(x) >$$

$$= \int_0^1 {}_0^C D_x^{\alpha} u(x) . \psi_{n,m}(x) \, dx$$

$$= \int_{\frac{n-1}{2^{k-1}}}^{\frac{n}{2^{k-1}}} {}_0^C D_x^{\alpha} u(x) . 2^{\frac{k-1}{2}} . \tilde{A} . E_m(2^{k-1} x - n + 1) \, dx,$$

where

$$\tilde{A} = \frac{1}{\sqrt{\frac{2(-1)^{m-1}(m!)^2}{(2m)!} E_{2m+1}(0)}}.$$

Now, changing the variable $2^{k-1} x - n + 1 = t$, then

$$\left|\tilde{c}_{n,m}\right| = \frac{1}{2^{\frac{k-1}{2}}}\left|\tilde{A}\int_0^1 {}_0^C D_t^{\alpha} u\left(\frac{t+n-1}{2^{k-1}}\right).E_m(t)\,dt\right|$$
$$\leq \frac{1}{2^{\frac{k-1}{2}}}\tilde{A}\int_0^1 \left|{}_0^C D_t^{\alpha} u\left(\frac{t+n-1}{2^{k-1}}\right).E_m(t)\right| dt \tag{8.26}$$
$$\leq \frac{1}{2^{\frac{k-1}{2}}}\tilde{A}\tilde{\mathfrak{R}}\int_0^1 \left|E_m(t)\right| dt.$$

By using the relation between Euler polynomials and Bernoulli polynomials [50], we have

$$\int_0^1 \left|E_m(t)\right| dt = \int_0^1 \left|\frac{2}{m+1}B_{m+1}(t) - \frac{2^{m+2}}{m+1}B_{m+1}\left(\frac{t}{2}\right)\right| dt$$
$$\leq \frac{2}{m+1}\int_0^1 \left|B_{m+1}(t)\right| dt + \frac{2^{m+2}}{m+1}\int_0^1 \left|B_{m+1}\left(\frac{t}{2}\right)\right| dt \tag{8.27}$$
$$\leq \frac{2}{m+1}\frac{16(m+1)!}{(2\pi)^{m+2}} + \frac{2^{m+2}}{m+1}\frac{16(m+1)!}{(2\pi)^{m+2}}, \quad \text{applying the property of Bernoulli polynomials [51]}$$
$$= \frac{32(m)!(1+2^{m+1})}{(2\pi)^{m+2}}.$$

Now by substituting Eq. (8.27) into Eq. (8.26), it follows

$$\tilde{c}_{n,m} \leq \frac{1}{2^{\frac{k-1}{2}}}\tilde{A}\tilde{\mathfrak{R}}\frac{32(m)!(1+2^{m+1})}{(2\pi)^{m+2}}. \tag{8.28}$$

Now from Eq. (8.25) and Eq. (8.28), it follows

$$\left\|{}_0^C D_x^{\alpha} u(x) - {}_0^C D_x^{\alpha}\tilde{u}_{k,M}(x)\right\|_2^2 \leq \sum_{n=2^{k-1}+1}^{\infty}\sum_{m=M}^{\infty}\left(\frac{\tilde{A}}{2^{\frac{k-1}{2}}}\tilde{\mathfrak{R}}\frac{32(m!)(1+2^{m+1})}{(2\pi)^{m+2}}\right)^2.\tilde{D}.$$

Therefore, $\left\|{}_0^C D_x^{\alpha} u(x) - {}_0^C D_x^{\alpha}\tilde{u}_{k,M}(x)\right\|_2 \to 0$ when M is fixed and $k \to \infty$.

8.3.5 Numerical Experiments

This section deals with the performance of the proposed method for solving FVFIDEs. To establish the applicability and efficiency of the method, the following examples have been presented here. The accuracy of the approached technique can be demonstrated by using absolute error function definition, i.e.,

$$\hat{e}(x) = |u(x) - \tilde{u}(x)|,$$

where $\tilde{u}(x)$ and $u(x)$ are the approximate solution and exact solution, respectively.

Furthermore, the L^2-error norm for the proposed wavelet is defined as follows:

$$L^2_{\text{error}} = \left[\int_0^1 |u(x) - \tilde{u}(x)|^2\, dx\right]^{\frac{1}{2}},$$

where $u(x)$ and $\tilde{u}(x)$ are the exact and approximate solutions, respectively. The L^2-error norm of the numerical experiments has been presented in Table 8.16.

Again, to display the accuracy of the proposed method, the numerical convergence rate (NCR) analysis has been obtained by using the following formula:

$$\text{NCR} = \log_2\left(\frac{L^\infty \text{error}_{k=\alpha, M=3}}{L^\infty \text{error}_{k=\alpha+1, M=3}}\right), \quad \alpha = 2,3,4,5,$$

where

$$L^\infty \text{error} = \underset{1\le p\le 2^{k-1}M}{\text{Max}} |u(x_p) - \tilde{u}(x_p)|, \quad x_p = \frac{2.p-1}{2.2^{k-1}.M}, \quad p = 1,2,\ldots,2^{k-1}.M,$$

where $\tilde{u}(x_p)$ and $u(x_p)$ are the approximate and exact solutions, respectively, and $x_p, p = 1,2,\ldots,2^{k-1}M$ are the collocation points.

Example 8.3.1

Consider the following FVFIDE with a weakly singular kernel [43]:

$$ {}_0^C D_x^{0.25} u(x) = f(x) + \frac{1}{2}\int_0^x \frac{u(z)}{(x-z)^{0.5}}\, dz + \frac{1}{3}\int_0^1 (x-z)u(z)\, dz, \tag{8.29}$$

$$u(0) = 0.$$

The exact solution is $u(x) = x^2 + x^3$. In this problem,

$$f(x) = \frac{\Gamma(3)}{\Gamma(2.75)}x^{1.75} + \frac{\Gamma(4)}{\Gamma(3.75)}x^{2.75} - \frac{\sqrt{\pi}\Gamma(3)}{2\Gamma(3.5)}x^{2.5} - \frac{\sqrt{\pi}\Gamma(4)}{2\Gamma(4.5)}x^{3.5} - \frac{7x}{36} + \frac{3}{20}.$$

The values of $\hat{e}(x)$ obtained by the Euler wavelet method (EWM) are presented in Table 8.1 for a fixed M and different values of k. Tables 8.2–8.4 show the comparison of absolute errors between EWM, second Chebyshev wavelets method (SCWM) [29], Taylor wavelets method (TWM) [37], and CAS wavelets method (CASWM) [29] for $k=2, M=3$, $k=3, M=3$, and $k=4, M=3$, respectively. Figures 8.1–8.8 display the numerical solutions and absolute error analysis by EWM. Table 8.5 lists the L^∞-error norm and NCR for different values of k and M. As we increase the size of k, we have an approximately third-order convergence rate.

TABLE 8.1
Absolute Error Values by EWM for Example 8.3.1

x	$k=2, M=3$	$k=3, M=3$	$k=4, M=3$	$k=5, M=3$	$k=6, M=3$
0	6.92622E-03	8.16823E-04	9.34293E-05	9.76036E-06	7.59016E-07
1/6	2.80869E-03	3.08178E-04	4.17109E-05	4.86887E-06	6.4575E-07
2/6	2.35849E-03	3.46757E-04	3.8555E-05	5.26614E-06	6.11718E-07
3/6	6.77794E-03	8.22022E-04	1.01106E-05	1.25124E-05	1.55389E-06
4/6	2.85561E-03	3.04049E-04	4.27003E-05	4.86705E-06	6.50901E-07
5/6	2.36678E-03	3.47707E-04	3.86022E-05	5.25658E-06	6.1186E-07

TABLE 8.2
Comparison of Absolute Errors for Example 8.3.1 with $k=2$, $M=3$

x	EWM	SCWM [29]	TWM [37]	CASWM [29] $(m'=6)$
0	6.92622E-03	8.3542E-03	6.9313E-03	3.0586E-02
1/6	2.80869E-03	1.2599E-03	2.8075E-03	4.4076E-02
2/6	2.35849E-03	9.3654E-03	2.3585E-03	3.8707E-02
3/6	6.77794E-03	2.2406E-02	6.7790E-03	1.5703E-02
4/6	2.85561E-03	1.9585E-02	2.8546E-03	2.8551E-02
5/6	2.36678E-03	3.2596E-02	2.3680E-03	9.8881E-02

TABLE 8.3
Comparison of Absolute Errors for Example 8.3.1 with $k=3$, $M=3$

x	EWM	SCWM [29]	TWM [37]	CASWM [29] $(m'=12)$
0	8.16823E-04	1.0620E-03	8.1806E-04	1.4328E-02
1/6	3.08178E-04	1.1189E-03	3.0810E-04	2.2762E-02
2/6	3.46757E-04	1.8879E-03	3.4663E-04	1.9409E-02
3/6	8.22022E-04	4.7945E-03	8.2215E-04	6.4173E-02
4/6	3.04049E-04	5.9543E-03	3.0420E-04	1.8012E-02
5/6	3.47707E-04	8.0256E-03	3.4753E-04	5.6450E-02

TABLE 8.4
Comparison of Absolute Errors for Example 8.3.1 with $k = 4$, $M = 3$

x	EWM	SCWM [29]	TWM [37]	CASWM [29] $(m' = 24)$
0	9.34293E-05	1.4395E-04	9.3735E-05	6.3491E-03
1/6	4.17109E-05	2.2617E-04	4.1690E-05	1.1460E-02
2/6	3.8555E-05	5.9826E-04	3.8572E-05	9.6982E-03
3/6	1.01106E-05	1.1274E-03	1.0113E-05	2.9504E-03
4/6	4.27003E-05	1.4961E-03	4.2679E-05	9.6733E-03
5/6	3.86022E-05	2.2056E-03	3.8628E-05	2.9484E-02

Figures 8.1–8.4 represent the numerical results obtained by EWM for $k = 2,3,4,5$ and $M = 3$.

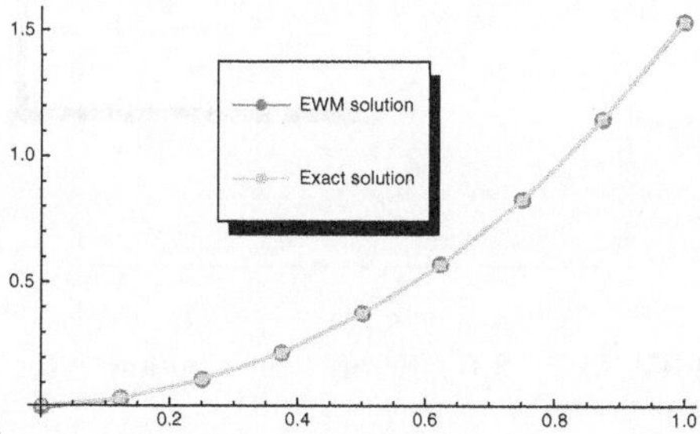

FIGURE 8.1 Numerical results for $k = 2$, $M = 3$.

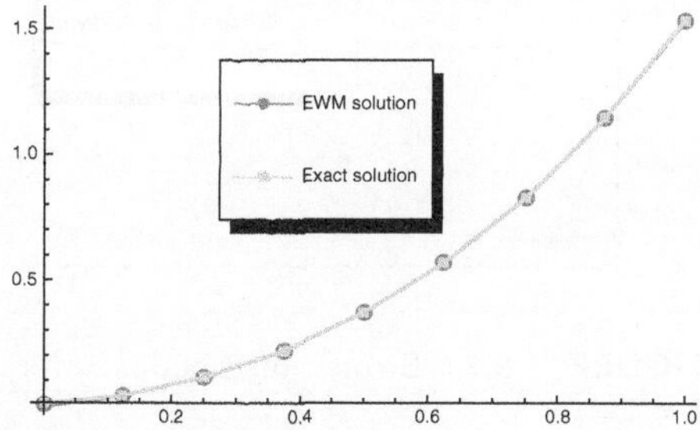

FIGURE 8.2 Numerical results for $k = 3$, $M = 3$.

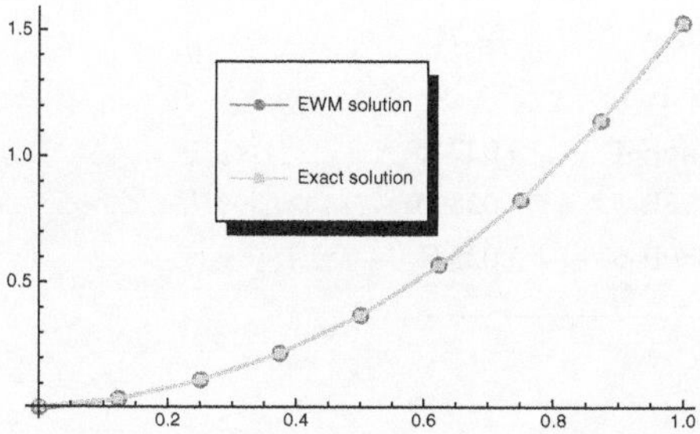

FIGURE 8.3 Numerical results for $k = 4$, $M = 3$.

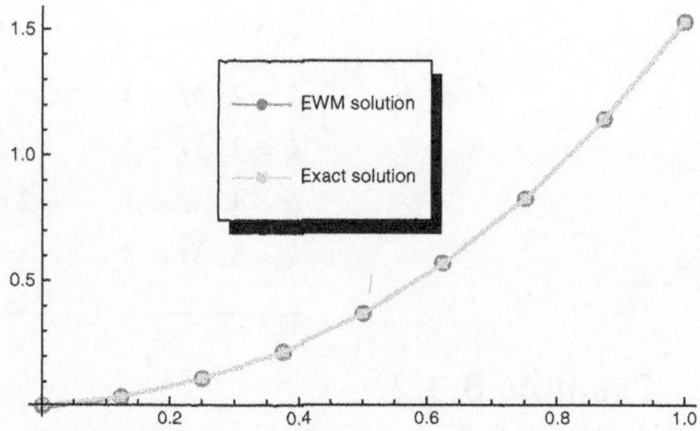

FIGURE 8.4 Numerical results for $k = 5$, $M = 3$.

Figures 8.5–8.8 represent the error analysis by EWM for $k = 2,3,4,5$ and $M = 3$.

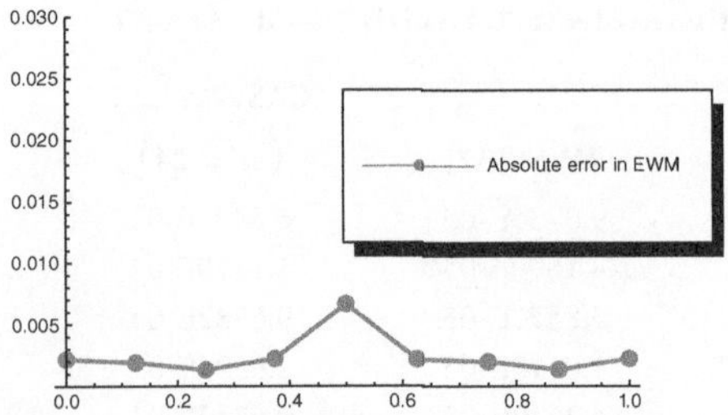

FIGURE 8.5 Error distribution for $k = 2$, $M = 3$.

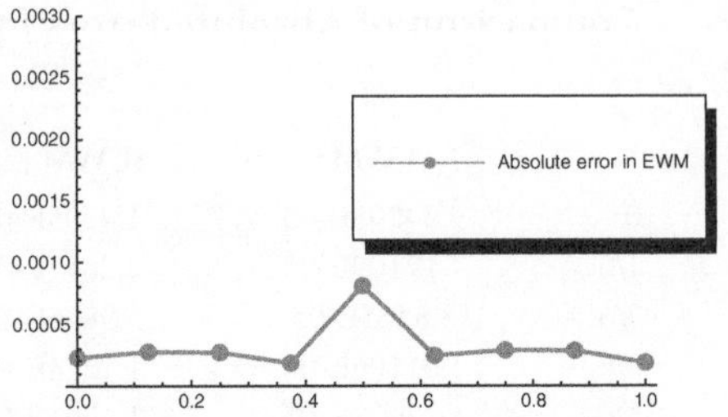

FIGURE 8.6 Error distribution for $k = 3$, $M = 3$.

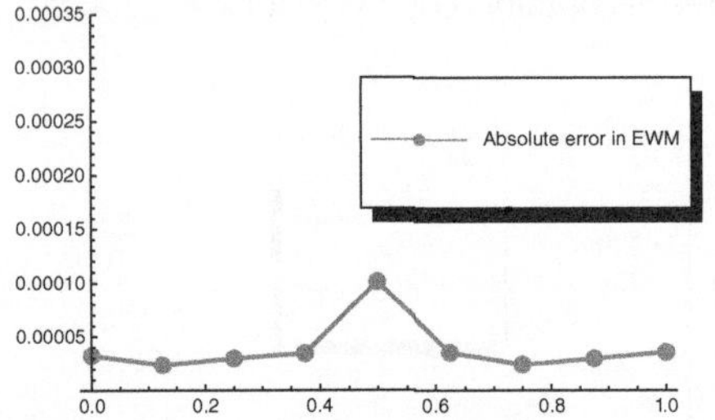

FIGURE 8.7 Error distribution for $k = 4$, $M = 3$.

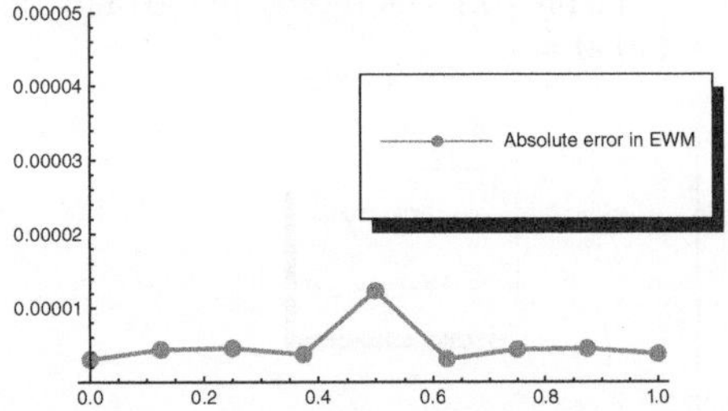

FIGURE 8.8 Error distribution for $k = 5$, $M = 3$.

TABLE 8.5
NCR for Example 8.3.1

k, M	L^∞ *error*	*NCR*
$k = 2$, $M = 3$	1.8684E-3	–
$k = 3$, $M = 3$	2.29146E-4	3.02746
$k = 4$, $M = 3$	2.80798E-5	3.02866
$k = 5$, $M = 3$	3.44869E-6	3.02541

Example 8.3.2

Consider the following FVFIDE [43]:

$$ {}_0^C D_x^{0.15} u(x) = f(x) + \frac{1}{4}\int_0^x \frac{u(z)}{(x-z)^{0.5}}\,dz + \frac{1}{7}\int_0^1 e^{x+z} u(z)\,dz, \tag{8.30}$$

$$u(0) = 0.$$

The exact solution is $u(x) = x^2 - x$. In this problem,

$$f(x) = \frac{\Gamma(3)}{\Gamma(2.85)} x^{1.85} - \frac{\Gamma(2)}{\Gamma(1.85)} x^{0.85} - \frac{\sqrt{\pi}\Gamma(3)}{4\Gamma(3.5)} x^{2.5} + \frac{\sqrt{\pi}\Gamma(2)}{4\Gamma(2.5)} x^{1.5} - \frac{e^{x+1} - 3e^x}{7}.$$

The values of $\hat{e}(x)$ obtained by the EWM are presented in Table 8.6 for a fixed M and different values of k. Tables 8.7–8.9 show the comparison of numerical solutions obtained by EWM, SCWM [29], TWM [37], and CASWM [29] for $k = 2, M = 3$, $k = 3, M = 3$, and $k = 4, M = 3$, respectively. Figures 8.9–8.16 display the numerical solutions and absolute error analysis by EWM. Table 8.10 lists the L^{∞}-error norm and NCR for different values of k and M. As we increase the size of k, we have an approximately first-order convergence rate.

TABLE 8.6
Absolute Error Values by EWM for Example 8.3.2

x	$k = 2, M = 3$	$k = 3, M = 3$	$k = 4, M = 3$	$k = 5, M = 3$	$k = 6, M = 3$
0	6.48787E-04	2.84609E-04	1.3234E-04	6.35677E-05	3.10869E-05
1/6	1.7182E-04	4.80499E-05	8.40833E-06	5.50901E-07	9.29938E-08
2/6	9.75513E-05	1.59967E-05	1.94159E-06	3.99176E-07	1.46062E-07
3/6	1.78743E-04	4.0093E-06	2.14527E-06	6.1413E-07	1.6785E-07
4/6	1.68513E-05	9.15943E-06	2.12246E-06	6.79011E-07	1.84612E-07
5/6	6.82207E-05	8.4428E-06	2.86997E-06	7.69137E-07	2.19725E-07

TABLE 8.7
Numerical Solutions for Example 8.3.2 with $k = 2$, $M = 3$

x	EWM	SCWM [29]	TWM [37]	CASWM [29] ($m' = 6$)	Exact Solution
0	−6.48787E-04	−1.5509E-02	−4.7843E-04	6.1480E-03	0
1/6	−1.38717E-01	−1.4412E-01	−0.1387E-01	−1.1115E-01	−1.38889E-01
2/6	−2.22125E-01	−2.2290E-01	−0.2221E-01	−1.8713E-01	−2.2222E-01
3/6	−2.49821E-01	−2.4994E-01	−0.2498E-01	−2.1160E-01	−2.5000E-01
4/6	−2.22205E-01	−2.2128E-01	−0.2222E-01	−1.8346E-01	−2.2222E-01
5/6	−1.38957E-01	−1.3718E-01	−0.1389E-01	−1.0222E-01	−1.38889E-01

TABLE 8.8
Numerical Solutions for Example 8.3.2 with $k = 3$, $M = 3$

x	EWM	SCWM [29]	TWM [37]	CASWM [29] ($m' = 12$)	Exact Solution
0	−2.84609E-04	−7.2939E-03	−2.0437E-04	2.3540E-03	0
1/6	−1.38841E-01	−1.3954E-01	−1.3885E-01	−1.2653E-01	−1.38889E-01
2/6	−2.22206E-01	−2.2252E-01	−2.2220E-01	−2.0680E-01	−2.2222E-01
3/6	−2.50004E-01	−2.4997E-01	−2.5000E-01	−2.3341E-01	−2.5000E-01
4/6	−2.22231E-01	−2.2196E-01	−2.2222E-01	−2.0590E-01	−2.2222E-01
5/6	−1.38897E-01	−1.3843E-01	−1.3889E-01	−1.2407E-01	−1.38889E-01

TABLE 8.9
Numerical Solutions for Example 8.3.2 with $k = 4$, $M = 3$

x	EWM	SCWM [29]	TWM [37]	CASWM [29] ($m' = 24$)	Exact Solution
0	−1.3234E-04	−3.4910E-03	−9.3603E-05	9.0900E-04	0
1/6	−1.3888E-01	−1.3912E-01	−1.3888E-01	−1.3345E-01	−1.38889E-01
2/6	−2.22224E-01	−2.2230E-01	−2.2222E-01	−2.1534E-01	−2.2222E-01
3/6	−2.50002E-01	−2.4999E-01	−2.5000E-01	−2.4266E-01	−2.5000E-01
4/6	−2.22224E-01	−2.2216E-01	−2.2222E-01	−2.1514E-01	−2.2222E-01
5/6	−1.38892E-01	−1.3878E-01	−1.3889E-01	−1.3269E-01	−1.38889E-01

Figures 8.9–8.12 represent the numerical results obtained by EWM for $k = 2,3,4,5$ and $M = 3$.

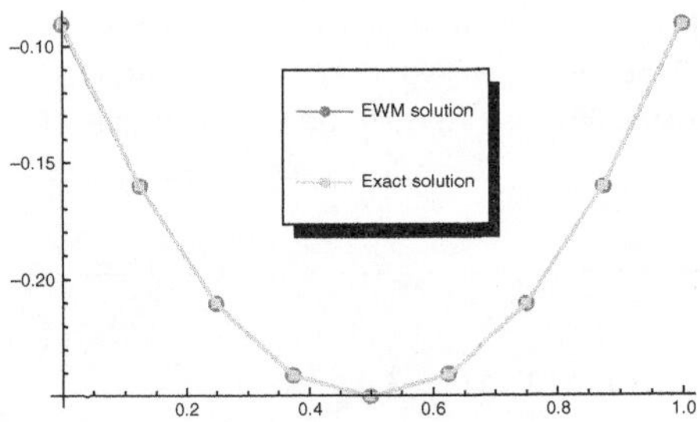

FIGURE 8.9 Numerical results for $k = 2$, $M = 3$.

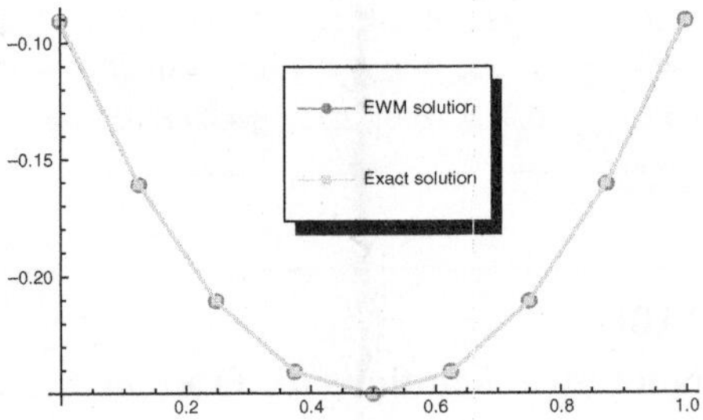

FIGURE 8.10 Numerical results for $k = 3$, $M = 3$.

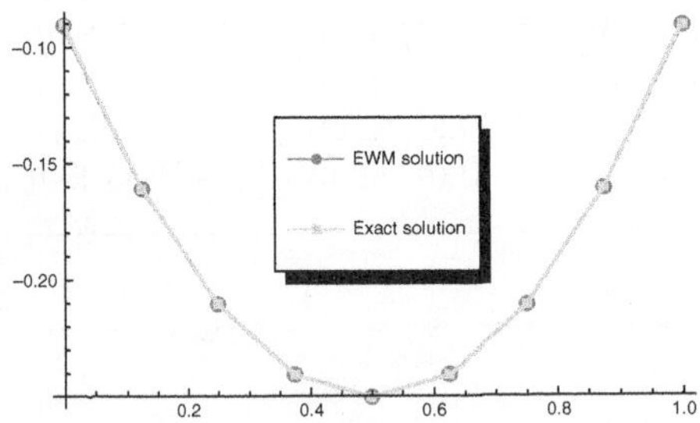

FIGURE 8.11 Numerical results for $k = 4$, $M = 3$.

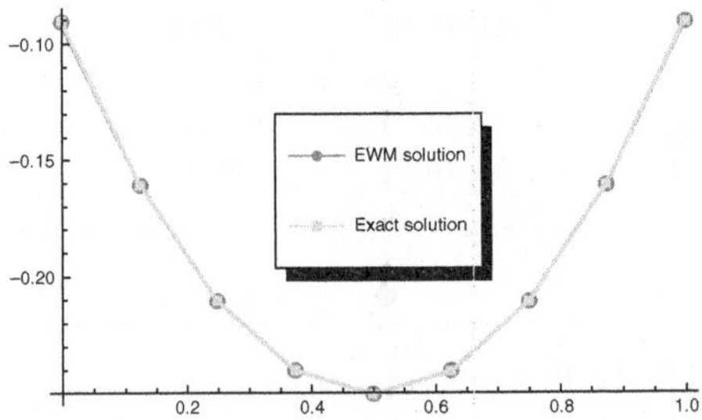

FIGURE 8.12 Numerical results for $k = 5$, $M = 3$.

Figures 8.13–8.16 represent the error analysis by EWM for $k = 2,3,4,5$ and $M = 3$.

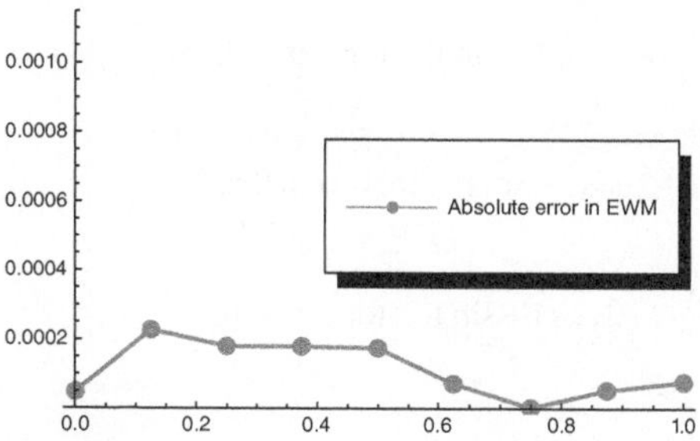

FIGURE 8.13 Error distribution for $k = 2$, $M = 3$.

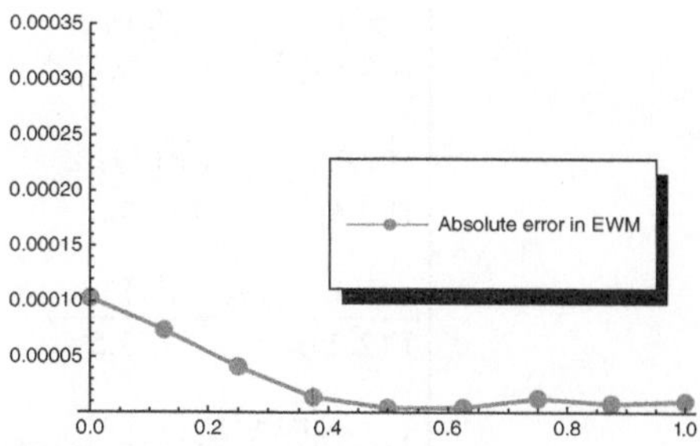

FIGURE 8.14 Error distribution for $k = 3$, $M = 3$.

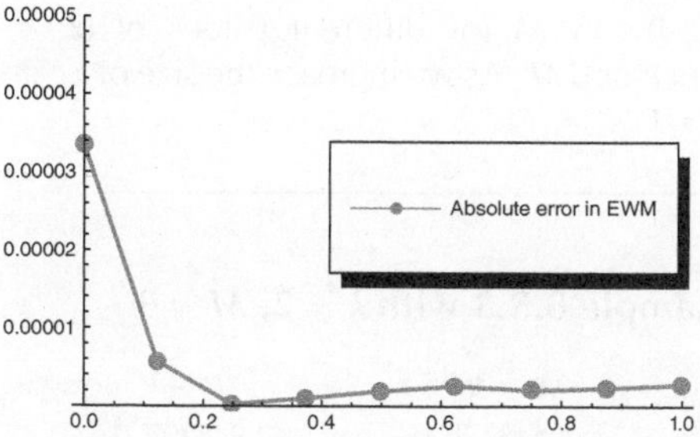

FIGURE 8.15 Error distribution for $k = 4$, $M = 3$.

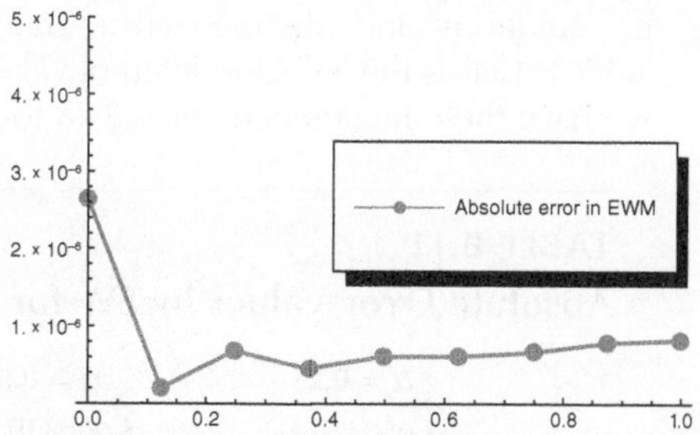

FIGURE 8.16 Error distribution for $k = 5$, $M = 3$.

TABLE 8.10
NCR for Example 8.3.2

k, M	L^∞ error	NCR
$k = 2, M = 3$	2.75161E-4	–
$k = 3, M = 3$	1.14711E-4	1.26227
$k = 4, M = 3$	5.29568E-5	1.11512
$k = 5, M = 3$	2.56104E-5	1.04809
$k = 6, M = 3$	1.25764E-5	1.02601

Example 8.3.3

Consider the following equation [29]:

$$ {}_0^C D_x^\alpha u(x) = f(x) + \int_0^x \frac{u(z)}{(x-z)^{0.5}} dz + \int_0^1 (x^2 + \cos z) u(z) \, dz, \tag{8.31}$$

$$u(0) = 0.$$

The exact solution is $u(x) = x^2$. In this problem,

$$f(x) = \begin{cases} \frac{\Gamma(3)}{\Gamma(2.75)} x^{1.75} - \frac{\sqrt{\pi}\Gamma(3)}{\Gamma(3.5)} x^{2.5} - \frac{x^2}{3} - 2\cos 1 + \sin 1, & \text{for} \quad \alpha = 0.25, \\ \frac{\Gamma(3)}{\Gamma(2.5)} x^{1.5} - \frac{\sqrt{\pi}\Gamma(3)}{\Gamma(3.5)} x^{2.5} - \frac{x^2}{3} - 2\cos 1 + \sin 1, & \text{for} \quad \alpha = 0.5, \\ \frac{\Gamma(3)}{\Gamma(2.25)} x^{1.25} - \frac{\sqrt{\pi}\Gamma(3)}{\Gamma(3.5)} x^{2.5} - \frac{x^2}{3} - 2\cos 1 + \sin 1, & \text{for} \quad \alpha = 0.75, \\ 2x - \frac{\sqrt{\pi}\Gamma(3)}{\Gamma(3.5)} x^{2.5} - \frac{x^2}{3} - 2\cos 1 + \sin 1, & \text{for} \quad \alpha = 1. \end{cases}$$

For different values of k and M, the values of $\hat{e}(x)$ obtained by the EWM are presented in Tables 8.11–8.14, respectively. Figures 8.17–8.40 display the numerical solutions and absolute error analysis by EWM for different values of α. Table 8.15 lists the NCR for different values of k and M. As we increase the size of k, we have the convergence rate ≈ 3.48 for $\alpha = 1$.

TABLE 8.11
Absolute Error Values by EW for Example 8.3.3 with $k = 2$, $M = 3$

x	$\alpha = 0.25$	$\alpha = 0.5$	$\alpha = 0.75$	$\alpha = 1$
0	3.05091E-04	5.60683E-04	4.54377E-04	2.94829E-08
1/6	1.1048E-04	2.0592E-04	9.13356E-05	7.49166E-08
2/6	7.82017E-05	1.63892E-04	1.83636E-05	1.02231E-07
3/6	4.99381E-05	9.24936E-05	1.61817E-04	1.42428E-07
4/6	5.08208E-05	9.89087E-05	2.35156E-04	2.00326E-07
5/6	2.30023E-05	7.88461E-05	3.77999E-04	2.82934E-07

TABLE 8.12
Absolute Error Values by EW for Example 8.3.3 with $k = 3$, $M = 3$

x	$\alpha = 0.25$	$\alpha = 0.5$	$\alpha = 0.75$	$\alpha = 1$
0	6.81629E-05	1.32553E-04	4.54377E-04	2.57343E-09
1/6	1.60419E-05	3.4961E-05	2.17437E-05	5.88687E-09
2/6	1.01368E-05	1.52965E-05	1.08809E-05	8.18578E-09
3/6	6.04943E-07	1.65824E-07	3.07033E-05	1.20108E-08
4/6	1.83265E-06	6.59213E-06	4.13088E-05	1.71341E-08
5/6	3.85197E-06	1.00839E-05	6.12942E-05	2.40453E-08

TABLE 8.13
Absolute Error Values by EW for Example 8.3.3 with $k = 4$, $M = 3$

x	$\alpha = 0.25$	$\alpha = 0.5$	$\alpha = 0.75$	$\alpha = 1$
0	1.59279E-05	3.21708E-05	2.69928E-05	2.28033E-10
1/6	2.48705E-06	3.55884E-06	2.34334E-07	4.63915E-10
2/6	2.24185E-07	6.58855E-07	3.1964E-06	7.11178E-10
3/6	2.10342E-07	1.29129E-08	5.86677E-06	1.04075E-09
4/6	4.25204E-07	1.05054E-06	8.32606E-06	1.48465E-09
5/6	2.20177E-07	1.24921E-06	1.23274E-05	2.08417E-09

TABLE 8.14
Absolute Error Values by EW for Example 8.3.3 with $k = 5$, $M = 3$

x	$\alpha = 0.25$	$\alpha = 0.5$	$\alpha = 0.75$	$\alpha = 1$
0	3.82692E-06	7.92295E-06	6.70722E-06	2.01885E-11
1/6	4.30019E-08	1.14471E-07	3.24046E-07	4.07056E-11
2/6	8.72765E-08	1.20946E-07	6.97159E-07	6.23041E-11
3/6	2.75789E-08	2.58172E-08	1.18368E-06	9.11047E-11
4/6	2.08818E-08	1.37744E-07	1.74634E-06	1.29933E-10
5/6	4.55665E-08	2.1481E-07	2.56511E-06	1.82343E-10

For $k = 2$, $M = 3$, Figures 8.17–8.20 represent the numerical results obtained by EWM for $\alpha = 0.25, 0.5, 0.75, 1$.

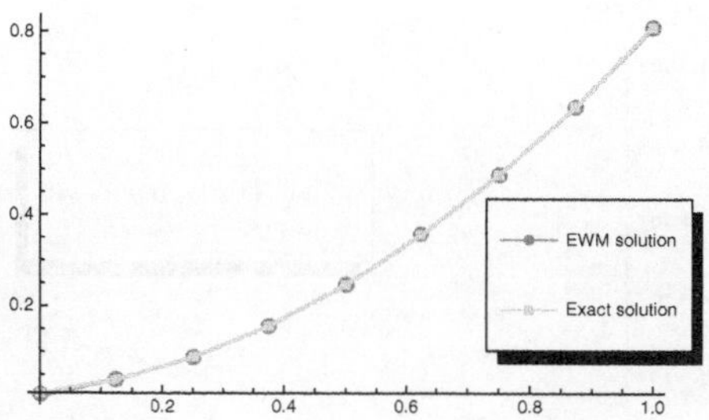

FIGURE 8.17 Numerical results for $\alpha = 0.25$.

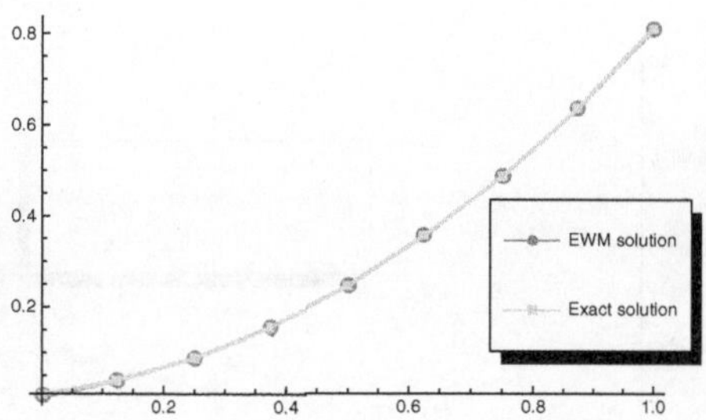

FIGURE 8.18 Numerical results for $\alpha = 0.5$.

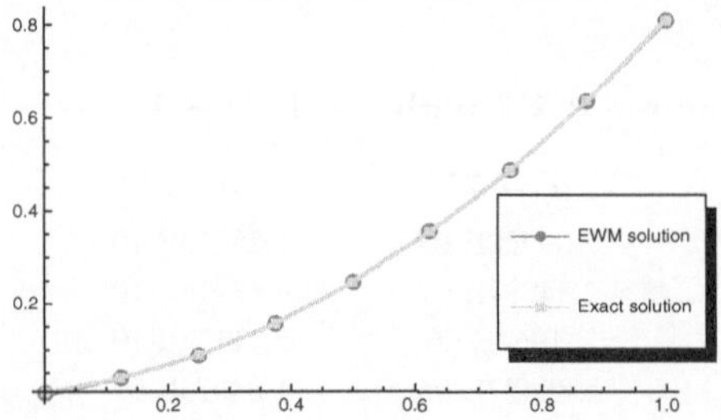

FIGURE 8.19 Numerical results for $\alpha = 0.75$.

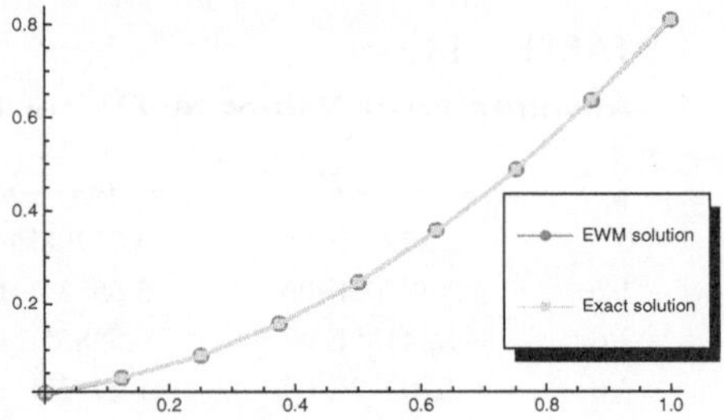

FIGURE 8.20 Numerical results for $\alpha = 1$.

For $k = 2$, $M = 3$, Figures 8.21–8.24 represent the error analysis by EWM for $\gamma = 0.25, 0.5, 0.75, 1$.

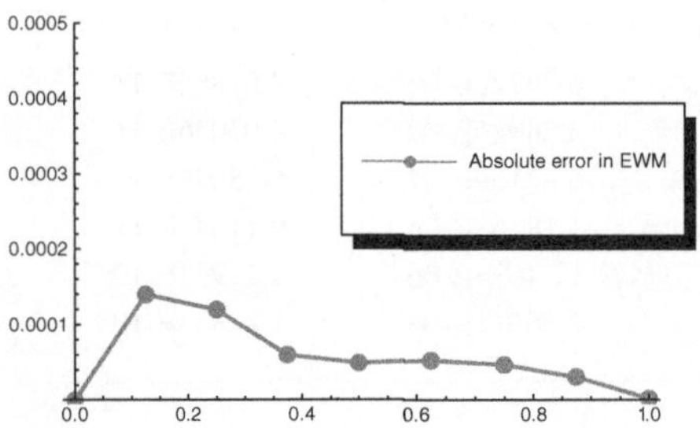

FIGURE 8.21 Error distribution for $\alpha = 0.25$.

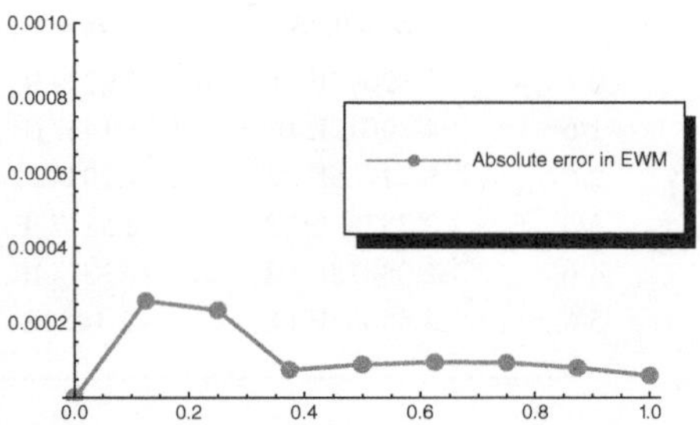

FIGURE 8.22 Error distribution for $\alpha = 0.5$.

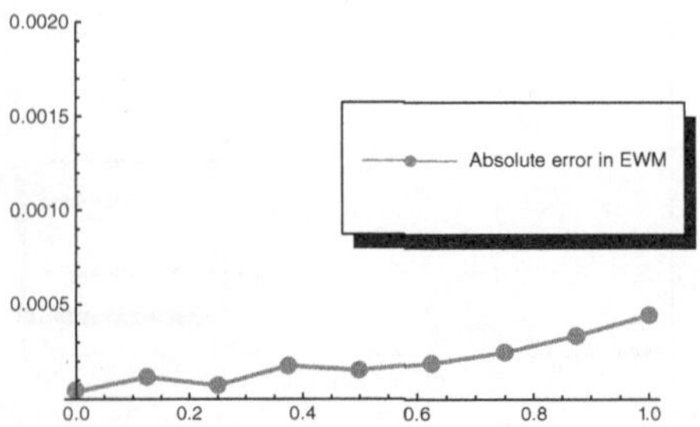

FIGURE 8.23 Error distribution for $\alpha = 0.75$.

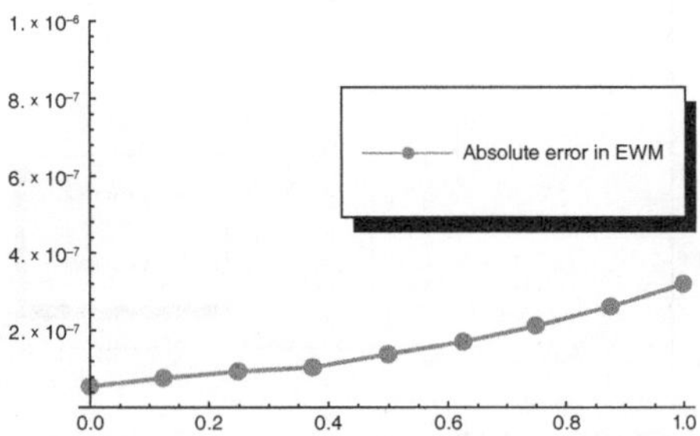

FIGURE 8.24 Error distribution for $\alpha = 1$.

For $k = 3$, $M = 3$, Figures 8.25–8.28 represent the numerical results obtained by EWM for $\alpha = 0.25, 0.5, 0.75, 1$.

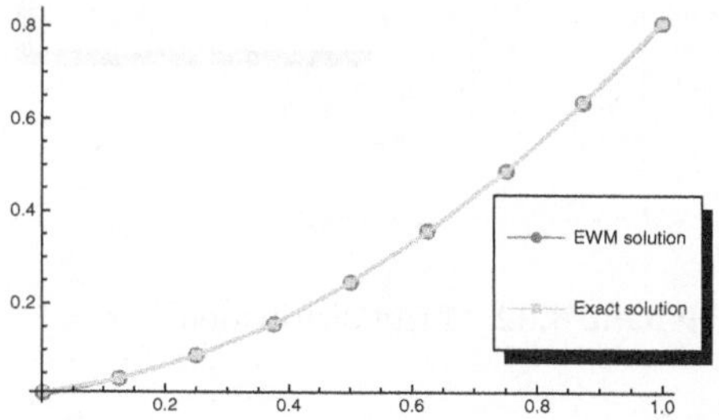

FIGURE 8.25 Numerical results for $\alpha = 0.25$.

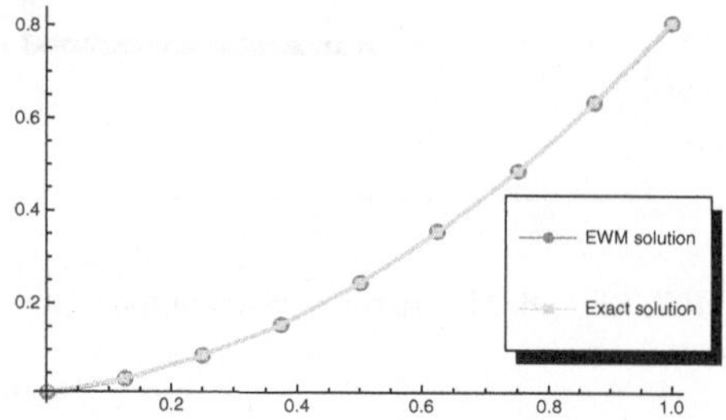

FIGURE 8.26 Numerical results for $\alpha = 0.5$.

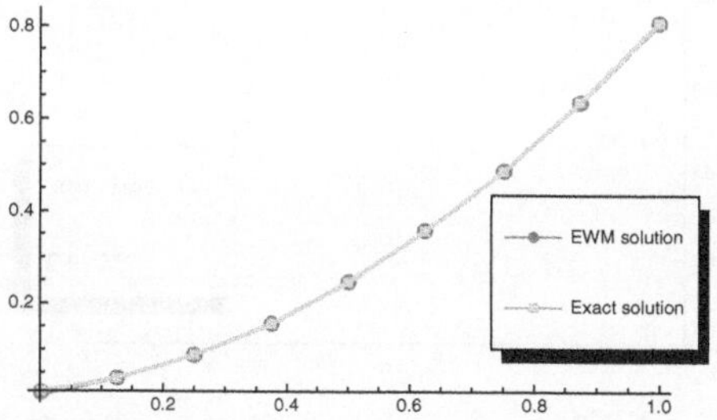

FIGURE 8.27 Numerical results for $\alpha = 0.75$.

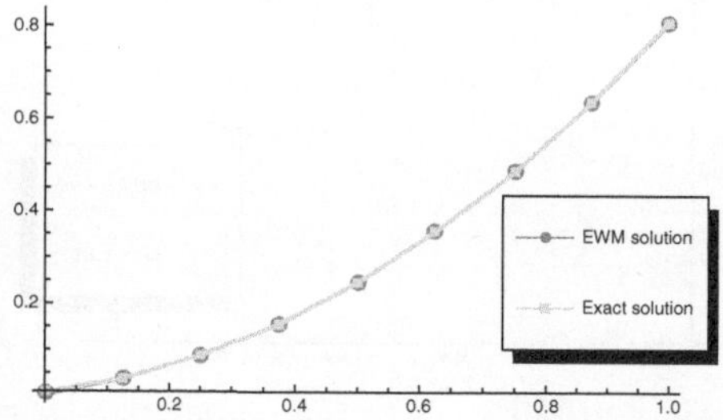

FIGURE 8.28 Numerical results for $\alpha = 1$.

For $k = 3$, $M = 3$, Figures 8.29–8.32 represent the error analysis by EWM for $\alpha = 0.25, 0.5, 0.75, 1$.

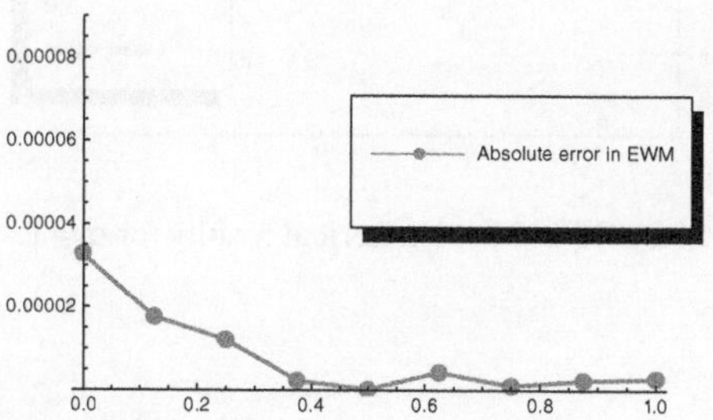

FIGURE 8.29 Error distribution for $\alpha = 0.25$.

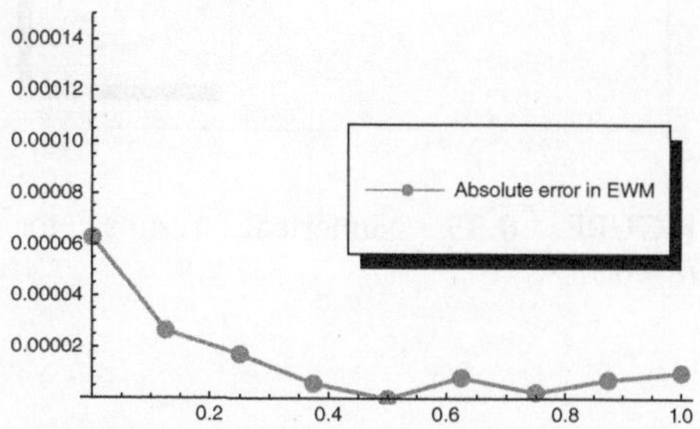

FIGURE 8.30 Error distribution for $\alpha = 0.5$.

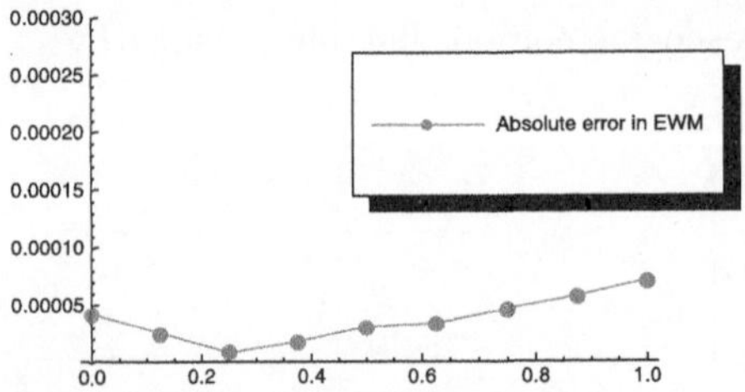

FIGURE 8.31 Error distribution for $\alpha = 0.75$.

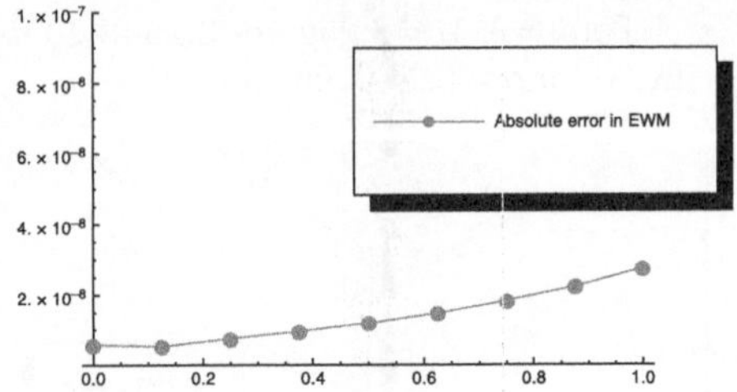

FIGURE 8.32 Error distribution for $\alpha = 1$.

For $k = 4$, $M = 3$, Figures 8.33–8.36 represent the numerical results obtained by EWM for $\alpha = 0.25, 0.5, 0.75, 1$.

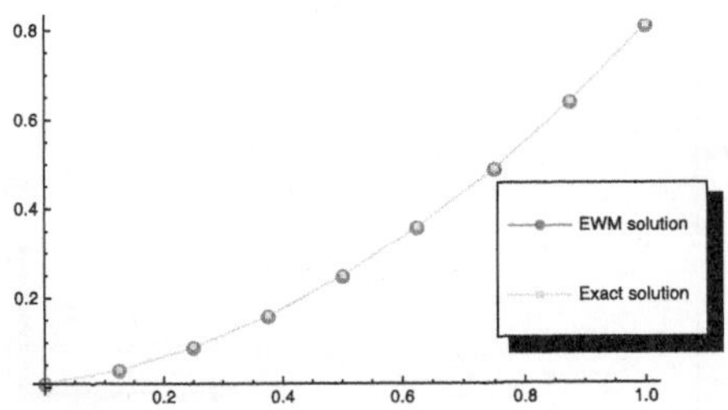

FIGURE 8.33 Numerical results for $\alpha = 0.25$.

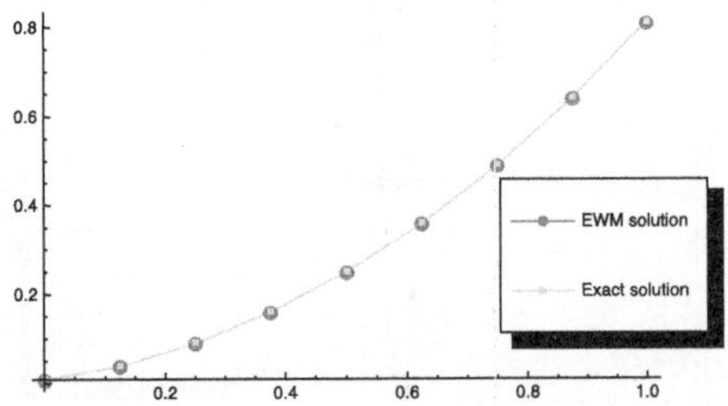

FIGURE 8.34 Numerical results for $\alpha = 0.5$.

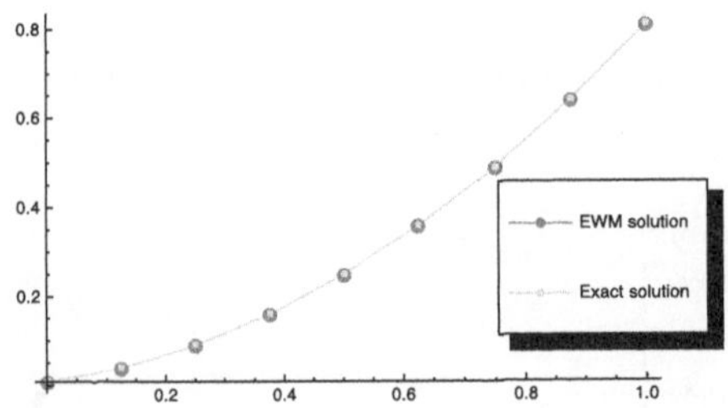

FIGURE 8.35 Numerical results for $\alpha = 0.75$.

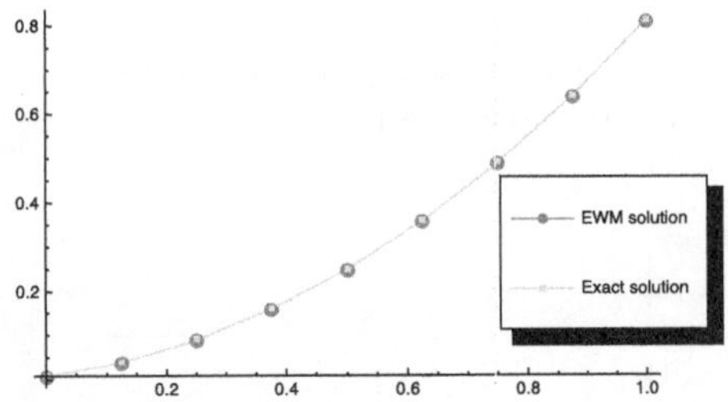

FIGURE 8.36 Numerical results for $\alpha = 1$.

For $k = 4$, $M = 3$, Figures 8.37–8.40 represent the error analysis by EWM for $\alpha = 0.25, 0.5, 0.75, 1$.

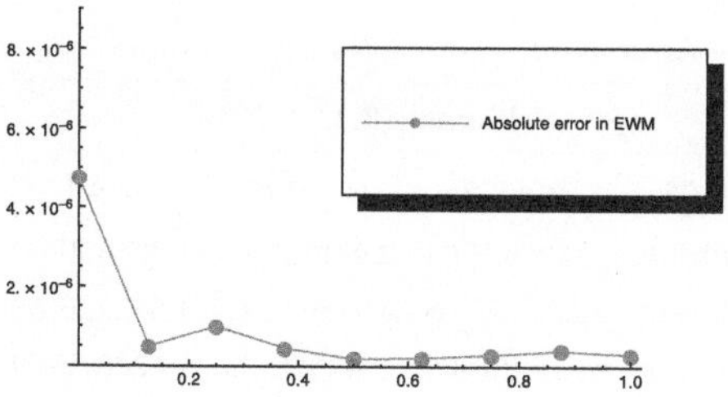

FIGURE 8.37 Error distribution for $\alpha = 0.25$.

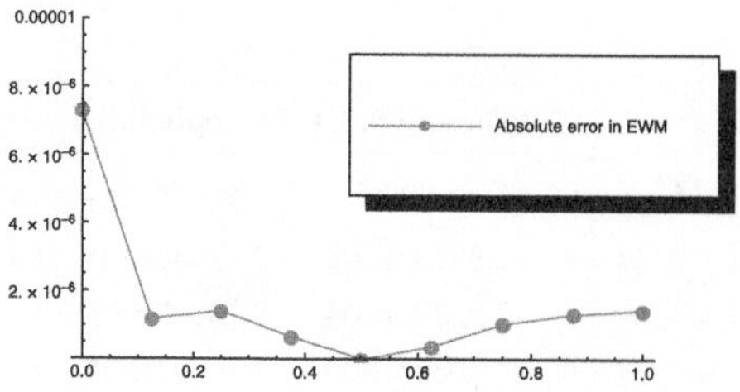

FIGURE 8.38 Error distribution for $\alpha = 0.5$.

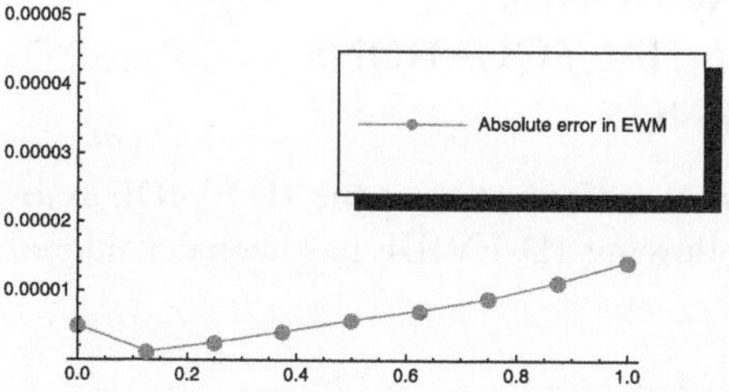

FIGURE 8.39 Error distribution for $\alpha = 0.75$.

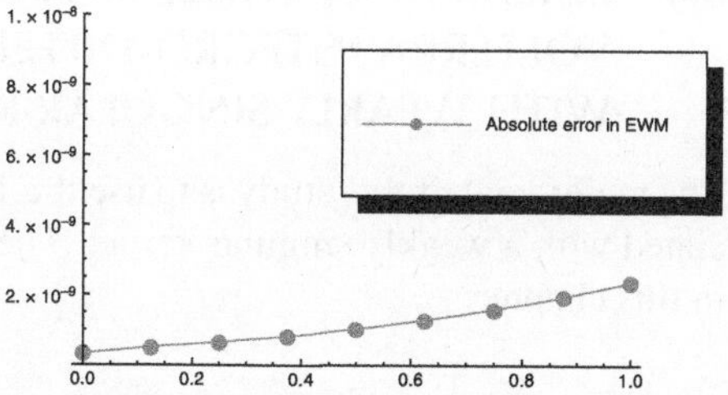

FIGURE 8.40 Error distribution for $\alpha = 1$.

TABLE 8.15
Numerical Convergence Rate Analysis for Example 8.3.3

	NCR			
k, M	$\alpha = 0.25$	$\alpha = 0.5$	$\alpha = 0.75$	$\alpha = 1$
$k = 2$, $M = 3$	–	–	–	–
$k = 3$, $M = 3$	2.11372	2.07856	2.524	3.4377
$k = 4$, $M = 3$	2.06946	2.03003	2.27601	3.47094
$k = 5$, $M = 3$	2.04374	2.01526	2.23861	3.48583

TABLE 8.16
L^2-Error Norm of Numerical Experiments for $M = 3$ and Different Values of k

	L^2 – *Error norm*					
	Example 8.3.1	**Example 8.3.2**	**Example 8.3.3**			
k, M	$\alpha = 0.25$	$\alpha = 0.15$	$\alpha = 0.25$	$\alpha = 0.5$	$\alpha = 0.75$	$\alpha = 1$
$k = 2, M = 3$	2.38737E-03	2.32688E-04	1.12279E-04	2.02894E-04	2.98172E-04	1.92685E-07
$k = 3, M = 3$	2.96777E-04	7.20007E-05	1.93008E-05	3.52575E-05	4.91567E-05	1.64222E-08
$k = 4, M = 3$	3.70385E-05	2.36702E-05	3.34661E-06	6.19676E-06	9.4758E-06	1.42267E-09
$k = 5, M = 3$	4.63808E-06	8.04426E-06	5.84334E-07	1.09259 E-06	1.90286 E-06	1.24458E-10

8.4 LINEAR AND NONLINEAR FRACTIONAL VOLTERRA INTEGRO-DIFFERENTIAL EQUATIONS WITH WEAKLY SINGULAR KERNELS

The major goal of this study is to use the TWs technique to solve the 1D FVIDE associated with a weakly singular kernel. The following 1D-FVIDE has been considered in this chapter:

$$\begin{cases} {}_0^C D_x^\alpha u(x) = f(x) + a(x)u(x) + \displaystyle\int_0^x \frac{\tilde{\Pi}(u(z))}{(x-z)^\beta} dz, \quad 0 < \beta < 1, \\ u(0) = 0, \end{cases} \tag{8.32}$$

where $u(x) \in L^2[0,1)$ and $a(x), f(x) \in L^2[0,1)$ are unknown and known functions, respectively. Here, the term $\tilde{\Pi}(u(z))$ is considered as $\tilde{\Pi}(u(z)) = u(z)$ and $\tilde{\Pi}(u(z)) = u^\tau(z)$, $\tau > 1$ for linear and nonlinear cases, respectively, and ${}_0^C D_x^\alpha$ denotes the Caputo fractional derivative of fractional order α, $0 < \alpha \leq 1$.

8.4.1 Function Approximation by TWs

If $h(x)$ is any function defined on $L^2[0,1)$, it can be approximated as follows:

$$h(x) \approx \sum_{n=1}^{2^{k-1}} \sum_{m=0}^{M-1} \tilde{h}_{n,m} \psi_{n,m}(x) = H^T \Psi(x), \tag{8.33}$$

where H and $\Psi(x)$ are $(2^{k-1}M \times 1)$-dimensional vectors given by

$$H = [\tilde{h}_{1,0}, \tilde{h}_{1,1}, \ldots, \tilde{h}_{1,M-1}, \ldots, \tilde{h}_{2^{k-1},0}, \tilde{h}_{2^{k-1},1}, \ldots, \tilde{h}_{2^{k-1},M-1}]^T \tag{8.34}$$

and

$$\Psi(x) = [\psi_{1,0}(x), \psi_{1,1}(x), \ldots, \psi_{1,M-1}(x), \ldots, \psi_{2^{k-1},0}(x), \psi_{2^{k-1},1}(x), \ldots \psi_{2^{k-1},M-1}(x)]^T, \quad (8.35)$$

where $\psi_{n,m}(x)$, $n = 1, 2, \ldots, 2^{k-1}$, $m = 0, 1, \ldots, M-1$ is the TWs defined on $[0,1)$ as follows:

$$\psi_{n,m}(x) = \begin{cases} 2^{\frac{k-1}{2}} \tilde{T}_m(2^{k-1}x - n + 1), & \frac{n-1}{2^{k-1}} \le x < \frac{n}{2^{k-1}}, \\ 0, & \text{otherwise.} \end{cases} \quad (8.36)$$

Using Eq. (8.33), the value of H can be calculated as follows:

$$H^T = \left(\int_0^1 h(x)\Psi^T(x)dx \right) . \tilde{D}^{-1}, \quad (8.37)$$

where $\tilde{D}$ is a $(2^{k-1}M \times 2^{k-1}M)$-dimensional matrix and is given by

$$\tilde{D} = \int_0^1 \Psi(x)\Psi^T(x)\,dx. \quad (8.38)$$

For $k = 2$, $M = 3$, we have

$$\tilde{D} = \begin{bmatrix} 1 & \frac{\sqrt{3}}{2} & \frac{\sqrt{5}}{3} & 0 & 0 & 0 \\ \frac{\sqrt{3}}{2} & 1 & \frac{\sqrt{15}}{4} & 0 & 0 & 0 \\ \frac{\sqrt{5}}{3} & \frac{\sqrt{15}}{4} & 1 & 0 & 0 & 0 \\ 0 & 0 & 0 & 1 & \frac{\sqrt{3}}{2} & \frac{\sqrt{5}}{3} \\ 0 & 0 & 0 & \frac{\sqrt{3}}{2} & 1 & \frac{\sqrt{15}}{4} \\ 0 & 0 & 0 & \frac{\sqrt{5}}{3} & \frac{\sqrt{15}}{4} & 1 \end{bmatrix}.$$

8.4.2 Operational Matrices

The construction of the OM of FI for the proposed scheme to solve FVIDEs has been explained in this section.

8.4.2.1 Operational Matrix of the Fractional Integration

The integration of the TWs vector $\Psi(x)$ is given by

$$\int_0^x \Psi(z)\,dz = \hat{P}_1\Psi(x), \tag{8.39}$$

where $\hat{P}_1$ is the $(2^{k-1}M \times 2^{k-1}M)$-dimensional OM of integration [48].

The FI of the vector $\Psi(x)$ is defined as follows:

$$\tilde{I}^{\alpha}\Psi(x) = P^{\alpha}\Psi(x), \tag{8.40}$$

where P^{α} is the OM of the FI of dimension $(2^{k-1}M \times 2^{k-1}M)$.

The OM P^{α} can be determined from Eq. (8.40) as follows:

$$P^{\alpha} = \left(\int_0^1 \tilde{I}^{\alpha}\Psi(x).\Psi^T(x)dx\right).\tilde{D}^{-1}. \tag{8.41}$$

8.4.2.2 Operational Matrix of the Integration with Singularity

For $0 < \beta < 1$, the OM of integration with singularity of the TWs vector $\Psi(x)$ is defined as [41]

$$\int_0^x \frac{\Psi(z)}{(x-z)^{\beta}}dz = \hat{P}_2\Psi(x), \tag{8.42}$$

where $\hat{P}_2$ is the OM of integration with singularity of dimension $(2^{k-1}M \times 2^{k-1}M)$.

8.4.3 Taylor Wavelet Method for FVIDE

In this work, the 1D-FVIDE Eq. (8.32) has been solved by implementing the OM approach derived by 1D TWs.

8.4.3.1 Linear Case

In the linear case, let $\tilde{\Pi}(u(z)) = u(z)$ in Eq. (8.32). Therefore, we have

$${}_0^C D_x^{\alpha}u(x) = f(x) + a(x)u(x) + \int_0^x \frac{u(z)}{(x-z)^{\beta}}dz. \tag{8.43}$$

To discuss the proposed scheme for solving Eq. (8.43), consider the approximation of following functions:

$${}_0^C D_x^{\alpha}u(x) \approx \Psi^T(x)\tilde{U}_1 = \tilde{U}_1^T\Psi(x), \tag{8.44}$$

$$u(0) \approx \Psi^T(x)\tilde{U}_2 = \tilde{U}_2^T\Psi(x), \tag{8.45}$$

$$f(x) \approx \Psi^T(x)F = F^T\Psi(x), \tag{8.46}$$

$$a(x) \approx \Psi^T(x)\tilde{A} = \tilde{A}^T\Psi(x). \tag{8.47}$$

Now, by integrating Eq. (8.44) with respect to x and utilizing Eq. (8.41), it follows

$$\begin{aligned} u(x) &= u(0) + \tilde{U}_1^T P^{\alpha}\Psi(x) \\ &= \tilde{U}_2^T\Psi(x) + \tilde{U}_1^T P^{\alpha}\Psi(x) \\ &= (\tilde{U}_2^T + \tilde{U}_1^T P^{\alpha})\Psi(x) \\ &= \Psi^T(x)\tilde{U}_3, \end{aligned} \tag{8.48}$$

where $\tilde{U}_3 = \tilde{U}_2 + P^{\alpha T}\tilde{U}_1$.

Now approximating the term $a(x)u(x)$ of Eq. (8.43) as follows:

$$\begin{aligned} a(x)u(x) &= (\tilde{A}^T\Psi(x)).(\tilde{U}_3^T\Psi(x)) \\ &= \tilde{A}^T\Psi(x).\Psi^T(x)\tilde{U}_3 \\ &= \tilde{A}^T\bar{U}_3\Psi(x) \\ &= \tilde{U}_4^T\Psi(x), \end{aligned} \tag{8.49}$$

where $\bar{U}_3 = \left(\int_0^1 \Psi(x)\Psi^T(x)\tilde{U}_3\Psi^T(x)\,dx\right).\tilde{D}^{-1}$ and $\tilde{U}_4 = \bar{U}_3^T\tilde{A}$.

Now, in Eq. (8.43), the integral part can be expressed as follows:

$$\begin{aligned} \int_0^x \frac{u(z)}{(x-z)^{\beta}}dz &= \int_0^x \frac{\tilde{U}_3^T\Psi(z)}{(x-z)^{\beta}}dz \\ &= \tilde{U}_3^T.\int_0^x \frac{\Psi(z)}{(x-z)^{\beta}}dz \\ &= \tilde{U}_3^T.\hat{P}_2.\Psi(x) \\ &= \tilde{U}_5^T.\Psi(x), \end{aligned} \tag{8.50}$$

where $\tilde{U}_5 = \hat{P}_2^T\tilde{U}_3$.

Now by plugging Eqs. (8.44)–(8.50) into Eq. (8.43), it follows

$$\Psi^T(x)\tilde{U}_1 = \Psi^T(x)F + \Psi^T(x)\tilde{U}_4 + \Psi^T(x)\tilde{U}_5. \tag{8.51}$$

Collocating Eq. (8.51) at $2^{k-1}M$-points x_i yields

$$\Psi^T(x_i)\tilde{U}_1 = \Psi^T(x_i)F + \Psi^T(x_i)\tilde{U}_4 + \Psi^T(x_i)\tilde{U}_5, \tag{8.52}$$

where x_i are the collocation points defined as follows:

$$x_i = \frac{2i-1}{2.2^{k-1}M}, \quad i = 1,2,...,2^{k-1}M.$$

The value of $\tilde{U}_1$ can be computed by simplifying the above linear algebraic system of Eq. (8.52), and then $\tilde{U}_3$ is calculated by using the value of $\tilde{U}_1$.

Thus, the linear FVIDE Eq. (8.43) can be numerically solved by substituting $\tilde{U}_3$ in Eq. (8.48).

8.4.3.2 Nonlinear Case

For the nonlinear case, let $\tilde{\Pi}(u(z)) = u^\tau(z)$, $\tau > 1$ in Eq. (8.32). Therefore, we have

$${}_0^C D_x^\alpha u(x) = f(x) + a(x)u(x) + \int_0^x \frac{u^\tau(z)}{(x-z)^\beta} dz. \tag{8.53}$$

Now in the case of $\tau = 2$, it follows

$$\begin{aligned} u^2(z) &= (\tilde{U}_3^T\Psi(z)).(\tilde{U}_3^T\Psi(z)) \\ &= \tilde{U}_3^T\Psi(z)\Psi^T(z)\tilde{U}_3 \\ &= \tilde{U}_3^T\bar{U}_3\Psi(z) \\ &= \tilde{U}_6^T\Psi(z), \end{aligned} \tag{8.54}$$

where $\tilde{U}_6 = \bar{U}_3^T.\tilde{U}_3$.

Now, using Eq. (8.54), the integral part of Eq. (8.53) can be expressed as follows:

$$\begin{aligned} \int_0^x \frac{u^2(z)}{(x-z)^\beta} dz &= \int_0^x \frac{\tilde{U}_6^T\Psi(z)}{(x-z)^\beta} dz \\ &= \tilde{U}_6^T.\int_0^x \frac{\Psi(z)}{(x-z)^\beta} dz \\ &= \tilde{U}_6^T.\hat{P}_2.\Psi(x) \\ &= \tilde{U}_7^T.\Psi(x), \end{aligned} \tag{8.55}$$

where $\tilde{U}_7 = \hat{P}_2^T\tilde{U}_6$.

Now by substituting Eqs. (8.44)–(8.49) and Eqs. (8.54) and (8.55) into Eq. (8.53), it follows

$$\Psi^T(x)\tilde{U}_1 = \Psi^T(x)F + \Psi^T(x)\tilde{U}_4 + \Psi^T(x)\tilde{U}_7. \tag{8.56}$$

Collocating Eq. (8.56) at $2^{k-1}M$-points x_i yields

$$\Psi^T(x_i)\tilde{U}_1 = \Psi^T(x_i)F + \Psi^T(x_i)\tilde{U}_4 + \Psi^T(x_i)\tilde{U}_7, \tag{8.57}$$

where x_i are the collocation points defined as follows:

$$x_i = \frac{2i-1}{2.2^{k-1}M}, \quad i = 1,2,...,2^{k-1}M.$$

The value of $\tilde{U}_1$ can be computed by solving the above nonlinear algebraic system of Eq. (8.57), and then $\tilde{U}_3$ is calculated by using the value of $\tilde{U}_1$.

Thus, the nonlinear FVIDE Eq. (8.53) can be numerically solved by substituting $\tilde{U}_3$ in Eq. (8.48).

8.4.4 Convergence Analysis

Theorem 8.4.1

Consider $u(x)$ be a continuous function in $L^2([0,1))$ and $|u(x)| \le \Re$, where $\Re$ is a real and positive constant. Then, the TWs expansion of $u(x)$ converges uniformly to $u(x)$, and also

$$\left|\tilde{f}_{n,m}\right| \le \frac{\Re\sqrt{2m+1}}{2^{\frac{k-1}{2}}.(m+1)}.$$

Proof:

The TWs coefficients can be calculated as follows:

$$\begin{aligned}\tilde{f}_{n,m} &= <u(x), \psi_{n,m}(x)> \\ &= \int_0^1 u(x)\psi_{n,m}(x)\,dx \\ &= 2^{\frac{k-1}{2}}\sqrt{2m+1}\int_{\frac{n-1}{2^{k-1}}}^{\frac{n}{2^{k-1}}} u(x)T_m(2^{k-1}x-n+1)\,dx.\end{aligned}$$

Now, changing the variable $2^{k-1}x - n + 1 = t$, it follows

$$= 2^{\frac{k-1}{2}}\sqrt{2m+1}\frac{1}{2^{k-1}}\int_0^1 u\left(\frac{t+n-1}{2^{k-1}}\right)T_m(t)\,dt.$$

This implies

$$\left|\tilde{f}_{n,m}\right| = \left|\frac{\sqrt{2m+1}}{2^{\frac{k-1}{2}}}\int_0^1 u\left(\frac{t+n-1}{2^{k-1}}\right)T_m(t)\,dt\right|$$

$$\leq \frac{\sqrt{2m+1}}{2^{\frac{k-1}{2}}}\int_0^1 \left|u\left(\frac{t+n-1}{2^{k-1}}\right)\right|\left|T_m(t)\right|dt$$

$$\leq \frac{\sqrt{2m+1}}{2^{\frac{k-1}{2}}}\Re\int_0^1 \left|T_m(t)\right|dt.$$

By the definition of the Taylor polynomial of degree m, $\int_0^1 \left|T_m(t)\right|dt = \frac{1}{m+1}$, $\text{Re}[m] > -1$, i.e.,

$$\left|\tilde{f}_{n,m}\right| \leq \frac{\Re\sqrt{2m+1}}{2^{\frac{k-1}{2}}.(m+1)}.$$

Hence, the series $\sum_{n=1}^{2^{k-1}}\sum_{m=0}^{M-1}\tilde{f}_{n,m}$ is absolutely convergent, and it follows that the series $\sum_{n=1}^{2^{k-1}}\sum_{m=0}^{M-1}\tilde{f}_{n,m}\,\psi_{n,m}(x)$ is uniformly convergent.

8.4.5 Error Analysis

In this part, the estimation of error bound of the TWM for solving FVIDE has been derived.

Theorem 8.4.2

Let $\tilde{u}_{2^{k-1}M}(x)$ be the 1D TWs expansion of the real sufficiently smooth function $u(x) \in L^2([0,1))$ and u is M times differentiable. Then,

$$\left\|u(x) - \tilde{u}_{2^{k-1}M}(x)\right\|_2 \leq \frac{2\Im}{M!\,2^{M(k+1)}},$$

where $\Im = \max_{x\in[0,1]}\left|u^{(M)}(x)\right|$.

Proof:

Assume that $u(x)$ be a sufficiently smooth function on $L^2([0,1))$ and $\tilde{u}_{2^{k-1}M}(x)$ is the best approximation of u from $\Theta_{2^{k-1}M}$, where

$$\Theta_{2^{k-1}M} = \mathrm{Span}\left\{\psi_{n,m}(x) : n = 1,2,\ldots,2^{k-1}, m = 0,1,\ldots,M-1\right\}.$$

Now dividing the closed interval [0,1] into 2^{k-1}-subintervals $\tilde{I}_n = \left[\frac{n-1}{2^{k-1}}, \frac{n}{2^{k-1}}\right]$ with $n = 1,2,\ldots,2^{k-1}$.

Then, for any $P_{2^{k-1}M}[u] \in \Theta_{2^{k-1}M}$, the following inequality is held:

$$\left\|u(x) - \tilde{u}_{2^{k-1}M}(x)\right\|_2 \le \left\|u(x) - P_{2^{k-1}M}[u]\right\|_2, \tag{8.58}$$

where $P_{2^{k-1}M}[u]$ is the interpolating polynomial of u at the Chebyshev nodes in the interval $\tilde{I}_n$. According to a study [52], the interpolation error is

$$\begin{aligned} \left|u(x) - P_{2^{k-1}M}[u]\right| &\le \frac{1}{2^{M-1}M!}\left(\frac{\left|\tilde{I}_n\right|}{2}\right)^M \max_{x\in\tilde{I}_n}\left|u^{(M)}(x)\right| \\ &\le \frac{2\Im}{M!2^{M(k+1)}}. \end{aligned} \tag{8.59}$$

In particular, from Eq. (8.58),

$$\begin{aligned} \left\|u(x) - \tilde{u}_{2^{k-1}M}(x)\right\|_2^2 &\le \left\|u(x) - P_{2^{k-1}M}[u]\right\|_2^2 \\ &= \int_0^1 \left|u(x) - P_{2^{k-1}M}[u]\right|^2 dx \\ &\le \int_0^1 \left(\frac{2\Im}{M!2^{M(k+1)}}\right)^2 dx \\ &= \left(\frac{2\Im}{M!2^{M(k+1)}}\right)^2. \end{aligned}$$

Therefore,

$$\left\|u(x) - \tilde{u}_{2^{k-1}M}(x)\right\|_2 \le \frac{2\Im}{M!2^{M(k+1)}}. \tag{8.60}$$

Hence, the theorem is proved. ■

Theorem 8.4.3

Let $\tilde{u}_{2^{k-1}M}(x)$ be the 1D TWs expansion of the real sufficiently smooth function $u(x) \in L^2([0,1))$ and u is M times differentiable. Then,

$$\left\|\tilde{I}^{\alpha}u(x)-\tilde{I}^{\alpha}\tilde{u}_{2^{k-1}M}(x)\right\|_2 \le \frac{2\Im}{\Gamma(\alpha)\sqrt{2\alpha(2\alpha-1)}M!2^{M(k+1)}},$$

where $\Im = \max_{x\in[0,1]}\left|u^{(M)}(x)\right|$.

Proof:

It may be referred to Ref. [53].

8.4.6 Numerical Experiments

The effectiveness of the presented approach for solving FVIDEs has been explored in this section. To verify the efficiency of the method, some numerical tests have been presented here. Using the absolute error definition and the L^2-error norm, the accuracy of the proposed method has been described as follows:

$$\tilde{\rho}(x) = \left|u(x)-\tilde{u}(x)\right|,$$

where $u(x)$ and $\tilde{u}(x)$ represent the exact and approximate solutions, respectively.

Furthermore, the L^2-error norm and L^∞-error norm for the proposed wavelet are defined as follows:

$$L^2_{\text{error}} = \left[\int_0^1 \left|u(x)-\tilde{u}(x)\right|^2 dx\right]^{\frac{1}{2}}$$

and

$$L^\infty_{\text{error}} = \underset{1\le j\le 2^{k-1}M}{\text{Max}}\left|u(x_j)-\tilde{u}(x_j)\right|,\quad x_j = \frac{2j-1}{2.2^{k-1}M},\; j=1,2,\ldots,2^{k-1}M,$$

where $u(x_j)$ and $\tilde{u}(x_j)$ signify the exact and approximate solutions, respectively, and $x_j, j=1,2,\ldots,2^{k-1}M$ are the collocation points.

Example 8.4.1

Consider a linear FVIDE with a weakly singular kernel [45]:

$${}_0^C D_x^{\frac{1}{3}}u(x) = f(x)+a(x)u(x)+\int_0^x \frac{u(z)}{(x-z)^{0.5}}dz,\quad x\in[0,1], \tag{8.61}$$

$$u(0)=0.$$

The exact solution is $u(x) = x^2 + x^{\frac{4}{3}}$. In this problem,

$$f(x)=\frac{6}{\Gamma\left(\frac{11}{3}\right)}x^{\frac{8}{3}}+\left(\frac{32}{35}-\frac{\Gamma\left(\frac{1}{2}\right)\Gamma\left(\frac{7}{3}\right)}{\Gamma\left(\frac{17}{6}\right)}\right)x^{\frac{11}{6}}+\Gamma\left(\frac{7}{3}\right)x \text{ and } a(x)=-\frac{32}{35}x^{\frac{1}{2}}.$$

Table 8.17 demonstrates the $\tilde{\rho}(x)$ values obtained by the TWM. Table 8.18 displays the comparison of $\tilde{\rho}(x)$ values between TWM and modification of hat functions (MHFs) method [45]. The numerical results and $\tilde{\rho}(x)$ error distributions by TWM are depicted in Figures 8.41–8.48. The L^2 and L^∞-error norms for $k = 2,3,4,5$ and $M = 3$ are presented in Table 8.19.

TABLE 8.17
$\tilde{\rho}(x)$ Errors by TWM for Example 8.4.1

	$\tilde{\rho}(x)$			
x	$k=2, M=3$	$k=3, M=3$	$k=4, M=3$	$k=5, M=3$
0.1	1.92952E-04	1.54211E-05	1.47331E-05	6.46664E-06
0.2	1.71535E-03	1.78429E-04	2.39499E-07	1.75919E-06
0.3	5.03695E-04	1.97697E-04	2.23936E-05	2.75911E-06
0.4	1.12881E-03	1.2374E-04	2.95712E-05	3.07349E-06
0.5	6.59046E-03	7.43476E-04	8.81926E-05	1.06951E-05
0.6	1.96597E-03	2.32328E-04	3.16054E-05	2.72005E-06
0.7	1.93434E-03	2.62484E-04	2.19312E-05	4.21854E-06
0.8	1.30101E-03	2.72845E-04	2.8982E-05	4.1025E-06
0.9	2.22347E-03	1.77051E-04	3.48077E-05	3.58307E-06

TABLE 8.18
Comparison of $\tilde{\rho}(x)$ Errors for Example 8.4.1

x	MHFs ($n=16$) [45]	TWM ($k=3, M=4$)	MHFs ($n=32$)[45]	TWM ($k=4, M=4$)	MHFs ($n=64$)[45]	TWM ($k=5, M=4$)
0.125	6.81733E-05	2.03701E-04	1.25593E-05	1.07178E-05	2.39892E-06	6.91388E-07
0.250	6.21373E-05	3.33379E-05	1.23046E-05	1.74466E-06	2.43485E-06	1.46628E-07
0.375	6.11849E-05	6.36651E-06	1.27213E-05	7.35124E-07	2.57556E-06	6.25335E-08
0.500	6.23334E-05	4.40477E-06	1.34327E-05	4.19375E-07	2.76220E-06	3.81931E-08
0.625	6.47637E-05	2.39489E-06	1.43297E-05	2.92725E-07	2.97850E-06	2.91264E-08
0.750	6.81214E-05	2.16435E-06	1.53667E-05	2.35559E-07	3.21794E-06	2.54765E-08
0.875	7.22219E-05	1.59285E-06	1.65207E-05	2.09184E-07	3.47739E-06	2.41397E-08

Figures 8.41–8.44 manifest the numerical results by TWM for $k = 2,3,4,5$ and $M = 3$.

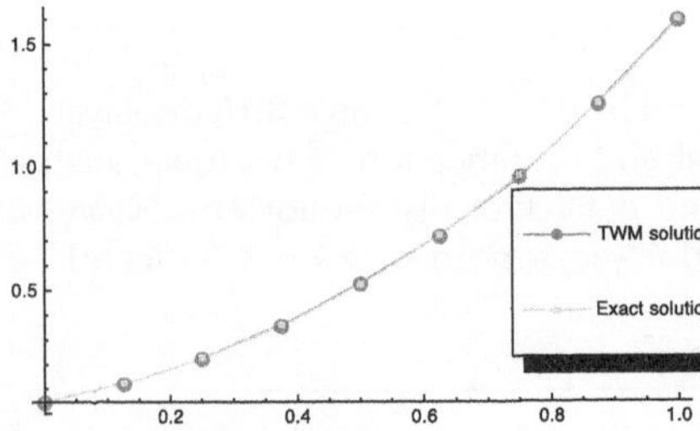

FIGURE 8.41 Numerical solution by TWM for $k = 2, M = 3$.

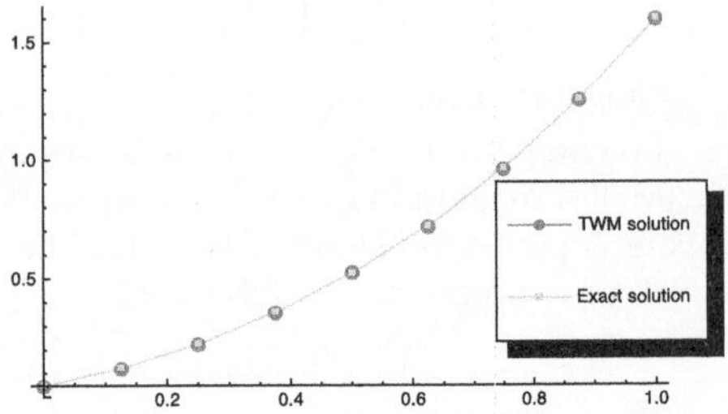

FIGURE 8.42 Numerical solution by TWM for $k = 3, M = 3$.

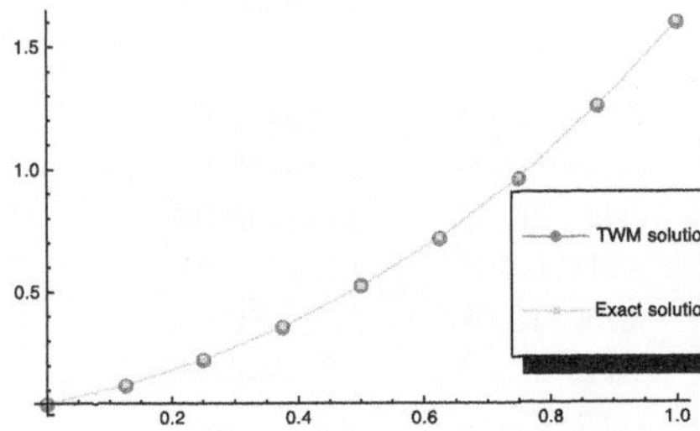

FIGURE 8.43 Numerical solution by TWM for $k = 4, M = 3$.

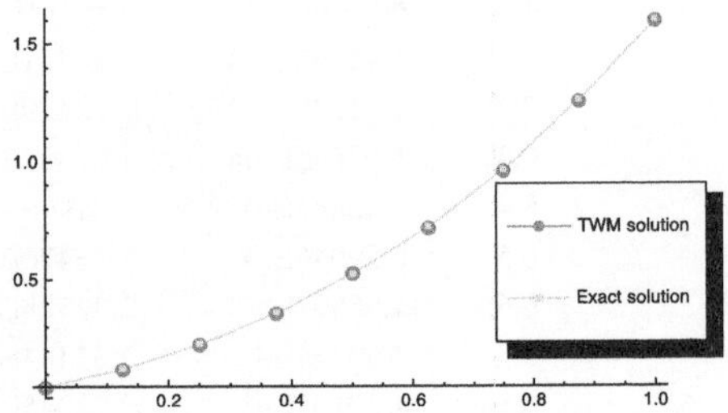

FIGURE 8.44 Numerical solution by TWM for $k = 5, M = 3$.

Figures 8.45–8.48 represent the error distribution by TWM for $k = 2,3,4,5$ and $M = 3$.

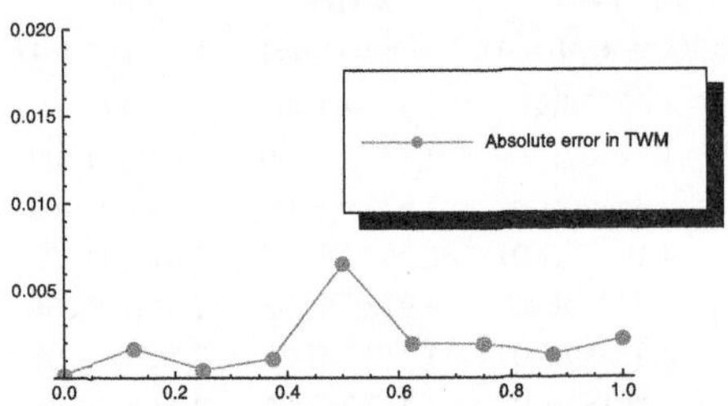

FIGURE 8.45 Error estimation by TWM for $k = 2, M = 3$.

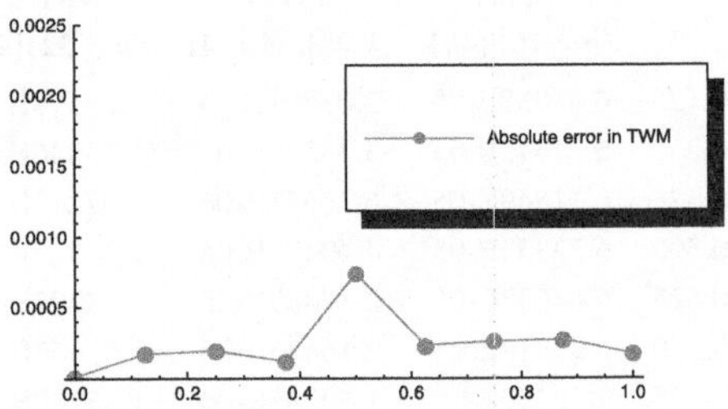

FIGURE 8.46 Error estimation by TWM for $k = 3, M = 3$.

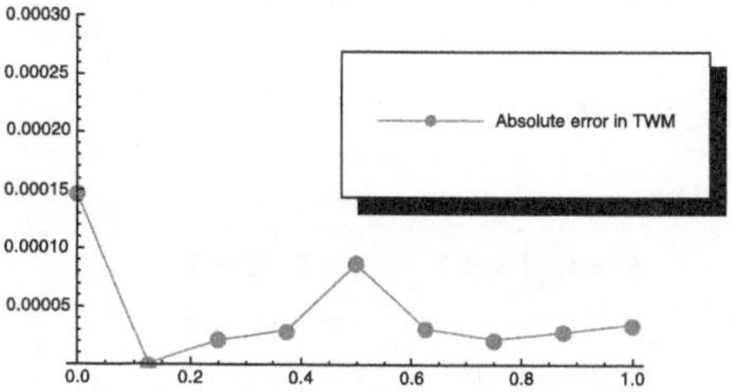

FIGURE 8.47 Error estimation by TWM for $k=4, M=3$.

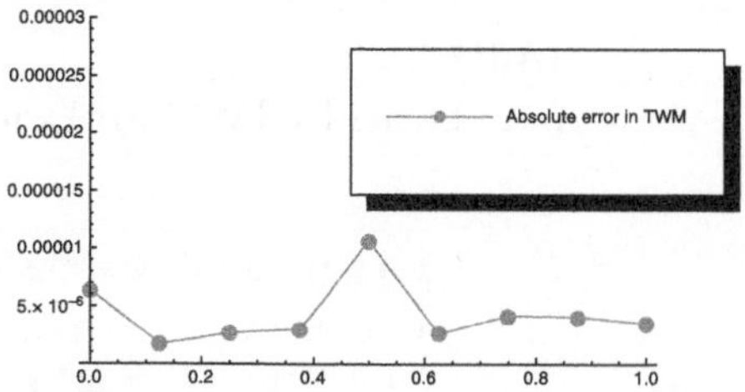

FIGURE 8.48 Error estimation by TWM for $k=5, M=3$.

TABLE 8.19
L^2-Error Norm and L^∞-Error Norm for Example 8.4.1

k, M	L^2 – Error norm	L^∞ – Error norm
$k=2, M=3$	1.77081E-03	1.75963E-03
$k=3, M=3$	3.21167E-04	6.20501E-04
$k=4, M=3$	9.3518E-05	2.90676E-04
$k=5, M=3$	2.76674E-05	1.21136E-04

Example 8.4.2

Consider the following linear FVIDE with a weakly singular kernel [54]:

$$ {}_0^C D_x^\alpha u(x) = f(x) + a(x)u(x) + \int_0^x \frac{u(z)}{(x-z)^{0.5}}\,dz, \quad x \in [0,1], \tag{8.62} $$

$$ u(0) = 0, $$

where

$$ f(x) = 2x \text{ and } a(x) = -\frac{16}{15}x^{\frac{1}{2}}. $$

The exact solution is $u(x) = x^2$, when $\alpha = 1$.

Table 8.20 demonstrates the $\tilde{\rho}(x)$ values obtained by the TWM. The numerical results and $\tilde{\rho}(x)$ error distributions by TWM are plotted in Figures 8.49–8.56. For different values of α, Figures 8.57 and 8.58 display the comparison of exact solution with the TWM solution. The L^2 and L^∞ -error norms for $k = 2,3,4,5$ and $M = 3$ are presented in Table 8.21.

TABLE 8.20
$\tilde{\rho}(x)$ Errors by TWM for Example 8.4.2

	$\tilde{\rho}(x)$			
x	$k=2, M=3$	$k=3, M=3$	$k=4, M=3$	$k=5, M=3$
0.1	1.96104E-05	2.33174E-06	5.8589E-07	1.97807E-09
0.2	2.43755E-05	6.5277E-06	2.59819E-08	1.40943E-10
0.3	5.44247E-05	3.20437E-07	1.04202E-08	1.49787E-09
0.4	7.0537E-05	3.99678E-07	1.05897E-08	1.09318E-09
0.5	1.14826E-05	1.19764E-08	1.37947E-08	1.06149E-09
0.6	4.19023E-06	3.65882E-07	3.5776E-08	1.70957E-09
0.7	1.99218E-06	6.79901E-07	3.57278E-08	1.72876E-09
0.8	7.06465E-06	4.861E-07	3.73598E-08	2.1964E-09
0.9	1.10272E-05	7.17788E-07	3.99176E-08	2.2628E-09

Figures 8.49–8.52 manifest the numerical results by TWM for $k = 2,3,4,5$ and $M = 3$.

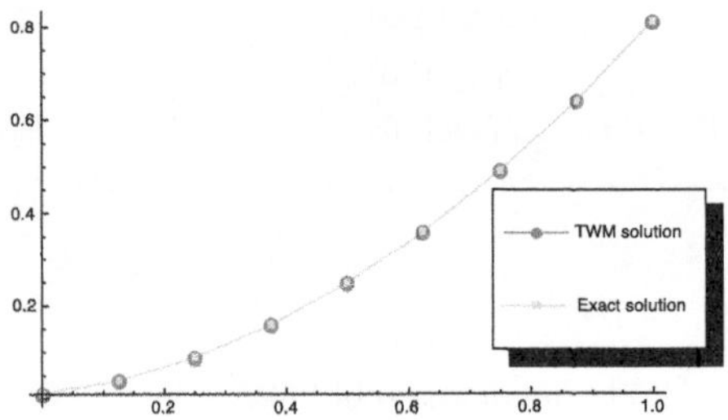

FIGURE 8.49 Numerical solution by TWM for $k = 2, M = 3$.

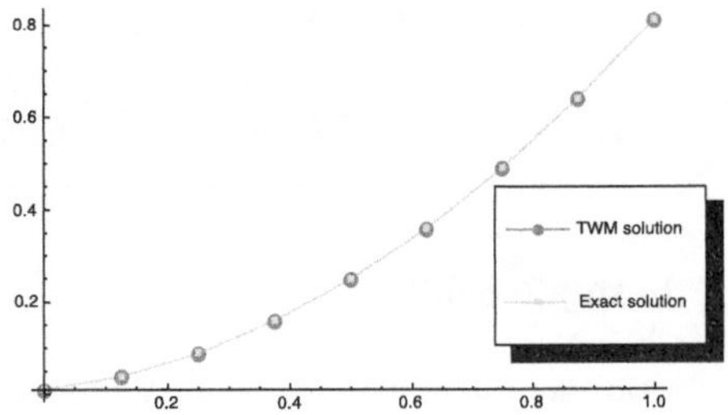

FIGURE 8.50 Numerical solution by TWM for $k = 3, M = 3$.

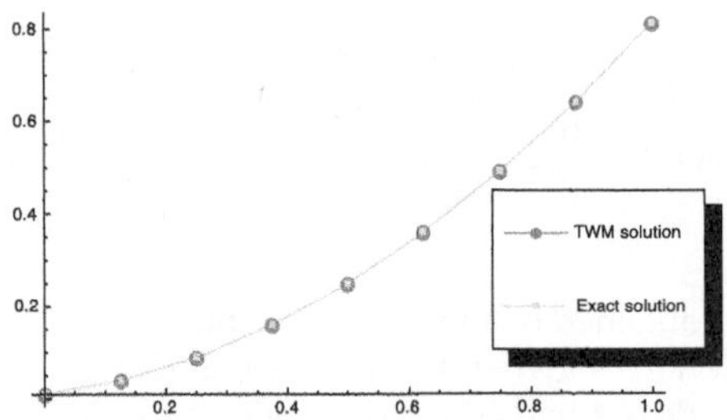

FIGURE 8.51 Numerical solution by TWM for $k = 4, M = 3$.

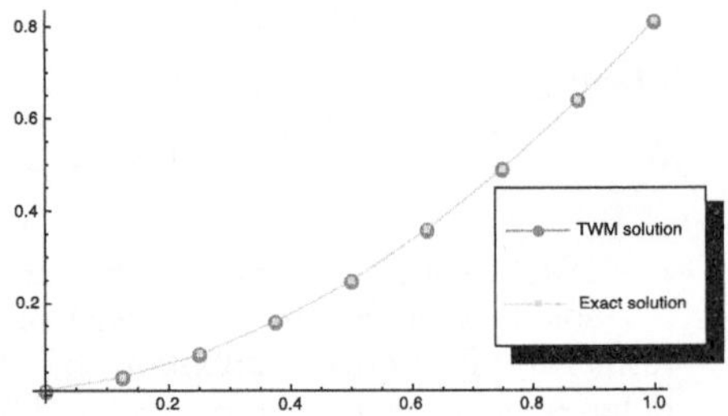

FIGURE 8.52 Numerical solution by TWM for $k = 5, M = 3$.

Figures 8.53–8.56 represent the error distribution by TWM for $k = 2,3,4,5$ and $M = 3$.

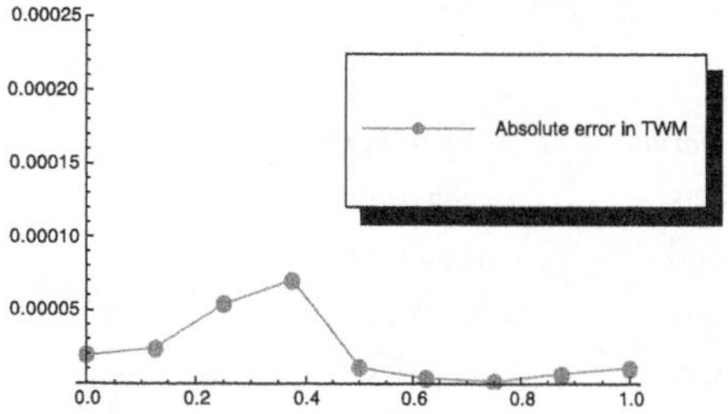

FIGURE 8.53 Error estimation by TWM for $k = 2, M = 3$.

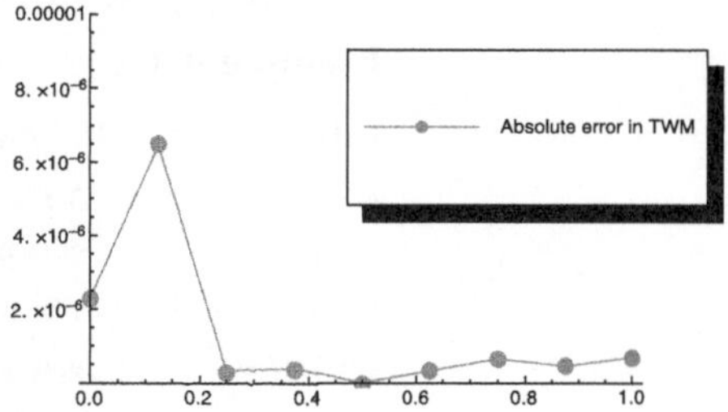

FIGURE 8.54 Error estimation by TWM for $k = 3, M = 3$.

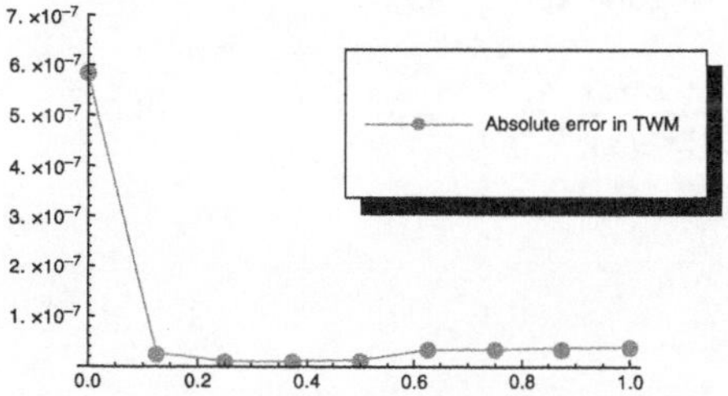

FIGURE 8.55 Error estimation by TWM for $k = 4, M = 3$.

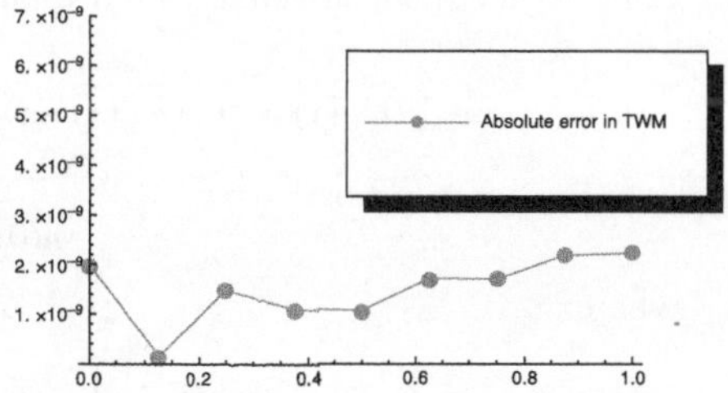

FIGURE 8.56 Error estimation by TWM for $k = 5, M = 3$.

Figures 8.57 and 8.58 depict the comparison of TWM solutions and exact solution for different values of α for $k = 2, M = 3$ and $k = 3, M = 3$.

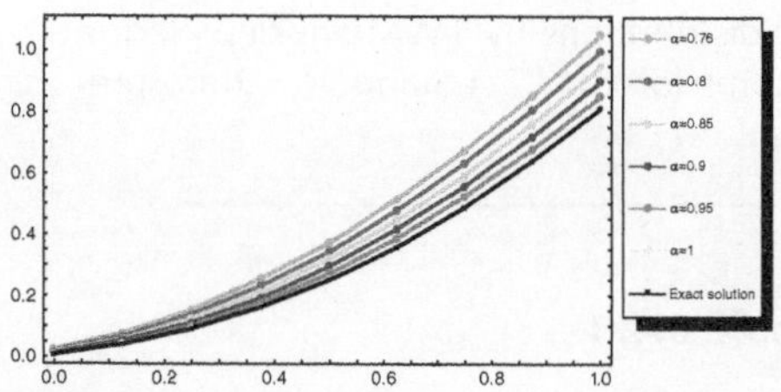

FIGURE 8.57 Exact solution and TWM solutions with different values of α for $k = 2, M = 3$.

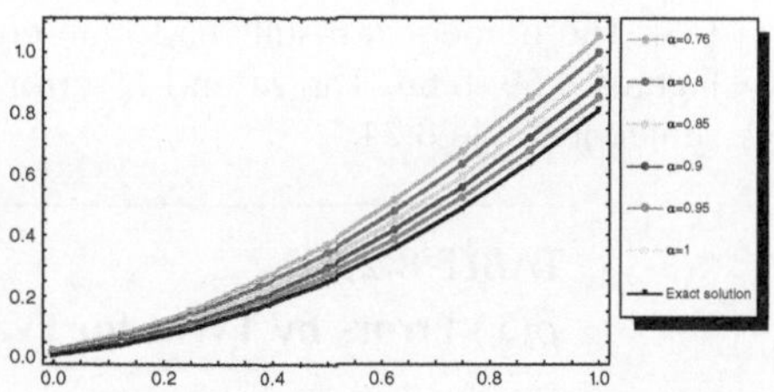

FIGURE 8.58 Exact solution and TWM solutions with different values of α for $k = 3, M = 3$.

TABLE 8.21
L^2-Error Norm and L^∞-Error Norm for Example 8.4.2

k, M	L^2 – *Error norm*	L^∞ – *Error norm*
$k = 2, M = 3$	3.75874E-05	7.18675E-05
$k = 3, M = 3$	2.46389E-06	6.65064E-06
$k = 4, M = 3$	1.56189E-07	5.96895E-07
$k = 5, M = 3$	9.76685E-09	5.30385E-08

Example 8.4.3

Consider a nonlinear FVIDE with a weakly singular kernel [45]:

$$ {}_0^C D_x^{\frac{2}{3}} u(x) = f(x) + a(x)u(x) + \int_0^x \frac{u^2(z)}{(x-z)^{0.5}} dz, \quad x \in [0,1], \tag{8.63} $$

$$ u(0) = 0, $$

where

$$ f(x) = \frac{3\Gamma\left(\frac{1}{2}\right)}{4\Gamma\left(\frac{11}{6}\right)} x^{\frac{5}{6}} - x^{\frac{5}{2}} - \frac{32}{35} x^{\frac{7}{2}} \text{ and } a(x) = x. $$

The exact solution is $u(x) = x^{\frac{3}{2}}$.

Table 8.22 demonstrates the $\tilde{\rho}(x)$ values obtained by the TWM. Table 8.23 displays the comparison of $\tilde{\rho}(x)$ values between TWM and MHFs methods [45]. The numerical results and $\tilde{\rho}(x)$ error distributions by TWM are depicted in Figures 8.59–8.66. The L^2 and L^∞-error norms for $k = 2,3,4,5$ and $M = 3$ are presented in Table 8.24.

TABLE 8.22
$\tilde{\rho}(x)$ Errors by TWM for Example 8.4.3

	$\tilde{\rho}(x)$			
x	$k = 2, M = 3$	$k = 3, M = 3$	$k = 4, M = 3$	$k = 5, M = 3$
0.1	2.10056E-03	1.73642E-04	1.27237E-04	8.29372E-06
0.2	4.14167E-04	3.66526E-04	2.37128E-05	2.74939E-06
0.3	1.75759E-03	7.2966E-05	8.59675E-06	1.7821E-06
0.4	1.15201E-03	7.1469E-05	7.94223E-06	7.77587E-07
0.5	4.9759E-04	8.96836E-05	1.39122E-05	1.93515E-06
0.6	2.51306E-04	2.20743E-05	5.15041E-06	4.94353E-07
0.7	6.98625E-05	3.66779E-05	3.3797E-06	4.63909E-07
0.8	3.18232E-04	2.52292E-05	2.09779E-06	3.97193E-07
0.9	3.30742E-04	2.27912E-05	2.65159E-06	2.35123E-07

TABLE 8.23
Comparison of $\tilde{\rho}(x)$ Errors for Example 8.4.3

x	MHFs ($n = 16$) [45]	TWM ($k = 3, M = 4$)	MHFs ($n = 32$) [45]	TWM ($k = 4, M = 4$)	MHFs ($n = 64$)[45]	TWM ($k = 5, M = 4$)
0.125	1.54710E-04	1.41518E-04	5.78251E-05	1.5561E-06	1.70440E-05	5.22423E-07
0.250	1.84256E-04	4.70073E-06	5.27913E-05	1.47661E-06	1.49408E-05	1.212E-07
0.375	1.93902E-04	6.45532E-06	5.38168E-05	6.41759E-07	1.50946E-05	4.82605E-08
0.500	2.20346E-04	4.14E-06	6.05455E-05	3.42001E-07	1.69314E-05	2.46619E-08
0.625	2.72394E-04	1.59144E-06	7.47583E-05	2.06709E-07	2.08999E-05	1.45147E-08
0.750	3.67651E-04	1.77439E-06	1.01259E-04	1.35814E-07	2.83408E-05	9.34954E-09
0.875	5.43833E-04	6.497E-07	1.50631E-04	9.4586E-08	4.22319E-05	6.57401E-09

Figures 8.59–8.62 manifest the numerical results by TWM for $k = 2,3,4,5$ and $M = 3$.

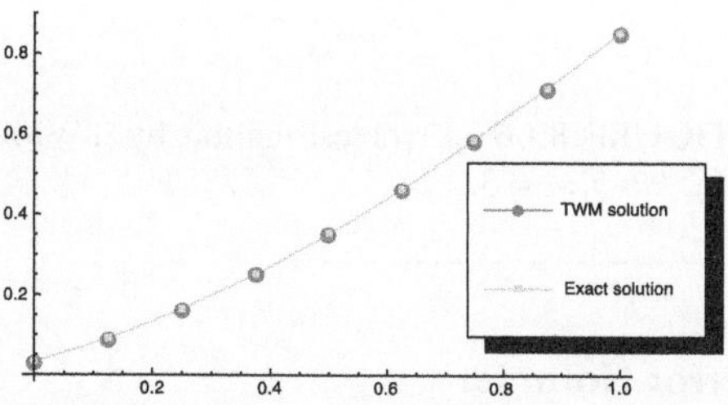

FIGURE 8.59 Numerical solution by TWM for $k = 2, M = 3$.

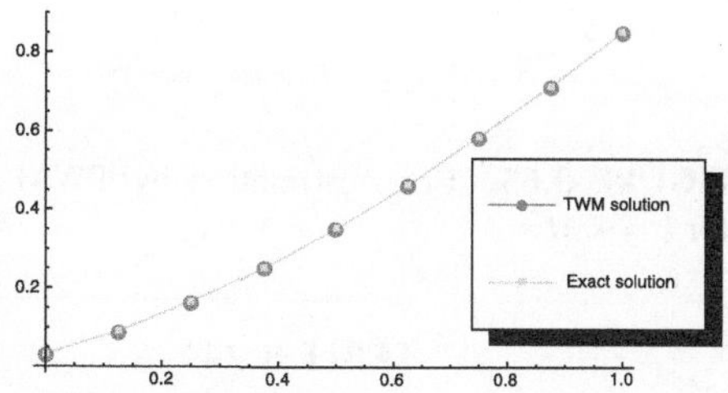

FIGURE 8.60 Numerical solution by TWM for $k = 3, M = 3$.

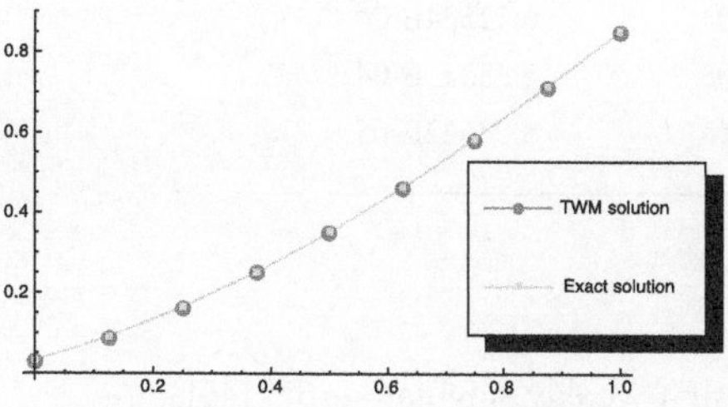

FIGURE 8.61 Numerical solution by TWM for $k = 4, M = 3$.

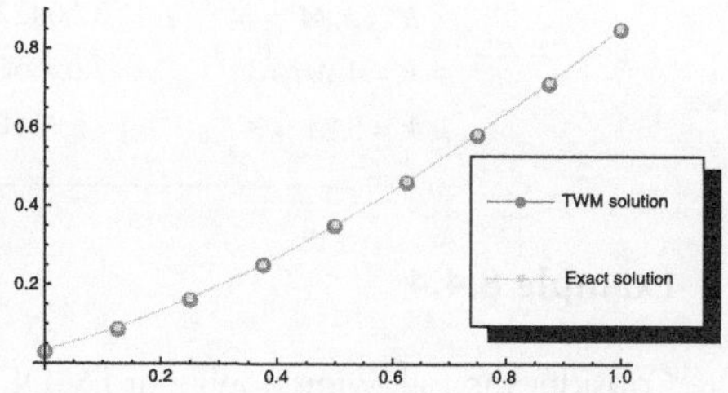

FIGURE 8.62 Numerical solution by TWM for $k = 5, M = 3$.

Figures 8.63–8.66 represent the error distribution by TWM for $k = 2,3,4,5$ and $M = 3$.

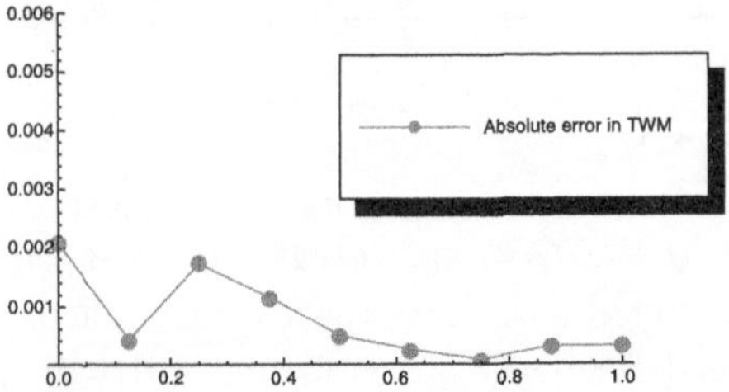

FIGURE 8.63 Error estimation by TWM for $k = 2, M = 3$.

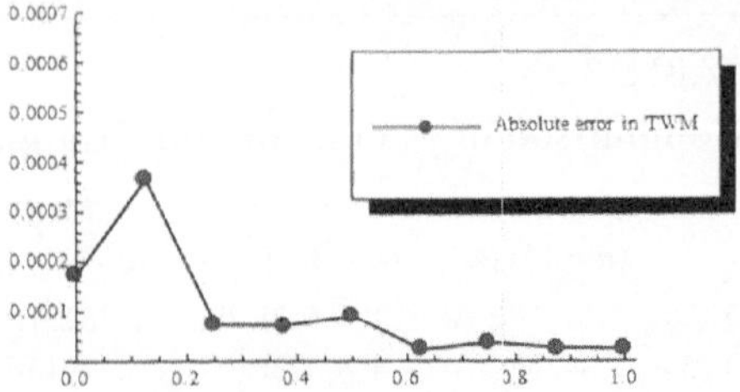

FIGURE 8.64 Error estimation by TWM for $k = 3, M = 3$.

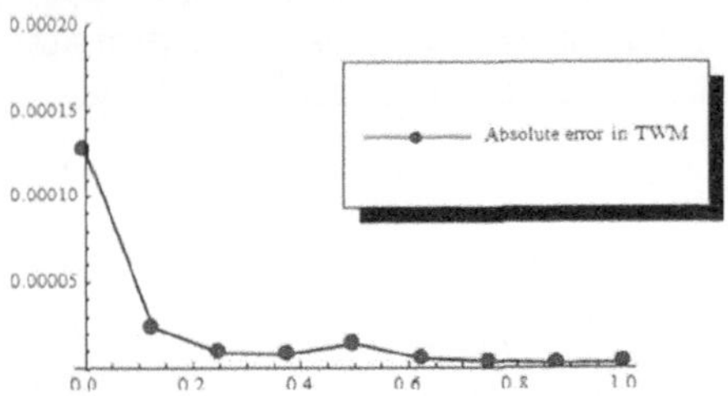

FIGURE 8.65 Error estimation by TWM for $k = 4, M = 3$.

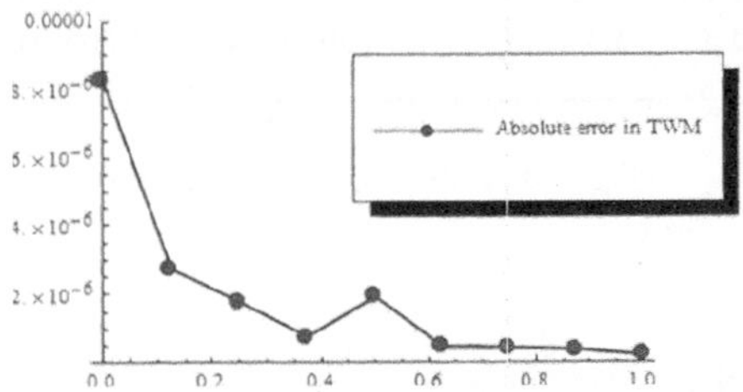

FIGURE 8.66 Error estimation by TWM for $k = 5, M = 3$.

TABLE 8.24
L^2-Error Norm and L^∞-Error Norm for Example 8.4.3

k, M	L^2 – Error norm	L^∞ – Error norm
$k = 2, M = 3$	1.234E-03	1.92233E-03
$k = 3, M = 3$	3.06422E-04	6.52524E-04
$k = 4, M = 3$	7.65656E-05	2.28842E-04
$k = 5, M = 3$	1.914E-05	8.07622E-05

Example 8.4.4

Consider the following nonlinear FVIDE with a weakly singular kernel [45]:

$$ {}_0^C D_x^\alpha u(x) = f(x) + \int_0^x \frac{u^2(z)}{(x-z)^{0.5}} dz, \quad x \in [0,1], \tag{8.64} $$

$$ u(0) = 0, $$

where

$$ f(x) = 3x^2 - \frac{\sqrt{\pi}\,\Gamma(7)}{\Gamma\left(\frac{15}{2}\right)} x^{\frac{13}{2}}. $$

The exact solution is $u(x) = x^3$, when $\alpha = 1$.

Table 8.25 demonstrates the $\tilde{\rho}(x)$ values obtained by the TWM. Table 8.26 displays the comparison of $\tilde{\rho}(x)$ values between TWM and MHFs methods [45]. The numerical results and $\tilde{\rho}(x)$ error distributions by TWM are plotted in Figures 8.67–8.74. For different values of α, Figures 8.75 and 8.76 display the comparison of exact solution with the TWM solution. The L^2 and L^∞-error norms for $k = 2,3,4,5$ and $M = 3$ are presented in Table 8.27.

TABLE 8.25
$\tilde{\rho}(x)$ Errors by TWM for Example 8.4.4

	$\tilde{\rho}(x)$			
x	$k = 2, M = 3$	$k = 3, M = 3$	$k = 4, M = 3$	$k = 5, M = 3$
0.1	2.24447E-03	2.18759E-04	3.51562E-05	3.41794E-06
0.2	1.75155E-03	2.81246E-04	2.7338E-05	4.39415E-06
0.3	1.74636E-03	2.80466E-04	2.73527E-05	4.39352E-06
0.4	2.24926E-03	2.17716E-04	3.50867E-05	3.41862E-06
0.5	6.70415E-03	7.93812E-04	9.80992E-05	1.22247E-05
0.6	2.10203E-03	2.20314E-04	3.49729E-05	3.40986E-06
0.7	1.81437E-03	2.7572E-04	2.70591E-05	4.37792E-06
0.8	1.56714E-03	2.68409E-04	2.74575E-05	4.37263E-06
0.9	2.0425E-03	2.05799E-05	3.43591E-05	3.42386E-06

TABLE 8.26
Comparison of $\tilde{\rho}(x)$ Errors for Example 8.4.4

x	MHFs ($n = 16$)[45]	TWM ($k = 3, M = 4$)	MHFs ($n = 32$)[45]	TWM ($k = 4, M = 4$)	MHFs ($n = 64$) [45]	TWM ($k = 5, M = 4$)
0.125	4.75615E-09	1.00842E-08	8.08279E-10	1.24458E-09	6.89080E-11	1.7489E-11
0.250	1.46318E-07	2.25235E-07	1.24726E-08	3.16579E-09	9.01380E-10	5.18944E-11
0.375	7.47778E-07	8.25911E-08	5.67297E-08	5.88184E-09	3.92295E-09	1.03856E-10
0.500	2.25373E-06	5.72476E-07	1.62864E-07	9.39111E-09	1.10301E-08	1.73354E-10
0.625	5.19464E-06	2.27175E-07	3.65499E-07	1.36914E-08	2.44518E-08	2.60383E-10
0.750	1.01464E-05	1.06135E-06	7.02044E-07	1.87794E-08	4.65856E-08	3.64913E-10
0.875	1.76618E-05	4.43716E-07	1.20829E-06	2.46501E-08	7.97163E-08	4.86903E-10

Figures 8.67–8.70 manifest the numerical results by TWM for $k = 2,3,4,5$ and $M = 3$.

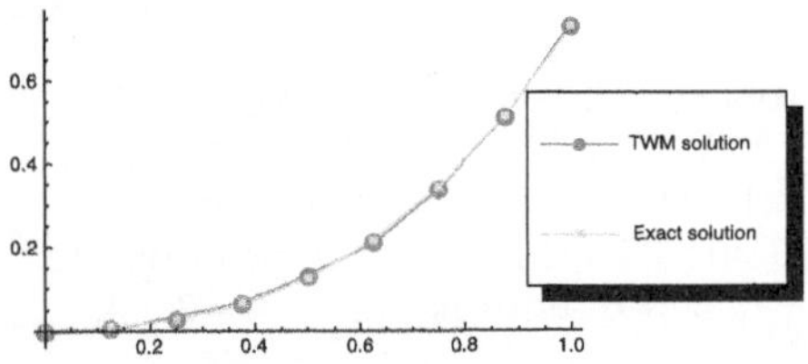

FIGURE 8.67 Numerical solution by TWM for $k = 2, M = 3$.

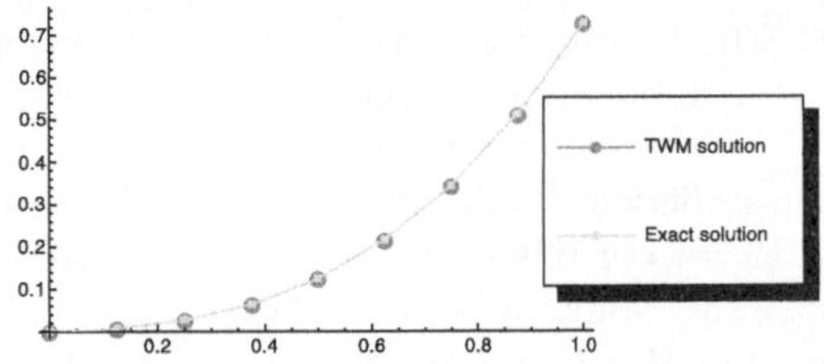

FIGURE 8.68 Numerical solution by TWM for $k = 3, M = 3$.

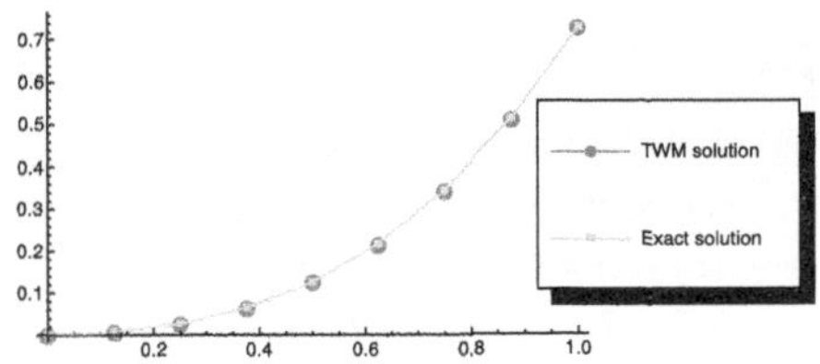

FIGURE 8.69 Numerical solution by TWM for $k = 4, M = 3$.

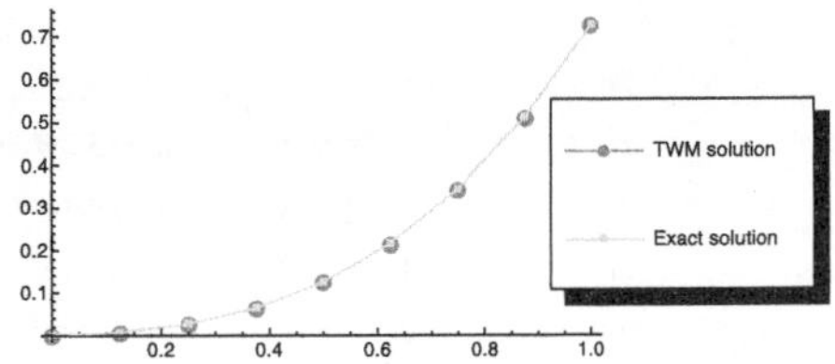

FIGURE 8.70 Numerical solution by TWM for $k = 5, M = 3$.

Figures 8.71–8.74 display the error distribution by TWM for $k = 2,3,4,5$ and $M = 3$.

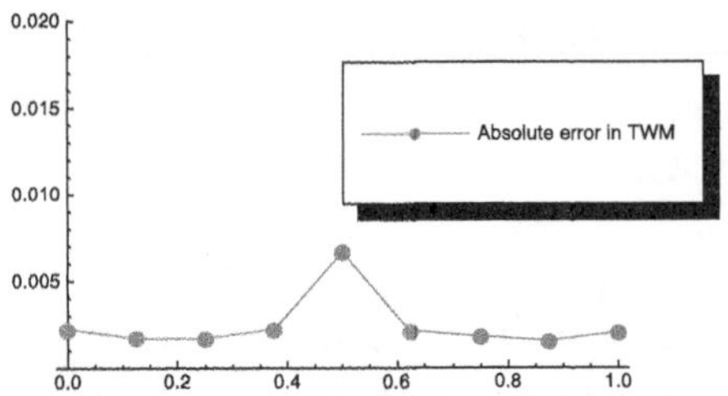

FIGURE 8.71 Error estimation by TWM for $k = 2, M = 3$.

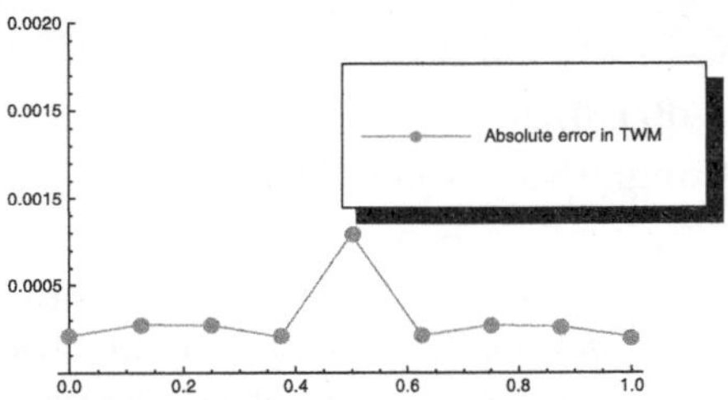

FIGURE 8.72 Error estimation by TWM for $k = 3, M = 3$.

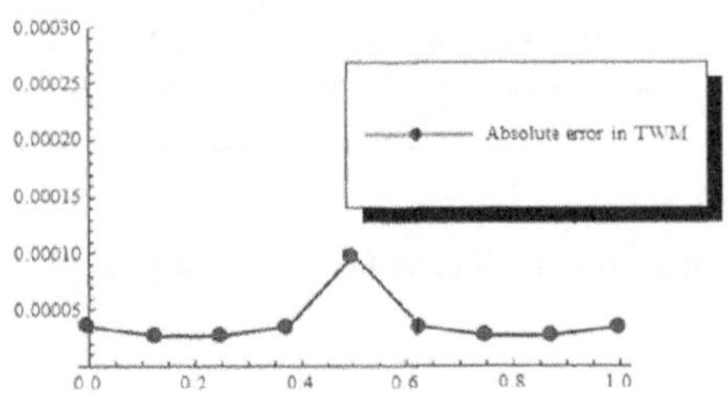

FIGURE 8.73 Error estimation by TWM for $k = 4, M = 3$.

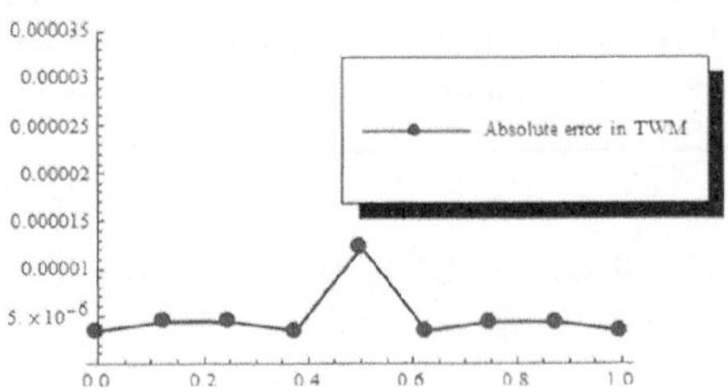

FIGURE 8.74 Error estimation by TWM for $k = 5, M = 3$.

Figures 8.75 and 8.76 depict the comparison of TWM solutions and exact solution for different values of α for $k = 2, M = 3$ and $k = 3, M = 3$.

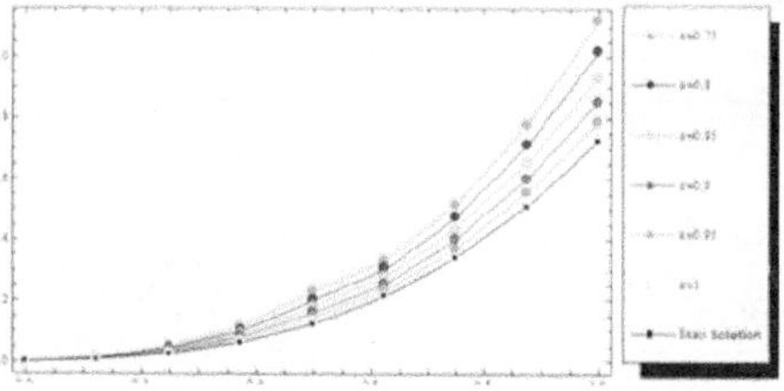

FIGURE 8.75 Exact solution and TWM solutions with different values of α for $k = 2, M = 3$.

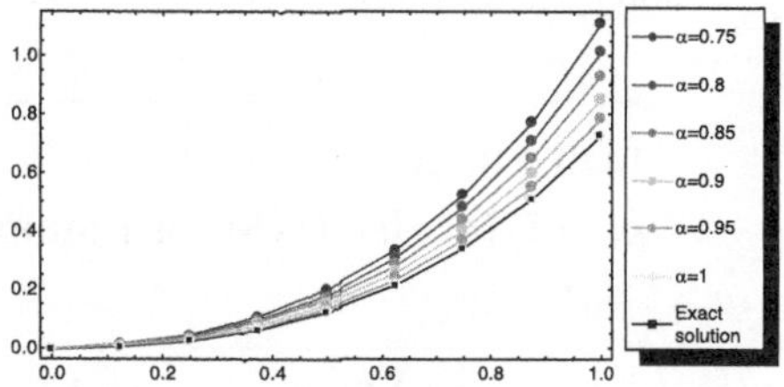

FIGURE 8.76 Exact solution and TWM solutions with different values of α for $k = 3, M = 3$.

TABLE 8.27
L^2-Error Norm and L^∞-Error Norm for Example 8.4.4

k, M	L^2 – Error norm	L^∞ – Error norm
$k = 2, M = 3$	2.36633E-03	1.6206E-03
$k = 3, M = 3$	2.95395E-04	2.02548E-04
$k = 4, M = 3$	3.69127E-05	2.53183E-05
$k = 5, M = 3$	4.61386E-06	3.16479E-06

Example 8.4.5

Consider the following nonlinear FVIDE with a weakly singular kernel [55]:

$$ {}_0^C D_x^{\frac{4}{5}} u(x) = h(x) + u^2(x) + \int_0^x \frac{u^2(z)}{(x-z)^{0.5}} dz, \quad x \in [0,1], \tag{8.65}$$

$$u(0) = 0,$$

where

$$h(x) = \frac{\Gamma(4)}{\Gamma\left(\frac{16}{5}\right)} x^{\frac{11}{5}} + \frac{\Gamma\left(\frac{7}{3}\right)}{\Gamma\left(\frac{23}{15}\right)} x^{\frac{8}{15}} - x^6 - 2x^{\frac{13}{3}} - x^{\frac{8}{3}}$$

$$-\sqrt{\pi}\left(\frac{\Gamma(7)}{\Gamma\left(\frac{15}{2}\right)} x^{\frac{13}{2}} + \frac{\Gamma\left(\frac{11}{3}\right)}{\Gamma\left(\frac{25}{6}\right)} x^{\frac{19}{6}} + \frac{2\Gamma\left(\frac{16}{3}\right)}{\Gamma\left(\frac{35}{6}\right)} x^{\frac{29}{6}}\right).$$

The exact solution is $u(x) = x^3 + x^{\frac{4}{3}}$.

Table 8.28 demonstrates the $\tilde{\rho}(x)$ values obtained by the TWM. The numerical results and $\tilde{\rho}(x)$ error distributions by TWM are plotted in Figures 8.77–8.84. The L^2 and L^∞-error norms for $k = 2,3,4,5$ and $M = 3$ are presented in Table 8.29.

TABLE 8.28
$\tilde{\rho}(x)$ Errors by TWM for Example 8.4.5

	$\tilde{\rho}(x)$			
x	$r = 2, S = 3$	$r = 3, S = 3$	$r = 4, S = 3$	$r = 5, S = 3$
0.1	5.32715E-05	2.96661E-04	1.07225E-04	7.11313E-06
0.2	2.18864E-03	8.53069E-05	2.50808E-07	1.60854E-06
0.3	9.16545E-04	2.27581E-04	2.03575E-05	2.6756E-06
0.4	1.17047E-03	1.39258E-04	2.75902E-05	2.75227E-06
0.5	7.26097E-03	7.66983E-04	8.87363E-05	1.06863E-05
0.6	1.59762E-03	2.12831E-04	3.04117E-05	2.93482E-06
0.7	1.88767E-03	2.40784E-04	2.32284E-05	3.915E-06
0.8	1.00647E-03	2.09673E-04	2.61167E-05	4.0494E-06
0.9	1.56821E-03	1.79943E-04	2.89596E-05	3.2059E-06

Figures 8.77–8.80 represent the numerical results by TWM for $k = 2,3,4,5$ and $M = 3$.

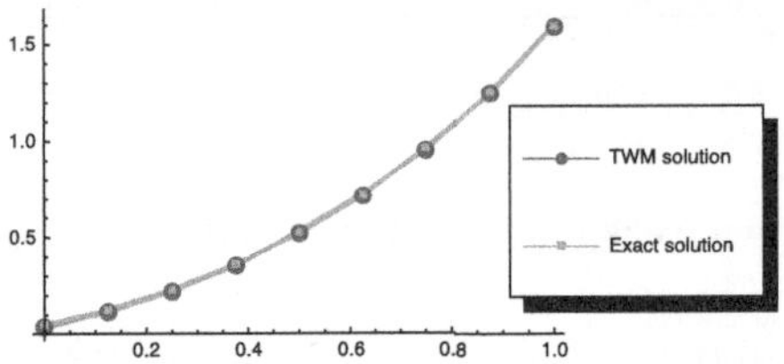

FIGURE 8.77 Numerical solution by TWM for $k = 2, M = 3$.

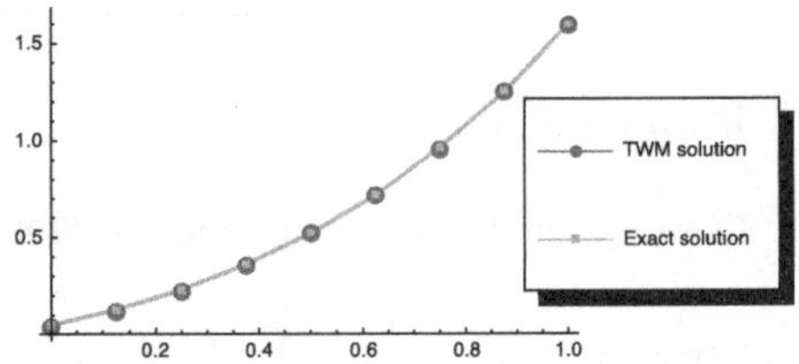

FIGURE 8.78 Numerical solution by TWM for $k = 3, M = 3$.

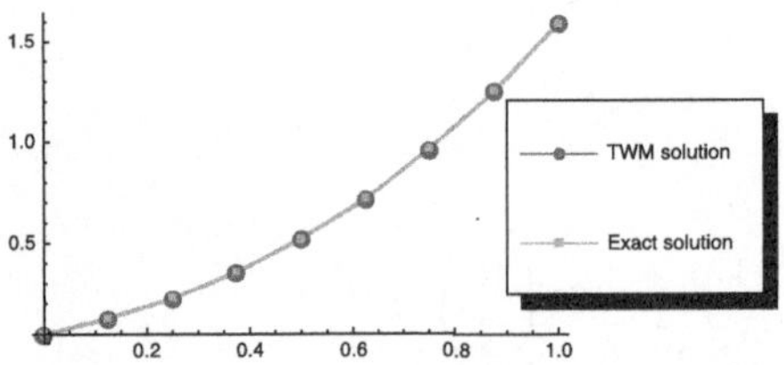

FIGURE 8.79 Numerical solution by TWM for $k = 4, M = 3$.

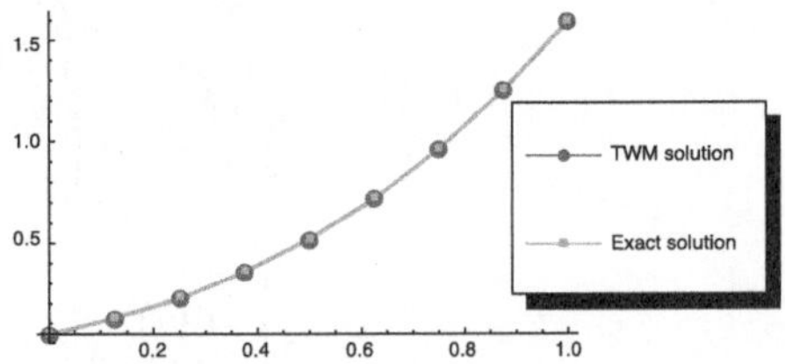

FIGURE 8.80 Numerical solution by TWM for $k = 5, M = 3$.

Figures 8.81–8.84 represent the absolute error estimation by TWM for $k = 2,3,4,5$ and $M = 3$.

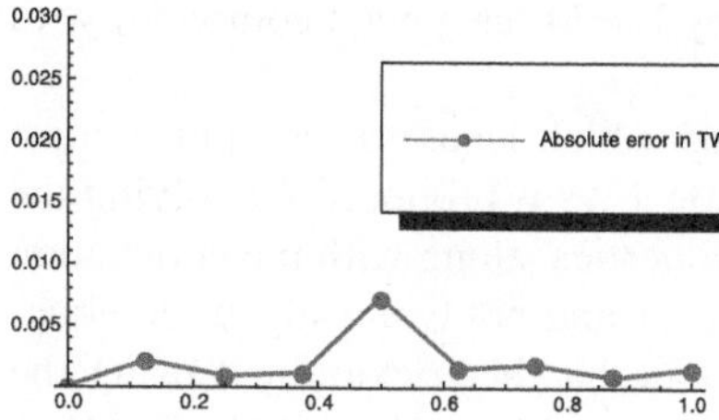

FIGURE 8.81 Error estimation by TWM for $k = 2, M = 3$.

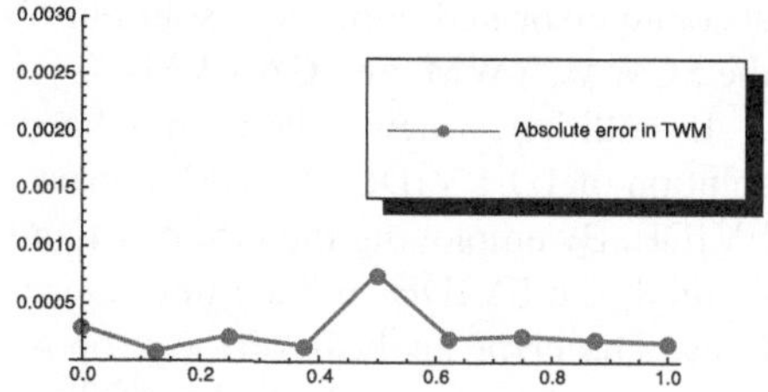

FIGURE 8.82 Error estimation by TWM for $k = 3, M = 3$.

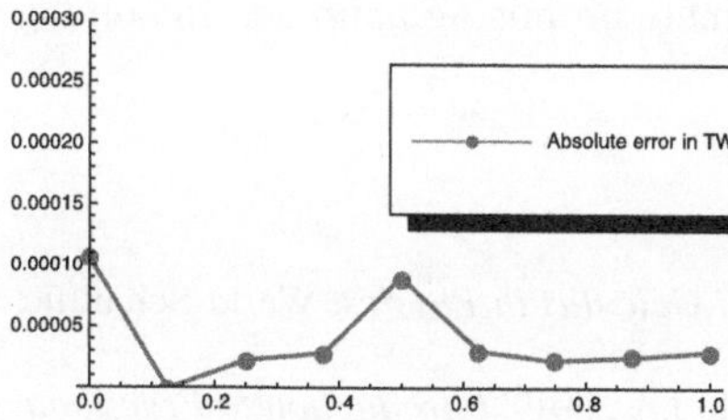

FIGURE 8.83 Error estimation by TWM for $k = 4, M = 3$.

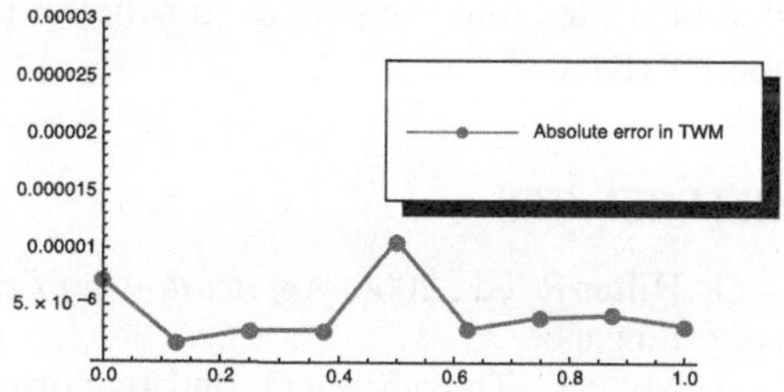

FIGURE 8.84 Error estimation by TWM for $k = 5, M = 3$.

TABLE 8.29
L^2-Error Norm and L^∞-Error Norm for Example 8.4.5

k, M	L^2 – Error norm	L^∞ – Error norm
$k = 2, M = 3$	1.89803E-03	1.87722E-03
$k = 3, M = 3$	3.73763E-04	6.9623E-04
$k = 4, M = 3$	1.07575E-04	3.28378E-04
$k = 5, M = 3$	3.13953E-05	1.3692E-04

8.5 CONCLUSION

In this chapter, a new scheme has been applied for the numerical solution of FVFIDE with a weakly singular kernel. The approached scheme is based on EWs for obtaining the approximate solution of FVFIDE. In the proposed method, by applying the OM of FI and properties of EWs, FVFIDEs convert to a system of linear algebraic equations, and that algebraic system has been solved by an appropriate numerical method. Moreover, the error analysis for the given function approximation has been studied by the sense of L^2-norm. The NCR analysis is presented to display the accuracy of the proposed method. Furthermore, some numerical experiments are solved to show the accuracy and applicability of the approached scheme. Moreover, the absolute error and numerical solution obtained by EWM have been compared with the SCWM, TWM, and CASWM.

In addition, a new scheme has been proposed which leads to the approximate solution of 1D-FVIDE. An OM approach based on TWs is proposed for solving the FVIDE. By employing the OM of FI and their properties, along with the collocation method, the FVIDEs reduce into a system of linear and nonlinear algebraic equations. The error analysis of the proposed technique has been examined using the definition of L^2-norm. Several numerical examples have been solved to demonstrate the accuracy and usefulness of the suggested numerical technique. Moreover, the NCR analysis of the numerical experiments is presented. Furthermore, a comparison of the error norms obtained by TWM with those obtained by other existing methods demonstrates that the proposed scheme is extremely promising approach in solving the FVIDEs.

REFERENCES

1. Hilfer R. ed., 2000, *Applications of Fractional Calculus in Physics*, World Scientific, Singapore.
2. Milici C., Draganescu G., and Tenreiro Machado J.A., 2019, *Introduction to Fractional Differential Equations*, Springer International Publishing, Cham, Switzerland.
3. Saha Ray S., 2020, *Nonlinear Differential Equations in Physics*, Springer Nature, Singapore.
4. Oldham K., and Spanier J., 1974, *The Fractional Calculus Theory and Applications of Differentiation and Integration to Arbitrary Order*, Elsevier, Amsterdam, Netherlands.
5. Kilbas A.A., Srivastava H.M., and Trujillo J.J., 2006, *Theory and Applications of Fractional Differential Equations, 204*, Elsevier, Amsterdam, Netherlands.
6. Koeller R.C., 1984, "Applications of fractional calculus to the theory of viscoelasticity", *Journal of Applied Mechanics*, 51, pp. 299–307.
7. Tang B.Q., and Li X.F., 2008, "Solution of a class of Volterra integral equations with singular and weakly singular kernels", *Applied Mathematics and Computation*, 199(2), pp. 406–413.
8. Zozulya V.V., and Gonzalez-Chi P.I., 1999, "Weakly singular, singular and hypersingular integrals in 3-D elasticity and fracture mechanics", *Journal of the Chinese Institute of Engineers*, 22(6), pp. 763–775.
9. Kythe P., and Puri P., 2011, *Computational Methods for Linear Integral Equations*, Springer Science & Business Media, Singapore.
10. Saha Ray S., 2021, "Similarity solutions for Keller-Segel model with fractional diffusion of cells", *Mathematical Methods in the Applied Sciences*, 44(10), pp. 8379–8396.
11. Mesgarani H., Aghdam Y.E., and Tavakoli H., 2021, "Numerical simulation to solve two-dimensional temporal-space fractional Bloch–Torrey equation taken of the spin magnetic moment diffusion", *International Journal of Applied and Computational Mathematics*, 7(3), pp. 1–14.
12. Baleanu D., Jajarmi A., Asad J.H., and Blaszczyk T., 2017, "The motion of a bead sliding on a wire in fractional sense", *Acta Physica Polonica A*, 131, pp. 1561–1564.

13. Ichise M., Nagayanagi Y., and Kojima T., 1971, "An analog simulation of non-integer order transfer functions for analysis of electrode processes", *Journal of Electroanalytical Chemistry and Interfacial Electrochemistry*, 33(2), pp. 253–265.
14. Jajarmi A., Hajipour M., Mohammadzadeh E., and Baleanu D., 2018, "A new approach for the nonlinear fractional optimal control problems with external persistent disturbances", *Journal of the Franklin Institute*, 355(9), pp. 3938–3967.
15. Bagley R.L., and Torvik P.J., 1985, "Fractional calculus in the transient analysis of viscoelastically damped structures", *AIAA Journal*, 23(6), pp. 918–925.
16. Ali K.K., and Gómez-Aguilar J.F., 2021, "New solitary wave solutions of the space–time coupled equal width wave equation (CEWE) and coupled modified equal width wave equation (CMEWE)", *International Journal of Applied and Computational Mathematics*, 7(4), pp. 1–19.
17. He J.H., 1999, "Some applications of nonlinear fractional differential equations and their approximations", *Bulletin of Science, Technology & Society*, 15(2), pp. 86–90.
18. Attia N., Akgül A., Seba D., and Nour A., 2021, "Numerical solutions to the time-fractional Swift–Hohenberg equation using reproducing kernel Hilbert space method", *International Journal of Applied and Computational Mathematics*, 7(5), pp. 1–35.
19. Golbabai A., Nikan O., and Nikazad T., 2019, "Numerical investigation of the time fractional mobile-immobile advection-dispersion model arising from solute transport in porous media", *International Journal of Applied and Computational Mathematics*, 5(3), p. 50.
20. Baleanu D., Jajarmi A., and Hajipour M., 2018, "On the nonlinear dynamical systems within the generalized fractional derivatives with Mittag–Leffler kernel", *Nonlinear Dynamics*, 94(1), pp. 397–414.
21. Saghali S., Javidi M., and Saei F.D., 2019, "Analytical solution of a fractional differential equation in the theory of viscoelastic fluids", *International Journal of Applied and Computational Mathematics*, 5(3), pp. 1–13.
22. Zhang X., Tang B., and He Y., 2011, "Homotopy analysis method for higher-order fractional integro-differential equations", *Computers & Mathematics with Applications*, 62(8), pp. 3194–3203.
23. Jiang W., and Tian T., 2015, "Numerical solution of nonlinear Volterra integro-differential equations of fractional order by the reproducing kernel method", *Applied Mathematical Modelling*, 39(16), pp. 4871–4876.
24. Nawaz Y., 2011, "Variational iteration method and homotopy perturbation method for fourth-order fractional integro-differential equations", *Computers & Mathematics with Applications*, 61(8), pp. 2330–2341.
25. Song L., and Wang W., 2013, "A new improved Adomian decomposition method and its application to fractional differential equations", *Applied Mathematical Modelling*, 37(3), pp. 1590–1598.
26. Zhao J., Xiao J., and Ford N.J., 2014, "Collocation methods for fractional integro-differential equations with weakly singular kernels", *Numerical Algorithms*, 65(4), pp. 723–743.
27. Mokhtary P., 2016, "Discrete Galerkin method for fractional integro-differential equations", *Acta Mathematica Scientia*, 36(2), pp. 560–578.
28. Wang Y., and Zhu L., 2017, "Solving nonlinear Volterra integro-differential equations of fractional order by using Euler wavelet method", *Advances in Difference Equations*, 1, pp. 1–16.
29. Wang Y., and Zhu L., 2016, "SCW method for solving the fractional integro-differential equations with a weakly singular kernel", *Applied Mathematics and Computation*, 275, pp. 72–80.

30. Eslahchi M.R., Dehghan M., and Parvizi M., 2014, "Application of the collocation method for solving nonlinear fractional integro-differential equations", *Journal of Computational and Applied Mathematics*, 257, pp. 105–128.
31. Momani S., and Noor M.A., 2006, "Numerical methods for fourth-order fractional integro-differential equations", *Applied Mathematics and Computation*, 182(1), pp. 754–760.
32. Mokhtary P., 2017, "Numerical analysis of an operational Jacobi Tau method for fractional weakly singular integro-differential equations", *Applied Numerical Mathematics*, 121, pp. 52–67.
33. Koundal R., Kumar R., Kumar R., Srivastava K., and Baleanu D., 2021, "A novel collocated-shifted Lucas polynomial approach for fractional integro-differential equations", *International Journal of Applied and Computational Mathematics*, 7(4), pp. 1–19.
34. Saeedi H., and Moghadam M.M., 2011, "Numerical solution of nonlinear Volterra integro-differential equations of arbitrary order by CAS wavelets", *Communications in Nonlinear Science and Numerical Simulation*, 16(3), pp. 1216–1226.
35. Heydari M.H., Hooshmandasl M.R., Cattani C., and Li M., 2013, "Legendre wavelets method for solving fractional population growth model in a closed system", *Mathematical Problems in Engineering*, 161030, p. 8.
36. Maleknejad K., Sahlan M.N., and Ostadi A., 2013, "Numerical solution of fractional integro-differential equation by using cubic B-spline wavelets", *Proceedings of the World Congress on Engineering*, 1, pp. 1–6.
37. Keshavarz E., and Ordokhani Y., 2019, "A fast numerical algorithm based on the Taylor wavelets for solving the fractional integro-differential equations with weakly singular kernels", *Mathematical Methods in the Applied Sciences*, 42(13), pp. 4427–4443.
38. Sahu P.K., and Saha Ray S., 2016, "A numerical approach for solving nonlinear fractional Volterra–Fredholm integro-differential equations with mixed boundary conditions", *International Journal of Wavelets, Multiresolution and Information Processing*, 14(05), p. 1650036.
39. Heydari M.H., Hooshmandasl M.R., Mohammadi F., and Cattani C., 2014, "Wavelets method for solving systems of nonlinear singular fractional Volterra integro-differential equations", *Communications in Nonlinear Science and Numerical Simulation*, 19(1), pp. 37–48.
40. Saha Ray S., and Bera R.K., 2005, "An approximate solution of a nonlinear fractional differential equation by Adomian decomposition method", *Applied Mathematics and Computation*, 167(1), pp. 561–571.
41. Behera S., and Saha Ray S., 2021, "Euler wavelets method for solving fractional-order linear Volterra–Fredholm integro-differential equations with weakly singular kernels", *Computational and Applied Mathematics*, 40(6), pp. 1–30.
42. Zhao J., Xiao J., and Ford N.J., 2014, "Collocation methods for fractional integro-differential equations with weakly singular kernels", *Numerical Algorithms*, 65(4), pp. 723–743.
43. Yi M., and Huang J., 2015, "CAS wavelet method for solving the fractional integro-differential equation with a weakly singular kernel", *International Journal of Computer Mathematics*, 92(8), pp. 1715–1728.
44. Yi M., Wang L., and Huang, J., 2016, "Legendre wavelets method for the numerical solution of fractional integro-differential equations with weakly singular kernel", *Applied Mathematical Modelling*, 40(4), pp. 3422–3437.
45. Nemati S., and Lima P.M., 2018, "Numerical solution of nonlinear fractional integro-differential equations with weakly singular kernels via a modification of hat functions", *Applied Mathematics and Computation*, 327, pp. 79–92.
46. Pedas A., Tamme E., and Vikerpuur M., 2016, "Spline collocation for fractional weakly singular integro-differential equations", *Applied Numerical Mathematics*, 110, pp. 204–214.

47. Avazzadeh Z., Heydari M.H., and Cattani C., 2019, "Legendre wavelets for fractional partial integro-differential viscoelastic equations with weakly singular kernels", *The European Physical Journal Plus*, 134(7), p. 368.
48. Behera S., and Saha Ray S., 2020, "An operational matrix based scheme for numerical solutions of nonlinear weakly singular partial integro-differential equations", *Applied Mathematics and Computation*, 367, p. 124771.
49. Podlubny I., 1998, *Fractional Differential Equations: An Introduction to Fractional Derivatives, Fractional Differential Equations, to Methods of Their Solution and Some of Their Applications*, Elsevier, Amsterdam, Netherlands.
50. Srivastava H.M., and Pinter A., 2004, "Remarks on some relationships between the Bernoulli and Euler polynomials", *Applied Mathematics Letters*, 17(4), pp. 375–380.
51. Saha Ray S., and Behera S., 2019, "A novel technique based on Bernoulli wavelets for numerical solutions of two-dimensional Fredholm integral equation of second kind", *Engineering Computations*, 36(6), pp. 1798–1819.
52. Stewart G.W., 1993, *Afternotes on Numerical Analysis*, University of Maryland, College Park, MD.
53. Toan P.T., Vo T.N., and Razzaghi M., 2019, "Taylor wavelet method for fractional delay differential equations", *Engineering with Computers*, 37, pp. 1–10.
54. Nemati S., Sedaghat S., and Mohammadi I., 2016, "A fast numerical algorithm based on the second kind Chebyshev polynomials for fractional integro-differential equations with weakly singular kernels", *Journal of Computational and Applied Mathematics*, 308, pp. 231–242.
55. Shi G., Gong Y., and Yi M., 2021, "Alternative Legendre polynomials method for nonlinear fractional integro-differential equations with weakly singular kernel", *Journal of Mathematics*, 2021, pp. 1–13.

9 Numerical Solutions of Fractional-Order Integro-Differential Equations

9.1 INTRODUCTION

One of the most significant components of mathematical analysis is fractional calculus. The theory of fractional calculus was developed in a conversation between Leibnitz and L'Hopital, and after its development, it drew the attention of physicists, mathematicians, and applied sciences researchers. In 1823, Abel was the first who applied fractional derivatives in finding the solution of an integral equation in the formulation of the Tautochrone problem. Fractional calculus plays an important role in the differentiation and integration of functions with non-integer orders [1,2]. The concept of fractional derivative has been introduced by two mathematicians, Leibnitz and L'Hopital, in 1695 [3]. In previous studies [4–7], fractional integral and fractional derivatives of the Riemann–Liouville concept have been used to solve numerous mathematical problems. Fractional calculus has applications in many disciplines, including electromagnetic waves [8], signal processing [9], image processing [10], viscoelasticity [11], fluid dynamic traffic [12], nonlinear dynamical systems [13], optimal control [14], economics [15], population growth [16], and continuum and statistical mechanics [17]. Due to the enormous contribution of fractional calculus to various disciplines of engineering, many researchers have proposed several approximate techniques such as Adomian decomposition method [18], operational Tau method [19], spline collocation method [20], and B-spline method [21] for the approximate solution of fractional equations.

In recent years, various polynomial operational matrices (OMs) have been derived using shifted Jacobi polynomials [22], shifted Chebyshev polynomials [23], shifted Legendre polynomials [24], and wavelet OMs have been derived using Chebyshev wavelets [25], Haar wavelets [26], and Legendre wavelets [27] for the solution of fractional differential equations (FDEs) and fractional integro-differential equations (FIDEs). Apart from that, different wavelet techniques [28–30] have been implemented for the solution of FIDEs.

Numerical solutions of VFIDEs have been studied by several authors such as Alkan and Hatipoglu [31], Khajehnasiri [32], Hamoud et al. [33], and Ali et al. [34]. Sahu and Saha Ray [35] have applied the Sinc–Galerkin technique for the numerical solution of fractional VFIDEs with weakly singular kernels. Zhou et al. [36] have solved fractional VFIDEs with mixed boundary conditions by using the Chebyshev

DOI: 10.1201/9781003453451-9

wavelet. Das et al. [37] have implemented the Homotopy perturbation method for solving Caputo-type fractional-order VFIDEs. The existence and uniqueness of solutions of fractional VFIDEs have been discussed in previous studies [38,39].

9.2 OUTLINE OF THE PRESENT STUDY

In this chapter, a wavelet-based OM scheme has been introduced to obtain the approximate solution of the linear and nonlinear fractional-order Volterra–Fredholm integro-differential equations (VFIDEs). For the suggested approach, the OM of the fractional integral for Taylor wavelets (TWs) has been constructed. Then, the fractional integral OM is utilized to reduce the solutions of linear and nonlinear VFIDEs to systems of linear and nonlinear algebraic equations, respectively. Moreover, the convergence and error estimation of the proposed technique has been analysed in this study. The numerical convergence rate is calculated to describe the accuracy of the presented method. Several illustrative experiments are included to verify the efficiency and validity of the proposed method. Moreover, the numerical results obtained by the suggested wavelet scheme have been compared with the other existing methods.

9.3 LINEAR AND NONLINEAR VOLTERRA–FREDHOLM INTEGRO-DIFFERENTIAL EQUATIONS

The main goal of this study is to implement the TWs method to solve the one-dimensional (1D) fractional linear and nonlinear VFIDEs. The fractional VFIDEs are considered as follows:

$$\begin{cases} {}_0^C D_x^{\alpha} u(x) = f(x) + \int_0^x k_1(x,z)\tilde{H}_1(u(z))\,dz + \int_0^1 k_2(x,z)\tilde{H}_2(u(z))\,dz, \\ u^{\rho}(0) = 0, \ \ \rho = 0,1,\ldots r-1, \ \ r-1 < \alpha \le r, \ \ r \in \mathbb{N}, \end{cases} \tag{9.1}$$

where $x \in \Omega := [0, L]$ and $u^{\rho}(x)$ stands for the ρ-th-order derivative of $u(x)$.

Here, $f(x)$ is a known function defined on Ω and $u(x)$ is the unknown function to be determined. The kernel functions $k_1(x,z)$ and $k_2(x,z)$ are defined on $\Lambda_1 := \{(x,z): 0 < x < L, 0 < z < x\}$ and $\Lambda_2 := \{(x,z): 0 < x < L, 0 < z < 1\}$, respectively. In this chapter, the terms $\tilde{H}_1(u(z))$ and $\tilde{H}_2(u(z))$ are considered as $\tilde{H}_1(u(z)) = \tilde{H}_2(u(z)) = u(z)$ and $\tilde{H}_1(u(z)) = \tilde{H}_2(u(z)) = u^p(z)$, $p > 1$ for linear and nonlinear cases, respectively, and ${}_0^C D_x^{\alpha}$ denotes the fractional derivative of fractional order α, $0 < \alpha \le 1$.

9.3.1 Function Approximation by TWs

If $g(x)$ is any function defined on $L^2[0,1)$, it can be approximated as follows:

$$g(x) \approx \sum_{n=1}^{2^{k-1}} \sum_{m=0}^{M-1} \tilde{g}_{n,m} \psi_{n,m}(x) = \tilde{G}^T \Psi(x), \tag{9.2}$$

where $\tilde{G}$ and $\Psi(x)$ are $(2^{k-1}M\times 1)$-dimensional vectors given by

$$\tilde{G}=[\tilde{g}_{1,0},\tilde{g}_{1,1},\ldots,\tilde{g}_{1,M-1},\ldots,\tilde{g}_{2^{k-1},0},\tilde{g}_{2^{k-1},1},\ldots,\tilde{g}_{2^{k-1},M-1}]^T \tag{9.3}$$

and

$$\Psi(x)=[\psi_{1,0}(x),\psi_{1,1}(x),\ldots,\psi_{1,M-1}(x),\ldots,\psi_{2^{k-1},0}(x),\psi_{2^{k-1},1}(x),\ldots\psi_{2^{k-1},M-1}(x)]^T, \tag{9.4}$$

where $\psi_{n,m}(x)$, $n=1,2,\ldots,2^{k-1}$, $m=0,1,\ldots,M-1$ is the TWs defined on $[0,1)$ as follows [40]:

$$\psi_{n,m}(x)=\begin{cases} 2^{\frac{k-1}{2}}\tilde{T}_m(2^{k-1}x-n+1), & \dfrac{n-1}{2^{k-1}}\le x<\dfrac{n}{2^{k-1}}, \\ 0, & \text{otherwise,} \end{cases} \tag{9.5}$$

with $\tilde{T}_m(x)=\sqrt{2m+1}.T_m(x)$, where $T_m(x)=x^m$ are the well-known Taylor polynomials of degree m.

Using Eq. (9.2), the value of $\tilde{G}$ can be calculated as follows:

$$\tilde{G}^T=\left(\int_0^1 g(x)\Psi^T(x)dx\right).\tilde{D}^{-1}, \tag{9.6}$$

where $\tilde{D}$ is a $(2^{k-1}M\times 2^{k-1}M)$-dimensional matrix and is given by

$$\tilde{D}=\int_0^1 \Psi(x)\Psi^T(x)\,dx. \tag{9.7}$$

For $k=2$, $M=3$, we have

$$\tilde{D}=\begin{bmatrix} 1 & \frac{\sqrt{3}}{2} & \frac{\sqrt{5}}{3} & 0 & 0 & 0 \\ \frac{\sqrt{3}}{2} & 1 & \frac{\sqrt{15}}{4} & 0 & 0 & 0 \\ \frac{\sqrt{5}}{3} & \frac{\sqrt{15}}{4} & 1 & 0 & 0 & 0 \\ 0 & 0 & 0 & 1 & \frac{\sqrt{3}}{2} & \frac{\sqrt{5}}{3} \\ 0 & 0 & 0 & \frac{\sqrt{3}}{2} & 1 & \frac{\sqrt{15}}{4} \\ 0 & 0 & 0 & \frac{\sqrt{5}}{3} & \frac{\sqrt{15}}{4} & 1 \end{bmatrix}.$$

Similarly, $k(x,z)$ can be approximated as follows:

$$k(x,z) = \Psi^T(x).\tilde{\Pi}.\Psi(z), \tag{9.8}$$

where $\tilde{\Pi}$ is a square matrix of dimension-$2^{k-1}M$ and is given by

$$\tilde{\Pi} = \tilde{D}^{-1}.\left(\int_0^1 \Psi(x).\left(\int_0^1 k(x,z)\Psi^T(z)dz\right)dx\right).\tilde{D}^{-1}.$$

9.4 OPERATIONAL MATRICES

This section deals with the construction of OM of the fractional integration (FI) for the proposed method to solve the fractional VFIDEs.

9.4.1 OM of FI

Let $\Psi(x)$ be the TWs vector introduced in Eq. (9.4). Then, the fractional integral OM is given by

$$\int_0^x \Psi(\zeta)d\zeta = P_1\Psi(x), \tag{9.9}$$

where P_1 is the $(2^{k-1}M \times 2^{k-1}M)$-dimensional OM of integration [41].

Similarly, the integration from 0 to qx $(0 < q < 1)$ of TWs vector $\Psi(x)$ is given by

$$\int_0^{qx} \Psi(\zeta)d\zeta = P_2\Psi(x), \tag{9.10}$$

where P_2 is the $(2^{k-1}M \times 2^{k-1}M)$-dimensional OM of integration.

Now, the FI of the TWs vector $\Psi(x)$ is defined as follows:

$$\tilde{I}^\alpha \Psi(x) = P^\alpha \Psi(x), \tag{9.11}$$

where P^α is the OM of the FI of dimension $(2^{k-1}M \times 2^{k-1}M)$.

Now using the Riemann–Liouville fractional integral definition, i.e.,

$$\tilde{J}^\alpha u(x) = \frac{1}{\Gamma(\alpha)}\int_0^\eta (x-z)^{\alpha-1}u(z)\,dz, \quad x > 0, \alpha \in \mathbb{R}^+,$$

and from Eq. (9.11), the matrix P^α can be obtained as follows:

$$P^\alpha = \left(\int_0^1 \tilde{I}^\alpha \Psi(x).\Psi^T(x)dx\right).\tilde{D}^{-1}. \tag{9.12}$$

For $k = 2$, $M = 2$ and $\alpha = 0.5$, we have

$$P^{0.5} = \begin{bmatrix} 0.531923 & 0.319154 & 0.440659 & -0.154879 \\ -0.106385 & 0.227967 & 0.0516265 & -0.0561003 \\ 0 & 0 & 0.531923 & 0.319154 \\ 0 & 0 & -0.106385 & 0.227967 \end{bmatrix}.$$

9.5 TAYLOR WAVELET METHOD FOR VFIDE

In this analysis, the linear and nonlinear VFIDEs have been solved by employing the OM approach derived by 2D wavelets.

9.5.1 Linear Case

In the linear case, let $\tilde{H}_1(u(z)) = \tilde{H}_2(u(z)) = u(z)$ in Eq. (9.1). Therefore, we have

$${}_0^C D_x^\alpha u(x) = f(x) + \int_0^x k_1(x,z)u(z)\,dz + \int_0^1 k_2(x,z)u(z)\,dz. \tag{9.13}$$

To discuss the proposed scheme for solving Eq. (9.13), consider the approximation of following functions:

$${}_0^C D_x^\alpha u(x) \approx \Psi^T(x)\tilde{A}_1 = \tilde{A}_1{}^T \Psi(x), \tag{9.14}$$

$$u(0) \approx \Psi^T(x)\tilde{U} = \tilde{U}^T \Psi(x), \tag{9.15}$$

$$f(x) \approx \Psi^T(x)\tilde{F} = \tilde{F}^T \Psi(x), \tag{9.16}$$

$$k_1(x,z) \approx \Psi^T(x)\tilde{K}_1\Psi(z) = \Psi^T(z)\tilde{K}_1^T\Psi(x), \tag{9.17}$$

$$k_2(x,z) \approx \Psi^T(x)\tilde{K}_2\Psi(z) = \Psi^T(z)\tilde{K}_2^T\Psi(x). \tag{9.18}$$

Integrating Eq. (9.14) with respect to x and utilizing Eq. (9.11), it follows

$$\begin{aligned} u(x) &= u(0) + \tilde{A}_1^T P^\alpha \Psi(x) \\ &= \tilde{U}^T \Psi(x) + \tilde{A}_1^T P^\alpha \Psi(x) \\ &= (\tilde{U}^T + \tilde{A}_1^T P^\alpha)\Psi(x) \\ &= \Psi^T(x)\tilde{A}_2, \end{aligned} \tag{9.19}$$

where $\tilde{A}_2 = \tilde{U} + P^{\alpha T}\tilde{A}_1$.

Now, in Eq. (9.13), the integral part can be expressed as follows:

$$\int_0^x k_1(x,z)u(z)\,dz = \int_0^x \Psi^T(x)\tilde{K}_1\Psi(z)\tilde{A}_2^T\Psi(z)\,dz$$

$$= \Psi^T(x)\tilde{K}_1\int_0^x \Psi(z)\Psi^T(z)\tilde{A}_2\,dz$$

$$= \Psi^T(x)\tilde{K}_1\int_0^x \bar{A}_2\Psi(z)\,dz,$$

where $\bar{A}_2 = \left(\int_0^1 \Psi(z)\Psi^T(z)\tilde{A}_2\Psi^T(z)\,dz\right).\tilde{D}^{-1}$.

Thus,

$$\begin{aligned}\int_0^x k_1(x,z)u(z)\,dz &= \Psi^T(x)\tilde{K}_1\bar{A}_2P_1\Psi(x)\\ &= \tilde{A}_3^T\Psi(x),\end{aligned} \tag{9.20}$$

where $\tilde{A}_3^T = \left(\int_0^1 \Psi^T(x)\tilde{K}_1\bar{A}_2P_1\Psi(x)\Psi^T(x)\,dx\right).\tilde{D}^{-1}$.

Similarly,

$$\int_0^1 k_2(x,z)u(z)\,dz = \int_0^1 \Psi^T(x)\tilde{K}_2\Psi(z)\tilde{A}_2^T\Psi(z)\,dz$$

$$= \Psi^T(x)\tilde{K}_2\int_0^1 \Psi(z)\Psi^T(z)\tilde{A}_2\,dz$$

$$= \Psi^T(x)\tilde{K}_2\int_0^1 \Psi(z)\Psi^T(z)\,dz.\tilde{A}_2 \tag{9.21}$$

$$= \Psi^T(x)\tilde{K}_2.\tilde{D}.\tilde{A}_2$$

$$= \Psi^T(x)\tilde{A}_4,$$

where $\tilde{A}_4 = \tilde{K}_2.\tilde{D}.\tilde{A}_2$.

Now by plugging Eqs. (9.14)–(9.21) into Eq. (9.13), it follows

$$\Psi^T(x)\tilde{A}_1 = \Psi^T(x)\tilde{F} + \Psi^T(x)\tilde{A}_3 + \Psi^T(x)\tilde{A}_4. \tag{9.22}$$

Collocating Eq. (9.22) at $2^{k-1}M$-points x_i yields

$$\Psi^T(x_i)\tilde{A}_1 = \Psi^T(x_i)\tilde{F} + \Psi^T(x_i)\tilde{A}_3 + \Psi^T(x_i)\tilde{A}_4, \tag{9.23}$$

where the collocation points x_i are given as follows:

$$x_i = \frac{2i-1}{2.2^{k-1}.M}, \quad i = 1, 2, \ldots, 2^{k-1}.M.$$

The value of $\tilde{A}_1$ can be computed using the above algebraic system of Eq. (9.23), and then $\tilde{A}_2$ is calculated by using the value of $\tilde{A}_1$.

Thus, the linear VFIDE (9.13) can be numerically solved by substituting $\tilde{A}_2$ in Eq. (9.19).

9.5.2 Nonlinear Case

For the nonlinear case, let $\tilde{H}_1(u(z)) = u^p(z)$, $p > 1$ and $\tilde{H}_2(u(z)) = u^q(z)$, $q > 1$ in Eq. (9.1). Therefore, we have

$${}_0^C D_x^{\alpha} u(x) = f(x) + \int_0^x k_1(x,z) u^p(z)\,dz + \int_0^1 k_2(x,z) u^q(z)\,dz. \tag{9.24}$$

Now in the case of $p = q = 2$, it follows

$$\begin{aligned} u^2(z) &= (\tilde{A}_2^T \Psi(z)).(\tilde{A}_2^T \Psi(z)) \\ &= \tilde{A}_2^T \Psi(z) \Psi^T(z) \tilde{A}_2 \\ &= \tilde{A}_2^T \bar{A}_2 \Psi(z) \\ &= \tilde{A}_5^T \Psi(z), \end{aligned} \tag{9.25}$$

where $\tilde{A}_5 = \bar{A}_2{}^T.\tilde{A}_2$.

Now, in Eq. (9.24), the integral part can be expressed as follows:

$$\begin{aligned} \int_0^x k_1(x,z) u^2(z)\,dz &= \int_0^x \Psi^T(x) \tilde{K}_1 \Psi(z) \tilde{A}_5^T \Psi(z)\,dz \\ &= \Psi^T(x) \tilde{K}_1 \int_0^x \Psi(z) \Psi^T(z) \tilde{A}_5\,dz \\ &= \Psi^T(x) \tilde{K}_1 \int_0^x \bar{A}_5 \Psi(z)\,dz, \end{aligned}$$

where $\bar{A}_5 = \left(\int_0^1 \Psi(z)\Psi^T(z)\tilde{A}_5\Psi^T(z)\,dz \right).\tilde{D}^{-1}$.

Thus,

$$\begin{aligned} \int_0^x k_1(x,z)u^2(z)\,dz &= \Psi^T(x)\tilde{K}_1\bar{A}_5P_1\Psi(x) \\ &= \tilde{A}_6^T\Psi(x), \end{aligned} \tag{9.26}$$

where $\tilde{A}_6^T = \left(\int_0^1 \Psi^T(x)\tilde{K}_1\bar{A}_5P_1\Psi(x)\Psi^T(x)\,dx \right).\tilde{D}^{-1}$.

Similarly,

$$\begin{aligned} \int_0^1 k_2(x,z)u^2(z)\,dz &= \int_0^1 \Psi^T(x)\tilde{K}_2\Psi(z)\tilde{A}_5^T\Psi(z)\,dz \\ &= \Psi^T(x)\tilde{K}_2 \int_0^1 \Psi(z)\Psi^T(z)\,dz.\tilde{A}_5 \\ &= \Psi^T(x)\tilde{K}_2.\tilde{D}.\tilde{A}_5 \\ &= \Psi^T(x)\tilde{A}_7, \end{aligned} \tag{9.27}$$

where $\tilde{A}_7 = \tilde{K}_2.\tilde{D}.\tilde{A}_5$.

Now by substituting Eqs. (9.14)–(9.19) and Eqs. (9.25)–(9.27) into Eq. (9.24), it follows

$$\Psi^T(x)\tilde{A}_1 = \Psi^T(x)\tilde{F} + \Psi^T(x)\tilde{A}_6 + \Psi^T(x)\tilde{A}_7. \tag{9.28}$$

Collocating Eq. (9.28) at $2^{k-1}M$-points x_i yields

$$\Psi^T(x_i)\tilde{A}_1 = \Psi^T(x_i)\tilde{F} + \Psi^T(x_i)\tilde{A}_6 + \Psi^T(x_i)\tilde{A}_7, \tag{9.29}$$

where the collocation points x_i are given as follows:

$$x_i = \frac{2i-1}{2.2^{k-1}.M}, \quad i = 1,2,\ldots,2^{k-1}.M.$$

The value of $\tilde{A}_1$ can be computed using the above algebraic system of Eq. (9.29), and then, $\tilde{A}_2$ is calculated by using the value of $\tilde{A}_1$.

Thus, the nonlinear VFIDE (9.24) can be numerically solved by substituting $\tilde{A}_2$ in Eq. (9.19).

9.6 CONVERGENCE ANALYSIS

In the present analysis, we established that the expansion of any continuous function $u(x) \in L^2([0,1))$, with respect to 1D TWs, as given in Eq. (9.2), converges uniformly to u.

Theorem 9.6.1

Consider $u(x)$ be a continuous function in $L^2([0,1))$ and $|u(x)| \leq \aleph$, where $\aleph$ is a real and positive constant. Then, the TWs expansion of $u(x)$ converges uniformly to $u(x)$, and also

$$\left|\tilde{f}_{n,m}\right| \leq \frac{\aleph\sqrt{2m+1}}{2^{\frac{k-1}{2}}.(m+1)}.$$

Proof:

The TWs coefficients can be calculated as follows:

$$\begin{aligned}
\tilde{f}_{n,m} &= < u(x), \psi_{n,m}(x) > \\
&= \int_0^1 u(x)\psi_{n,m}(x)\,dx \\
&= 2^{\frac{k-1}{2}}\sqrt{2m+1}\int_{\frac{n-1}{2^{k-1}}}^{\frac{n}{2^{k-1}}} u(x)T_m(2^{k-1}x-n+1)\,dx,
\end{aligned}$$

Now, changing the variable $2^{k-1}x-n+1=s$, it follows

$$= 2^{\frac{k-1}{2}}\sqrt{2m+1}\frac{1}{2^{k-1}}\int_0^1 u\left(\frac{s+n-1}{2^{k-1}}\right)T_m(s)\,ds.$$

This implies

$$\begin{aligned}
\left|\tilde{f}_{n,m}\right| &= \left|\frac{\sqrt{2m+1}}{2^{\frac{k-1}{2}}}\int_0^1 u\left(\frac{s+n-1}{2^{k-1}}\right)T_m(s)\,ds\right| \\
&\leq \frac{\sqrt{2m+1}}{2^{\frac{k-1}{2}}}\int_0^1 \left|u\left(\frac{s+n-1}{2^{k-1}}\right)\right| |T_m(s)|\,ds \\
&\leq \frac{\sqrt{2m+1}}{2^{\frac{k-1}{2}}}\aleph\int_0^1 |T_m(s)|\,ds.
\end{aligned}$$

By the definition of TP of degree m, $\int_0^1 |T_m(s)|\, ds = \frac{1}{m+1}$, $\mathrm{Re}[m] > -1$.

That is,

$$\left|\tilde{f}_{n,m}\right| \le \frac{\aleph\sqrt{2m+1}}{2^{\frac{k-1}{2}}.(m+1)}.$$

Hence, the series $\sum_{n=1}^{2^{k-1}} \sum_{m=0}^{M-1} \tilde{f}_{n,m}$ is absolutely convergent, and it follows that the series $\sum_{n=1}^{2^{k-1}} \sum_{m=0}^{M-1} \tilde{f}_{n,m} \psi_{n,m}(x)$ is uniformly convergent [42].

9.7 ERROR ANALYSIS

In this part, the estimation of error bound of the Taylor wavelet method (TWM) for solving VFIDE has been derived.

Theorem 9.7.1

Let $\tilde{u}_{2^{k-1}M}(x)$ be the 1D TWs expansion of the real sufficiently smooth function $u(x) \in L^2([0,1))$ and u is M times differentiable. Then,

$$\left\| u(x) - \tilde{u}_{2^{k-1}M}(x) \right\|_2 \le \frac{2\Re}{M!\,2^{M(k+1)}}, \tag{9.30}$$

where $\Re = \max_{x\in[0,1]} \left| u^{(M)}(x) \right|$.

Proof:

Assume that $u(x)$ be a sufficiently smooth function on $L^2([0,1))$ and $\tilde{u}_{2^{k-1}M}(x)$ is the best approximation of u from $\Lambda_{2^{k-1}M}$, where

$$\Lambda_{2^{k-1}M} = \mathrm{Span}\left\{ \psi_{n,m}(x) : n = 1,2,\ldots,2^{k-1}, \quad m = 0,1,\ldots,M-1 \right\}.$$

Now dividing the closed interval [0,1] into 2^{k-1}-subintervals $\tilde{I}_n = \left[\frac{n-1}{2^{k-1}}, \frac{n}{2^{k-1}} \right]$ with $n = 1,2,\ldots,2^{k-1}$.

Then, for any $P_{2^{k-1}M}[u] \in \Lambda_{2^{k-1}M}$, the following inequality is held:

$$\left\| u(x) - \tilde{u}_{2^{k-1}M}(x) \right\|_2 \le \left\| u(x) - P_{2^{k-1}M}[u] \right\|_2, \tag{9.31}$$

where $P_{2^{k-1}M}[u]$ is the interpolating polynomial of u at the Chebyshev nodes in the interval $\tilde{I}_n$. According to a study [43], the interpolation error is

$$\begin{aligned}\left|u(x)-P_{2^{k-1}M}[u]\right| &\le \frac{1}{2^{M-1}M!}\left(\frac{|\tilde{I}_n|}{2}\right)^M \max_{x\in\tilde{I}_n}\left|u^{(M)}(x)\right| \\ &\le \frac{2\Re}{M!2^{M(k+1)}}.\end{aligned} \tag{9.32}$$

In particular, from Eq. (9.31),

$$\begin{aligned}\left\|u(x)-\tilde{u}_{2^{k-1}M}(x)\right\|_2^2 &\le \left\|u(x)-P_{2^{k-1}M}[u]\right\|_2^2 \\ &= \int_0^1 \left|u(x)-P_{2^{k-1}M}[x]\right|^2 dx \\ &\le \int_0^1 \left(\frac{2\Re}{M!2^{M(k+1)}}\right)^2 dx \\ &= \left(\frac{2\Re}{M!2^{M(k+1)}}\right)^2.\end{aligned}$$

Therefore,

$$\left\|u(x)-\tilde{u}_{2^{k-1}M}(x)\right\|_2 \le \frac{2\Re}{M!2^{M(k+1)}}. \tag{9.33}$$

Hence, the theorem is proved. ■

Theorem 9.7.2

Let $\tilde{u}_{2^{k-1}M}(x)$ be the 1D TWs expansion of the real sufficiently smooth function $u(x)\in L^2([0,1))$ and u is M times differentiable. Then,

$$\left\|\tilde{I}^{\alpha}u(x)-\tilde{I}^{\alpha}\tilde{u}_{2^{k-1}M}(x)\right\|_2 \le \frac{2\Re}{\Gamma(\alpha)\sqrt{2\alpha(2\alpha-1)}M!2^{M(k+1)}},$$

where $\Re = \max_{x\in[0,1]}\left|u^{(M)}(x)\right|$.

Proof:

It may be referred to Ref. [44].

9.8 NUMERICAL EXPERIMENTS

In this part, some linear and nonlinear examples are provided to show the applicability of the presented technique. The absolute error definition and L^2-error norm are specified as follows to justify the accuracy and efficiency of the proposed method:

$$\tilde{E}(x) = |u(x) - \tilde{u}(x)|$$

and

$$L^2_{\text{error}} = \left[\int_0^1 |u(x) - \tilde{u}(x)|^2 \, dx \right]^{\frac{1}{2}},$$

where $\tilde{u}(x)$ is the approximate solution and $u(x)$ is the exact solution.

The numerical convergence rate (NCR) formula is proposed to demonstrate the efficacy of the presented scheme and is given by

$$\text{NCR} = \log_2 \left(\frac{\text{MAE}_{k=\beta, M=3}}{\text{MAE}_{k=\beta+1, M=3}} \right), \quad \beta = 2,3,4,$$

where

$$\text{MAE} = \underset{1 \le j \le 2^{k-1}M}{\text{Max}} |u(x_j) - \tilde{u}(x_j)|, \quad x_j = \frac{2.j-1}{2.2^{k-1}M}, \quad j = 1,2,\ldots,2^{k-1}M.$$

Example 9.8.1

Consider the following linear fractional VFIDE [45]:

$${}_0^C D_x^{1.7} u(x) = f(x) + \int_0^x (x-z)u(z)\,dz + \int_0^1 (x+z)u(z)\,dz, \tag{9.34}$$

$$u(0) = u'(0) = 0,$$

and

$$f(x) = \frac{x^{0.3}\Gamma(3)}{\Gamma(1.3)} + \frac{x^{1.3}\Gamma(4)}{\Gamma(2.3)} - \frac{x^4}{12} - \frac{x^5}{20} - \frac{7x}{12} - \frac{9}{20}.$$

The exact solution is $u(x) = x^2 + x^3$.

For $k = 2,3,4$ and $M = 3$, the absolute error values obtained by the TWM are presented in Table 9.1, and it shows that the increase in the size of k and M provides a better absolute error result. Figures 9.1–9.3 display the comparison graphically between the numerical solutions obtained by TWM and exact solution, and the absolute errors distribution by TWM are displayed in Figures 9.4–9.6. The NCR for different values of k, M is presented in Table 9.2, and it indicates that the increase in the size of k leads to an approximately third-order convergence rate. Table 9.3 shows the comparison of approximate solutions between TWM, Legendre wavelets method (LWM) [45], and CAS wavelets method (CASWM) [45] for $k = 4, M = 2$ and $k = 5, M = 2$, respectively.

TABLE 9.1
$\tilde{E}(x)$ Errors Obtained by TWM for Example 9.8.1

	Error function $\tilde{E}(x)$		
x	$k = 2, M = 3$	$k = 3, M = 3$	$k = 4, M = 3$
0.1	2.12268E-03	2.51741E-04	2.40341E-05
0.2	1.87704E-03	2.39392E-04	2.75011E-05
0.3	1.53126E-03	2.72783E-04	2.71873E-05
0.4	2.10221E-03	2.19481E-04	3.50755E-05
0.5	6.30135E-03	7.82384E-04	9.77012E-05
0.6	2.23298E-03	2.18371E-04	3.51817E-05
0.7	1.75309E-03	2.81609E-04	2.7359E-05
0.8	1.74102E-03	2.81238E-04	2.73355E-05
0.9	2.24936E-03	2.18702E-04	3.5152E-05

Figures 9.1–9.3 represent the comparison of TWM solutions with exact solution for $k = 2,3,4$ and $M = 3$.

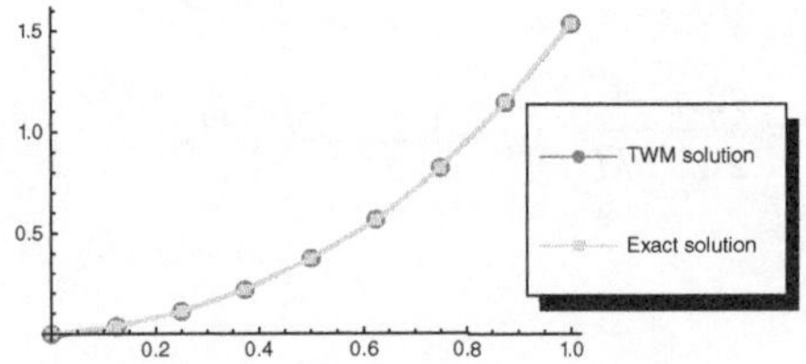

FIGURE 9.1 Comparison of TWM solution with exact solution for $k = 2, M = 3$.

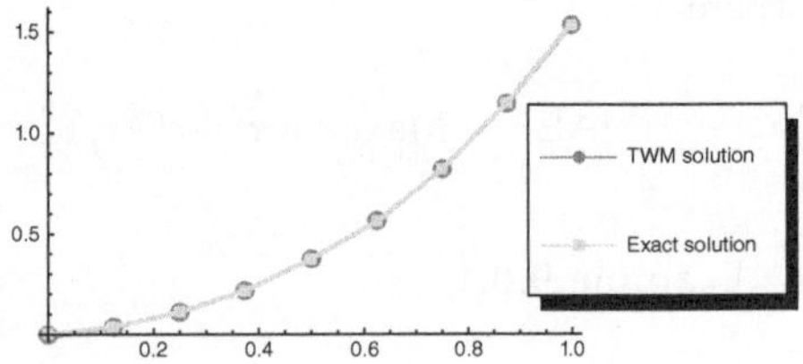

FIGURE 9.2 Comparison of TWM solution with exact solution for $k = 3, M = 3$.

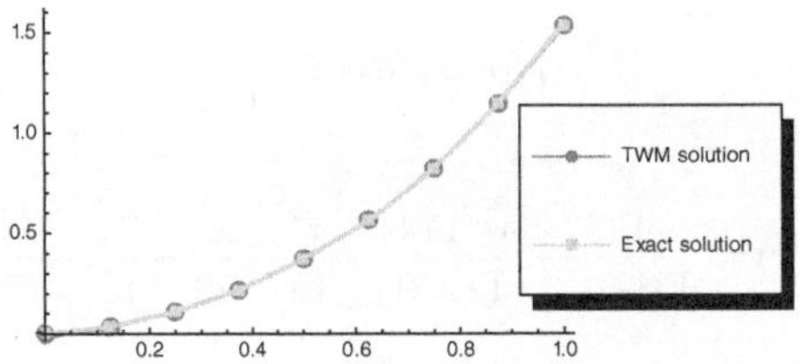

FIGURE 9.3 Comparison of TWM solution with exact solution for $k = 4, M = 3$.

Figures 9.4–9.6 represent the error distribution by TWM for $k = 2,3,4$ and $M = 3$.

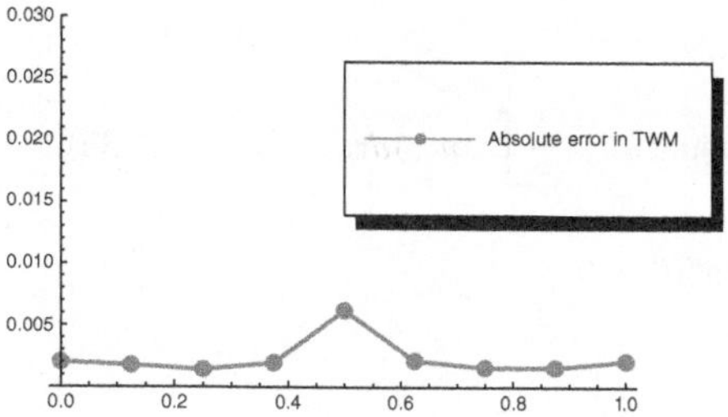

FIGURE 9.4 Absolute error for $k = 2, M = 3$.

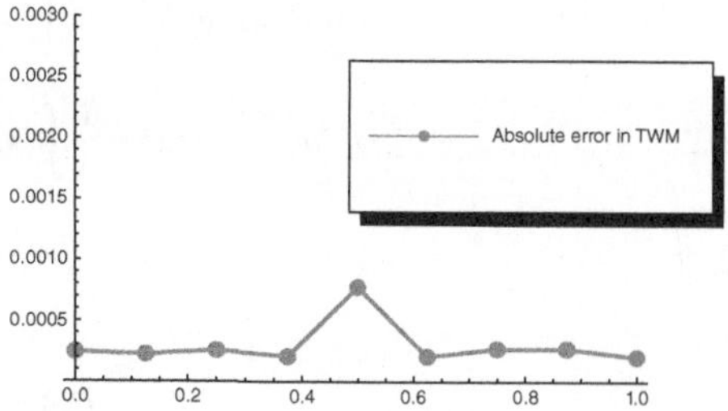

FIGURE 9.5 Absolute error for $k = 3, M = 3$.

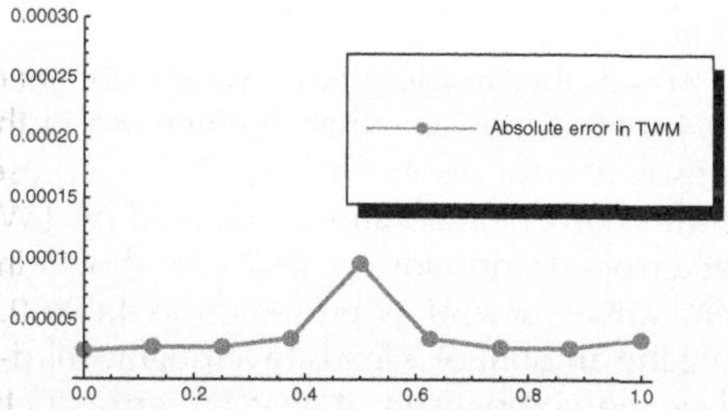

FIGURE 9.6 Absolute error for $k = 4, M = 3$.

TABLE 9.2
NCR for Example 9.8.1

k, M	*MAE*	*NCR*
$k = 2, M = 3$	1.6225E-03	–
$k = 3, M = 3$	2.03347E-04	2.9962
$k = 4, M = 3$	2.54671E-05	2.99724

TABLE 9.3
Comparison of Approximate Solutions for Example 9.8.1

	$k = 4, M = 2$			$k = 5, M = 2$			
x	LWM [45]	CASWM [45]	TWM	LWM [45]	CASWM [45]	TWM	Exact Solution
1/8	0.015822	0.023571	0.0135838	0.016551	0.021438	0.0166323	0.017578
2/8	0.075531	0.075531	0.0731687	0.077098	0.084365	0.0769361	0.078125
3/8	0.192880	0.216487	0.18743	0.193159	0.204620	0.191927	0.193359
4/8	0.361498	0.398469	0.368096	0.368505	0.392246	0.373323	0.375000
5/8	0.622950	0.652830	0.626886	0.626293	0.651032	0.632845	0.634765
6/8	0.930901	1.134367	0.97552	0.963194	0.128764	0.98221	0.984375
7/8	1.391340	1.525654	1.42572	1.408939	1.498832	1.43314	1.435546

Example 9.8.2

Consider the linear fractional VFIDE [45]:

$$ {}_{0}^{C}D_x^{2.3}u(x) = f(x) + \frac{1}{4}\int_0^x (x-z)u(z)\,dz + \frac{1}{2}\int_0^1 zxu(z)\,dz, \quad (9.35)$$

$$u(0) = u'(0) = u''(0) = 0,$$

and

$$f(x) = \frac{x^{1.2}\Gamma(4.5)}{\Gamma(2.2)} - \frac{x^{5.5}}{99} - \frac{x^5}{20} - \frac{x}{11}.$$

The exact solution is $u(x) = x^{\frac{7}{2}}$.

For $k = 2,3,4$ and $M = 3$, the absolute error values obtained by the TWM are presented in Table 9.4, and it indicates that the increase in the size of k and M yields an improved absolute error result. Figures 9.7–9.9 represent the comparison graphically between the numerical solutions obtained by TWM and exact solution, and the absolute errors distribution by TWM are shown in Figures 9.10–9.12. The NCR for different values of k, M is presented in Table 9.5, and it is clearly evident that increasing the amount of k leads to a roughly third-order convergence rate. Table 9.6 displays the comparison of absolute errors between TWM, LWM [45], and Genocchi polynomials method (GPM) [46].

TABLE 9.4
$\tilde{E}(x)$ Errors Obtained by TWM for Example 9.8.2

	Error function $\tilde{E}(x)$		
x	$k = 2, M = 3$	$k = 3, M = 3$	$k = 4, M = 3$
0.1	2.23832E-03	1.79631E-04	2.06389E-05
0.2	2.0325E-03	2.33502E-04	2.53669E-05
0.3	1.77031E-03	3.65651E-04	3.38191E-05
0.4	2.64159E-03	2.86988E-04	5.02611E-05
0.5	1.15401E-02	1.33089E-03	1.58912E-04
0.6	4.13742E-03	3.82631E-04	5.81896E-05
0.7	3.37762E-03	4.94064E-04	4.93255E-05
0.8	3.24563E-03	5.68652E-04	5.41586E-05
0.9	4.37211E-03	4.43736E-04	7.40552E-05

Figures 9.7–9.9 represent the comparison of TWM solutions with exact solution for $k = 2,3,4$ and $M = 3$.

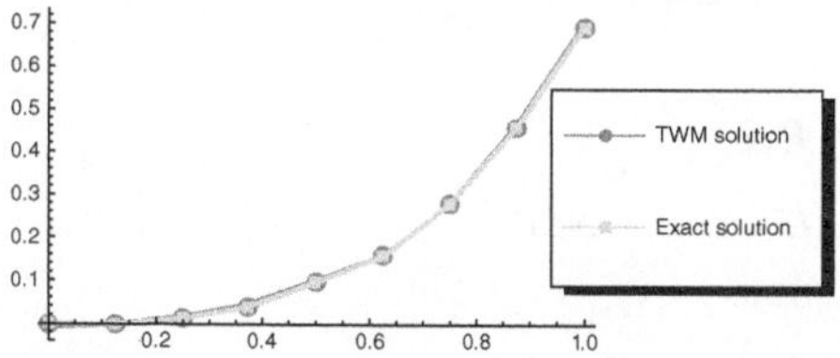

FIGURE 9.7 Comparison of TWM solution with exact solution for $k = 2, M = 3$.

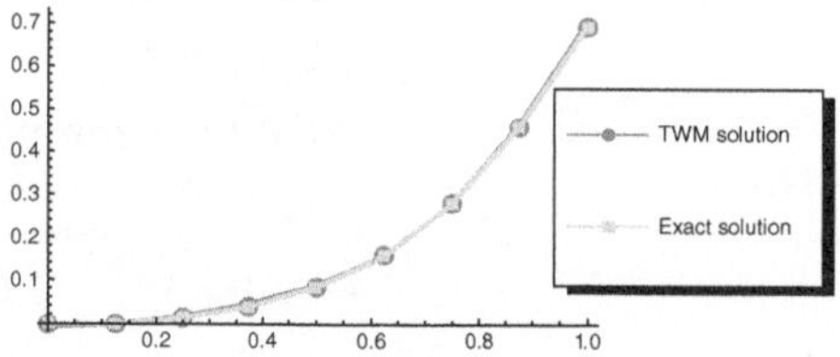

FIGURE 9.8 Comparison of TWM solution with exact solution for $k = 3, M = 3$.

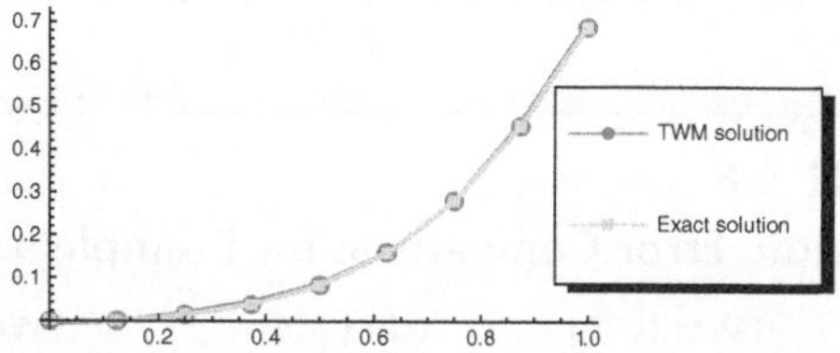

FIGURE 9.9 Comparison of TWM solution with exact solution for $k = 4, M = 3$.

Figures 9.10–9.12 represent the error distribution by TWM for $k = 2,3,4$ and $M = 3$.

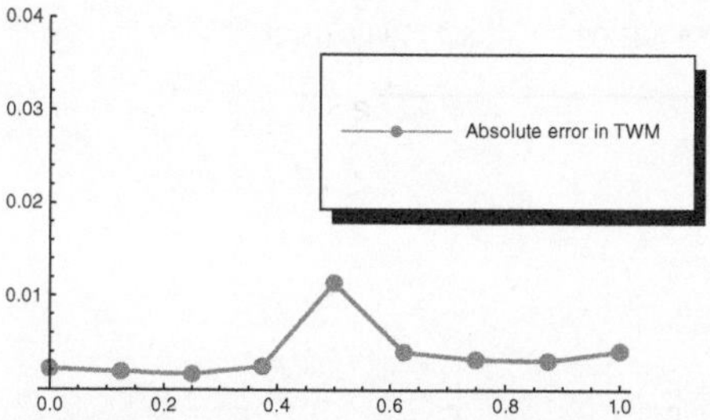

FIGURE 9.10 Absolute error for $k = 2, M = 3$.

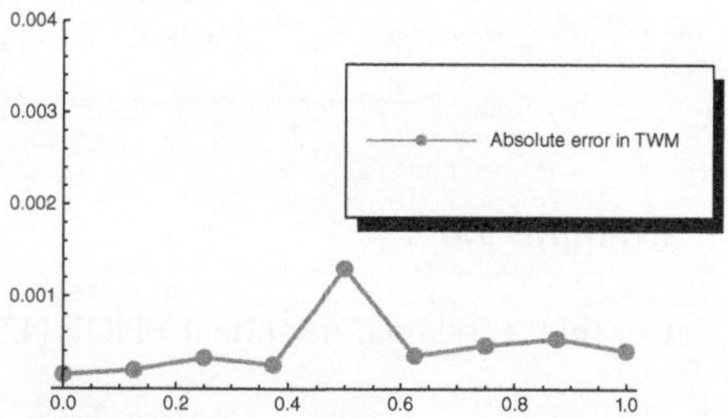

FIGURE 9.11 Absolute error for $k = 3, M = 3$.

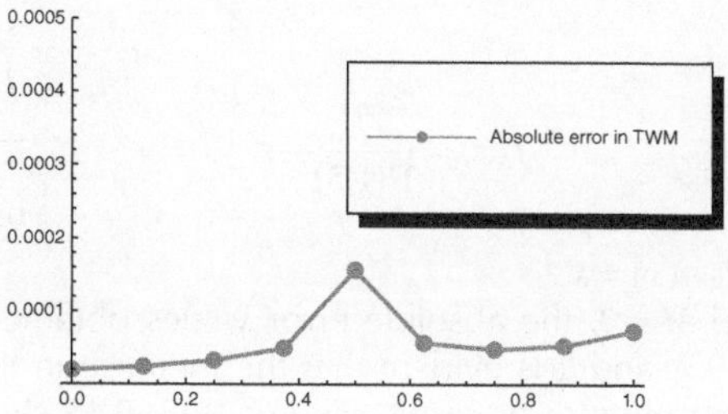

FIGURE 9.12 Absolute error for $k = 4, M = 3$.

TABLE 9.5
NCR for Example 9.8.2

k, M	MAE	NCR
$k = 2, M = 3$	3.18649E-03	–
$k = 3, M = 3$	4.21326E-04	2.91896
$k = 4, M = 3$	5.40431E-05	2.96275
$k = 5, M = 3$	6.83973E-06	2.9821

TABLE 9.6
Absolute Error Comparison for Example 9.8.2

	LWM [435]	GPM [46]	TWM
x	$k = 3, M = 2$	$e_8(x)$	$k = 2, M = 4$
1/8	2.2475E-04	1.60950E-04	1.41434E-04
2/8	5.6882E-04	6.47020E-04	1.85755E-04
3/8	8.0634E-04	1.26848E-03	1.51394E-04
4/8	2.3712E-03	1.94127E-03	2.89421E-04
5/8	2.8084E-03	2.66168E-03	8.10252E-05
6/8	3.1612E-03	3.43505E-03	1.0603E-04
7/8	3.3599E-03	4.25649E-03	8.21934E-05

Example 9.8.3

Consider the linear fractional VFIDE [47]:

$$ {}_0^C D_x^{0.5} u(x) = f(x) - \frac{x^2 e^x}{3} u(x) + \int_0^x e^x z u(z)\, dz + \int_0^1 x^2 u(z)\, dz, \tag{9.36} $$

$$ u(0) = 0, $$

and

$$ f(x) = \frac{x^{0.5}}{\Gamma(1.5)} - \frac{x^2}{2}. $$

The exact solution is $u(x) = x$.

For $k = 2,3,4$ and $M = 3$, the absolute error values obtained by the TWM are presented in Table 9.7, and it is evident that the increase in the size of k and M generates a better absolute error result. Figures 9.13–9.15 show the comparison graphically between the numerical solutions obtained by TWM and exact solution, and the absolute errors distribution by TWM are displayed in Figures 9.16–9.18.

The NCR for different values of k, M is presented in Table 9.8. As we increase the size of k, it approximately converges linearly to 1.

TABLE 9.7
$\tilde{E}(x)$ Errors Obtained by TWM for Example 9.8.3

	Error function $\tilde{E}(x)$		
x	$k = 2, M = 3$	$k = 3, M = 3$	$k = 4, M = 3$
0.1	8.34604E-05	6.57633E-04	1.57072E-04
0.2	1.30596E-03	3.11533E-04	8.0678E-06
0.3	1.08406E-03	1.68119E-04	3.86273E-06
0.4	5.82223E-04	1.56731E-05	1.09704E-06
0.5	6.473E-04	3.95617E-06	3.60557E-07
0.6	3.26605E-04	7.94219E-06	2.60205E-07
0.7	1.08968E-04	1.56032E-06	3.11285E-07
0.8	5.61131E-06	1.18918E-06	2.86823E-07
0.9	1.71324E-05	3.06618E-06	4.24706E-08

Figures 9.13–9.15 represent the comparison of TWM solutions with the exact solution for $k = 2,3,4$ and $M = 3$.

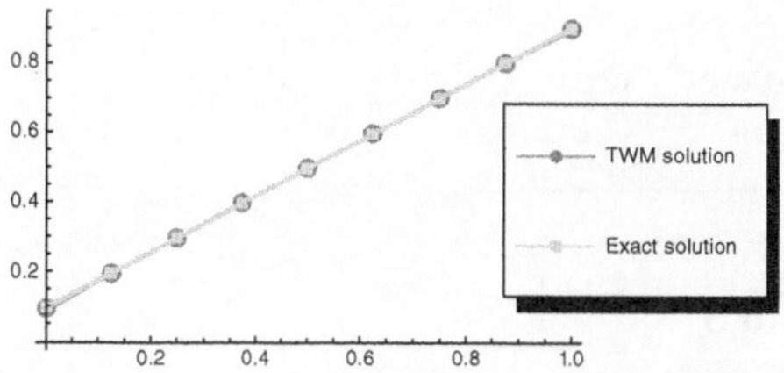

FIGURE 9.13 Comparison of TWM solution with exact solution for $k = 2, M = 3$.

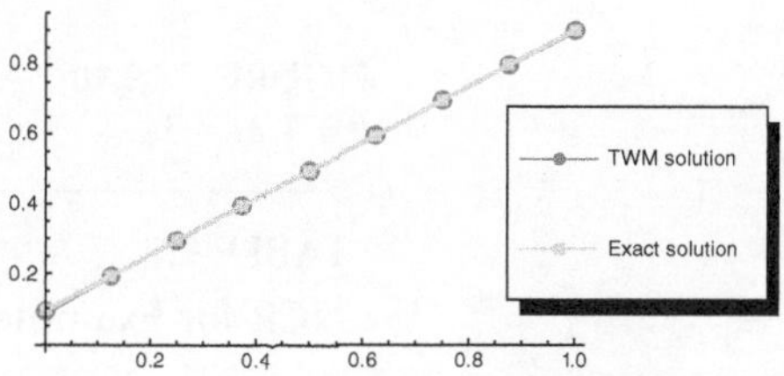

FIGURE 9.14 Comparison of TWM solution with exact solution for $k = 3, M = 3$.

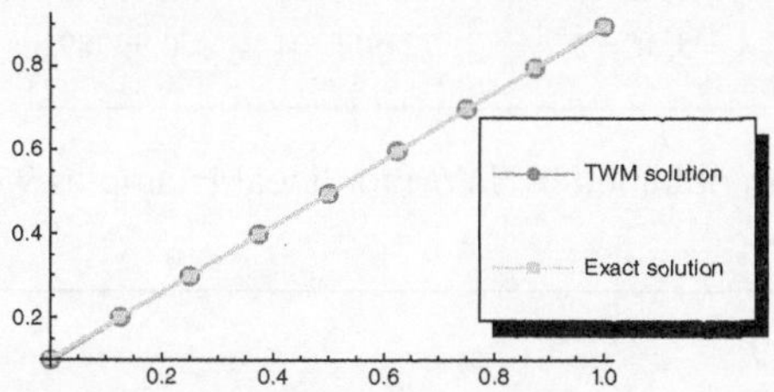

FIGURE 9.15 Comparison of TWM solution with exact solution for $k = 4, M = 3$.

Figures 9.16–9.18 represent the error distribution by TWM for $k = 2,3,4$ and $M = 3$.

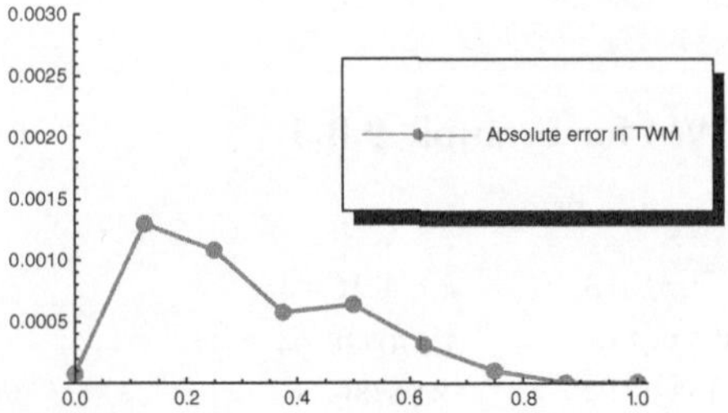

FIGURE 9.16 Absolute error for $k = 2, M = 3$.

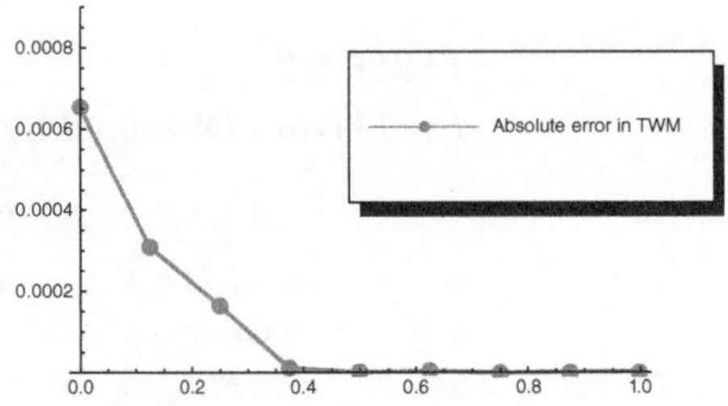

FIGURE 9.17 Absolute error for $k = 3, M = 3$.

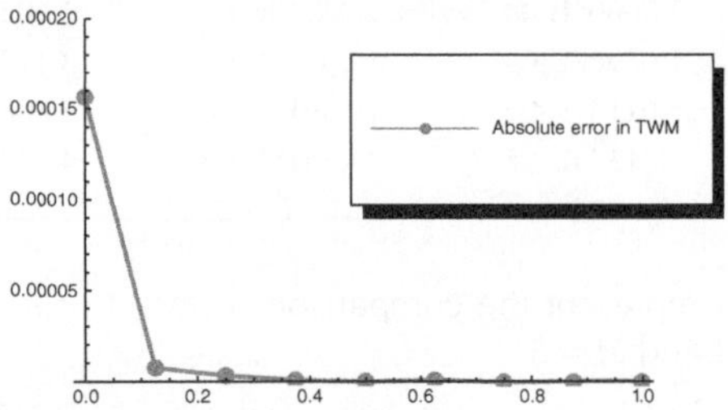

FIGURE 9.18 Absolute error for $k = 4, M = 3$.

TABLE 9.8
NCR for Example 9.8.3

k, M	*MAE*	*NCR*
$k = 2, M = 3$	1.37556E-03	–
$k = 3, M = 3$	6.8884E-04	0.997778
$k = 4, M = 3$	3.44566E-04	0.999389
$k = 5, M = 3$	1.72296E-04	0.999891

The L^2-error norm obtained by TWM for linear Examples 9.8.1–9.8.3 are tabulated in Table 9.9.

TABLE 9.9
L^2_{error}-Norm for Examples 9.8.1–9.8.3

	L^2 – *Error norm*		
k, M	**Example 9.8.1**	**Example 9.8.2**	**Example 9.8.3**
$k = 2, M = 3$	2.36665E-03	3.63829E-03	9.97033E-04
$k = 3, M = 3$	2.96656E-04	4.56139E-04	3.57983E-04
$k = 4, M = 3$	3.72922E-05	5.70706E-05	1.26819E-04

Example 9.8.4

Consider the following nonlinear fractional VFIDE [48]:

$$ {}_{0}^{C}D_{x}^{\frac{3}{4}}u(x) = f(x) + \frac{1}{20}\int_{0}^{x}(3x+4z)u(z)\,dz + \frac{1}{20}\int_{0}^{1}(3x-z)u^{2}(z)\,dz, \tag{9.37} $$

$$ u(0) = 0, $$

and

$$ f(x) = \frac{2x^{\frac{1}{4}}}{\Gamma\left(\frac{5}{4}\right)} + \frac{3\sqrt{\pi}x^{\frac{3}{4}}}{4\Gamma\left(\frac{7}{4}\right)} - \frac{41x^{\frac{7}{2}}}{350} - \frac{17x^{3}}{60} - \frac{229x}{260} + \frac{47}{450}. $$

The exact solution is $u(x) = x^{\frac{3}{2}} + 2x$.

For $k = 2,3,4$ and $M = 3$, the absolute error values obtained by the TWM are presented in Table 9.10, and it shows that the increase in the size of k and M provides a better absolute error result. Figures 9.19–9.21 represent the comparison graphically between the numerical solutions obtained by TWM and exact solution, and the absolute errors distribution by TWM are plotted in Figures 9.22–9.24. The NCR for different values of k, M is presented in Table 9.11. As we increase the size of k, it approximately converges linearly to 1.

TABLE 9.10
$\tilde{E}(x)$ Errors Obtained by TWM for Example 9.8.4

	Error function $\tilde{E}(x)$		
x	$k = 2, M = 3$	$k = 3, M = 3$	$k = 4, M = 3$
0.1	2.03269E-03	1.31438E-03	8.9718E-05
0.2	2.5251E-03	7.68784E-05	2.49415E-05
0.3	4.33826E-03	1.02668E-04	3.90265E-06
0.4	1.44078E-04	7.01869E-05	6.8445E-06
0.5	1.39003E-03	1.04029E-04	1.39742E-05
0.6	1.60527E-04	1.45904E-05	4.94297E-06
0.7	6.39555E-06	3.32297E-05	3.50675E-06
0.8	2.03951E-04	1.98808E-05	1.95501E-06
0.9	1.70957E-04	2.07776E-05	2.39365E-06

Figures 9.19–9.21 represent the comparison of TWM solutions with exact solution for $k = 2,3,4$ and $M = 3$.

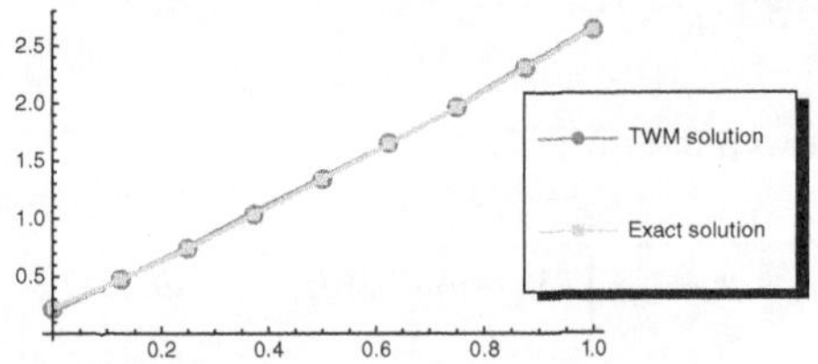

FIGURE 9.19 Comparison of TWM solution with exact solution for $k = 2, M = 3$.

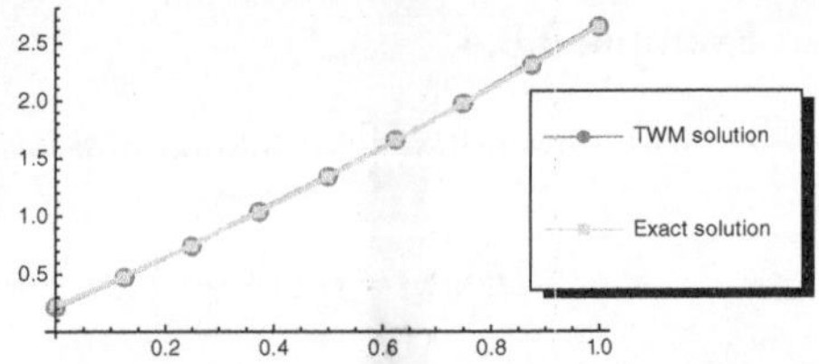

FIGURE 9.20 Comparison of TWM solution with exact solution for $k = 3, M = 3$.

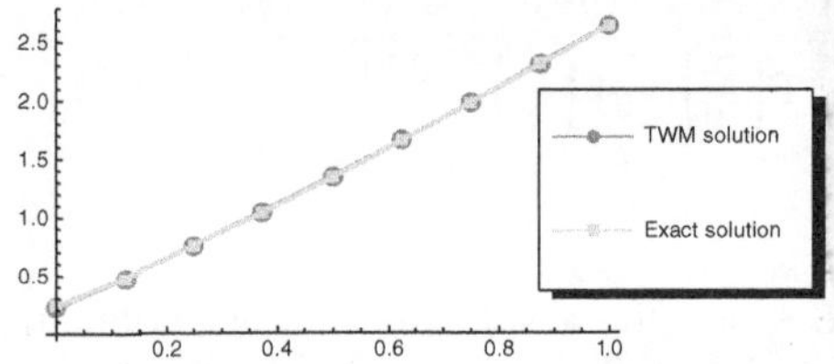

FIGURE 9.21 Comparison of TWM solution with exact solution for $k = 4, M = 3$.

Figures 9.22–9.24 represent the error distribution by TWM for $k = 2,3,4$ and $M = 3$.

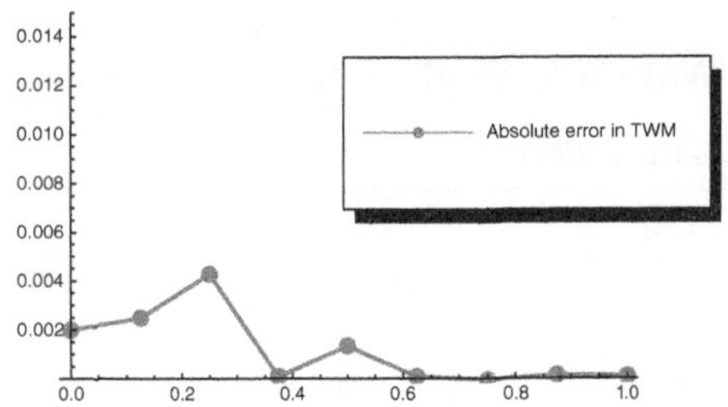

FIGURE 9.22 Absolute error for $k = 2, M = 3$.

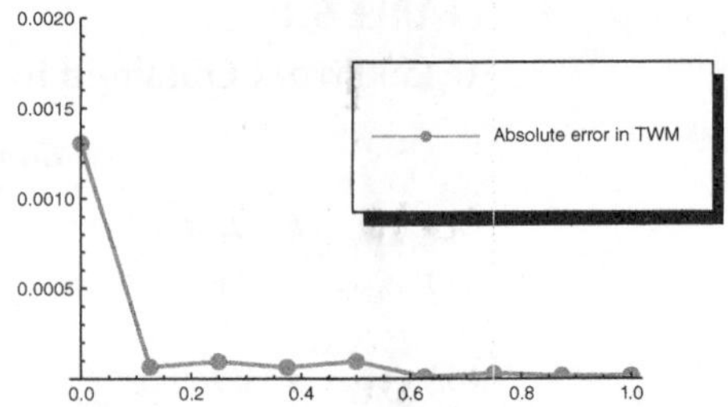

FIGURE 9.23 Absolute error for $k = 3, M = 3$.

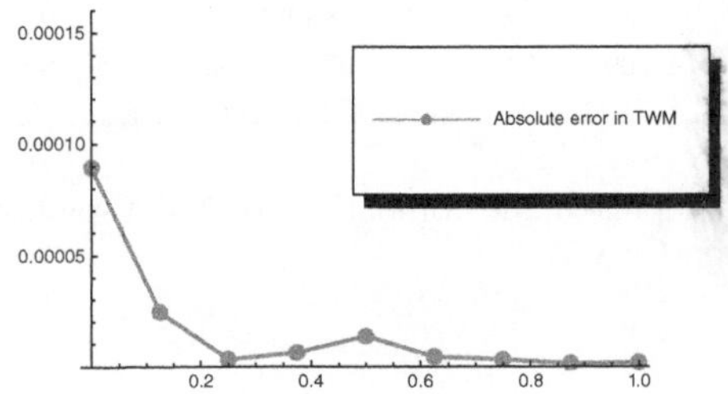

FIGURE 9.24 Absolute error for $k = 4, M = 3$.

TABLE 9.11
NCR for Example 9.8.4

k, M	MAE	NCR
$k=2, M=3$	4.02678E-03	–
$k=3, M=3$	1.88148E-03	1.09776
$k=4, M=3$	8.94564E-04	1.07261
$k=5, M=3$	4.30975E-04	1.05358

Example 9.8.5

Consider the following nonlinear fractional VFIDE [45]:

$$ {}_0^C D_x^{2.2} u(x) = f(x) + \frac{1}{3}\int_0^x (x+z)[u(z)]^2\,dz + \frac{1}{4}\int_0^1 (x-z)[u(z)]^3\,dz, \tag{9.38} $$

$$ u(0) = u'(0) = u''(0) = 0, $$

and

$$ f(x) = \frac{x^{0.8}\Gamma(4)}{\Gamma(1.8)} - \frac{5x^8}{56} - \frac{x}{40} + \frac{1}{44}. $$

The exact solution is $u(x) = x^3$.

For $k = 2,3,4$ and $M = 3$, the absolute error values obtained by the TWM are presented in Table 9.12, and it illustrates that the increase in the size of k and M provides a better absolute error result. Figures 9.25–9.27 represent the comparison graphically between the numerical solutions obtained by TWM and exact solution, and the absolute errors distribution by TWM is shown in Figures 9.28–9.30.

TABLE 9.12
$\tilde{E}(x)$ Errors Obtained by TWM for Example 9.8.5

	Error function $\tilde{E}(x)$		
x	$k=2, M=3$	$k=3, M=3$	$k=4, M=3$
0.1	2.20702E-03	2.15789E-04	3.46467E-05
0.2	1.72591E-03	2.77181E-04	2.73203E-05
0.3	1.74949E-03	2.81704E-04	2.7351E-05
0.4	2.21917E-03	2.18596E-04	3.51596E-05
0.5	6.24442E-03	7.81164E-04	9.76538E-05
0.6	2.24777E-03	2.18714E-04	3.51524E-05
0.7	1.74313E-03	2.81227E-04	2.73437E-05
0.8	1.75835E-03	2.81108E-04	2.73418E-05
0.9	2.25668E-03	2.18957E-04	3.51536E-05

The NCR for different values of k, M is presented in Table 9.13, and it is evident that increasing the size of k leads to a third-order convergence rate. Table 9.14 displays the comparison of maximum absolute errors (MAEs) between TWM, LWM [45], and CASWM [45].

TABLE 9.13
NCR for Example 9.8.5

k, M	MAE	NCR
$k = 2, M = 3$	1.62646E-03	–
$k = 3, M = 3$	2.03062E-04	3.00174
$k = 4, M = 3$	2.5385E-05	2.99987

TABLE 9.14
Comparison of MAE Norm for Example 9.8.5

MAE	LWM [45]	CASWM [45]	TWM
$k = 4, M = 2$	4.432652E-03	7.844735E-02	9.61684E-04
$k = 5, M = 2$	2.143872E-03	5.243756E-03	2.42081E-04

Figures 9.25–9.27 represent the comparison of TWM solutions with exact solution for $k = 2,3,4$ and $M = 3$.

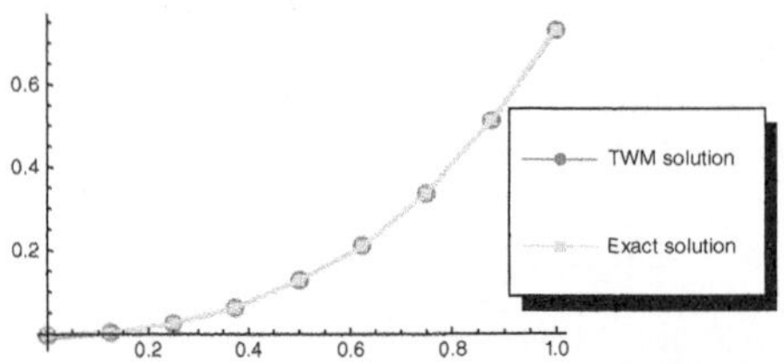

FIGURE 9.25 Comparison of TWM solution with exact solution for $k = 2, M = 3$.

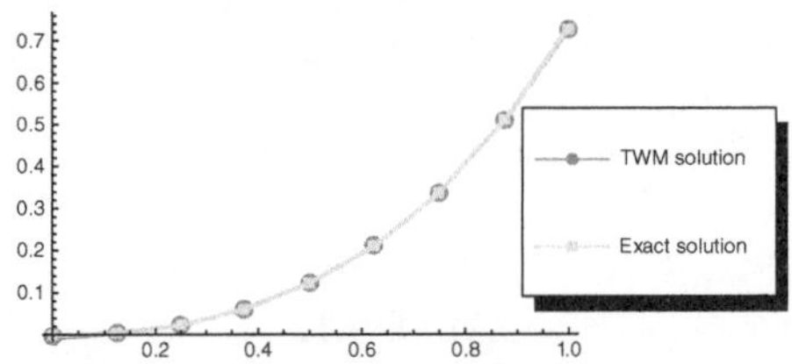

FIGURE 9.26 Comparison of TWM solution with exact solution for $k = 3, M = 3$.

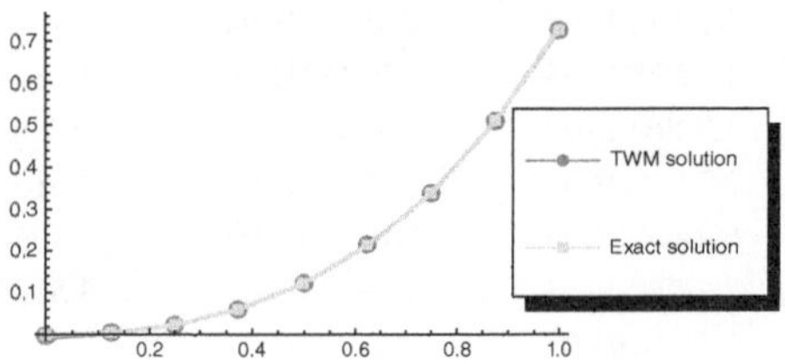

FIGURE 9.27 Comparison of TWM solution with exact solution for $k = 4, M = 3$.

Figures 9.28–9.30 represent the error distribution by TWM for $k = 2,3,4$ and $M = 3$.

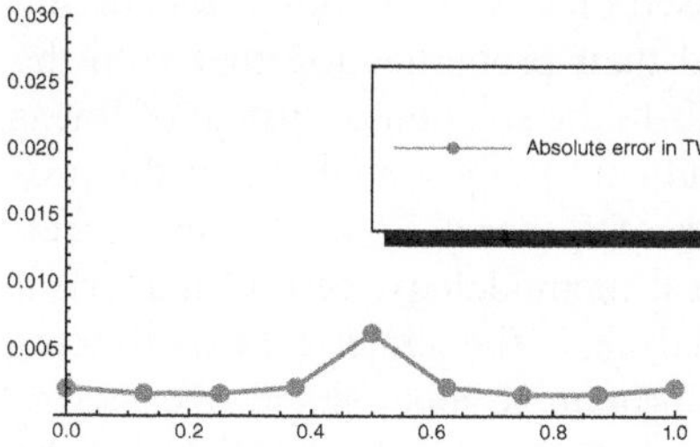

FIGURE 9.28 Absolute error for $k = 2, M = 3$.

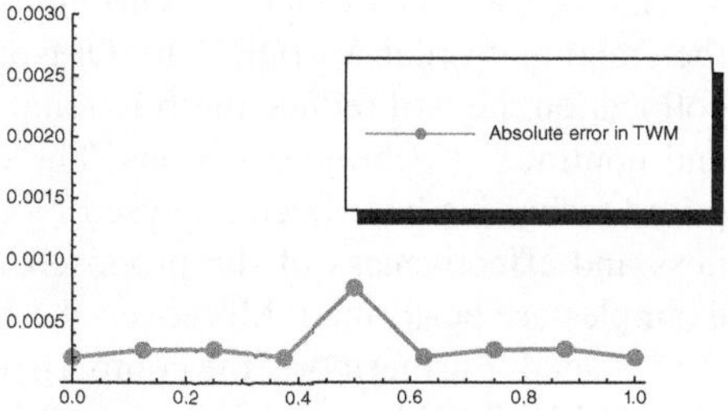

FIGURE 9.29 Absolute error for $k = 3, M = 3$.

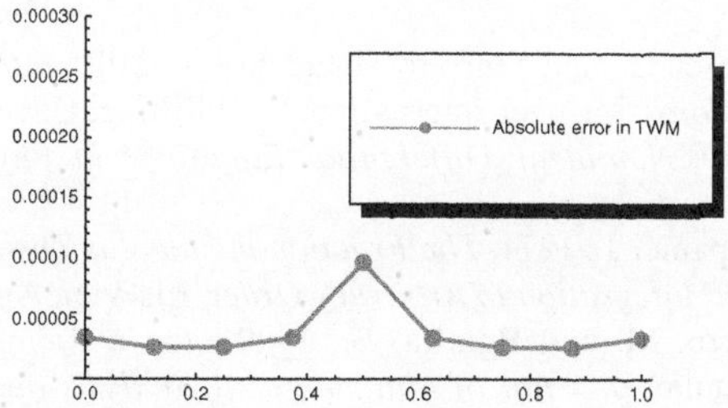

FIGURE 9.30 Absolute error for $k = 4, M = 3$.

Moreover, the L^2-error norm obtained by TWM for linear Examples 9.8.4 and 9.8.5 is tabulated in Table 9.15.

TABLE 9.15
L^2_{error}-Norm for Examples 9.8.4 and 9.8.5

	L^2 – *Error norm*	
k, M	**Example 9.8.4**	**Example 9.8.5**
$k = 2, M = 3$	2.71166E-03	2.36242E-03
$k = 3, M = 3$	8.79703E-04	2.95294E-04
$k = 4, M = 3$	2.93745E-04	3.69112E-05

9.9 CONCLUSION

The present work provides a new scheme that leads to the numerical solution of 1D VFIDE of fractional order. An OM approach based on TWs is proposed for solving the fractional-order VFIDE. The OM of FI and their properties together with the collocation method reduce the fractional VFIDEs to the solution of system of linear and nonlinear algebraic equations. The error and convergence analyses of the proposed technique have been analysed by the sense of L^2-norm. To prove the correctness and effectiveness of the proposed numerical methodology, certain numerical examples are performed. Moreover, the NCR analysis of the numerical experiments is presented. Furthermore, the comparisons of approximate solution and error norms obtained by TWM with the results of the other existing methods establish that the proposed method is very much effective to solve the fractional VFIDEs.

REFERENCES

1. Milici C., Draganescu G., and Tenreiro Machado J.A., 2019, *Introduction to Fractional Differential Equations*, Springer International Publishing, Cham, Switzerland.
2. Saha Ray S., 2020, *Nonlinear Differential Equations in Physics*, Springer Nature, Singapore.
3. Oldham K., and Spanier J., 1974, *The Fractional Calculus Theory and Applications of Differentiation and Integration to Arbitrary Order*, Elsevier, Amsterdam, Netherlands.
4. Trujillo, J.J., Rivero, M. and Bonilla, B., 1999, "On a Riemann–Liouville generalized Taylor's formula", *Journal of Mathematical Analysis and Applications*, **231**(1), pp. 255–265.
5. Anastassiou, G.A., 2008, "Opial type inequalities involving Riemann–Liouville fractional derivatives of two functions with applications", *Mathematical and Computer Modelling*, **48**(3–4), pp. 344–374.
6. Liang, Y.S., 2010, "Box dimensions of Riemann–Liouville fractional integrals of continuous functions of bounded variation", *Nonlinear Analysis: Theory, Methods & Applications*, **72**(11), pp. 4304–4306.
7. Almeida, R. and Torres, D.F., 2009, "Calculus of variations with fractional derivatives and fractional integrals", *Applied Mathematics Letters*, **22**(12), pp. 1816–1820.
8. Tarasov V.E., 2009, "Fractional integro-differential equations for electromagnetic waves in dielectric media", *Theoretical and Mathematical Physics*, **158**(3), pp. 355–359.
9. Povstenko Y., 2010, "Signaling problem for time-fractional diffusion-wave equation in a halfspace in the case of angular symmetry", *Nonlinear Dynamics*, **59**(4), pp. 593–605.
10. Ghanbari, B., and Atangana, A., 2020, "A new application of fractional Atangana–Baleanu derivatives: Designing ABC-fractional masks in image processing", *Physica A: Statistical Mechanics and Its Applications*, **542**, p. 123516.
11. Saghali S., Javidi M., and Saei F.D., 2019, "Analytical solution of a fractional differential equation in the theory of viscoelastic fluids", *International Journal of Applied and Computational Mathematics*, **5**(3), pp. 1–13.
12. He J.H., 1999, "Some applications of nonlinear fractional differential equations and their approximations", *Bulletin of Science, Technology & Society*, **15**(2), pp. 86–90.
13. Baleanu D., Jajarmi A., and Hajipour M., 2018, "On the nonlinear dynamical systems within the generalized fractional derivatives with Mittag–Leffler kernel", *Nonlinear Dynamics*, **94**(1), pp. 397–414.
14. Rabiei K., Ordokhani Y., and Babolian E., 2018, "Numerical solution of 1D and 2D fractional optimal control of system via Bernoulli polynomials", *International Journal of Applied and Computational Mathematics*, **4**(1), pp. 1–17.

15. Tarasov, V.E., 2019, On history of mathematical economics: Application of fractional calculus. *Mathematics*, **7**(6), p. 509.
16. Yang, X.J. and Tenreiro Machado, J.A., 2017, A new insight into complexity from the local fractional calculus view point: Modelling growths of populations. *Mathematical Methods in the Applied Sciences*, **40**(17), pp. 6070–6075.
17. Mainardi F., 1997, *Fractional Calculus: Some Basic Problems in Continuum and Statistical Mechanics*, Springer, New York.
18. Saha Ray S., 2008, "A new approach for the application of Adomian decomposition method for the solution of fractional space diffusion equation with insulated ends", *Applied Mathematics and Computation*, **202**(2), pp. 544–549.
19. Vanani S.K., and Aminataei, A., 2011, "Operational Tau approximation for a general class of fractional integro-differential equations", *Computational & Applied Mathematics*, **30**(3), pp. 655–674.
20. Pedas A., and Tamme E., 2006, "Spline collocation method for integro-differential equations with weakly singular kernels", *Journal of Computational and Applied Mathematics*, **197**(1), pp. 253–269.
21. Majeed A., Kamran M., and Rafique M., 2020, "An approximation to the solution of time fractional modified Burgers' equation using extended cubic B-spline method", *Computational and Applied Mathematics*, **39**(4), pp. 1–21.
22. Bhrawy A.H., Tharwat M.M., and Alghamdi M.A., 2014, "A new operational matrix of fractional integration for shifted Jacobi polynomials", *Bulletin of the Malaysian Mathematical Sciences Society*, **37**(4), pp. 983–995.
23. Bhrawy A.H., and Alofi A.S., 2013, "The operational matrix of fractional integration for shifted Chebyshev polynomials", *Applied Mathematics Letters*, **26**(1), pp. 25–31.
24. Akrami M.H., Atabakzadeh M.H., and Erjaee G.H., 2013, "The operational matrix of fractional integration for shifted Legendre polynomials", *Iranian Journal of Science and Technology*, **37**(4), pp. 439–444.
25. Zhu L., and Fan Q., 2012, "Solving fractional nonlinear Fredholm integro-differential equations by the second kind Chebyshev wavelet", *Communications in Nonlinear Science and Numerical Simulation*, **17**(6), pp. 2333–2341.
26. Aziz I., and Fayyaz M., 2013, "A new approach for numerical solution of integro-differential equations via Haar wavelets", *International Journal of Computer Mathematics*, **90**(9), pp. 1971–1989.
27. Sahu P.K., and Saha Ray S., 2016, "A numerical approach for solving nonlinear fractional Volterra–Fredholm integro-differential equations with mixed boundary conditions", *International Journal of Wavelets, Multiresolution and Information Processing*, **14**(05), p. 1650036.
28. Zhu L., and Fan Q., 2013, "Numerical solution of nonlinear fractional-order Volterra integro-differential equations by SCW", *Communications in Nonlinear Science and Numerical Simulation*, **18**(5), pp. 1203–1213.
29. Wang Y., and Zhu L., 2017, "Solving nonlinear Volterra integro-differential equations of fractional order by using Euler wavelet method", *Advances in Difference Equations*, **2017**(1), pp. 1–16.
30. Saeedi H., Moghadam M.M., Mollahasani N., and Chuev G.N., 2011, "A CAS wavelet method for solving nonlinear Fredholm integro-differential equations of fractional order", *Communications in Nonlinear Science and Numerical Simulation*, **16**(3), pp. 1154–1163.
31. Alkan S., and Hatipoglu V.F., 2017, "Approximate solutions of Volterra-Fredholm integro-differential equations of fractional order", *Tbilisi Mathematical Journal*, **10**(2), pp. 1–13.
32. Khajehnasiri A.A., 2016, "Numerical solution of nonlinear 2D Volterra–Fredholm integro-differential equations by two-dimensional triangular function", *International Journal of Applied and Computational Mathematics*, **2**(4), pp. 575–591.

33. Hamoud A.A., and Ghadle K.P., 2018, "Modified Laplace decomposition method for fractional Volterra-Fredholm integro-differential equations", *Journal of Mathematical Modeling*, **6**(1), pp. 91–104.
34. Ali M.R., Hadhoud A.R., and Srivastava H., 2019, "Solution of fractional Volterra–Fredholm integro-differential equations under mixed boundary conditions by using the HOBW method", *Advances in Difference Equations*, **2019**(1), pp. 1–14.
35. Sahu P.K., and Saha Ray S., 2016, "Sinc-Galerkin technique for the numerical solution of fractional Volterra-Fredholm integro-differential equations with weakly singular kernels", *International Journal of Nonlinear Sciences & Numerical Simulation*, **17**(6), pp. 315–323.
36. Zhou F., and Xu X., 2019, "Numerical solution of fractional Volterra-Fredholm integro-differential equations with mixed boundary conditions via Chebyshev wavelet method", *International Journal of Computer Mathematics*, **96**(2), pp. 436–456.
37. Das, P., Rana, S. and Ramos, H., 2019, Homotopy perturbation method for solving Caputo-type fractional-order Volterra-Fredholm integro-differential equations. *Computational and Mathematical Methods*, **1**(5), p. e1047.
38. Younis, M.F., Abed, A.M. and Hamoud, A.A., 2021, Existence and uniqueness results for bvp of nonlinear fractional Volterra-Fredholm integro-differential equation. *Advances in Dynamical Systems and Applications*, **16**(2), pp. 457–467.
39. Hamoud, A.A. and Ghadle, K.P., 2019, Some new existence, uniqueness and convergence results for fractional Volterra-Fredholm integro-differential equations. *Journal of Applied and Computational Mechanics*, **5**(1), pp. 58–69.
40. Keshavarz E., and Ordokhani Y., 2019, "A fast numerical algorithm based on the Taylor wavelets for solving the fractional integro-differential equations with weakly singular kernels", *Mathematical Methods in the Applied Sciences*, **42**(13), pp. 4427–4443.
41. Behera S., and Saha Ray S., 2020, "An operational matrix based scheme for numerical solutions of nonlinear weakly singular partial integro-differential equations", *Applied Mathematics and Computation*, **367**, p. 124771.
42. Liu N., and Lin E.B., 2010, "Legendre wavelet method for numerical solutions of partial differential equations", *Numerical Methods for Partial Differential Equations: An International Journal*, **26**(1), pp. 81–94.
43. Stewart G.W., 1993, *Afternotes on Numerical Analysis*, University of Maryland, College Park, MD.
44. Toan P.T., Vo T.N., and Razzaghi M., 2019, "Taylor wavelet method for fractional delay differential equations", *Engineering with Computers*, **37**, pp. 1–10.
45. Meng Z., Wang L., Li H., and Zhang W., 2015, "Legendre wavelets method for solving fractional integro-differential equations", *International Journal of Computer Mathematics*, **92**(6), pp. 1275–1291.
46. Loh J.R., Phang C., and Isah A., 2017, "New operational matrix via Genocchi polynomials for solving Fredholm-Volterra fractional integro-differential equations", *Advances in Mathematical Physics*, **2017**, pp. 1–13.
47. Hamoud A.A., and Ghadle K.P., 2019, "Some new existence, uniqueness and convergence results for fractional Volterra-Fredholm integro-differential equations", *Journal of Applied and Computational Mechanics*, **5**(1), pp. 58–69.
48. Das P., Rana S., and Ramos H., 2020, "A perturbation-based approach for solving fractional-order Volterra–Fredholm integro differential equations and its convergence analysis", *International Journal of Computer Mathematics*, **97**(10), pp. 1994–2014.

Index

Note: **Bold** page numbers refer to tables and *italic* page numbers refer to figures.

G

H

I

J

K

L

M

N

O

P

Q

R

S

W